Oceanographic and Marine Cross-Domain Data Management for Sustainable Development

Paolo Diviacco
Istituto Nazionale di Oceanografia e di Geofisica Sperimentale (OGS), Italy

Adam Leadbetter
Marine Institute, Ireland

Helen Glaves
British Geological Survey, UK

A volume in the Advances in Environmental Engineering and Green Technologies (AEEGT) Book Series

IGI GLOBAL
DISSEMINATOR OF KNOWLEDGE

www.igi-global.com

Published in the United States of America by
IGI Global
Information Science Reference (an imprint of IGI Global)
701 E. Chocolate Avenue
Hershey PA, USA 17033
Tel: 717-533-8845
Fax: 717-533-8661
E-mail: cust@igi-global.com
Web site: http://www.igi-global.com

Library of Congress Cataloging-in-Publication Data

Names: Diviacco, Paolo, editor. | Leadbetter, Adam, 1979- editor. | Glaves, Helen, 1966- editor.
Title: Oceanographic and marine cross-domain data management for sustainable development / Paolo Diviacco, Adam Leadbetter, and Helen Glaves, editors.
Description: Hershey, PA : Information Science Reference, 2017. | Series: Advances in environmental engineering and green technologies | Includes bibliographical references and index.
Identifiers: LCCN 2016023682| ISBN 9781522507000 (hardcover) | ISBN 9781522507017 (ebook)
Subjects: LCSH: Marine resources--Management--Data processing. | Marine resources conservation--Data processing. | Sustainable aquaculture--Data processing. | Sustainable marine
Classification: LCC GC1018.5 .O25 2017 | DDC 333.91/64--dc23 LC record available at https://lccn.loc.gov/2016023682

This book is published in the IGI Global book series Advances in Environmental Engineering and Green Technologies (AEEGT) (ISSN: 2326-9162; eISSN: 2326-9170)

British Cataloguing in Publication Data
A Cataloguing in Publication record for this book is available from the British Library.

Advances in Environmental Engineering and Green Technologies (AEEGT) Book Series

ISSN: 2326-9162
EISSN: 2326-9170

Mission

Growing awareness and an increased focus on environmental issues such as climate change, energy use, and loss of non-renewable resources have brought about a greater need for research that provides potential solutions to these problems. Research in environmental science and engineering continues to play a vital role in uncovering new opportunities for a "green" future.

The **Advances in Environmental Engineering and Green Technologies (AEEGT)** book series is a mouthpiece for research in all aspects of environmental science, earth science, and green initiatives. This series supports the ongoing research in this field through publishing books that discuss topics within environmental engineering or that deal with the interdisciplinary field of green technologies.

Coverage

- Policies Involving Green Technologies and Environmental Engineering
- Air Quality
- Water Supply and Treatment
- Electric Vehicles
- Radioactive Waste Treatment
- Alternative Power Sources
- Waste Management
- Biofilters and Biofiltration
- Cleantech
- Pollution Management

IGI Global is currently accepting manuscripts for publication within this series. To submit a proposal for a volume in this series, please contact our Acquisition Editors at Acquisitions@igi-global.com or visit: http://www.igi-global.com/publish/.

The Advances in Environmental Engineering and Green Technologies (AEEGT) Book Series (ISSN 2326-9162) is published by IGI Global, 701 E. Chocolate Avenue, Hershey, PA 17033-1240, USA, www.igi-global.com. This series is composed of titles available for purchase individually; each title is edited to be contextually exclusive from any other title within the series. For pricing and ordering information please visit http://www.igi-global.com/book-series/advances-environmental-engineering-green-technologies/73679. Postmaster: Send all address changes to above address.

Table of Contents

Foreword xvi

Preface xviii

Chapter 1
Oceanographic Data Management: Quills and Free Text to the Digital Age and "Big Data" 1
Justin J. H. Buck, British Oceanographic Data Centre, National Oceanography Centre, UK
Roy K. Lowry, British Oceanographic Data Centre, National Oceanography Centre, UK

Chapter 2
Balancing Formalization and Representation in Cross-Domain Data Management for Sustainable Development 23
Paolo Diviacco, Istituto Nazionale di Oceanografia e di Geofisica Sperimentale - OGS, Italy
Adam Leadbetter, Marine Institute, Ireland

Chapter 3
Developing a Common Global Framework for Marine Data Management 47
Helen M. Glaves, British Geological Survey, UK

Chapter 4
Linked Ocean Data 2.0 69
Adam Leadbetter, Marine Institute, Ireland
Michelle Cheatham, Wright State University, USA
Adam Shepherd, Woods Hole Oceanographic Institution, USA
Rob Thomas, British Oceanographic Data Centre, National Oceanography Centre, UK

Chapter 5
Documenting Provenance for Reproducible Marine Ecosystem Assessment in Open Science 100
Xiaogang Ma, Rensselaer Polytechnic Institute, & University of Idaho, USA
Stace E. Beaulieu, Woods Hole Oceanographic Institution, USA
Linyun Fu, Rensselaer Polytechnic Institute, USA
Peter Fox, Rensselaer Polytechnic Institute, USA
Massimo Di Stefano, Rensselaer Polytechnic Institute, USA & University of New Hampshire, USA
Patrick West, Rensselaer Polytechnic Institute, USA

Chapter 6
Semantic Search Engine for Data Management and Sustainable Development: Marine Planning Service Platform 127
Giuseppe Manzella, ETT SpA, Italy
Roberto Bartolini, CNR ILC, Italy
Franco Bustaffa, Delta Progetti, Italy
Paolo D'Angelo, ETT SpA, Italy
Maurizio De Mattei, Delta Progetti, Italy
Francesca Frontini, CNR ILC, Italy
Maurizio Maltese, Delta Progetti, Italy
Daniele Medone, Delta Progetti, Italy
Monica Monachini, CNR ILC, Italy
Antonio Novellino, ETT SpA, Italy
Andrea Spada, SyO srl, Italy

Chapter 7
SeaDataNet: Towards a Pan-European Infrastructure for Marine and Ocean Data Management 155
Dick Schaap, Marine Information Service (MARIS), The Netherlands

Chapter 8
Repositioning Data Management Near Data Acquisition 178
Paolo Diviacco, Istituto Nazionale di Oceanografia e di Geofisica Sperimentale - OGS, Italy
Jordi Sorribas, CSIC, Spain
Karien De Cauwer, Royal Belgian Institute of Natural Sciences, Belgium
Jean Marc Sinquin, Ifremer, France
Raquel Casas, CSIC, Spain
Alessandro Busato, Istituto Nazionale di Oceanografia e di Geofisica Sperimentale - OGS, Italy
Yvan Stoyanov, Royal Belgian Institute of Natural Sciences, Belgium
Serge Scory, Royal Belgian Institute of Natural Sciences, Belgium

Chapter 9
Interoperability in Marine Sensor Networks through SWE Services: The RITMARE Experience ... 200
Alessandro Oggioni, Institute for Electromagnetic Sensing of the Environment, National Research Council (IREA-CNR), Italy
Paolo Tagliolato, Institute of Marine Sciences, National Research Council (ISMAR-CNR), Italy
Cristiano Fugazza, Institute for Electromagnetic Sensing of the Environment, National Research Council (IREA-CNR), Italy
Monica Pepe, Institute for Electromagnetic Sensing of the Environment, National Research Council (IREA-CNR), Italy
Stefano Menegon, Institute of Marine Sciences, National Research Council (ISMAR-CNR), Italy
Fabio Pavesi, Institute for Electromagnetic Sensing of the Environment, National Research Council (IREA-CNR), Italy
Paola Carrara, Institute for Electromagnetic Sensing of the Environment, National Research Council (IREA-CNR), Italy

Chapter 10
Building a 4D Voxel-Based Decision Support System for a Sustainable Management of Marine Geological Resources 224
Vera Van Lancker, Royal Belgian Institute of Natural Sciences, Belgium
Frederic Francken, Royal Belgian Institute of Natural Sciences, Belgium
Lars Kint, Royal Belgian Institute of Natural Sciences, Belgium
Nathan Terseleer, Royal Belgian Institute of Natural Sciences, Belgium
Dries Van den Eynde, Royal Belgian Institute of Natural Sciences, Belgium
Lies De Mol, FPS Economy, Belgium
Guy De Tré, Ghent University, Belgium
Robin De Mol, Ghent University, Belgium
Tine Missiaen, Ghent University, Belgium
Vasileios Hademenos, Ghent University, Belgium
Marcel Bakker, Geological Survey of the Netherlands, The Netherlands
Denise Maljers, Geological Survey of the Netherlands, The Netherlands
Jan Stafleu, Geological Survey of the Netherlands, The Netherlands
Sytze van Heteren, Geological Survey of the Netherlands, The Netherlands

Chapter 11
Providing Oceanographic Data and Information to Pacific Island Communities 253
James Potemra, University of Hawai'i, USA
John Maurer, University of Hawai'i, USA
Echelle S. Burns, University of Hawai'i, USA

Chapter 12
Marine Environment Data Management Related to the Human Activity in the South-Eastern Baltic Sea (The Lithuanian Sector) 282
Algimantas Grigelis, Institute of Geology and Geography, Lithuania
Nerijus Blažauskas, Klaipėda University, Lithuania
Leonora Živilė Gelumbauskaitė, Institute of Geology and Geography, Lithuania
Saulius Gulbinskas, Klaipėda University, Lithuania
Sergej Suzdalev, Klaipėda University, Lithuania
Christian Ferrarin, Institute of Marine Sciences-National Research Council (ISMAR-CNR), Italy

Chapter 13
Data and Operational Oceanography: A Review in Support of Responsible Fisheries and Aquaculture 303
Enrique Wulff, Andalusian Institute for Marine Sciences (CSIC), Spain

Chapter 14
Integrated Monitoring in the Voordelta, The Netherlands: Monitoring and Data Management for Evaluation of Nature Compensation Measures 325
Niels Kinneging, Deltares, The Netherlands
Meinte Blaas, Deltares, The Netherlands
Arjen Boon, Deltares, The Netherlands
Kees Borst, Rijkswaterstaat, The Netherlands
Gerrit Hendriksen, Deltares, The Netherlands
Gerard Van der Kolff, Deltares, The Netherlands
Theo Prins, Deltares, The Netherlands
Willem Stolte, Deltares, The Netherlands

Chapter 15
Analysis of Ocean in Situ Observations and Web-Based Visualization: From Individual Measurements to an Integrated View 345
Alexander Barth, University of Liege, Belgium
Sylvain Watelet, University of Liege, Belgium
Charles Troupin, Balearic Islands Coastal Ocean Observing and Forecasting System (SOCIB), Spain
Aida Alvera-Azcárate, University of Liege, Belgium
Jean-Marie Beckers, University of Liege, Belgium

Compilation of References 372

About the Contributors 408

Index 421

Detailed Table of Contents

Foreword xvi

Preface xviii

Chapter 1
Oceanographic Data Management: Quills and Free Text to the Digital Age and "Big Data" 1
Justin J. H. Buck, British Oceanographic Data Centre, National Oceanography Centre, UK
Roy K. Lowry, British Oceanographic Data Centre, National Oceanography Centre, UK

In this chapter the authors look at technological issues within and the influence of technological developments outside oceanography on two very different, but interrelated, facets of ocean data: the labelling of parameters and the autonomous collection of oceanographic data. In particular, consideration is given to the influence of standards developed within computer science, such as Sensor Web Enablement (SWE), Resource Description Framework (RDF) and the Simple Knowledge Organisation System (SKOS). It is shown that the development of autonomy in oceanographic data collection is highly dependent on semantically sound data and metadata with presentation of related issues such as unambiguous data citation, resolution of systematic biases and integration of data from complementary but independently governed oceanographic observation programmes.

Chapter 2
Balancing Formalization and Representation in Cross-Domain Data Management for Sustainable Development 23
Paolo Diviacco, Istituto Nazionale di Oceanografia e di Geofisica Sperimentale - OGS, Italy
Adam Leadbetter, Marine Institute, Ireland

Sustainable development is grounded in complex systems where phenomena and processes are so interrelated and intermingled that they can only be studied in a multiple disciplinary manner. Limitiations in understanding such systems can be identified that are either epistemic and cognitive or socio-economic. Both cases are problematic for collaborative and multidisciplinary studies in sustainable development. This chapter explores these difficulties with a particular focus on data management, an activity that is at the same time very important and very sensitive to such problems. To allow users to find their way through data and information spaces, solutions are proposed that take advantage of either the formalization of knowledge; or of representations, such as schemas, mind maps or graphs. Through a careful balance of these two pathways this chapter demonstrates that it is possible to mitigate the difficulties mentioned above and allow a community to work together at profit.

Chapter 3
Developing a Common Global Framework for Marine Data Management..47
Helen M. Glaves, British Geological Survey, UK

The paradigm shift in marine research moving from the traditional discipline based methodology to a multidisciplinary, ecosystem level approach is being driven by changes in both the policies for the management and exploitation of the ocean, and the scientific method itself. The availability of large volumes of good quality data is fundamental to this increasingly holistic approach to ocean research but there are significant barriers to its re-use. The Ocean Data Interoperability Platform (ODIP) project has been funded in parallel by the European Commission, National Science Foundation in the USA and the Australian Government to promote the development of a common framework for marine data management that leverages the existing marine e-infrastructures which have been created in response to the need for greater sharing of marine data at a regional level.

Chapter 4
Linked Ocean Data 2.0 69
Adam Leadbetter, Marine Institute, Ireland
Michelle Cheatham, Wright State University, USA
Adam Shepherd, Woods Hole Oceanographic Institution, USA
Rob Thomas, British Oceanographic Data Centre, National Oceanography Centre, UK

Within the theme of sustainable development, it is not desirable to either have data siloed in one location where it cannot be reused for purposes beyond which it was originally collected, or in a state where it cannot be integrated into a holistic view of the marine environment. As such, the links between datasets should be formally documented and exploited as best as possible. Given this, the use of Semantic Web technology and information modelling patterns are explored in this chapter with reference to the marine domain. Further, new strategies for adding semantic annotation to data in real-time are discussed and prototyped.

Chapter 5
Documenting Provenance for Reproducible Marine Ecosystem Assessment in Open Science 100
Xiaogang Ma, Rensselaer Polytechnic Institute, & University of Idaho, USA
Stace E. Beaulieu, Woods Hole Oceanographic Institution, USA
Linyun Fu, Rensselaer Polytechnic Institute, USA
Peter Fox, Rensselaer Polytechnic Institute, USA
Massimo Di Stefano, Rensselaer Polytechnic Institute, USA & University of New Hampshire, USA
Patrick West, Rensselaer Polytechnic Institute, USA

Open Science not only means the openness of various resources involved in a scientific study but also the connections among those resources that demonstrate the origin, or provenance, of a scientific finding or derived dataset. In this chapter, the authors used the PROV Ontology, a community standard for representing and exchanging machine-readable provenance information in the Semantic Web, and extended it for capturing provenance in the IPython Notebook, a software platform that enables transparent workflows. The developed work was used in conjunction with scientists' workflows in the Ecosystem Assessment Program of the U.S. NOAA Northeast Fisheries Science Center. This work provides a pathway towards formal, well-annotated provenance in an electronic notebook. Not only will the use of such technologies and standards facilitate the verifiability and reproducibility of ecosystem assessments, their use will also provide solid support for Open Science at the interface of science and ecosystem management for sustainable marine ecosystems.

Chapter 6
Semantic Search Engine for Data Management and Sustainable Development: Marine Planning Service Platform 127
Giuseppe Manzella, ETT SpA, Italy
Roberto Bartolini, CNR ILC, Italy
Franco Bustaffa, Delta Progetti, Italy
Paolo D'Angelo, ETT SpA, Italy
Maurizio De Mattei, Delta Progetti, Italy
Francesca Frontini, CNR ILC, Italy
Maurizio Maltese, Delta Progetti, Italy
Daniele Medone, Delta Progetti, Italy
Monica Monachini, CNR ILC, Italy
Antonio Novellino, ETT SpA, Italy
Andrea Spada, SyO srl, Italy

This chapter presents a computer platform supporting a Marine Information and Knowledge System based on a repository that gathers, classify and structures marine scientific literature and data, guaranteeing their accessibility by means of standard protocols. This requires the access to quality controlled data and to information that is provided in grey literature and/or in relevant scientific literature. There exist efforts to develop search engines to find author's contributions to scientific literature or publications. This implies the use of persistent identifiers. However very few efforts are dedicated to link publications to data that was used, or cited in them or that can be of importance for the published studies. Full-text technologies are often unsuccessful since they assume the presence of specific keywords in the text; to fix this problem, it is suggested to use different semantic technologies for retrieving the text and data and thus getting much more complying results.

Chapter 7
SeaDataNet: Towards a Pan-European Infrastructure for Marine and Ocean Data Management...... 155
Dick Schaap, Marine Information Service (MARIS), The Netherlands

SeaDataNet is actively operating and further developing a Pan-European infrastructure for managing, indexing and providing access to ocean and marine data sets and data products, acquired from research cruises and other observational activities in European marine waters and global oceans. It is undertaken by the National Oceanographic Data Centres (NODCs), and marine information services of major research institutes, from 35 coastal states bordering the European seas. Its data centres have the essential capabilities and facilities for data quality control, long term stewardship, retrieval and distribution. Development and provision of standards, services and tools for marine and ocean data management has proven instrumental for wider uptake. This is further encouraged by cooperation and synergy with EU marine research projects, and by adoption of the SeaDataNet infrastructure in the EMODnet (European Marine European Marine Observation and Data Network) development. EMODnet is an initiative in the framework of the EU Marine Directive.

Chapter 8
Repositioning Data Management Near Data Acquisition... 178
Paolo Diviacco, Istituto Nazionale di Oceanografia e di Geofisica Sperimentale - OGS, Italy
Jordi Sorribas, CSIC, Spain
Karien De Cauwer, Royal Belgian Institute of Natural Sciences, Belgium
Jean Marc Sinquin, Ifremer, France
Raquel Casas, CSIC, Spain
Alessandro Busato, Istituto Nazionale di Oceanografia e di Geofisica Sperimentale - OGS, Italy
Yvan Stoyanov, Royal Belgian Institute of Natural Sciences, Belgium
Serge Scory, Royal Belgian Institute of Natural Sciences, Belgium

This chapter intends to propose a solution to the progressive paradigmatic drift that emerges when data is part of a workflow where processes and contexts are not accessible. In this case end users can have difficulties in identifying possible anomalies, or events that might be very important for the dataset and its interpretation. Since sustainable development is based on cross-disciplinary studies, this can revert in misunderstanding and difficulties to work collaboratively. To address this issue it is proposed to fill the gap between the data and its acquisition through a logging system named EARS that records underway data such as for example: meteo or swell, and events such as: anomalies or acquisition milestones. Once all these information are stored, they can be linked to the data through an OGC compliant metadata model and gathered as summaries as required by several data management initiatives.

Chapter 9
Interoperability in Marine Sensor Networks through SWE Services: The RITMARE Experience... 200
Alessandro Oggioni, Institute for Electromagnetic Sensing of the Environment, National Research Council (IREA-CNR), Italy
Paolo Tagliolato, Institute of Marine Sciences, National Research Council (ISMAR-CNR), Italy
Cristiano Fugazza, Institute for Electromagnetic Sensing of the Environment, National Research Council (IREA-CNR), Italy
Monica Pepe, Institute for Electromagnetic Sensing of the Environment, National Research Council (IREA-CNR), Italy
Stefano Menegon, Institute of Marine Sciences, National Research Council (ISMAR-CNR), Italy
Fabio Pavesi, Institute for Electromagnetic Sensing of the Environment, National Research Council (IREA-CNR), Italy
Paola Carrara, Institute for Electromagnetic Sensing of the Environment, National Research Council (IREA-CNR), Italy

The rapid growth and development in different fields related to sensors has, together with the huge increase of devices due to the decrease of device costs, led to a shift from traditional monitoring, where the data collected is not subject to any management actions, to sensor/processing networks, where in the life cycle more stages are devoted to make the data accessible. Data integration is the first step in advanced environmental monitoring, but assuring that heterogeneous systems can interoperate is still a challenge. The Sensor Web Enablement (SWE) initiative defines a framework to address this issue, offering a set of standard models and interfaces to improve sensor interoperability and to face quality issues in the reliability of sensors. The need for seamless access to observations from marine sensors has been the focus of several research projects. This chapter presents the actions taken in the development of the Spatial Data Infrastructure for project RITMARE to ease the adoption of SWE within the Italian marine community overcoming the main constraints in SWE adoption.

Chapter 10
Building a 4D Voxel-Based Decision Support System for a Sustainable Management of Marine Geological Resources 224
Vera Van Lancker, Royal Belgian Institute of Natural Sciences, Belgium
Frederic Francken, Royal Belgian Institute of Natural Sciences, Belgium
Lars Kint, Royal Belgian Institute of Natural Sciences, Belgium
Nathan Terseleer, Royal Belgian Institute of Natural Sciences, Belgium
Dries Van den Eynde, Royal Belgian Institute of Natural Sciences, Belgium
Lies De Mol, FPS Economy, Belgium
Guy De Tré, Ghent University, Belgium
Robin De Mol, Ghent University, Belgium
Tine Missiaen, Ghent University, Belgium
Vasileios Hademenos, Ghent University, Belgium
Marcel Bakker, Geological Survey of the Netherlands, The Netherlands
Denise Maljers, Geological Survey of the Netherlands, The Netherlands
Jan Stafleu, Geological Survey of the Netherlands, The Netherlands
Sytze van Heteren, Geological Survey of the Netherlands, The Netherlands

For sustainable management of marine geological resources, a geological knowledge base is being built for the Belgian and southern Netherlands part of the North Sea. Voxel models of the subsurface are used for predictions on sand and gravel quantities and qualities, to ensure long-term resource use. The voxels are filled with geological data from boreholes and seismic lines, but other information can be added also. The geology provides boundary conditions needed to run environmental impact models that calculate resource depletion and regeneration under various scenarios of aggregate extraction. Such analyses are important in monitoring progress towards good environmental status, as outlined in the Marine Strategy Framework Directive. By including uncertainty, data products can be generated with confidence limits, which is critical for assessing the significance of changes in the habitat or in any other resource-relevant parameter. All of the information is integrated into a cross-domain, multi-criteria decision support system optimised for user-friendliness and online visualisation.

Chapter 11
Providing Oceanographic Data and Information to Pacific Island Communities 253
James Potemra, University of Hawai'i, USA
John Maurer, University of Hawai'i, USA
Echelle S. Burns, University of Hawai'i, USA

The Pacific Islands Ocean Observing System (PacIOOS) is a partnership of Federal, State and local agencies, academia, industry and various other groups that provides data and information about the ocean environment to a wide array of users in the insular Pacific. By collecting measurements and running numerical model forecasts, conducting scientific analyses and then providing these to the public through common interfaces, PacIOOS enables science-based decision making with data services and products. This chapter describes how data collection from a wide-array of sensors can support decision-making in the region if the data are properly (easily, effectively, consistently) served.

Chapter 12
Marine Environment Data Management Related to the Human Activity in the South-Eastern Baltic Sea (The Lithuanian Sector) ... 282

Algimantas Grigelis, Institute of Geology and Geography, Lithuania
Nerijus Blažauskas, Klaipėda University, Lithuania
Leonora Živilė Gelumbauskaitė, Institute of Geology and Geography, Lithuania
Saulius Gulbinskas, Klaipėda University, Lithuania
Sergej Suzdalev, Klaipėda University, Lithuania
Christian Ferrarin, Institute of Marine Sciences-National Research Council (ISMAR-CNR), Italy

The chapter provides an overview of the activities fostering the blue growth and measures supporting the sustainable development and maintaining the good status of the marine environment in Lithuania. The recent findings and developed practical management tools facilitating the effective management of the marine resources support the implementation of EU Strategy for the Baltic Sea Region. Preparation of the first Lithuanian integrated maritime spatial plan, development of innovative solutions for handling dredged soil in ports, the new approach for selection of new dumping sites and prediction of possible adverse effects are among the activities that provided the most valuable results for the further development of marine activities also in the wider region.

Chapter 13
Data and Operational Oceanography: A Review in Support of Responsible Fisheries and Aquaculture ... 303

Enrique Wulff, Andalusian Institute for Marine Sciences (CSIC), Spain

Data obtained within the operational oceanography infrastructures denote a possibility of significant analysis of fisheries in the progress of the implementation of the 1995 FAO Code for Responsible Fisheries. Data products for fisheries and aquaculture cater for research needs, thereby ensuring a capacity to meet temporal resolution, time frame, reporting frequency and interoperability formats requirements. The existence of this legislation and the costs of its enforcement make that the present data infrastructures are employed under relatively precarious circumstances. This chapter provides with an overview, not exhaustive but demonstrative, of what has been achieved for data within operational oceanography addressing the needs of fisheries and aquaculture scientists.

Chapter 14
Integrated Monitoring in the Voordelta, The Netherlands: Monitoring and Data Management for Evaluation of Nature Compensation Measures ... 325

Niels Kinneging, Deltares, The Netherlands
Meinte Blaas, Deltares, The Netherlands
Arjen Boon, Deltares, The Netherlands
Kees Borst, Rijkswaterstaat, The Netherlands
Gerrit Hendriksen, Deltares, The Netherlands
Gerard Van der Kolff, Deltares, The Netherlands
Theo Prins, Deltares, The Netherlands
Willem Stolte, Deltares, The Netherlands

Monitoring of the environmental effects of a harbour extension and the compensation measures is a very complex task. The Voordelta area has high natural values, but is also of high economic importance. To implement a monitoring strategy for this area a multidisciplinary consortium has been formed, consisting of a number of institutes and companies. A central data management facility was set up for data storage and management. This chapter illustrates the data management approach using the Voordelta monitoring programme for the years 2004 to 2013. A central data management facility was set up for data storage and management. A repository gives access to raw data files to all team members. From the analysis of the raw data a number of information products have been developed and disseminated to the authorities and the public through Google Earth. It will be shown, that the presence of a strong multidisciplinary team and good collaboration is the key to success in this complex programme. The way the data have been managed supports this process enormously.

Chapter 15
Analysis of Ocean in Situ Observations and Web-Based Visualization: From Individual Measurements to an Integrated View 345
Alexander Barth, University of Liege, Belgium
Sylvain Watelet, University of Liege, Belgium
Charles Troupin, Balearic Islands Coastal Ocean Observing and Forecasting System (SOCIB), Spain
Aida Alvera-Azcárate, University of Liege, Belgium
Jean-Marie Beckers, University of Liege, Belgium

The sparsity of observations poses a challenge common to various ocean disciplines. Even for physical parameters where the spatial and temporal coverage is higher, current observational networks undersample a broad spectrum of scales. This situation is generally more severe for chemical and biological parameters because such sensors are less widely deployed. The present chapter describes the analysis tool DIVA (Data-Interpolating Variational Analysis) which is designed to generate gridded fields from in situ observations. DIVA has been applied to various physical (temperature and salinity), chemical (concentration of nitrate, nitrite and phosphate) and biological parameters (abundance of a species). The chapter also shows the technologies used to visualize the gridded fields. Visualization of analyses from in situ observations provide a unique set of challenges since the accuracy of the analysed field is not spatially uniform as it strongly depends on the location of the observations. In addition, an adequate treatment of the depth and time dimensions is essential.

Compilation of References 372

About the Contributors 408

Index 421

Foreword

Within the earth and environmental sciences, the oceanography and marine science communities have been exemplary in the systematic development and deployment of joined-up data systems and open standards covering a range of disciplines. This is driven partly by the scale of the application domain - the oceans are a single connected global system, so the data required to explain it must be as well - but also by the related social organization of science programs, which involve a small number of expensive platforms (ships, satellites) managed collectively. The principal scientist knows from the outset that they are only the current occupant of that seat, and that the data describing the observations made on their watch will be fed immediately into a system that will (ultimately) make it available beyond the immediate scientific goals. Alongside this, the longstanding utilisation of the oceans for food and for transportation has led to a tradition of hydrographic and meteorology standards that also underpin the other technical applications.

Location is a key aspect of ocean data, so the oceanography community has been an important user of geospatial technologies, and also developer of techniques in the geospatial sciences, particularly in observational systems. Geospatial has been a relatively well-organised cross-domain horizontal for a while (the Open Geospatial Consortium celebrated its 20th anniversary in 2014), exemplified through collaborative design and publication of data and interface standards, and vigorous development in both closed and open source software in support. Mainstreaming of mapping and location-based technologies through consumer applications on the web has accelerated this. Meanwhile, the web has also breathed new life into the long-promised artificial intelligence project in a variety of ways, including the easy availability of a large corpus of data, along with the emergence of linking as a key enabler. While very much still a minority sport, the semantic web uses linking as a platform which can support a range of both advanced (reasoning) and relatively mundane capabilities (lookups).

The contributions in this volume lean on a number of these elements as they address a cross-domain data management challenge. All of the papers in the first section (theory) build on the key web paradigm, hypertext, in its machine processable guise of 'linked data'. However, the main effort is now at the next layer up the semantic stack, and in the development of consensus and establishment of suitable governance arrangements around the shared agreements (a more functional name for 'standards'), which may in practice be informal and emergent. In the second section (technologies) the focus is mainly on architectures, as no-one is fretting about data formats and structures any more - geospatial data and services are in the background, a mature technology. The use of coordinated metadata indexes and controlled vocabularies is still seen as a key enabler, but with more efficient approaches to tagging. The final section (case studies) provides a set of illustrations of the effective use of joined-up marine and oceanographic data in a variety of geographic and thematic settings. As applications, there is a particular emphasis on visualization and user interactions.

Joining the mature geospatial technologies with the more modern semantic technologies is perhaps a key theme threaded through this volume. It highlights a renewal underway in the geospatial community, but also a challenge. The geospatial standards community were early and enthusiastic adopters of a range of web 'standards', some of which barely saw the light of day anywhere else. They now confront an 'innovator's dilemma' as the mass market bypassed XML technologies, web-services, and complex metadata standards in favour of JSON, REST, and content mining supplemented by lightweight tagging. The oceanographic and marine community appear to be demonstrating a pragmatic combination, taking elements from both sides of this divide, and thereby maintaining their reputation as leaders in theory and in practice.

Simon J. D. Cox
CSIRO, Australia
Melbourne, 2016-03-23

Preface

SUSTAINABLE DEVELOPMENT

Following the Brundtland Commission's report (Brundtland, 1987) sustainable development is defined as *"development which meets the needs of current generations without compromising the ability of future generations to meet their own needs"*. Here, development is intended to be restricted to economic and social development, in particular for people with a low economic standard of living.

A popular model that describes how this may occur in practice is what is generally referred to as the *"triple bottom line"* (Elkington, 1994). This is an accounting framework that incorporates three dimensions of performance: social, environmental and financial. It differs from traditional reporting frameworks as it includes ecological (or environmental) and social measures. Needless to say, it can be difficult to assign appropriate means of measurement to this framework.

Stakeholder participation and acceptance determine perceived trends and policy implementation. Socio-economic factors are, therefore, very important but at the same time very difficult to handle since they are dependent on the context and on the capacity of each country to respond to each of its social problems.

A certain number of authors such as Murphy (2012) complained that since sustainable development became a matter for discussion, a preferential focus has been put on the environmental side, while socio-economic factors were considered less relevant. The reason for this is generally ascribed to the difficulties in quantitatively handling social matters.

In the definition of sustainable development given by the Brundtland Commission's report a very important emphasis is given to intragenerational and intergenerational equity (Holden et al., 2014). For some authors this can include parameters such as criminality, obesity or internet users. To contrast this approach other authors, such as Hopwood, Mellor, and O'Brien (2005), have highlighted that there is a risk that by over-extending the scope of the definition of sustainable development the concept could be diluted and become irrelevant. In addition, mixing social and environmental factors in the field of sustainable development can make it difficult for researchers to analyze phenomena.

If socio-economic factors are considered difficult to study, this is contrasted by the general assumption that the human impact on the environment is easier to analyse. One of the aims of this book is to understand whether this is the case or whether within the study of human impact on the environment as well as in socio-economics, issues might rise that could result in a further imbalance of the "triple bottom line".

SUSTAINABILITY

The term "sustainability" has been defined in the ecological sciences to mean the conditions that must be present for the ecosystem to sustain itself over the long term (Holden et al., 2014). The Brundtland report details that *"sustainable development must not endanger the natural systems that support life on Earth"*. A fundamental assumption then is that human development tends to disrupt the balance of ecosystems, and since preservation of ecosystems is of paramount importance all impacts should be monitored and in case mitigated.

To implement this it is necessary to understand environmental systems and how human pressure impacts on them.

Environmental systems are complex systems where multiple factors interact and intermingle at the same time. In these cases it is very difficult to isolate *a posteriori* the contribution that each of these factors might have exerted.

The study of pressures and environmental reactions is a highly cross-domain research field where scientists from different backgrounds, with different interests and different ways of thinking, meet. The interaction between the different methods of study, cognitive processes, practices and sometimes also human attitudes, becomes itself a complex system, adding to the complicacies of the already intricate environmental systems

ISSUES OF THE ENVIRONMENTAL SCIENCES

One of the main problems of environmental sciences is the intrinsic difficulty to devise experiments to study environmental phenomena. Decomposing these latter and reconstructing the physical laws that govern them can be difficult because factors are interdependent and often nonlinear, meaning they are very sensitive to small variations (the so-called "butterfly effect"). In addition, in the environmental sciences, correlation often does not mean causation, highlighting the role of an underestimated parameter (Lawlor, Davey Smith, & Ebrahim, 2004). Besides, laws are often built considering a low number of cases, or cases that only resemble what is under observation, which creates problems in finding analogies (Engelhardt & Zimmerman, 1982). This undermines the classic approach of the scientific method where theories need to be verified (or falsified), opening the possibility for multiple visions and theories to co-exist. This contrasts one of the foundations of scientific research, which is that science is supposed to be objective. This perspective paves the way for a different approach where scientific visions result from the convergence of communities.

CROSS DISCIPLINARY ISSUES

The convergence of a community implies that initially partners might have different positions. Convergence implies communication. Three levels of communication can be identified:

- **Syntactic:** following Shannon and Weaver (1949), accurate transmission of data between sender and receiver across a boundary needs to be based on a shared and stable syntax/technology. The syntactic level does not consider the meaning of what is transmitted.

- **Semantic:** is a higher level where, besides the ability to transfer data, the focus is on its use, introducing now the need to consider the meaning of terms.
- **Pragmatics:** is related to the significance of information referring also to the state of the surrounding world excluded from the utterance, including cultural and personal factors.

Full understanding among partners is achieved only if all three levels are accounted for. A suitable level of work can be reached even if not all levels are met but mechanisms are implemented that can fill the gap between different conceptualizations and cognitive models. These mechanism can be of various nature: from learning, to semantic agents, to knowledge bases, to computer supported collaborative work tools.

One of the most important aim of this book is to explore how these gaps between domains, scientific fields and also between mental habits, practices or, using the specific term coined by Kuhn (1962), paradigms, can be bridged, in order that the scientific community working on an environmental problem related to sustainable development can converge on a sound and shared solution.

TRENDS

To fill the gaps between different conceptualizations, two strategies can be invoked. One possibility is to formalize knowledge. This allows to understand where possible inconsistencies can occur and account for them. The other strategy relies on representation of knowledge. This can be less detailed and can be used as a road-map that does not mandates which actual path to use between two points, rather offers multiple solutions to be chosen by users upon their needs.

In the first case, knowledge must be explicit. Following Polanyi (1966) this is a difficult result to obtain. In fact, part of knowledge cannot be made explicit because obscured or embedded in practices or even tools. At the same time knowledge formalization cannot be performed by a single person, rather it should be the result of a community effort. This somehow can be seen as moving the problem of concurrent conceptualization from collaborative use of terms and concepts, to the collaborative design of knowledge formalization.

Representation is a projection of knowledge. Using the terminology of Calender and Cohen (2006) it is prone to the *omissions* of possible factors and to their *commission*, meaning the deliberate act of changing the relations between them. Strictly speaking all representations are false (Chakravartty,1995). What is interesting in our case is that omissions and commissions allow building and sharing pragmatic codes between cultures and paradigms. These take place in artifacts called *boundary objects*, which aim to bridge the gaps between different cognitive models being at the same time generic for a community and detailed for another.

In the study of human impact on the environment, boundary objects are often built creating geographic maps of physical and chemical parameters. These, essentially, are forms of superpositions, that can become a base for mediation among communities and paradigms when the original data is processed to result in a new product. Every party contributes to it while, at the same time, the result is not under the control of only one party.

Geographic maps do not have explicit semantics. This can be added through annotation. However, the introduction of text or symbols reintroduce the problem of possible contrasting cognitive models, paradigms or practices.

Maps can also be non geographic. Concept or mind maps are an example of this. In these cases conceptualization is made explicit but at the same time granularity can vary. In the case of mind maps for example details are generally nested, so that users can stop before entering details they are not interested in. This allows the simultaneous use of such maps by different users as a boundary object.

ORGANISATION OF THIS BOOK

Within this perspective, the contents of this book have been arranged into three sections, The first deals with theoretical and historical aspects of cross-domain data management as applied to the marine sciences; the second section details specific technologies which have been utilised or developed in order to apply these theories; and the third and final section expounds on use cases illustrating the practical application of both the theory and technology.

Section 1: Theory

The first chapter of this introductory section, "Oceanographic Data Management: Quills and Free Text to the Digital Age and 'Big Data" by Justin Buck and Roy Lowry paints a picture showing how marine data management has evolved from the days of paper based recording and reporting, through the use of early internet and web based systems and projects forward ideas which are fundamental in other chapters of this book on pervasive computing and future internet and web applications for oceanographic data delivery and integration.

Following this scene setting chapter is a contribution by Paolo Diviacco and Adam Leadbetter, "Balancing Formalization and Representation in Cross-Domain Data Management for Sustainable Development" which looks into the issues of the philosophy of data interpretation and how the tools introduced by Lowry and Buck may be used to represent formalised data structures, while non-formalised data representation is also introduced. A key phrase from John Delaney is introduced, which underpins several ideas in later chapters: "There are emergent technologies throughout the fields around oceanography which we will incorporate into oceanography, and through that convergence we will make oceanography into something even more magical" (Delaney, 2010).

The third chapter, "Developing a Common Global Framework for Marine Data Management," by Helen Glaves, then looks to how the shift in marine research from a siloed, discipline specific activity to a multidisciplinary Earth System activity is driving data practitioners to collaborate on a global scale. This chapter has particular focus on ongoing collaborations between marine data custodians in member states of the European Commission, the United States and Australia to use common frameworks for a global data infrastructure under the auspices of the Ocean Data Interoperability Platform.

One of the technologies being championed within the Ocean Data Interoperability Platform is Linked Data, a World Wide Web Consortium (W3C) standard for publishing structured data online. Chapter 4, "Linked Ocean Data 2.0," by Adam Leadbetter, Michelle Cheatham, Adam Shepherd and Robert Thomas, surveys progress within the application of this technology to oceanographic research datasets including the development of reusable publication patterns and techniques for visualising the interconnections between datasets. Foreshadowed by Buck and Lowry in Chapter one, the application of Linked Data techniques to data streams from pervasive sensors (Linked Big Ocean Data) is introduced and explored.

The reproducibility of data is a fundamental issue in epistemology, and there has been much activity in Earth and space science informatics in recent years in applying the W3C Provenance Data Model to datasets to better understand where the data came from and how it has been processed. Xiaogang Ma, Linyun Fu, Peter Fox, Massimo Di Stefano and Patrick West discuss this topic in Chapter 5, "Documenting Provenance for Reproducible Marine Ecosystem Assessment in Open Science." Their specific focus is on the use of reproducible research, using processing notebooks published online and the Semantic Web and Linked Data techniques introduced in earlier chapters with respect to ecosystem assessment in fisheries activities.

Section 2: Technologies

With the theory of the Semantic Web and Linked Data now described to readers over the course of a number of chapters the opening chapter of the Technologies section of this book. Chapter 6, "Semantic Search Engine for Data Management and Sustainable Development," with Giuseppe Manzella as lead author, applies these techniques in a knowledge management system. Underlying this chapter is the epistemological understanding that data can be organised and presented as information and that integration of and conversations about information leads to knowledge. In the scientific method knowledge is often encapsulated in academic journal articles. The system Manzella and his co-authors present collates marine science data and knowledge and allows access to them through standard interfaces, while utilising semantic technologies to provide more compelling results to users than traditional full-text based searches.

The theme of data integration from a range of sources is continued by Dick Schaap in Chapter 7, "SeaDataNet: Towards a Pan-European Infrastructure for Marine and Ocean Data Management." As it stands, the technological infrastructure of SeaDataNet connects together many national oceanographic data centres from around Europe, the Middle East and north Africa to provide a coherent metadata model for those data centres, and to provide users with federated access to their data through a single web portal. Coherence in the semantics of fields within the metadata is provided via the controlled vocabularies described earlier in the book and information products generated from the SeaDataNet data stock have been supplied to the EMODnet programme which is described in subsequent chapters.

Chapter 8, "Repositioning Data Management Near Data Acquisition," by Paolo Diviacco, Jordi Sorribas, Karien de Cauwer, Jean-Marc Sinquin, Raquel Casas, Alessandro Busato, Yvan Stojanov and Serge Scory propose the movement of the annotation of data files with controlled vocabulary terms closer to the point of acquisition. In the case of this chapter the technology is an automated event logging software to be deployed on research vessels which allows direct connection of data into, amongst other projects, SeaDataNet.This echoes both the "fog computing" paradigm (Bonomi, Milito, Zhu, & Addepalli, 2012) and the "Born Connected" approach described in Chapters 2 and 4 which will be increasingly important as the number of data collecting platforms in the ocean increases through programmes such as Argo and Everyone's Gliding Observatories.

Alongside Linked Data, a key technology platform in making the "Born Connected" dream a reality is Sensor Web Enablement (SWE) which is an Open Geospatial Consortium standard for publishing measurements from sensors and is useful for real time data delivery. Chapter 9, "Interoperability in Marine Sensor Networks through SWE Services," led by Alessandro Oggioni describes the application of SWE within the Italian national marine research programme, RITMARE. Of note in their advances is the bridge between the traditional file store or relational database data management systems of the national oceanographic data centres and the expanding web services world.

Section 3: Case Studies

Having established the theories and technologies of use in marine science data management as it applies to sustainable development, we now turn our attention to describing some practical applications of these areas. Chapter 10, led by Vera Van Lancker, "Building a 4D Voxel-Based Decision Support System for a Sustainable Management of Marine Geological Resources," describes the need for marine geological resource management and associated tools to present holistic solutions to those needs. Again, an emerging theme is the epistemological need to integrate data from multiple sources into information products which must be interpreted to create new knowledge applicable to a domain for which the original data sources may not have been collected.

James Potemra, John Maurer, and Echelle Burns continue the theme of decision support tools based upon scientific data and information in Chapter 11, "Providing Oceanographic Data and Information to Pacific Island Communities." The range of stakeholders to which their Pacific Islands Ocean Observing System speaks is large and includes: research scientists, operational agencies, planners, local agencies, risk managers and the general public. These groups collaborate to identify new data sets for incorporation in the system and new data products, developing the awareness and utility of the data system.

Chapter 12, "Marine Environment Data Management Related to Human Activity in the South-Eastern Baltic Sea (Lithuanian Segment)," by Algimantas Grigelis, describes the integration of environmental, economic and social needs in marine spatial planning in the Baltic Sea region. As noted in the chapter, the goal of marine spatial planning is to identify and control potential conflicts between these various needs at as early a stage as possible in a development process. This favourably compares with our earlier definition of sustainable development. The marine spatial planning tools presented here have been used to sustainably develop ports and disposal sites at sea.

Chapter 13, "Data and Operational Oceanography: A Review in Support of Responsible Fisheries and Aquaculture," by Enrique Wulff-Barreiro analyses the mismatch between the data user and the data provider with reference to researchers in the fisheries and aquaculture domain. The large infrastructure systems of operational oceanography data providers and the high quality time series provided in a more delayed mode are shown to often be disconnected. These observations leave room for many future developments in marine data management which could be addressed by approaches outlined in earlier chapters.

Nils Kinnegin and his co-authors present how appropriate collaboration and integration of data has been achieved for a specific marine monitoring and evaluation plan in Chapter 14, "Integrated Monitoring in the Voordelta, The Netherlands". They describe both policy drivers and information needs driving monitoring programmes for the creation of observational data. The authors of this chapter observe that the human elements of collaboration and frequent interaction are as important, if not more so, than the technological aspects of building such an integrated, cross-disciplinary monitoring system.

The final chapter of the book, Chapter 15, "Analysis of Ocean In Situ Observations and Web-Based Visualization: From Individual Measurements to an Integrated View," by Alexander Barth, Sylvain Watelet, Charles Troupin, Aida Alvera-Azcárate and Jean-Marie Beckers returns to the theme of representation, in this case through Web-based techniques for the display of profiles and maps of ocean data. These tools are increasingly important in the visualisation of missions undertaken by autonomous underwater vehicles and gridded data fields such as from models and remotely sensed datasets.

CONCLUSION

Within the tripartite structure of this book we have been able to provide a solid insight into the issues and the possible solutions related to the analysis of pressures on the marine environment and how these can be mitigated. We have highlighted that scientific research in this area is not easy to perform since the intrinsic cross-disciplinary nature of this field introduces the need to manage concurrent cognitive models and paradigms.

If traditionally the triple bottom line perspective of sustainable development described above was considered biased only by the difficulties in quantifying social factors, we demonstrate in this book that the situation is not straightforward in studying and handling the environment as well.

The message, from first chapter to last, is clear: data from the oceans are proliferating and the tools to integrate the data to allow for interpretation, in particular in support of sustainable development, are being actively developed. To make the most of these endeavours this book shows that in order to account for the above mentioned difficulties we must move beyond managing data into managing meaning. Results to-date show that, although a lot still of work is still to be done, balancing formalization and representation of knowledge can be the key to address such issues. This idea is discussed theoretically and put in practice in several examples within this book.

REFERENCES

Bonomi, F., Milito, R., Zhu, J., & Addepalli, S. (2012, August). Fog computing and its role in the internet of things. In *Proceedings of the first edition of the MCC workshop on Mobile cloud computing* (pp. 13-16). ACM. doi:10.1145/2342509.2342513

Brundtland Commission Report. (1987). *Lawlor*. Retrieved from https://en.wikisource.org/wiki/Brundtland_Report on 7/10/2015

Callender, C., & Cohen, J. (2005). There is no special problem about scientific representation. *Theoria*, *55*, 67–81.

Chakravartty, A. (1995). Truth and representation in science: Two inspirations from art. In R. Frigg & M. Hunter (Eds.), *Beyond Mimesis and Convention: Representation in Art and Science* (pp. 33-50). Berlin: Springer.

Delaney, J. (2010, April). *Wiring an interactive ocean* [video file]. Retrieved October 13, 2015 from https://www.ted.com/talks/john_delaney_wiring_an_interactive_ocean?language=en

Elkington, J. (1994). Towards the sustainable corporation: Win-win-win business strategies for sustainable development. *California Management Review*, *36*(2), 90–100. doi:10.2307/41165746

Engelhardt, W., & Zimmerman, J. (1982). *Theory of earth science*. Cambridge, UK: Cambridge University Press.

Holden, E., Linnerud, K., & Banister, D. (2014). Sustainable development: Our common future revisited. *Global Environmental Change*, *26*, 130–139. doi:10.1016/j.gloenvcha.2014.04.006

Hopwood, B., Mellor, M., & O'Brien, G. (2005). Sustainable development: Mapping different approaches. *Sustainable Development*, *13*(1), 38–52. doi:10.1002/sd.244

Lawlor, D. A., Davey Smith, G., & Ebrahim, S. (2004). Commentary: The hormone replacement-coronary heart disease conundrum: Is this the death of observational epidemiology? *International Journal of Epidemiology*, *33*(3), 464–467. doi:10.1093/ije/dyh124 PMID:15166201

Murphy, K. (2012). The social pillar of sustainable development: A literature review and framework for policy analysis. *Sustainability: Science, Practice, & Policy, 8*(1),15-29. Retrieved from http://www.google.it/archives/vol8iss1/1008-041.murphy.html

Polanyi, M. (1966). *The tacit dimension*. New York, NY: Anchor Day Books.

Chapter 1

Oceanographic Data Management:
Quills and Free Text to the Digital Age and "Big Data"

Justin J. H. Buck
British Oceanographic Data Centre, National Oceanography Centre, UK

Roy K. Lowry
British Oceanographic Data Centre, National Oceanography Centre, UK

ABSTRACT

In this chapter the authors look at technological issues within and the influence of technological developments outside oceanography on two very different, but interrelated, facets of ocean data: the labelling of parameters and the autonomous collection of oceanographic data. In particular, consideration is given to the influence of standards developed within computer science, such as Sensor Web Enablement (SWE), Resource Description Framework (RDF) and the Simple Knowledge Organisation System (SKOS). It is shown that the development of autonomy in oceanographic data collection is highly dependent on semantically sound data and metadata with presentation of related issues such as unambiguous data citation, resolution of systematic biases and integration of data from complementary but independently governed oceanographic observation programmes.

INTRODUCTION

Oceanographic data collection started some 500 years ago with scribes recording readings from tide poles using quill pens and parchment and data management was the curation of these parchments. During the first half of the last Century oceanographic data management comprised the curation of printed books. Very little had changed over hundreds of years.

However, from 1950 onwards the rate of change accelerated phenomenally with the birth of the digital age. Oceanographic data management became formalised as an international activity during the 1960s. Since then the technological environment has changed beyond all recognition, which has inevitably revolutionised data management practices in line with new usage of data.

DOI: 10.4018/978-1-5225-0700-0.ch001

The chapter will examine the history of a selection of data management threads, looking at the impacts on them of technological change and the consequences of this change in terms of data accessibility, data usability and the quality of science. There are many potential candidates for such a study. Some are obvious and directly linked to hardware developments, such as data volume constraints and data availability timescales. Others are more subtle involving the technological changes driving cultural changes, such as the development and adoption of standards (including controlled vocabularies) and the integration of data into the scientific literature. The ongoing increase of internet bandwidth has meant exposure and access of data via the web has become the norm in less than two decades.

The next stage in the journey will be to examine the current state of development of these threads, with particular reference to what data consumers are now able to do plus what they cannot. The needs of both operational and scientific users of data will be considered. Operational users readily accept crude data quality levels to achieve rapid delivery. This is fundamentally different to the high quality data needed for precise estimates of quantities to monitor climate change such as the heat content of the ocean where small errors in source data propagate into significant errors in scientific findings (Barker, Dunn, & Domingues, 2011; Pedro & Goni 2011).

Next the consequences of technological evolution will be addressed. When there are significant technology jumps in terms of storage or data delivery there is a risk that previously processed data will be left behind making it unavailable or inconsistent in terms of format or semantic description. Aggregate datasets spanning centuries or longer enable long term scientific analyses e.g. Roemmich, Gould, & Gilson (2012). Thus, avoiding the creation of legacy data issues should be part of the adoption of new technologies.

Finally, the Chapter will look at how these threads could develop further in the future using technologies such as the Semantic Web following Leadbetter, Cheatham, Shepherd and Thomas (2017, chapter 4 this book) and the solutions currently under development for the facets of the 'Big Data' problem. The 'Big Data' age will place stringent demands on data semantics, but if these demands are met it will make data accessible to new forms of science. However, if they are not then ambiguous or non-machine readable data descriptions may hinder progress. Further, there is the very real risk that unless cultural issues involving the move towards 'open data' are addressed then the technological progress and the resulting scientific progress might be stopped in its tracks.

This chapter will address these topics by examining two complementary case studies, the first being the development of oceanographic data semantics, before moving on to consider the rapid and ongoing revolution that is ocean observation by autonomous platforms. The latter of which identifies additional issues and topics such as data standards, lessons learnt and data citation developments.

BACKGROUND

Oceanographic and Operational Context

Data consumers for oceanographic data are broadening in scope beyond the research project based synoptic studies that tend to fund research cruises. A major additional consumer is the increasingly data hungry family of ocean forecast models that require low-grade versions of data less than 24 hours after observation. Up until the 1990s ocean data were only available once a research cruise had reached port and the physical media (tape, disk etc.) had been submitted to data centres. These data were then aggregated into products for dissemination to users. Satellite and mobile communications have enabled observations to be transmitted from observation platforms to shore in near-real-time. Through the advent of the internet and the Global Telecommunications System (GTS) it became possible to move data without moving physical media. The GTS was initially created to transfer meteorological data between meteorological centres globally and it now carries oceanographic data to enable ocean forecast models and the quantification of air-sea heat flux in metrological forecast models (Lea, Martinc & Oke, 2014). Progress continues with the UK Meteorological Office developing a new biogeochemical model that will require biogeochemical parameters such as phosphate, nitrate and chlorophyll in addition to physical oceanographic parameters like temperature and salinity (Met Office, 2015).

In addition to these near real time demands there is a pressing need for high quality climate grade data to enable fundamental oceanographic research questions such as those concerning oceanic heat content, sea level, the carbon and nitrate cycles, and ecosystems. These data need to be of a much higher precision (< 0.005°C in temperature and < 0.01 PSU in salinity for physics data). Any systematic biases found in lower grade scientific data such as historic issues in the eXpedible BathyThermograph (XBT) dataset of upper ocean temperature or pressure biases in the global Argo profiling float dataset have the potential significantly damage results and the reputation of science (Pedro, DiNezio & Gustavo, 2011; Barker et al., 2011).

Data Labelling

The labelling of data started with handwritten annotations. As the digital age dawned these developed into column headings in data files and spreadsheets. For closed communities working with limited numbers of parameters, this was relatively easy to manage as the whole community fully understood all of the concepts used. However, data exchange across communities and an ever-increasing number of parameters has required management using tools such as controlled vocabularies, supported by increasingly sophisticated technology. Through these it becomes possible for automated systems to determine whether parameters described by different communities in different ways are the same. More importantly, it allows superficially similar descriptions of significantly different parameters to be differentiated.

In parallel to the specification and development of consistent parameter naming conventions Linked Data (Bizer et al, 2009; Leadbetter, Cheatham, Shepherd & Thomas, 2017, chapter 4 this book) and OGC Sensor Web Enablement technologies (OGC, 2013; Diviacco et al., 2017, chapter 8 this book; Oggioni et al. 2017, chapter 9 this book) have evolved. These technologies are enabling a fundamental shift in the sharing, exposure, and linking of data via standardised web interfaces thus enabling ingestion and aggregation of data for 'Big Data' use cases.

Data Formats

Prior to 1990, data storage format design was dominated by problems caused by the lack of standardisation between computer manufacturers and the physical formatting requirements of 9-track magnetic tape. Several manufacturers such as DEC and UNIVAC pioneered the current character standard encoded in American Standard Code for Information Interchange (ASCII) mapped to 4 8-bit bytes in a 32-bit word. However, other major players didn't. IBM had 32-bit words and 8-bit bytes but encoded characters in Extended Binary Coded Decimal (EBCDIC) whilst Control Data had 60-bit words with 10 6-bit bytes with characters encoded in Binary Coded Decimal (BCD). Floating point binary data were even more of an issue with every manufacturer having their own ideas on how sign, exponent and mantissa could be mapped into a word. Magnetic tapes could be physically formatted in many different ways, with density, block size and various types of non-data block being some of the variable parameters.

As a result of these drivers data storage formats were designed by specific organisations to make optimal use of their chosen hardware, such as the British Oceanographic Data Centre's (BODC) PXF - a binary format designed to maximise data capacity of magnetic tape. Such a format worked well in internal systems, but could only be used for data exchange if the recipient had considerable technical expertise. One of the authors (RKL) spent two years of his early career full time writing software to allow BODC to handle the range of formats used by its data suppliers. There was a major international effort led by IODE to develop a data exchange format known as GF3 (UNESCO, 1987) that eliminated many of the degrees of freedom and could have worked. However, it was never widely adopted for political rather than technical reasons.

At the end of the 1980s the demise of the mainframe and rise of desktop workstations resulted in the removal of many of these degrees of freedom. The 32-bit word made up of 4 8-bit bytes with character information encoded in ASCII and floating point data encoding following a convention proposed by IEEE became universal. Desktop computing tools, especially spreadsheets, became popular and these brought further de facto standards for storage formats, such as ASCII-encoded comma-separated value (CSV). Exchange format development was driven by sets of conventions negotiated by user communities engaged in the large international environmental science projects of the 1990s, such as the World Ocean Circulation Experiment (WOCE) and Joint Global Ocean Flux Survey (JGOFS). One result was the Cooperative Ocean/Atmosphere Research Data Service (COARDS) conventions that provide the basis much of today's environmental data exchange.

Currently oceanic and atmospheric data tend to be stored in either ASCII or binary formats. ASCII formats are a compromise with the representation of complex multidimensional data being non-trivial and data stored in an inefficient manner. Non-ASCII binary formats offer a solution with the atmospheric and physical oceanographic scientific domains adopting the Network Common Data Format (NetCDF) (Unidata, 2016). NetCDF is a set of software libraries and self-describing, machine-independent data formats that support the creation, access, and sharing of array-oriented scientific data. The Climate Forecast (CF) community produced a version of NetCDF (CF-NetCDF) with standardised variable names, structures and attributes (CF community, 2015).

CASE STUDIES

Labelling Data Case Study

There are many pieces of information required in the labelling of data known as metadata. As data are collections of measurements, the most fundamental label required is a description of the phenomenon that was measured. When data were delivered on paper this information was delivered through written annotations. These were read by humans whose information processing capabilities meant that issues such as incorrect spelling or variable word order had minimal impact on understanding. The portfolio of oceanographic measurements made at that time was so small that even labelling in different languages didn't cause significant problems.

When the measurements migrated from paper to digital media the parameter labels obviously needed to accompany them. However, this brought problems. Technology, such as the FORTRAN66 programming language, dictated that the data be encoded in fields with a predetermined fixed size. Storage was also at an absolute premium - a punched card only held 80 byes - so format designers were driven to cut field sizes to the absolute minimum. These pressures led to the idea of encoding the parameter description as an alphanumeric code that mapped to a larger text string known as a parameter code. The controlled vocabulary was born.

Controlled Vocabularies

A controlled vocabulary is a set of labels that may be used to populate a field in a metadata document or a data file. For example, for data presented as a spreadsheet parameter codes may be used to label the data columns. It could be argued that simple human-readable labels could be used instead of parameter codes. However, such plaintext labels are unlikely to be understandable by software agents. It can even be the case that after the passage of surprisingly short periods of time they cease to be understandable by humans, including their author.

This simplest manifestation of a parameter code controlled vocabulary is an identifier corresponding to a text string. This addresses machine-readability issues such as the elimination of spelling mistakes and could possibly work for a small number of parameters. However, as scale and complexity increases problems are inevitable with this approach. Word order and the richness of language could easily result in a simple observed phenomenon being described by multiple terms, such as 'air temperature', 'atmospheric temperature', 'temperature of the atmosphere' and 'temperature of the air'.

This problem escalates as the information content of the parameter label increases. Many physical parameters may be described by a simple phrase relating to the observed phenomenon such as 'salinity'. However, when it comes to disciplines such as contaminants in biota the description needs to carry a lot more information, such as what was measured, the matrix in which it was measured, the observed phenomenon of the measurement. Data model limitations have even lead to information on how the measurement was made being included. Consider the number of different ways that cadmium levels in livers collected from female specimens of a given fish species may be expressed as a text string. If every laboratory making these measurements composed their own text string the richness of expression would be so great that semantic interoperability would be an impossible dream.

The approach to this problem adopted by the BODC Parameter Usage Vocabulary (PUV) used by SeaDataNet described in Schaap et al. (2017, chapter 7 this book) and EMODNet Chemistry Lot amongst others is to use a semantic model to break down the information content into discrete items, each of which is itself a controlled vocabulary. For example, a contaminant in biota description is broken down into nine elements thus:

- Measured phenomenon e.g. 'Concentration'
- Measured phenomenon statistic e.g. 'standard deviation'
- Chemical substance e.g. 'total iron'
- Measured phenomenon to matrix relationship e.g. 'per unit dry weight of'
- Matrix e.g. 'biota'
- Biological entity e.g. 'Asteroidea' (a semantic model within the semantic model including elements such as taxon name, gender and development stage)

- Sample preparation e.g. 'acid digestion'
- Analysis e.g. 'ICP-MS'
- Data processing e.g. 'calculation from multiple measurements'

A text generator is used to convert the labels of the elements into a single human-readable parameter label such as 'Concentration standard deviation of total iron per unit dry weight of biota {Asteroidea} by acid digestion and ICP-MS and 'calculation from multiple measurements'. The resulting labels are consistent in both word order and in the wording used to describe the embedded concepts.

However, the technique delivers more than consistent labels. Semantic interoperability inevitably involves mapping between parameter vocabularies developed independently by different laboratories. If each of these is described in terms of compatible semantic models then the mapping process is significantly simplified (Lowry, Bird & Haaring, 2007) to the extent where it can become semi-automated.

The semantic model also provides a mechanism for the location of the required parameter label from a large and complex vocabulary like the BODC PUV, which contains over 35,000 concepts. Consider a client in which each of the nine semantic elements is offered to the user as a drop-down list or search box. The user selects the appropriate entry from each of these – a little like spinning the wheels of a one-armed bandit. The selected combination is matched against the PUV model instances and if the combination is found the user 'wins' a parameter code.

However, what if the user loses? One response would be to add the user's combination to the PUV automatically. However, an element of caution is required because not all combinations of the model elements are semantically valid. Consider two elements, such as animal and colour. Combinations such as 'brown dog' and 'green snake' occur naturally. The combination 'green dog' requires hair dye. Consequently, proposed new combinations need to be quarantined and quality controlled before addition to the vocabulary.

Having described the BODC PUV, which has been under development since the 1980s, we will now look at how technology has influenced its management and development.

Vocabulary Governance

Controlled vocabularies require two types of governance, both of which need technological support. Content Governance controls the codes used and their labelling, making decisions such as whether new codes should be added and whether existing code descriptions should be modified. Technical Governance provides the mechanisms that make the outputs from Content Governance available to data encoders and users. This evolution of controlled vocabularies along with the broader development of information technology is shown in Figure 1 and will be further described in the subsequent section.

Parameter vocabulary governance in oceanographic data management had its foundation in the 1980s through the activities of the Intergovernmental Oceanographic Commission's (IOC) Group of Experts on the Technical Aspects of Data Exchange (GETADE). The technology available to this group for Content Governance activities were face-to-face meetings, desk telephones, the postal service and rudimentary telex-based electronic mail. The technology available for their Technical Governance was a printed book (UNESCO, 1987), which by IOC rules needed to be in five languages. A critical examination of their activities shows that technology provided adequate support for their Content Governance activities, but failed their needs for Technical Governance completely.

Figure 1. Timeline showing the evolution of vocabulary governance with the development of information technology

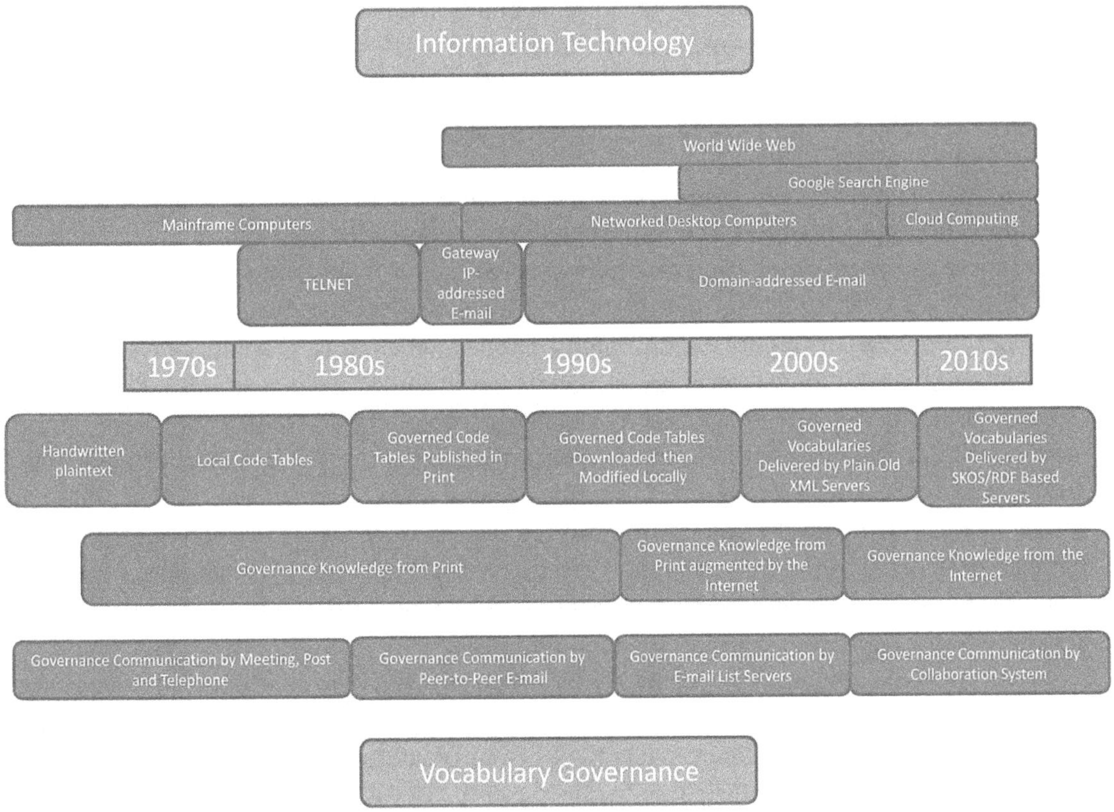

Technology and Vocabulary Content Governance

Until the end of the 1980s oceanographic data management focused on a very small number of parameters, most of which were physical oceanographic. Scientists were making other kinds of measurements, but these never troubled the data management community and consequently were never shared outside scientific publications. However, the move to collaborative multidisciplinary oceanographic projects such as NERC's North Sea Project in 1988-89 and the international Joint Global Ocean Flux Study (JGOFS) project that commenced fieldwork in 1989 changed this. Not only did these projects include a wide range of chemical and biological measurements but their collaborative nature meant the data required intra-project management and their cost brought an expectation of project data sets suitable for future re-use.

The influence of technology on parameter vocabulary content governance was seen in two areas: accessibility to knowledge and the tooling available for collaboration. Consider the position of one of the authors in 1988 when suddenly faced with the prospect of extending the BODC PUV to cover the requirements of the North Sea Project and UK JGOFS. Based in a physical oceanographic laboratory, accessible knowledge was restricted to the 1986 edition of the 'Rubber Book' (CRC, 1986) and inter-library loans.

The result of this restricted access to knowledge of the content governance for chemical and biological parameters was that total trust was placed in the scientists making the measurements. Some might say that this is the way things should be as they are the experts. However, years of experience have shown the parochial nature of data labelling by scientists. Incorrect spellings, shorthand and even 'pet names' are common in data files submitted to BODC. The labelling is only tidied up when the data are written up into publications. This has led to many errors in the BODC PUV that could only be corrected thanks to knowledge resources made available through advances in information technology.

Today it seems like we have had the World Wide Web forever. However, it didn't exist until 1994 and it wasn't until the turn of the Century that authoritative content and the tools to discover that content reached usable maturity. The Google search engine was born in 1996 and was used in BODC PUV content governance, particularly for checking spelling of taxon names, by 1998. The 'fuzzy searches' provided in those days were a little unpredictable – having 'Praunus' match to 'porn' produced some vey unexpected results. The sophistication of the text matching in search engines has improved significantly and most unexpected results now are usually helpful. Even so, an element of skepticism is still required, especially when searching for chemical names. Errors have resulted when a search for a little-known compound returned hits for a compound used by industry with a very similar spelling.

The Web and Search Engines provide the plumbing that gives access to knowledge resources beyond the wildest dreams of twenty years ago. Taxon naming ceased to be a problem when the World Register for Marine Species (WoRMS) (VLIZ, 2016) provided instant, easy access to knowledge provided by the best marine taxonomic brains in the world. Problems with chemical nomenclature have been greatly eased by sites such as the Chemical Entities of Biological Interest (EMBL-EBI,2016) and many others set up to provide the knowledge base for the chemical industry.

Improvements in communication technology have also benefited the development of collaboration in vocabulary content governance. Vocabulary development for the GF3 standard was supported by a well-funded international organisation that recruited a group of international experts and supported their collaboration through face-to-face meetings. In the case of the BODC PUV the content governance was initially the responsibility of a single individual who sought collaboration through ad-hoc support from the scientific community. This took the form of one-to-one meetings, telephone conversations and e-mail or opportunistic group discussions during research cruises and science conferences.

The first technical innovation that facilitated broader consultation was the development of e-mail list server systems, such as Mailman. This allowed collaborative content governance approaching the standard of the days of GF3, but at significantly lower cost in terms of both money and experts' time. E-mail list servers have supported vocabulary content governance in communities such as Climate and Forecast (CF) for over a decade and still operate in that role today.

However, list servers have their limitations as collaborative tools. For example, Mailman preserves discussion threads in an archive, but the location of specific items of information requires significant effort. Consequently, systems have been developed to enhance collaboration by simplifying access to this information. The CF community has developed a bespoke tool to track the progress of vocabulary concept development from initial proposal to acceptance (CEDA, 2016). In the EU Geo-Seas project Colla (Diviacco, 2016), a collaborative tool developed at OGS Trieste was used for vocabulary development discussions. More refined team collaboration systems, such as Atlassian Confluence (Atlassian, 2016), are now marketed commercially. The challenge facing the BODC PUV is to embrace these technologies in the development of effective collaborative content governance.

Technology and Vocabulary Technical Governance

Controlled vocabulary technical governance in oceanographic data management had its origins in the publication in the five official languages of the IOC of the GF3 Code Tables (UNESCO, 1987). This successfully delivered the efforts of the content governance authority to its potential user base. However, this was only one of their requirements. Even if science were static, no content governance authority can possibly deliver complete coverage. In reality, vocabulary content governance faces the situation amusingly portrayed when Edmund Blackadder encountered Dr. Johnson with his 'complete' English Dictionary. No matter how many concepts are created there will always be a need for more. This creates a vocabulary technical governance use case that the printed word cannot address within realistic budgets. A viewpoint put forward all too often in the late 1980s was that a new controlled vocabulary needed to be created because the GF3 code tables could not be maintained. It should be noted that this was an era when standards and interoperability were unpopular because many gave lip service to a desire to share data that they did not support in their hearts.

By the time the GF3 Code Tables rolled off the presses technology was developing that could have supported the maintenance use case. Desktop computers were becoming commonplace in offices. These ran software, such as word processors and spreadsheets that could be used to manage code table lists. The computers included floppy disk drives that had the potential to distribute updated code table lists by physical exchange. This potential was compromised to some extent by the facts that there were three physical floppy disk formats in widespread use and that the commercial software used to maintain the code lists used proprietary binary formats.

This level of technology could have provided GETADE with technical governance capable of addressing the code table maintenance use case economically by periodic distribution of mass-produced floppy disks. Sadly, by the time this became possible GETADE's ability to provide effective content governance had been destroyed by the political challenges and savage funding cuts that hit UNESCO. Even sadder was the fact that in the early 1990s further technical advances in the form of the World Wide Web could have provided GETADE with the means to maintain a single master copy of the code tables that was accessible to users 24/7, which is the Nirvana of vocabulary technical governance.

The 1990s could be regarded as the Dark Ages for controlled vocabularies in oceanographic data management. Ironically, the development of desktop computing and the web provided the means to accelerate the rate at which localised controlled vocabularies developed in the absence of effective overall content governance. Every project had the tools to develop vocabularies precisely matched to its requirements and then share them between project participants. These tools were used to full effect resulting in rapid semantic divergence between projects that created a large workload for those working towards semantic interoperability in more recent years by mapping vocabularies.

At the turn of this century the European Union SEASEARCH project (National Oceanography Centre, 2015) had the objective of producing fully interoperable discovery metadata across European and Mediterranean data centres. This required semantic interoperability, which was to be implemented through controlled vocabularies known as 'libraries'. Moreover, it recognised the need for vocabulary content governance and implemented it. This content governance was able to function effectively using available technologies in the form of master spreadsheet files maintained for download on web sites or FTP servers.

However, there were issues. In particular the technology available for inclusion of the controlled vocabulary terms into metadata documents comprised the copying and pasting of identifiers or even labels from one desktop office application into another. There was no mechanism in place to ensure that the copied document was the latest version from the content governance or that it had not been the subject of local extensions. This caused almost farcical situations when locally-validated metadata documents were submitted to holding centres only to be rejected because controlled fields failed to validate centrally.

Whilst SEASEARCH was running, one of the significant technology developments for fully interoperable metadata and data was reaching maturity. This was Extensible Markup Language (XML). SEASEARCH was based on ASCII files conforming to format and Microsoft Access database files. These were impossible to verify without the development of bespoke conformance checkers. However, XML delivered document structural conformance checking through generic tools. Furthermore, through Schematron technology it provided the means to verify field contents against a centrally-maintained controlled vocabulary.

SEASEARCH was followed by SeaDataNet (Schaap et al., 2017, chapter 7 this book) with the objective of building a fully interoperable pan-European distributed data system based on XML technology. The controlled vocabulary technical governance requirements of SeaDataNet were met through the development of a vocabulary server by BODC. The fundamental design of this system was that each controlled vocabulary term was encoded in metadata documents and data files as a URN. This URN resolved to a URL that in turn resolved to an XML document containing the attributes of the term such as preferred label and definition.

Since the establishment of these basic principles, technological developments have had a profound influence on the management of data labels through vocabulary server technology. Two standards developed through W3C have been particularly influential. The first of these is the Simple Knowledge Organisation System (W3C, 2012b), which provides both a basic framework for controlled vocabulary organisation with concepts mapping to the parameter codes of the 1980s aggregated into collections, which in turn map to controlled vocabularies. These SKOS elements have attributes that provide a rich XML document content model for controlled vocabularies. Furthermore, SKOS includes a simple set of mappings between concepts, which provides the basis for the development of multiple controlled vocabularies into basic ontologies.

This ability empowers the semantic modelling of parameter labelling described above to provide both functional semantic interoperability through model mapping and a 'Linked Data' experience for users reading parameter labels. Consider a typical contaminant in biota parameter label for cadmium concentrations in the edible mussel, which is represented by the URL http://vocab.nerc.ac.uk/collection/P01/current/CDDWCF14/ (NERC, 2015).

This delivers a human-readable document (Table 1) containing the URL, URN, preferred label, alternative label and definition for the concept. In addition it includes a set of mappings to a categorisation vocabulary (P02), a suggested unit of measure (P06) plus a five-element semantic model:

- S06 - Measured phenomenon (Concentration)
- S27 - Chemical substance (cadmium)
- S02 - Measured phenomenon to matrix relationship (per unit dry weight of)
- S26 - Matrix (biota)
- S25 - Biological entity (Mytilus edulis (ITIS: 79454: WoRMS 140480) [Subcomponent: flesh])

Table 1. Styled XHTML document for example URL

URI	http://vocab.nerc.ac.uk/collection/P01/current/CDDWCF14/_
Identifier ()	SDN:P01::CDDWCF14
Preferred label (en)	**Concentration of cadmium {Cd CAS 7440-43-9} per unit dry weight of biota {Mytilus edulis (ITIS: 79454: WoRMS 140480) [Subcomponent: flesh]}**
Alternative label (en)	DryWtCd_BE006336
Definition (en)	The amount (mass or number of moles) of the specified analyte per unit weight of the specified organism or part thereof after water has been removed.
Version Info ()	4
Deprecated()	false
Broader	http://vocab.nerc.ac.uk/collection/P02/current/BCMT/_
Broader	http://vocab.nerc.ac.uk/collection/S06/current/S0600045/_
Broader	http://vocab.nerc.ac.uk/collection/S25/current/BE006336/_
Broader	http://vocab.nerc.ac.uk/collection/S26/current/MAT01963/_
Broader	http://vocab.nerc.ac.uk/collection/S27/current/CS002363/_
Related	http://vocab.nerc.ac.uk/collection/P06/current/UMKG/_
Related	http://vocab.nerc.ac.uk/collection/S02/current/S041/_
Date ()	2015-07-02 09:01:52.0

The other four elements of the semantic model haven't been specified in this case and so are omitted.

Furthermore, following some of these links, such the chemical substance or biological entity leads to further mappings, including external resources such a ChEBI and WoRMS.

Functional semantic interoperability is provided through the second influential RDF standard, which is the Resources Description Framework (RDF). This provides an XML structure capable of presenting the SKOS content model above in a machine-readable form. If the document shown above has its stylesheet removed then it can be seen that it is an RDF document thus:

```
<?xml version="1.0" encoding="UTF-8" ?>
- <rdf:RDF xmlns:rdf="http://www.w3.org/1999/02/22-rdf-syntax-ns#"
xmlns:skos="http://www.w3.org/2004/02/skos/core#" xmlns:dc="http://purl.org/
dc/terms/" xmlns:dce="http://purl.org/dc/elements/1.1/" xmlns:rdfs="http://
www.w3.org/2000/01/rdf-schema#" xmlns:owl="http://www.w3.org/2002/07/owl#"
xmlns:void="http://rdfs.org/ns/void#">
- <skos:Concept rdf:about="http://vocab.nerc.ac.uk/collection/P01/current/CD-
DWCF14/">
  <skos:prefLabel xml:lang="en">Concentration of cadmium {Cd CAS 7440-43-9}
per unit dry weight of biota {Mytilus edulis (ITIS: 79454: WoRMS 140480) [Sub-
component: flesh]}</skos:prefLabel>
  <skos:altLabel xml:lang="en">DryWtCd_BE006336</skos:altLabel>
  <skos:definition xml:lang="en">The amount (mass or number of moles) of the
specified analyte per unit weight of the specified organism or part thereof
after water has been removed.</skos:definition>
```

```
<dc:identifier>SDN:P01::CDDWCF14</dc:identifier>
<dce:identifier>SDN:P01::CDDWCF14</dce:identifier>
<skos:notation>SDN:P01::CDDWCF14</skos:notation>
<owl:versionInfo>4</owl:versionInfo>
<dc:date>2015-07-02 09:01:52.0</dc:date>
<skos:note xml:lang="en">accepted</skos:note>
<owl:deprecated>false</owl:deprecated>
<skos:broader rdf:resource="http://vocab.nerc.ac.uk/collection/P02/current/BCMT/" />
<skos:related rdf:resource="http://vocab.nerc.ac.uk/collection/P06/current/UMKG/" />
<skos:related rdf:resource="http://vocab.nerc.ac.uk/collection/S02/current/S041/" />
<skos:broader rdf:resource="http://vocab.nerc.ac.uk/collection/S06/current/S0600045/" />
<skos:broader rdf:resource="http://vocab.nerc.ac.uk/collection/S25/current/BE006336/" />
<skos:broader rdf:resource="http://vocab.nerc.ac.uk/collection/S26/current/MAT01963/" />
<skos:broader rdf:resource="http://vocab.nerc.ac.uk/collection/S27/current/CS002363/" />
<void:inDataset rdf:resource="http://vocab.nerc.ac.uk/.well-known/void" />
</skos:Concept>
</rdf:RDF>
```

As with every useful technology, vocabulary serving is not monopolised by a single system. In addition to NVS, other servers have been implemented in the oceanographic domain such as the Marine Metadata Interoperability (MMI) Ontology Registry and Repository (http://mmisw.org/) in the USA and Spatial Information Services Stack Vocabulary Service (SISSVoc) (Cox et.al, in press). From a user perspective, it is highly desirable to be able to ask the same questions of each of these. A third W3C standard, SPARQL, is a query language for vocabularies that allows users to do this, providing a suitable endpoint is provided by the vocabulary server. Furthermore, it is possible to federate multiple endpoints such that a single user query delivers hits from multiple vocabulary servers.

In the case of NVS, the SPARQL endpoint is implemented using the Apache Jena Fuseki (2015) open source software. The master copies of the controlled vocabularies are held in an Oracle relational database schema, which is translated daily into a set of RDF triples that are then placed in a Fuseki triple store linked to a Fuseki SPARQL endpoint (http://vocab.nerc.ac.uk/sparql/).

It is becoming clear as semantic operability implementation develops in projects such as Ocean Data Interoperability Platform (ODIP) described in Glaves, H. (2017, chapter 3 this book) that the simple set of mapping predicates provided by SKOS is insufficient. Some parameter mappings between separately-governed vocabularies can be semantically challenging. For example, there is a concept in the BODC Parameter Usage Vocabulary for '2m air temperature', a standard in meteorology, but in the CF Standard Names there is only 'air_temperature' plus a co-ordinate variable 'height' in the data files with a value of 2 and the unit of measure 'metres'. Matching these automatically requires predicates with rich semantics plus sets of rules. In other words full ontologies managed using technologies such as OWL (W3C, 2012a) or even collaborative development of new semantic standards in projects like ODIP. This is where the future of labelling data now lies.

AUTONOMOUS OCEAN OBSERVATION CASE STUDY

Argo is an operational array of neutrally buoyant freely drifting Lagrangian floats deployed throughout the world's oceans. The floats have the capability to change their buoyancy and hence their depth below the ocean surface by pumping oil into an external bladder, changing the volume and thus the buoyancy. The project aims to deploy of one float in each 3 by 3 degree region of the ocean deeper than 2,000 m. The original scientific goal was to measure ocean currents and for this reason the core array primarily consists of floats that park at 1,000 m depth. Every ten days each float descends to 2,000 m then ascends to the surface measuring temperature and salinity. On reaching the surface it transmits the data using the ARGOS or Iridium satellite communication networks before returning to its park depth. Floats typically have a life span of 5 years with a few floats reaching 10 years or more (Roemmich et al. 2012).

Data management was central to the design of the Argo project at its inception. Data are initially processed by national-level data centres before transmission to the GTS and copying the data across to two mirrored Global Data Assembly Centres (GDAC) where they are made available to data users. The floats freely drift with the ocean currents so frequently pass through the Economic Exclusion Zones (EEZ) of multiple nations. The Law of the Sea dictates that data must be made available when an EEZ is crossed so the decision was made that the inception of Argo that all data would be publicly available.

The Argo project highlighted the need for archival of raw data and unambiguous metadata in 2009 when differences in the behavior of different pressure sensors in differing float types become apparent. The issue had significant effects on the quality of the data returned including biases in the depth of the oceanic thermocline (Barker et al. 2011). Early in the Argo project there were three manufacturers for the pressure sensors used in the Seabird SBE41 instrument that were installed on the majority of the floats. Each sensor type has different long term characteristics with Amatek and Paine sensors exhibiting a slow positive drift to +8 dBar over the float's life. Druck sensors had a pathology called "snow-flakes" where particles precipitated out of the oil shorting out the electrodes in the sensor and causing large positive data spikes. This was complicated by the manufacturer producing the most widely used float only transmitting any positive pressure sensor drifts. During 2008 the welding using in construction of Druck pressure sensors was changed without informing customers or data users. The new technique was fundamentally flawed causing the welds to crack so oil leaked from the sensors leading to a negative measurement drift and failure of the sensor within 2 years. On the floats that only transmitted positive pressure drifts this negative drift was uncorrectable.

At the same time these problems were identified calibrated data were urgently required by the scientific community for assimilation into models and data analyses for heat content and sea level estimates for the IPCC 5th assessment of global climate change. In order to provide this the following issues needed resolution:

- Multiple float types with different methodologies for transmitting and handling pressure sensor drift
- Argo project members and an external scientific community where no single group understood all the float types
- Three different pressure sensor pathologies
- Insufficient metadata to unambiguously identify affected sensors and thus consistently correct the Argo dataset

To resolve these issues over two years the Argo data community documented float and sensor characteristics with a controlled vocabulary, applied corrections to the dataset, flagged data where floats potentially had uncorrectable pressure drifts, and liaised with manufactures to ensure all the required engineering data was transmitted by the float for future deployments. The Argo pressure issues are not unique in oceanography and another example came from the global XBT dataset. XBTs are deployed as part of geophysics research cruises to determine the speed of sound in water for acoustic instrumentation, and by meteorological services to determine upper-ocean temperature. It was discovered that the assumed fall rate and thus depth measurement of XBTs was incorrect for many of the models. This led to a similar recalibration of the entire XBT dataset (Pedro et al, 2011). These lessons are being taken forward to fledgling observing programs such as Everyones Glider Observatories (EGO) which will be composed of autonomous submarine gliders and surface vehicles. Many of these platforms and sensors are novel with the final methodologies to produce a globally consistent dataset undetermined so it is not known which data will be crucial in future years. An example of such an issue from submarine gliders is the thermal lag in the conductivity sensor degrading the basic physics measurements. A solution exists for a particular sensor and platform configurations such as for the Slocum family of gliders with a free flushing sensor (Garau et al., 2011) but this is not generically applicable and research is ongoing for different glider and sensor configurations. To address these issues and uncertainty in future data needs key technical and engineering data are included in the exchange formats. Additionally, raw data are preserved in a long term archives. The goal is to reduce the effort required to resolve dataset wide issues as and when they are identified in coming years.

As of July 2015 Argo data have been used in 2123 publications and it has become essential that the data are accurately citable for the transparency of science and policy that the associated research influences. This means the version of data used in a publication is unambiguous thereby facilitating reproducible research. As a global collaboration usage of data needs to traceable in order to sustain funding. This has proved an issue for Argo as initial data citation efforts focused on static unchanging datasets. The Argo time series grows as the time series from individual floats grow and new floats are deployed. Argo data also change with time as data are screened and calibrations are updated annually. Also, major updates such as resolution of the pressure issues change the entire dataset. When the GDACs were established the need to be able to download historic versions of the data was not foreseen. US National Oceanographic Data Centre (NODC, now part of the National Centers for Environmental Information, NCEI) store weekly snapshots of the entire GDAC when they joined the program however to date only the most recent snapshot has been exposed to users.

Citation and publication of static datasets is now commonplace and described in the International Oceanographic Commission Data Publication Cookbook (Leadbetter et al, 2013) however the topic of dynamic data citation is an active area of research and the issue is not unique to the Argo project. Operational environmental observing systems from many disciplines have similar requirements. In 2013 DataCite defined the following approach for dynamic data citation (DataCite, 2015):

- Time slices
- Snap shots
- Cite the growing data with an access time

The lack of foresight to track changes in the dataset at the Argo GDACs means that in the case of Argo the third option does not guarantee data reproducibility for users. Consequently Digital Object Identifiers (DOI) have been assigned to monthly snapshots of data stored at the European Argo GDAC (the French Research Institute for Exploitation, Ifremer). This approach does mean that there is a growing number of DOIs for the dataset, which complicates the identification of data usage in the academic literature. Such measures are crucial for activities such as Argo to sustain funding. Efforts are ongoing to move to a single DOI that will encompass snapshots held at US NCEI and Ifremer. Identification of an individual snapshot will be via a substring extension to the single DOI. The use of persistent identifiers has become a core component of the design of e-infrastructures such as EUDAT, NEON and PANGAEA. In parallel to the Argo activity the Research Data Alliance data citation working group has been working on the citation of dynamic data via workshops and recommendations (Rauber, Asmi, van Uytvanck & Prőll, 2015). These recommendations will meet the data reducibility requirement. However, the proposed approach is not readily implementable on legacy data systems where resources are limited such as the Argo project and pursuing the citation of snapshots is the current strategy.

The Argo project is expanding its mission to sample in regional seas, at depths greater than 2,000 m, introducing biogeochemical sampling, deploying floats under ice, and extending the mission to cover near the diurnally changing ocean surface. This means the floats are being equipped with new sensors expanding to parameters traditionally only measured by research cruise. These immature sensors on autonomous platforms tend to be less accurate than the highly precise research ship based measurements and the established float sensors. In order for these data to be integrated into the operational data stream they need to be adequately described including parameter names, sensor metadata and indications of quality. Mapping Argo parameter names to CF standard names (CF Conventions and Metadata, 2015) and NERC PUV terms ensures they are unambiguously described. It also enables Argo to move away from using the existing vocabulary with 4 character names, which was desired to make the new parameter names human readable. An Argo vocabulary has been created for sensor models and the Argo formats support the inclusion of sensor precision and accuracy. To prevent data and engineering values where quality is not assured the Argo data management team made the decision to split biogeochemical data from core physics data in an ancillary file (the B-file). The core file is merged with proven variables from the B-File to produce a high quality merged data product for non-expert users. The B-file containing all the raw and derived data variables is also available to expert users. At depths greater than 2,000 m core physics data are of unknown quality because the existing conductivity-temperature-depth sensor is not proven or the new high precision variant is a an unproven prototype so the data are flagged as probably good to make data users aware that of the unknown quality level of the data. These strategies form a template for new sensors and variables to be introduced into the Argo project while ensuring the project continues to deliver high quality data.

The success of the Argo project meant it formed a template for activities that contribute to the Global Ocean Observing System (GOOS). Another component of GOOS is OceanSites, the international coordination group for fixed point ocean observatories. The OceanSites data format was adapted from the Argo format with adjustments to accommodate the different data type. Beyond current components of the GOOS the natural evolution for Argo floats are submarine gliders. By adding wings, movable ballast and hydrodynamic improvements to the float design the flow over the wings during vertical displacement in the water column produces horizontal motion introducing the ability to pilot the observation platform. The Gliders for Research, Ocean Observation and Management (GROOM) project was a design study to evaluate the requirements to set up a sustainable European glider infrastructure to safely operate individual and fleets of gliders in order to create a continuum of observations. GROOM included a task aiming to deliver a template for glider data management. It is envisaged that submarine gliders form part of the GOOS. A key deliverable of GROOM was to produce the Everyones Glider Observatories (EGO) format where EGO is the international submarine glider community. The EGO format was based on the OceanSites format reinstating elements of the Argo format to include metadata for autonomous platforms, and technical and engineering data after the lessons learnt from Argo pressure issues. To enable data interoperability the EGO format also includes mappings to the BODC PUV in addition to the CF standard names that were mapped to in the OceanSites and Argo formats. Although cosmetically similar the formats have one subtle but significant difference. The adoption of climate forecast conventions differs between the formats. Argo has the simplest possible implementation with additional attributes used on the data variables. The OceanSites community are actively working towards CF compliance. The EGO format is CF 1.6 compliant using a trajectory-profile feature type allowing the data to be used by different users where operational users prefer profile data and scientists need the trajectory in order to fully calibrate the data. The adoption of the Attribute Convention for Dataset Discovery (ACDD) (ESIP, 2015) also differs slightly between projects. The ACDD defines a set of NetCDF global attributes recommended for describing a NetCDF dataset to discovery systems such as Digital Libraries so inconsistencies in adoption hinders dataset interoperability.

Within the GOOS framework for ocean observing Essential Ocean Variables (EOV) and Essential Climate Variables (ECV) have been identified as the key observation to focus on as the GOOS develops (Ocean Observations Panel for Climate, 2015). The recently funded AtlantOS project is a trans-Atlantic collaboration aiming to bring together the data from multiple observing systems into a single data portal. One of the first elements of work is to map the EOVs and ECVs to the BODC PUV vocabularies and CF standard names to enable interoperability between datasets.

The GOOS is composed of multiple platforms from multiple manufacturers. This has led to numerous raw data formats. An example is that when the Argo dataset was reprocessed to enable a scientifically viable dataset of subsurface trajectory data over 150 data decoders were required. This proliferation of raw data formats is set to increase as the types of sensors and platforms increase placing an additional burden on data processors. A solution to this issue is to standardise the data at the earliest stage possible. The Open Geospatial Consortium (OGC) Sensor Web Enablement (SWE) initiative defines standards for retrieving and exchanging data from sensors (OGC, 2013; Diviacco et al., 2017, chapter 8 this book; Oggioni et al., 2017, chapter 9 this book). The current and pending standards are:

- Observations & Measurements (O&M) –The general models and XML encodings for observations and measurements.
- PUCK Protocol Standard – Defines a protocol to retrieve a SensorML description, sensor "driver" code, and other information from the device itself, thus enabling automatic sensor installation, configuration and operation.

- Sensor Model Language (SensorML) – Standard models and XML Schema for describing the processes within sensor and observation processing systems.
- Sensor Observation Service (SOS) – Open interface for a web service to obtain observations and sensor and platform descriptions from one or more sensors.
- Sensor Planning Service (SPS) – An open interface for a web service by which a client can 1) determine the feasibility of collecting data from one or more sensors or models and 2) submit collection requests.
- SWE Common Data Model – Defines low-level data models for exchanging sensor related data between nodes of the OGC® Sensor Web Enablement (SWE) framework.
- SWE Service Model – Defines data types for common use across OGC Sensor Web Enablement (SWE) services. Five of these packages define operation request and response types.

Implementation of OGC SWE has thus far been limited within the oceanographic community. One of the first implementations was by the Spanish national research council Consejo Superior de Investigaciones Científicas (CSIC) to retrieve metadata and data from ships' underway data as part of the SANY project (Bleier et al., 2009). Since then SWE has been implemented by several groups internationally using different software packages such as the database based 52North implementation (GmbH, 2016) and the THREDDS ncSOS (Unidata, 2015) extension that places a SOS service over a repository of NetCDF files. Many of these software packages are open source freely available from online repositories such as GitHub enabling community development. SWE uptake is increasing within the oceanographic community with all four successful project proposals in the 2013 Ocean of Tomorrow call focusing on new sensors for autonomous ocean observation including OGC SWE in their data formats.

Not only are the data types growing but physical data volumes are also growing with autonomous platforms capable of collecting as much data as the memory card attached to the platform can store. Types of data include video and passive acoustic data. The volumes are such that transmission of these data is not economical by satellite so they tend to be processed on platform recovery. Video data from a single research cruise deploying remotely operated vehicles can exceed 1TB in volume. Passive acoustic data volumes from autonomous vehicles are limited to the size of the media logging the data so can regularly exceed 16GB. These increases in data volume although small by scientific standards (the Large Hardon Collider at CERN collects petabytes per day when it is operational) are several orders of magnitude larger than traditional oceanographic data holdings. Efforts are ongoing to develop centralised processing, storage and delivery mechanisms on large computing infrastructure such as the UK JASMIN facility (STFC, 2014).

FUTURE RESEARCH DIRECTIONS

Over the past 40 years different communities have each developed their own controlled vocabulary to describe the parameters included in their data outputs. The challenge now faced is to build tools that are capable of automatically aggregating these data outputs, including working out which data streams may be merged without creating scientific nonsense. Techniques such as the description of parameters using a semantic model expressed in SKOS and knowledge capture through mapping goes some way towards this. However, it is clear that greater semantic richness in mapping operators and formalised machine-readable descriptions of the semantic models behind the parameter descriptions are required.

Once these are available communities will be in a position to develop tools that can make scientifically valid judgements and decisions.

The move into the digital age and ongoing increase in the number of platforms and environmental variables measured means that in order to make data readily ingestible by users data standards are required at the earliest stage possible in the data chain. The mapping of OGC SWE SensorML and Observations and Measurements standards for oceanographic data is ongoing in European Union funded research projects such as those in the 2013 Ocean of Tomorrow call. This development is including how data quality indicators can be unambiguously defined. This implementation is limited to the manufacturers and sensors linked to those projects but approach is readily extensible with openly available deliverables describing how the standards have been implemented making more widespread implementation possible.

Linked Data was an initially separate development to OGC SWE. Linked Data has introduced the Semantic Sensor Network and JSON-LD metadata formats akin to SensorML In order to maximise the number of users ideally metadata and data will be exposed in both OGC SWE and Linked Data standards. Elements to pursue this work are defined such as the PROV-O ontology (W3C, 2013) following Ma et al. (2017, chapter 5 this book) that will be key for linking SSN to O&M as it defines a standard for provenance or more precisely the entities, activities, and people involved in producing a piece of data or object. The exchange formats for oceanographic data that are part of the GOOS are predominately NetCDF. Linked Data attributes for NetCDF have been defined by Yu, Leadbetter, Simons and Cox (2015), which is a significant step towards the exposure of NetCDF data via Linked Data technology.

The issue of increasing data volumes in environmental science is being actively investigated. The UK Natural Environment Research Council (NERC) data centres are working on serving the data from high performance computing centres such as the NERC data analysis environment JASMIN. These data types also bring "new old issues" to the fore again though. One such example is passive acoustic data that are highly compressed in a non-standard format to allow the recording to be the maximum length possible. This means an exchange format needs to be agreed on and the data mapped to controlled vocabularies before it can be fully served to data users in an interoperable manner.

Unambiguous data citation for dynamically growing and mutating datasets or use in data products presents a continuing issue. Proposed methodologies have emerged from the work of DataCite and the Research Data Alliance. These now need implementing and evaluating. Internationally ratified approaches for how to make dynamic data from legacy data centre infrastructure where full implementation is not possible such as for Argo where development resource is limited are also required.

CONCLUSION

Controlled vocabulary Content Governance would be impossible without the knowledge base currently accessible through the internet. Its processes have been significantly enhanced though the availability of increasingly sophisticated collaboration tools.

Controlled vocabulary Technical Governance has been shown by historical example to be fatally flawed without the functionality provided by vocabulary servers based on the design principles of Linked Data and the Semantic Web.

The application of controlled vocabularies across the different components of the GOOS will enable interoperability between datasets aiding the integration of data from different platform types. The goal being to reduce the resource required to aggregate data for use in science, environmental monitoring, production of data products and commercial use of data.

Application of controlled vocabularies and common formats to data at source using standards such as OGC SWE and Linked Data will enable rapid (and automated cross discipline) ingestion and assimilation of data by users both within and beyond the oceanographic domain.

Both the Argo and XBT datasets have shown that storage of raw data and transfer of key engineering data to exchange formats is required for ocean observing networks. This will reduce the impact of systematic issues when they are identified years after data were initially collected and disseminated.

REFERENCES

Atlassian. (2016). *Confluence – Team collaboration Software*. Retrieved April 5, 2016, from https://www.atlassian.com/software/confluence/

Barker, P. M., Dunn, J. R., Domingues, C. M., & Wijffels, S. E. (2011). Pressure Sensor Drifts in Argo and Their Impacts. *Journal of Atmospheric and Oceanic Technology*, *28*(8), 1036–1049. doi:10.1175/2011JTECHO831.1

Bizer, C., Heath, T., & Berners-Lee, T. (2009). Linked Data—The Story So Far. *International Journal on Semantic Web and Information Systems, 5*(3), 1–22. doi:10.4018/jswis.2009081901

Bleier, T., Bozic'Roman, B., da Costa, A., Costes, S., Iosifescu, I., Martin, O., … Simonis, I. (2009). *SANY and open service architecture for sensors networks*. Retrieved October 30, 2015, from http://www.frisia-it.com/assets/images/SANY_Book.pdf

Centre for Environmental Data Archival (CEDA). (2016). *Proposals for CF*. Retrieved April 5, 2016, from http://cfeditor.ceda.ac.uk/proposals/1?status=active&namefilter=&proposerfilter=&descfilter=&filter+and+display=filter

CF Community. (2015). *CF Conventions and Metadata*. Retrieved October 30, 2015, from http://cfconventions.org/standard-names.html

CF Community. (2016). *NetCDF Conventions*. Retrieved October 30, 2015, from http://www.unidata.ucar.edu/software/netcdf/conventions.html

Claustre, H., Bishop, J., Boss, E., Bernard, S., Berthon, J. F., Coatanoan, C., . . . Uitz, J. (2009). Bio-optical profiling floats as new observational tools for biogeochemical and ecosystem studies: potential synergies with ocean color remote sensing. *Proceedings of OceanObs'09: Sustained Ocean Observations and Information for Society*. Retrieved February 10, 2016, from http://www.ioccg.org/groups/Claustre_et_al._OO09.pdf

Cox, S., Yu, J., & Rankine, T. (in press). SISSVoc: A Linked Data API for SKOS vocabularies. Submitted to Semantic Web - Interoperability, Usability. *Applicability*.

CRC. (1986). *CRC Handbook of Chemistry and Physics* (67th ed.). CRC.

DiNezio, P. N., & Goni, G. J. (2011). Direct Evidence of a Changing Fall-Rate Bias in XBTs Manufactured during 1986–2008. *Journal of Atmospheric and Oceanic Technology*, *28*(11), 1569–1578. doi:10.1175/JTECH-D-11-00017.1

Diviacco, P. (2016). *Collaborative toolkit for Scientific Project Development*. Retrieved February 12, 2016, from http://colla.ogs.trieste.it/COLLA.2/

Diviacco, P., Sorribas, J., De Cauwer, K., Sinquin, J. M., Casas, R., Busato, A., & Scory, S. et al. (2017). Repositioning data management near data acquisition. In *Oceanographic and Marine Cross-Domain Data Management for Sustainable Development*. Hershey, PA: IGI Global.

European Molecular Biology Laboratory European Bioinformatics Institute (EMBL-EBI). (2016). *Chemical Entities of Biological Interest (ChEBI)*. Retrieved April 5, 2016, from https://www.ebi.ac.uk/chebi/

Federation of Earth Science Information Partners (ESIP). (2015). *Category: Attribute Conventions Dataset Discovery*. Retrieved February 15, 2016, from http://wiki.esipfed.org/index.php/Category:Attribute_Conventions_Dataset_Discovery

Garau, B., Ruiz, S., Zhang, W. G., Pascual, A., Heslop, E., Kerfoot, J., & Tintoré, J. (2011). Thermal Lag Correction on Slocum CTD Glider Data. *Journal of Atmospheric and Oceanic Technology*, *28*(9), 1065–1071. doi:10.1175/JTECH-D-10-05030.1

Glaves, H. (2017). Developing a Common Global Framework for Marine Data Management. In *Oceanographic and Marine Cross-Domain Data Management for Sustainable Development*. Hershey, PA: IGI Global.

Jena, A. (2015), *A free and open source Java framework for building Semantic Web and Linked Data applications (2015)*. Retrieved October 30, 2015, from https://jena.apache.org/index.html

Lea, D. J., Martin, M. J., & Oke, P. R. (2014). Demonstrating the complementarity of observations in an operational ocean forecasting system. *Quarterly Journal of the Royal Meteorological Society*, *140*(683), 2037–2049. doi:10.1002/qj.2281

Leadbetter, A., Cheatham, M., Shepherd, A., & Thomas, R. (2017). Linked Ocean Data 2.0. In *Oceanographic and Marine Cross-Domain Data Management for Sustainable Development*. Hershey, PA: IGI Global.

Leadbetter, A., Raymond, L., Chandler, C., Pikula, L., Pissierssens, P., & Urban, E. (2013). *Ocean Data Publication Cookbook, IOC Manuals and guides No. 64*. Retrieved October 30, 2015, from http://www.iode.org/index.php?option=com_oe&task=viewDocumentRecord&docID=10574

Lowry, R., Bird, L., & Haaring, P. (2007): A semantic modelling approach to biological parameter interoperability. In *Proceedings Ocean Biodiversity Informatics:International Conference on Marine Biodiversity Data Management*. VLIZ.

Ma, X., Fu, L., Fox, P., Di Stefano, M., & West, P. (2017). Documenting Provenance for Reproducible Marine Ecosystem Assessment in Open Science. In *Oceanographic and Marine Cross-Domain Data Management for Sustainable Development*. Hershey, PA: IGI Global.

Met Office. (2015). *Earth system model development*. Retrieved February 15, 2016, from http://www.metoffice.gov.uk/research/areas/chemistry-ecosystems/earth-system

National Oceanography Centre. (2015). *SEA-SEARCH*. Retrieved April 5, 2016, from http://www.bodc.ac.uk/projects/european/seasearch/

Natural Environment Research Council (NERC). (2015). *Concentration of cadmium {Cd CAS 7440-43-9} per unit dry weight of biota {Mytilus edulis (ITIS: 79454: WoRMS 140480) [Subcomponent: flesh]}*. Retrieved April 5, 2016, from http://vocab.nerc.ac.uk/collection/P01/current/CDDWCF14/

52° North Initiative for Geospatial Open Source Software GmbH. (2016). *52norrth exploring horizons.* Retrieved February 14, 2016, from http://52north.org/

Ocean Observations Panel for Climate. (2015). *Essential variables.* Retrieved October 30, 2015, from http://ioc-goos-oopc.org/obs/ecv.php

Oggioni, A., Tagliolato, P., Fugazza, C., Pepe, M., Menegon, S., Pavesi, F., & Carrara, P. (2017). Interoperability in marine sensor networks through SWE services. In *Oceanographic and Marine Cross-Domain Data Management for Sustainable Development*. Hershey, PA: IGI Global.

Open Geospatial Consortium. (2013). *OGC® Sensor Web Enablement: Overview and High Level Architecture.* Retrieved October 30, 2015, from http://docs.opengeospatial.org/wp/07-165r1/

Rauber, A., Asmi, A., van Uytvanck, D., & Pröll, S. (2015). *Data Citation of Evolving Data, Recommendations of the Working Group on Data Citation (WGDC), Revision September 24th 2015.* Retrieved October 30, 2015, from https://rd-alliance.org/system/files/documents/RDA-DC-Recommendations_150924.pdf

Roemmich, D., Gould, W. J., & Gilson, J. (2012). 135 years of global ocean warming between the Challenger expedition and the Argo Programme. *Nature Climate Change*, *2*(6), 425–428. doi:10.1038/nclimate1461

Schaap, D. (2017). SeaDataNet – Towards a Pan-European Infrastructure for Marine and Ocean Data Management. In *Oceanographic and Marine Cross-Domain Data Management for Sustainable Development*. Hershey, PA: IGI Global.

Science and Technology Facilities Council (STFC). (2014). *JASMIN, Petascale storage and cloud computing for big data challenges in environmental science.* Retrieved April 5, 2016, from http://www.jasmin.ac.uk

Starr, J., Ammann, N., Ashton, J., Barton, A., Elliott, J., Jacquemot-Perba, M., … Ziedorn, F. (2015). *DataCite Metadata Schema for the Publication and Citation of Research Data, Version 3.1 August 2015.* doi:10.5438/0010

UNESCO. (1987). GF3: A General formatting system for geo-referenced data. Vol. 2. Technical description of the GF3 format and code tables. Intergovernmental Ocanographic Commission Manuals and Guides.

Unidata. (2015). *THREDDS Data Server (TDS).* Retrieved February 15, 2016, from http://www.unidata.ucar.edu/software/thredds/current/tds/

Unidata. (2016). *Network Common Data Form (NetCDF).* Retrieved February 15, 2016, from http://www.unidata.ucar.edu/software/netcdf/

Vlaams Instituut Voor De Zee (VLIZ). (2016). *WoRMS: World Register of Marine Species.* Retrieved April 5, 2016, from http://www.marinespecies.org/

W3C. (2012a). *Web Ontology Language (OWL)*. Retrieved April 5, 2016, from http://www.w3.org/2001/sw/wiki/OWL

W3C. (2012b). *SKOS Simple Knowledge Organization System - Home Page*. Retrieved April 5, 2016, from http://www.w3.org/2004/02/skos/

W3C. (2013). *PROV-O: The PROV Ontology*. Retrieved February 15, 2016, from https://www.w3.org/TR/prov-o/

Yu, J., Car, N. J., Leadbetter, A., Simons, B. A., & Cox, S. J. D. (2015). Towards Linked Data Conventions for Delivery of Environmental Data Using netCDF. Springer.

KEY TERMS AND DEFINITIONS

Argo: Argo is a global array of more than 3,000 free-drifting profiling floats that measures the temperature and salinity of the upper 2000 m of the ocean.

Autonomous Ocean Observation: Observation and measure of the ocean using fully or semi-autonomous robotic platforms equipped with sensor payloads.

Controlled Vocabulary: A collection of concepts that may populate a given field in a metadata document or data model.

Data Citation: The unambiguous citation of data by persistent universal resource identifiers allowing reproducible science and tracking of data usage in the scientific literature.

Global Ocean Observing System: a permanent global system for observations, modelling and analysis of marine and ocean variables to support operational ocean services worldwide.

Global Telecommunications System (GTS): The coordinated global system of telecommunication facilities and arrangements for the rapid collection, exchange and distribution of observations and processed information within the framework of the World Weather Watch.

Linked Data: The W3C recommended best practice for exposing, sharing, and connecting pieces of data, information, and knowledge on the Semantic Web using URIs and RDF.

Open Geospatial Consortium (OGC) Sensor Web Enablement (SWE): OGC's defined standards enable developers to make all types of sensors, transducers and sensor data repositories discoverable, accessible and useable via the Web.

Parameter: The entity that is quantified by a measurement.

Vocabulary Content Governance: The process of determining the concepts to be included in a controlled vocabulary and their descriptive labels, especially their formal definitions.

Vocabulary Technical Governance: The process of storing controlled vocabularies, changing them as instructed by the Content Governance and making them accessible to both human and software agent users.

Chapter 2
Balancing Formalization and Representation in Cross-Domain Data Management for Sustainable Development

Paolo Diviacco
Istituto Nazionale di Oceanografia e di Geofisica Sperimentale - OGS, Italy

Adam Leadbetter
Marine Institute, Ireland

ABSTRACT

Sustainable development is grounded in complex systems where phenomena and processes are so interrelated and intermingled that they can only be studied in a multiple disciplinary manner. Limitiations in understanding such systems can be identified that are either epistemic and cognitive or socio-economic. Both cases are problematic for collaborative and multidisciplinary studies in sustainable development. This chapter explores these difficulties with a particular focus on data management, an activity that is at the same time very important and very sensitive to such problems. To allow users to find their way through data and information spaces, solutions are proposed that take advantage of either the formalization of knowledge; or of representations, such as schemas, mind maps or graphs. Through a careful balance of these two pathways this chapter demonstrates that it is possible to mitigate the difficulties mentioned above and allow a community to work together at profit.

INTRODUCTION

Sustainable development is the possibility to benefit from natural resources without incurring in their depletion or triggering negative side-effects in other domains. This dualism is also reflected in the United Nation's (UN) Sustainable Development Goals (UN resolution A/RES/70/1, 2015), where on one hand the UN lists goals such as end poverty and hunger or ensure healthy lives, while, at the same

DOI: 10.4018/978-1-5225-0700-0.ch002

time, claiming that these have to be achieved ensuring conservation of the environment, ecosystems and combatting climate change and its impacts.

This balance is very hard to obtain because it is very difficult to understand all the aspects of the related processes and phenomena and it is very complicated to try to control them. Indeed in sustainable development, understanding is problematic because:

- The necessary skills and competences come from multiple disciplines such as biology, geology, sociology or economics
- Since multiple fields are involved, research projects in this area are very likely to be collaborative and cross-disciplinary
- Approaches and methods needed for sustainable development related to scientific fields are often not as rigorous as those needed in other domains such as for example computer science or engineering
- The need to collaborate with other people can introduce issues that range from simple misunderstanding, to lack of awareness, tacit knowledge, up to politics or rhetoric

It is important to understand that these effects are not due to insufficient efforts spent during coordination or managing of projects but to the intrinsic nature of knowledge, and its usage, in the domain of sustainable development. It is a cultural problem. To address such issues, therefore, instead of organizing the work within a "one-size-fits-all" manner that has been devised for other sectors, we need to develop strategies and perspectives focused on the specific case of sustainable development. In this chapter we will consider the case of data management, an activity where all the above mentioned issues very clearly emerge since data, and the information built on top of those data, are at the base of all sustainable development activity. In this sense it is possible to highlight the existence of a gap between what is required by stakeholders and what can be offered by experts. This is not only a gap but a structural difference. The main focus of stakeholders is on norms, decisions and protocols, while scientists offer observations and hypotheses.

The availability of data is therefore overwhelmingly important. It is necessary to enable information to be built from different perspectives and ways of thinking by actors with different point of views, approaches and aims. To facilitate this, data should be understandable by all the designated community.

To address all this we will, in this chapter:

- Describe the mechanisms that are at the base of the issues that limit and obstruct handling of such topics
- Introduce the concept of representation and its implication in data management and in particular in the case where contrasting cognitive models access the same representation
- Introduce the problem of formalization of knowledge that can be applied to assist data access and to frame collaborative research activities
- Discuss the advantages and disadvantages of formalization and how this can be put in practice and the tools needed
- Discuss whether knowledge can be exploited even without the need to formalize it; how this can be put in practice and the tools needed

BACKGROUND

Sustainable Development

The Brundtland Commission's report (Brundtland, 1987) defined sustainable development as "development which meets the needs of current generations without compromising the ability of future generations to meet their own needs". The main point of interest of such a definition is the connection between economic and social development, in particular for people with a low economic standard of living; and the importance of protecting the natural resource base and the environment. If the latter is jeopardized, no improvement can be obtained in economic and social well-being. In addition the exploitation of resources has to consider a long term perspective where the effects of development will condition the environment for future generations.

The study of processes and phenomena in this context is based on two pillars: the socio-economic; and the environmental. Authors such as for example Murphy (2012) highlighted that a specific focus has been put on the environmental side, while socio-economic factors were considered less relevant. On the contrary, they are very important and very complex to handle since they are dependent on the context and on the capacity of each country to respond to each its social problems.

The complexity of the matter was mirrored also in the progressive loss of enthusiasm and rise of scepticism that from the United Nations Conference on Environment and Development in Rio de Janeiro in 1992, through the Johannesburg Summit lead to the current situation, where the focus moved from an all-encompassing attempt to address all issues simultaneously towards a more pragmatic approach where local and simpler issues are under examination.

Sustainable development is therefore a cross-domain field where researchers and stakeholders from different backgrounds, with different interests and different ways of thinking, meet. Processes and phenomena in sustainable development are, in short, complex systems.

MAIN FOCUS OF THE CHAPTER

Limitations of Reasoning in Sustainable Development

Complex systems are entities where multiple factors interact and intermingle at the same time. In these cases it is very difficult to isolate *a posteriori* the contribution that each of these factors might have exerted, while on the other hand it is important to identify them and understand how they can influence the whole system.

As mentioned above scientific disciplines involved in the study of sustainable development span a wide range, from environmental to socio economic studies. Practices, theoretical frameworks, even the mathematics used are often very distinct between disciplines. However, there is one common aspect that can be found in all these domains: the difficulty in providing exact quantitative, unbiased measurements. As a consequence, it is very difficult in the domain of sustainable development to apply what is generally understood to be the standard view of the scientific method. It should be noted that this does not apply only for "environmental related" disciplines, but maybe even more so in the case of socio-economic ones.

The standard view of the scientific method prescribes that scientific experiments should be reproducible, meaning that having set the same conditions we should get the same results. This rarely happens in environmental and socio-economic sciences since the factors governing a process and their interdependence are very sensitive to small variations (the so-called "butterfly effect"); they are generally nonlinear systems, so that it is very difficult to exactly replicate the configuration of an experiment. In these domains correlation often does not mean causation, highlighting the role of an underestimated parameter (Lawlor, Davey Smith & Ebrahim, 2004). In addition, laws are often built considering a low number of cases, or cases that only resemble what is under observation, which creates problems in finding analogies (Engelhardt & Zimmerman, 1982).

Another important factor which conditions the possibility to handle reliable measurements is theory laden-ness. This is the unintentional bias that researchers impose on measurements and observations. These latter are based on planned experiments that are based on one or more theories; meaning essentially to what someone is expecting to see. Therefore, quoting Hanson (1958)

Observation of x is shaped by prior knowledge of x.

This generally results in highlighting or neglecting information (Bishop, 1992) (see the section on representation later in this chapter). It is important to say that this happens at all levels, even at the level of perception (Diviacco & Busato, 2015).

The standard view of the scientific method is based on an inductive-deductive loop that from initial observations drives the formulation of a law that later through an experiment could be validated (or following Popper (1963), "corroborated"). It is easy to understand that in the domain of sustainable development, considering what has been said so far, the possibility to validate any theory through observations is very limited. To avoid the short circuit that this situation can generate, environmental and socio-economic sciences tend to substitute to the inductive/deductive loop with what Peirce (1958) called abduction. Abduction is a form of reasoning where the resulting state of affairs and the law is known, while the controlling state of affairs is sought (retrodiction). Abductive methods are very useful in the creation of new ideas[1].

Since abductive reasoning merely suggest that something maybe (Peirce, 1958) it is possible, and is often the case, that multiple explanations can coexist. Of course, adding this to the difficulty in setting experiments for a process or phenomenon, it is easy to see how complex and contradictory the activity of the experts in specific domains can be.

An important point in evaluating the contribution of science in understanding phenomena in domains such as sustainable development is the problem of objectivity and the commonly misleading definition of what is a fact and what is an opinion. In Diviacco (2015), and Diviacco and Busato (2015) these issues are described in depth, but several ideas of note are restated here to show how this can further limit the reliability of the standard methods to understand processes and phenomena in sustainable development.

The standard view depicts scientific research as an objective activity. To understand if this can be true we need first to understand what the term objective actually means. Objectivity is contrasted to subjectivity, and since subjectivity reasonably means "from someone's point of view", the resulting popular idea is that objectivity could be associated with "truth" and "reality", rather than how things may appear or might be imagined. This is an oversimplification. The common definition of objectivity refers to what should be actually called "metaphysical objectivity" meaning a form of objectivity that does not depend on experience, but of course we know that science is based on observation, and therefore we should revert to what we better identify as "epistemic objectivity". This can be defined as the sets of claims or judgments which truth values are not determined privately. Objectivity can be explained only as "not just from someone's point of view", objectivity is always limited to the point of view of a certain number of people that agree and converge on a shared vision.

Summing up subjectivity is private and objectivity is public or, to put it another way, only shared points of view are objective (Diviacco, 2015; Searle, 1995).

Now, having defined the difference between metaphysical and epistemic objectivity we can make a further step and understand what a fact is. Facts are metaphysically objective, and as such practically we cannot access to them. What we can handle are merely possible situations that can be expressed by non-contradictory claims that we call "state of affairs."

The Sociologic

We described above the limitations of reasoning in research fields related to sustainable development that force a "relaxed" approach towards the possibility to access reality. We highlighted that these limitations lead to a model where multiple points of view are very likely to emerge, and that therefore multiple competing theories can coexist (without necessarily clashing).

Any explanation of a phenomenon or scientific model, following Woo (1999), have two forms of approximations and uncertainties: the intrinsic ones that depend on the limitations of the model and the aleatory uncertainties that are related to the beliefs of each researcher. In fact each researcher focuses only on what he thinks are the more likely causes of a phenomenon and excludes others. This is bias, a problem which we are not generally prepared to handle. Popper (1963) maintains that bias is unavoidable, and that we instead need to identify it and prevent its influence on research. In this perspective Pshenichny (2004) proposes that bias should be explicitly formulated in the theories and provides tools to take bias into account.

We showed that the state of affairs, meaning what experts consider in their everyday activities, can be seen as shared spaces on which a community needs to converge. It is therefore easy to understand how reasoning in the fields related to sustainable development is strongly conditioned by the social connection of the people working within the same project. Each actor brings his/her bias. It is easy to understand that if identifying the bias of a single actor could theoretically be possible, the interaction between biases of a community can be seen itself as a complex system.

Kuhn (1962) showed that scientists tend to isolate themselves in what he called "paradigms", meaning a mental environment defined by the personal history of the single researcher, their points of view, the tradition they are part of, their models and myths, in general their background. Paradigms condition their "normal" scientific life and tend to gather scientists in groups that, following Becher and Trowler (2001), resemble "tribes". They tend to evolve separately and eventually cannot understand each other (incommensurability). In this space, following Whitley (2000), control over facilities and resources of research define status and identity of researchers and tribes. Following Latour and Woolgar (1979), the lives of scientists and experts are conditioned by the network of allies and by the strategic position that the researcher is able to maintain in the community. In few words "Science is a social construct" (Latour & Woolgar, 1979).

Communication

In this landscape we have to consider a further complication, which is communication. Traditionally, this is layered into syntax, semantics and pragmatics (Morris, 1971, Diviacco 2013). Due to paradigms and incommensurability it is easy to understand that communication can be complex and prone to misunderstanding. These effects can be traced at all levels of communication, but while sometimes efforts are spent in trying to agree on a common syntax, those spent in trying to achieve a common understanding and meaning are generally very problematic, while within the scope of pragmatics they are commonly not even considered.

Shannon and Weaver (1949) provided methods to overcome communication difficulties. However these methods were mainly focused on the syntactic level and are more related to the technical implementation of communication. They do not address problems related to the human "contribution" to these difficulties.

Following Mirowski (2002) in Shannon's information theory, information is something that stands apart from and is independent of the processor of information and its internal characteristics, then it deals with encoding, transmission and decoding, not caring about the ability to understand of the payload.

In the case of sustainable development, instead, as we demonstrated, the role of the human-human interaction and therefore the need for a convergence of a community is paramount

We can identify essentially two classes of issues that can generate such difficulties, and namely:

- Epistemic and cognitive ones: that are generally structural, intrinsic, often innate and unconscious, such as tacit knowledge and theory laden-ness
- Socio-economic ones: that are conscious and voluntary, where actors aim to exert an influence on a process, project or activity in general

In the first case, we already described what theory laden-ness is; we will now briefly describe what tacit knowledge is. Tacit knowledge is knowledge that cannot be made explicit because obscured or embedded in practices or even tools. Following Polanyi (1966)

We know more than we can tell.

Tacit knowledge therefore remains often outside communication since cannot be captured.

Socio economic issues are also very relevant. In any community, in order to have control over the evolution of a project, partners are committed to the development and successful strategic positioning of their visions and of themselves as an "obligatory passage point" (Latour & Woolgar, 1979). Actors' contributions will then be ideological and political, which introduce another dimension to take under control in communication. Following David and Bunn (1988) who analysed the possible economic interests that can interfere with the quest for interoperability, this has to be considered at all communication levels: pragmatic and semantic of course, but even at the syntactic one.

SOLUTIONS AND RECOMMENDATIONS

Summing up, in the domain of sustainable development, on one hand we can have serious intrinsic reasoning issues in understanding processes and phenomena, on the other side we have complex social interactions that can affect knowledge building. How can we be optimistic about the possibility to devise any data management tool that can be used in this domain, where, as mentioned above, collaborative work is expected?

Knowledge

Knowledge can be associated with the acquisition and processing of information that under certain conditions can be used, expressed, shared, and transmitted. The syntax/semantics/pragmatics tri-partition we described in the section about communication, can be roughly associated with the distinction between data, information and knowledge (to delve more into the argument please see Diviacco, 2013 and Diviacco, 2015). The problem we have to deal with is that traditionally only a one-directional path from data towards knowledge was identified in knowledge building. On the contrary, as we demonstrated above, epistemic, cognitive and socio-economic factors, following the opposite direction condition severely information and data acquisition and management (Fox 2014, Diviacco 2013). How can we cope with this?

We can follow essentially two paths:

- Knowledge formalization (controlled vocabularies, ontologies, archetypes). This is based on the identification and definition of the concepts that constitute that particular abstraction of the world and on the construction of a map of their relations. This environment can be later used to classify and reason on the available data and information, and reconfigure that knowledge to adapt to other paradigms
- Non-formalized knowledge (boundary objects, representation). This is based on the introduction of artefacts that could be used to allow multiple paradigms to access the same data and information space from different point of view.

Knowledge Formalisation

In order to truly pursue sustainable development - which in the definition above takes place over the time scale of generations, when data, information or knowledge are captured there must be enough supporting information alongside the new data to allow future users to re-analyse, re-visualise and reinterpret the data without recourse to the data originator. Once this is achieved, environmental observations may be used to evaluate future problems of sustainable development, and even to assess the sustainability of certain development choices. It is also instructive to consider how data becomes knowledge.

A History of Knowledge Formalisation in Marine Data Management

Whilst this may seem at first to be a utopian vision for data managers, it is in fact the mantra that has been used within the International Oceanographic Data and Information Exchange (IODE) community for many years as a way of showing the data management process adds value to data (Lowry, Rickards, & Brown, 2005). This has required the formal definition of marine data models, which have evolved over many years and iterations. Whilst the history of the common vocabulary definitions used within these data models is explored in detail by Buck and Lowry (2016, see chapter 1 this volume) a brief overview of the formal data models produced in the marine science domain is of use in setting the scene for this, and subsequent chapters. Initially, knowledge formalisation in the marine sciences was through definitions of common metadata schema and file formats for exchange of information and results between partners in specific projects. The classical example of this is the GF3 format, developed by the Intergovernmental Oceanographic Commission's International Oceanographic Data and Information Exchange (IODE) and first proposed in 1979 as a generalized formatting system for general use in the transfer of marine science data (UNESCO, 1987). A key, indeed prescient, feature within GF3 was the publication of common vocabularies, in a paper form, which is a theme we will return to later in this section in a World Wide Web context. GF3 was eventually refined by the IODE Group of Experts on the Technical Aspects of Data Exchange (GETADE). The GETADE Data Format was developed in response to the perceived agedness of GF3, which had led to many national oceanographic data centres developing their own file formats proving to cause a barrier to the exchange of data between nations. However, the knowledge formalisation within GETADE remained within the GF3 code tables. The first real move away from this model was the Marine XML project. Marine XML was not a project designed to create a new XML standard encoding, but to provide the crosswalks between various standards through knowledge encapsulation. The results of the Marine XML project showed that there was still work to be done on mapping various metadata schema from the different national oceanographic data centres and domain projects; and that access to the GF3 codelists should be improved (Isenor & Lowry, 2005). By the time of the Marine XML project, the codelists had moved from paper to FTP site downloads, but follow-on work in the EnParDis and NETMAR projects and through the use of the successor, web-enabled controlled vocabularies in the SeaDataNet infrastructure has brought these to modern web standards (Leadbetter, Lowry, & Clements, 2014).

The Marine XML project began to bridge the gap between data and knowledge through its crosswalk paradigm, but how can we understand a more general flow of data to knowledge? The data ecosystem may be seen to encapsulate the data-information-knowledge-wisdom (DIKW) model of Shedroff (1994). An ecosystem grows, evolves and thrives on diversity. In the DIKW model, data may grow through presentation and organisation into information. Information itself may evolve into knowledge as consumers from a wider context than the data were originally collected in bring their experience to conversations about the data and information which can then tell new stories. Taking the example of marine science data above, the raw data straight from the research vessel cruise obviously fits into the data part of the model. Through organisation and presentation of the raw data into something cleaner and more usable, at the "beta" stage in the software release metaphor, the lines between data and information have become blurred and the data are evolving towards information. At the "release" stage where the data are organised and presented in such a way that they are totally reusable, and may even have a dataset description paper from an academic journal to accompany them, the evolution to information is complete. These "information-sets" can now provide the stimulus for researchers to converse about what they contain and to begin building knowledge, While many of the drivers for data sharing examined in this paper have come from the sciences, the idea of research data evolving into information, knowledge and wisdom has touchstones in the arts, particularly the field of literature. As Will Self (Guardian, 2008) notes, creative writing is "not building things. You are removing things, chipping away at language to reveal a living form." Indeed, as scholarly communication in all disciplines is currently predicated on the peer-reviewed, written research paper it could be seen that all synthesis of research data into the information and knowledge these papers contain is a creative writing process.

The Current State of the Art in Marine Data Management Knowledge Formalisation

There are emergent technologies throughout the fields around oceanography which we will incorporate into oceanography, and through that convergence we will make oceanography into something even more magical. (Delaney, 2010)

In the emerging Web Science (the study of large-scale socio-technical systems, such as the World Wide Web and the relationship between people and technology, the ways that society and technology co-constitute one another and the impact of this co-constitution on broader society, Shadbolt & Berners-Lee, 2008) the concepts of both the Semantic Web and Linked Data are very powerful tools in the formalisation of knowledge. Indeed the Semantic Web was born of Tim Berners-Lee's vision to create a Web of Data analogous to the World Wide Web of documents (Berners-Lee, 1998). The Semantic Web is a standards based approach to the Web of Data which promotes common data formats and exchange protocols on the Web, most fundamentally the Resource Description Framework (RDF). According to the World Wide Web Consortium, "the Semantic Web provides a common framework that allows data to be shared and reused across application, enterprise, and community boundaries" (Herman, 2008). The Semantic Web encapsulates knowledge in a spectrum of formalisation, from low to high. The points of this spectrum, often called the "ontology spectrum" or "semantic spectrum", are defined by McGuinness (2003). In overview, the simplest point on the spectrum is the catalogue (or glossary, as in Navigli & Velardi, 2008) that simply provides a list of terms used to annotate data and their definitions. A catalogue is not necessarily published to either the WWW or Semantic Web. Nor does a catalogue require a group of experts to provide governance over its content. Complexity is increased in the controlled vocabularies (Warner, 2002) and data dictionaries (Kumpati, 1985, IBM, 1993) which add, respectively, naming conventions and publication on top of the catalogue. A data model (van Renssen, 2005) then adds some definition of the relationship between the entities of a data dictionary. A taxonomy further adds a tree hierarchy with the concepts at the bottom of the tree inheriting the properties of those concepts at the top of the tree (Navigli & Velardi, 2008). A concept can be viewed as the definition of an idea or notion; a unit of thought (Miles & Bechhofer, 2009). A more complex point of the semantic spectrum is the ontology. An ontology describes all the concepts in a given domain and the relationships between them, and following best practice (Cox, 2012), is conceptualised using Uniform Resource Identifiers (URIs) for each of the data elements, relationships and property types. An ontology can be parsed by a reasoning engine and then used to infer knowledge about the domain.

Similarly, Linked Data makes use of these Semantic Web publication patterns, but then uses the expectations of a Web document (URIs are used to identify things; when looked-up those URIs resolve over HTTP; the resolution returns a document compliant with standards; links are provided to other URIs so users may discover other resources) to provide connections to other data published on the Semantic Web (Berners-Lee, 2011). Leadbetter, Cheatham, Shepherd & Thomas (2016, see chapter X this volume) further explore the use of Linked Data in the marine science domain. These approaches, as well as the emergence of real-time analytical applications for marine data streamed from instruments to data centres, have given rise to the concept of using the Web as the query and programming interface for marine data (Chandler & Arko, 2015). While born of the Linked Data paradigm, this is also seen in the Sensor Web Enablement activities of the Open Geospatial Consortium, and in products such as NOAA's ERDDAP server (Mendelssohn & Simons, 2008). What remains, however, is a suite of non-standard, domain specific interfaces without a common unifying "glue".

Semantic Web practitioners have developed the notion of the upper ontology in response to this problem of using data encoded with the knowledge of different domains. The upper ontology should describe concepts which are common across all domains, and in this regard bridges the difference between the information modelling ontology and the philosophical ontology. An important function of an upper ontology is to support very broad semantic interoperability between large numbers of ontologies which are accessible ranking "under" this upper ontology. As the rank metaphor suggests, it is usually a hierarchy of entities and associated rules (both theorems and regulations) that attempts to describe those general entities that do not belong to a specific problem domain. The World Wide Web Commission has published both a Provenance Data Model (Missier, Belhajjame, and Cheney 2013), and a Semantic Web encoded version of this model as the Provenance Ontology, or PROV-O (Lebo et al. 2013). PROV-O may be considered an upper-level ontology as it provides a very broad level of semantic interoperability between more domain specific ontologies, or it is "a domain-independent common-sense theory constructed by aggregating suitable contributions from areas such as descriptive metaphysics, philosophical logic, cognitive science and linguistics" (Guizzardi 2006). PROV-O is based around a simple model of three classes:

- Entity a physical, digital, conceptual, or other kind of thing with some fixed aspects; entities may be real or imaginary
- Activity something that occurs over a period of time and acts upon or with entities; it may include consuming, processing, transforming, modifying, relocating, using, or generating entities
- Agent something that bears some form of responsibility for an activity taking place, for the existence of an entity, or for another agent's activity

A number of attributes have been defined which describe the relationship between the classes by characterising the influence that an instance of one class has on another instance, either of the same class or of a different class. There are several examples of PROV-O being used as an upper ontology in the Earth Sciences (Compton, Corsar, & Taylor, 2014; Ma et al., 2014; Leadbetter & Vodden, 2015). Its limited number of classes mean that there is a clear pathway for ontology alignment by placing the classes of a domain-specific ontology into the class structure provided by PROV-O. Taking the alignment of the Semantic Sensor Network with PROV-O as a first case study (Compton, Corsar, & Taylor, 2014), it can be seen the Semantic Sensor Network Observation class defines an observation an interpretation of real-world events and the results of those events. For example, a stimulus (wind) spins the cups on a wind sensor, generating a current, and through an equation modelling the relationship between this and the physical property of wind speed the sensor outputs a value that we can choose to interpret as an observation of the wind speed at that moment. The observation is the social construct of the interpretation, not the act of the observing itself. Following the Semantic Sensor Network model, Observation would be aligned with the PROV-O class Entity. This however contrasts with the view of the ISO 19156 Observations and Measurements (O&M) model, which states that an Observation is an action whose result is an estimate of some property of a feature-of-interest, obtained through the use of a specific procedure (Cox, 2006). In O&M an Observation therefore aligns with a PROV-O Activity. Therefore, alignment can be achieved between the two lower-level ontologies as PROV-O Activity is not a disjoint class with PROV-O Entity. However, care needs to be taken in this regard as the Observation class from the two different lower-level ontologies also represents subtly different concepts the record of an observation in the Semantic Sensor Network and an Observation Event in O&M (Cox, 2015). It can therefore be seen that this use of PROV-O as an upper ontology helps to both align the ontologies for interoperability but also highlights the specific differences of the creation philosophy of the ontologies being aligned.

One ontology which describes, in general terms, oceanographic deployments, such as research vessel surveys, is the Ocean Data Ontology (ODO, http://www.ocean-data.org/). This ontology should be able to be aligned with other ontologies describing specific details of the sensors deployed (for example the Semantic Sensor Network Ontology) or the observations made in the environment during the deployment (for example, O&M). From ODO, the classes of interest are Dataset, which describes a collection of observations made during a Deployment and is linked to the specific research project making the deployment and the funding award which paid for the deployment; and Parameter which links an instrument on a deployment to the environment property it is observing as well as describing the observable properties recorded within a Dataset. It can be seen that these classes should be available for users from a range of disciplines who wish to use the data in decision making tools or other applications, and therefore these classes will be aligned with PROV-O. As shown by Leadbetter and Vodden (2015), the oceanographic domain often uses the term "parameter" where the O&M community use the term "observable property", therefore the "Parameter" class of ODO should be most closely aligned with O&M's Observable Property, showing that formalization is community dependent and that bridging between models of knowledge formalization must be done at a human level. This is an attribute of the O&M class Observation, so therefore in a PROV-O model could come from any of the three classes. It makes no sense for the Parameter to be an agent as it bears no responsibility for the Observation activity taking place or the creation of an Entity; likewise the Parameter is not an Activity as it does not act upon or with Entities. Therefore, a Parameter, or Observable Property, is an Entity and an Observation Activity can use said Entity. Similarly, within the Semantic Sensor Network a sensor, or instrument, observes a property, or an ODO Parameter. A sensor is known to be defined as both a PROV-O Agent and Entity (Compton, Corsar, & Taylor, 2014), therefore an ODO Instrument should also be defined in this way and a Parameter may be attributed to it. As we have seen, Observations are PROV-O Activities, resulting in PROV-O Entities (either O&M Results or Semantic Sensor Network Observations). From here, a group of results should also be a group of PROV-O Entities, and therefore a PROV-O Entity should sufficiently describe an ODO Dataset. Therefore, an ODO Dataset can be said to be derived from the PROV-O Entities which are the O&M Results of the Observations made using an ODO Parameter.

Archetypes as an Alternative Method in Marine Data Management Knowledge Formalisation

An alternative strategy for true knowledge capture and interoperability across domains both within and without the marine sciences to that of the upper ontology alignment method may be found by looking further afield than the Earth Science informatics domain. By taking Delaney's approach that there are interesting developments in other domains on the fringe of ocean sciences and looking to the field of health informatics, where much work has been done on the topic of separating information models and knowledge models we can see that archetypes (Beale, 2002) offer one solution to a true encoding of knowledge. As the volumes of observational data from Argo floats, gliders and ocean observatories increases and the desire to semantically annotate the data at source also grows, solutions have been designed as a strategy to compensate for the fact that data captures only a part of what researchers can expect from an observation (Diviacco et al. 2014, Fredericks, 2015), we are reaching an age of ubiquitous marine computing which should allow for more user friendly and comprehensive data capture (Garde, Hovenga, Buck, & Knaup, 2007). However, this precludes the heavy manual involvement of the data manager in the process of creating the "utopian" vision of users not wishing to continually return to the data originator in order to understand their datasets. Archetypes have been used in healthcare informatics to define integrity constraints; data required for the interpretation of a measurement; multilingual support; and replication of domain knowledge in ubiquitous computing environments to address this issue. Likewise, healthcare informatics experts have used archetypes successfully to provide interoperability between otherwise heterogeneous data systems (Lezcano, Sicilia, & Rodríguez-Solano, 2011). As Beale (2002) states, with large information models for domains it is often logistically and technically difficult for different data producers and users to agree on a common data model. This makes automated processing and decision support using observational data very difficult to achieve. In the typical archetype use-case, the archetype is used to describe knowledge while the ontology is used to describe information, designed to address issues where knowledge can be volatile (i.e. the definitions of concepts within the information model can change over time), or the number of knowledge entities for a given domain is vast. This second point is likely to be of most use in the marine science domain, particularly when considering the NERC Vocabulary Server (a commonly used Semantic Web resource in the marine data management community; Leadbetter, Lowry, & Clements, 2014) contains some 98,000 knowledge entities (as of October 21st, 2015).

Lezcano, Sicilia, & Rodríguez-Solano (2011) provide a mechanism to match the definitions of archetypes to ontologies and to provide archetypes with computational semantics by providing rules for the encoding of archetypes in the W3C's Web Ontology Language (OWL). This is particularly useful because the definition language used in producing archetypes does not allow for inferencing (Beale, Heard, Kalra, & Lloyd, 2005) but OWL based archetype definitions can be enhanced by the inclusion of Semantic Web Rule Language entries (Horrocks et al., 2004). Given the example of how the Semantic Sensor Network, Observations & Measurements, and Ocean Data Ontology all formalise descriptions of measurements within the environment in slightly different ways, and instructive use case for the application of archetypes to the marine science domain is to prototype an Observation archetype synthesising the knowledge encapsulated by these ontologies. This is expanded below using the Archetype Definition Language (Beale, Heard, Kalra. & Lloyd, 2005).

A prototype observation archetype should capture the following knowledge as already described in this chapter: that an observation event leads to a result, which contains a value and a unit of measure; an observation event requires the observation of a parameter, or observable property; and a process, including a sensor or instrument, acts on the observable property to generate the result. Observations may be combined to generate a set of observations for grouping. An archetype conforming to this definition is shown in Listing 1.

It can be seen that there are growing numbers of methods and results in formalizing cross-disciplinary knowledge, but there are still difficulties arising in achieving a utopian vision where all formalized knowledge is truly interoperable. Therefore, there are many advances in representing knowledge in less formalized ways. This will be discussed below.

METHODS BASED ON NON-FORMALIZED KNOWLEDGE

Representation

We defined above the notion of "epistemic objectivity" as the sets of claims or judgments which truth values are not determined privately. Judgments imply a form of intentional portrayal of things, such as beliefs, precepts, news reports, books, articles, sentences and pictures, which can defines "representation". Intentionality connects epistemic objectivity with representation. Representations can be, amongst other things, theories and models, linguistic and mathematical entities, computer simulations, concrete objects (Diviacco, 2015). Representation can take place in multiple ways: linguistically when something is explained; graphically in a map or a graph reporting the observation of a phenomenon; but also in other ways in which the linguistic aspects are less relevant or at least are only a medium for communication and do not necessarily contain all of the message. Diviacco (2015) gives examples of forms of representation where this de-literalization allows us to move the focus toward other forms of communication such as art where concurrent conceptual models can co-exist.

In light of the discussion in this chapter, representation is very interesting because it is communication that can be biased by cognitive and epistemic issues, by ideology or politics; or concurrently by all these factors. Strictly speaking, descriptions of entities and processes afforded by scientific representations are generally false (Chakravartty, 1995). Following Callender and Cohen (2006), representation has features reality does and does not have. It is characterized by omissions and commissions. Omission is the act of neglecting some possible causes among those that can explain a phenomenon; commission is the act of deliberately changing the network of possible causes and explanations of a phenomenon. Representation can then be used to change the parameters of the target system, representing it in a way that constitutes a distortion of its true nature (see the above section on the limitations of reasoning in sustainable development). Similarly in the case of painting or sculpture, it should be possible to distinguish realistic and non-realistic representations. Realism in representation is indicative of an easy to interpret display, even for the non-specialist. Non-realistic representation is such because it requires an interpretation phase where conventions need to be set. Accessibility depends on the ease with which representation can issue the same information and those viewing the representation can internalize conventions.

A key aspect of collaborative work is the possibility to access the same conventions. This can depend on the actual complexity of the phenomenon under study. In fact, using again the similitude with the case of painting; when simple objects are to be described, the iconic use of the image is often easy to grasp, meaning few conventions are needed. When objects are complex, the needed symbology becomes more abstract and therefore need to be managed in a specific way, incurring into the issues we described above.

Boundary Objects

The formalized knowledge paradigm requires large investment of resources in the governance of all the tools it can offer. In addition, the status of such an infrastructure cannot be considered stable, since new concepts continuously appear and have to be mapped in the environment.

An un-formalized paradigm allows a more relaxed approach since is intended essentially to offer access to data and information without imposing a specific viewpoint.

The approach is based on the ethnological studies done in nineties by Susan Leigh Star and James R. Griesemer (1989). They found that different communities, characterized by different languages, cultures and traditions can collaborate effectively notwithstanding the fact that they could have rather different cognitive models. In these cases they found that this process is facilitated by the presence of artifacts that can be used in different ways but that gather users in the same cognitive space bridging concurrent cognitive models through abstraction from all the domains of the partners. To do so artefacts must be weakly structured in common use, but strongly structured in individual use.

Boundary objects are artefacts related to sets of information, conversations, interests, rules, or plans, and are at the centre of communities of practices (Diviacco, 2012; Star, & Griesemer, 1989). Their application to bridge professional boundaries was studied by Kimble, Grenier and Goglio-Primard (2010), while their application in the field of the Arts can be seen in Van Saaze and Dekker (2013)

The idea of the boundary object can be put in practice through many kinds of artefacts. What is relevant is that, in practice, they are forms of representations. These can be for example: maps, graphs or work flows, to mention a few. These artefacts can be designed in order to expose "nodes" that can be used to gather data and information in the form of files, message threads, wikis or else. Each node can be labelled associating it to a concept, process or phenomenon, so that the artefact will gather all the knowledge of the designated community. On the other hand the artefact is intended as a passive entity, which performs no reasoning: the "work" is done mainly by the users.

Following Scardamalia and Bereiter (2006) knowledge building, in the perspective we are adopting here, can have two forms, and namely: (i) a weak form, where findings only become contributions to community knowledge when they are brought into public discourse, and (ii) a strong version, where the discourse is itself the state of public knowledge in a community.

We find very interesting that the state of knowledge of a community could be not only in the minds of the individual members of the community, but also in the discourse itself. In fact, if the discourse takes place in an artefact, this latter will become the state of knowledge of the designated community (Diviacco 2015) and will follow its evolution: practical implementations of this approach have been proposed by Diviacco (2015), Diviacco (2013), Diviacco and Pshenichny (2011) and Diviacco and Pshenichny (2010)

Limitations to such approach have also been highlighted (Diviacco, 2015). These relates mainly to the flexibility of the artefact itself. For example, a mind map already populated with data and information is very difficult to be changed in a later stage because that could possibly mean moving information across nodes, which of course cannot be automatized. A more flexible approach, which can be applied often, but not in all the cases, could be data annotation. In this case data is represented, for example plotted in a graph or projected in a map, while users are provided with tools to draw objects on the representation, such as lines, circles, arrows to capture the attention of the community on a feature that can be used as a node and therefore gather data, information, messages or else.

FUTURE RESEARCH DIRECTIONS

Considering the two paths identified, namely formalization and representation, we need to identify two groups of future research directions.

In the case of formalized knowledge, while Ontology Design Patterns, as explored by Leadbetter, Cheatham, Shepherd & Thomas (2016), offer one approach to integration of data by publishing data to a given pattern the considerations of this chapter offer two future research directions to use the patterns in different ways. The first would be to align the Ontology Design Patterns with PROV-O in order to use an upper ontology approach to combining data from different domains published under Ontology Design Patterns. The second would be to pull the Ontology Design Patterns up to the higher level in a two-level information model and to rewrite the knowledge captured in the patterns as archetypes.

In the case of representation based knowledge management, we think that a very promising future direction would be to explore the possibility to create on the fly the artefacts that help user to navigate the collaborative knowledge space. This could be possible for example analysing terms used in messaging or annotation. A positive aspect of this approach would be that possibly this could pose a bridge between the two worlds of formalization and representation since an interesting approach could be to ground this activity in controlled vocabularies or ontologies (Diviacco and Fox, in preparation).

CONCLUSION

This chapter aimed at exploring the difficulties behind data management in sustainable development and at proposing possible solutions to overcome them. We have highlighted the intrinsic limits of reasoning in this domain and the role of relationships and communication between the partners in the designated community. Tacit knowledge and contrasting cognitive models do not allow partners to understand each other's, and therefore knowledge, often can be formalized only with difficulties.

The resulting scenario is very complex. We claim that in spite of the difficulties we have described there are opportunities to address collaborative research in the domain of sustainable development through the exploitation of a careful balance of formalized knowledge and non-formalized knowledge representation. In many senses the practitioners of sustainable development in general, and ocean data management in particular, must look beyond their immediate domains in order to solve these issues.

Where knowledge can be formalized, tools such as controlled vocabularies, ontologies or archetypes can assist partners with different backgrounds to bridge the differences and incompatibilities of their original contexts. Where knowledge remains tacit, or difficult to agree upon different strategies should be introduced. These can be based on artefacts that allow the communication to take place without interfering with it. Such artefacts should be representations of knowledge that while being unique can be interpreted differently by the different communities.

We have introduced concepts from computer science; health informatics; philosophy; and art in order to provide an outline strategy to allow the management, integration and representation of marine data for sustainable development.

Tools based on such strategy will allow the end user to find and access data, information and knowledge on IT systems that eventually become themselves representations of the state of knowledge of a community.

REFERENCES

Beale, T. (2002, November). Archetypes: Constraint-based domain models for future-proof information systems. In OOPSLA 2002 workshop on behavioural semantics (Vol. 105).

Beale, T., Heard, S., Kalra, D., & Lloyd, D. (2005). *Archetype definition language (adl).* OpenEHR specification, the openEHR foundation.

Becher, T., & Trowler, P. R. (2001). *Academic tribes and territories intellectual enquiry and culture of disciplines*. Buckingham, UK: The Society for Research into Higher Education and Open University Press.

Berners-Lee, T. (1998). *Semantic web road map.* Retrieved from: https://www.w3.org/DesignIssues/Semantic

Berners-Lee, T. (2011). *Linked data-design issues (2006).* Retrieved from http://www.w3.org/DesignIssues/LinkedData.html

Buck, J., & Lowry, R. (2016). Oceanographic Data Management: Quills and Free Text to the digital age and 'Big Data. In P. Diviacco, A. Leadbetter, & H. Glaves (Eds.), *Oceanographic and Marine Cross-Domain Data Management for Sustainable Development*. Hershey, PA: IGI Global.

Bishop, M. A. (1992) Theory-Ladenness of Perception Arguments. In *Proceedings of the Biennial Meeting of the Philosophy of Science Association* (pp. 287-299). Academic Press.

Brundtland Commission Report. (1987). *Lawlor*. Retrieved from https://en.wikisource.org/wiki/Brundtland_Report on 7/10/2015

Chakravartty, A. (1995). Truth and Representation in Science: Two Inspirations from Art. In R. Frigg & M. Hunter (Eds.), *Beyond Mimesis and Convention: Representation in Art and Science* (pp. 33-50). Berlin: Springer.

Chandler, C., & Arko, R. (2015). Big Data Challenges: Variety and Veracity. Ocean Data Interoperability Platform Workshop Presentation, Paris, France.

Compton, M., Corsar, D., & Taylor, K. (2014). *Sensor data provenance: SSNO and PROV-O together at last*. In 7th International Semantic Sensor Networks Workshop (October 2014).

Cox, S. (2006). *Observations and measurements. Open Geospatial Consortium Best Practices Document*. Open Geospatial Consortium.

Cox, S. (2012). *Best practice in formalizing a SKOS vocabulary. Solid Earth and Environment GRID*. Commonwealth Scientific and Industrial Organisation (CSIRO). Retrieved from https://www.seegrid.csiro.au/wiki/Siss/VocabularyFormalizationInSKOS

Cox, S. (2015). *Ontology alignment – is PROV-O good enough?* Open Geospatial Consortium GeoSemantics Summit, Boulder, CO.

David, P. A., & Bunn, J. A. (1988). The economics of gateway technologies and network evolution. *Information Economics and Policy*, *3*(2), 165–202. doi:10.1016/0167-6245(88)90024-8

Delaney, J. (2010, April). *Wiring an interactive ocean* [video file]. Retrieved October 13, 2015 from https://www.ted.com/talks/john_delaney_wiring_an_interactive_ocean?language=en

Diviacco, P., & Busato, A. (2015). Maps, graphs and annotations as Boundary Objects in Knowledge Networks, Distributed Cognition and Collaborative e-Research. In P. Diviacco, P. Fox, C. Pshenichny, & A. Leadbetter (Eds.), *Collaborative Knowledge in Scientific Research Networks*. Hershey, PA: IGI Global. doi:10.4018/978-1-4666-6567-5.ch019

Diviacco, P. (2015). Reconciling Knowledge and Collaborative e-Research. In Collaborative Knowledge in Scientific Research Networks. Hershey, PA: IGI Global. doi:10.4018/978-1-4666-6567-5.ch001

Diviacco, P., Pshenichny, C. A., Carniel, R., Khrabrykh, Z., Shterkhun, V., Mouromtsev, D., & Pascolo, P. et al. (2014). Organization of a geophysical information space by using an event-bush-based collaborative tool. *Earth Sci Inform*, *10*. doi:10.1007/s12145-014-0182-2

Diviacco, P., DeCawer, K., Leadbetter, A., Sorribas, J., Stojanov, Y., Busato, A., & Cova, A. (2014). *Bridging semantically different paradigms in the field of marine acquisition event logging, Earth Science Informatics*. Springer. doi:10.1007/s12145-014-0192-0

Diviacco, P., & Busato, A. (2013). The Geo-Seas Seismic data viewer: a tool to facilitate control of data access. *Bollettino di Geofisica Teorica ed Applicata, 54*(3), 257-270.

Diviacco, P. (2013). Addressing Conflicting Cognitive Models in Collaborative e-Research: A Case Study in Exploration Geophysics. In *Collaborative and Distributed e-Research: Innovations in Technologies, Strategies and Applications*. Hershey, PA: IGI Global press.

Diviacco, P. (2012). Addressing Conflicting Cognitive Models in Collaborative E-Research: A Case Study in Exploration Geophysics. In *Collaborative and Distributed E-Research: Innovations in Technologies, Strategies and Applications*. Hershey, PA: IGI Global Press; doi:10.4018/978-1-4666-0125-3.ch012

Diviacco, P., & Pshenichny, C. A. (2011). A case study on the use of event bushes as a formal representation for Computer Supported Collaborative Work in the GeoSciences European Geosciences Union (EGU) General Assembly, Vienna (Austria), April 2010. *Geophysical Research Abstracts*, 12.

Diviacco, P., Lowry, R., & Leadbetter, A. (2011). Collaborative work and tools towards wide scientific community driven metadata model and vocabulary building: The case of the EU GeoSeas project. European Geosciences Union (EGU) General Assembly, Vienna (Austria), April 2010. *Geophysical Research Abstracts*, 12.

Diviacco, P., & Pshenichny, C. A. (2010). Concept-referenced spaces in computer-supported collaborative work. *Geophysical Research Abstracts*, 12.

Diviacco, P. (2005). An open source, web based, simple solution for seismic data dissemination and collaborative research. *Computers & Geosciences*, *31*(5), 599–605. doi:10.1016/j.cageo.2004.11.008

Eco, U. (1979). *A Theory of Semiotics*. Indiana University Press.

Engelhardt, W., & Zimmerman, J. (1982). *Theory of earth science*. Cambridge, UK: Cambridge University Press.

Fredericks, J. (2015). Persistence of Knowledge Across Layered Architectures. In P. Diviacco, P. Fox, C. Pshenichny, & A. Leadbetter (Eds.), *Collaborative Knowledge in Scientific Research Networks*. Hershey, PA: IGI Global. doi:10.4018/978-1-4666-6567-5.ch013

Fox, P. (2014). Data Science and Analytics Curriculum Development at Rensselaer (and the Tetherless World Constellation). In *Training Students to Extract Value from Big Data: Summary of a Workshop* (pp. 18–21). Washington, DC: National Academies Press.

Garde, S., Hovenga, E., Buck, J., & Knaup, P. (2007). Expressing clinical data sets with openEHR archetypes: A solid basis for ubiquitous computing. *International Journal of Medical Informatics*, *76*, S334–S341. doi:10.1016/j.ijmedinf.2007.02.004 PMID:17392019

Guardian. (2008, July 22). *Will Self* [Author biography]. Retrieved from http://www.theguardian.com/books/2008/jun/11/willself

Hanson, N. R. (1958). *Patterns of Discovery*. Cambridge, UK: Cambridge University Press.

Herman, I. (2008). *W3C semantic web activity*. World Wide Web Consortium.

Horrocks, I., Patel-Schneider, P. F., Boley, H., Tabet, S., Grosof, B., & Dean, M. (2004). *SWRL: A semantic web rule language combining OWL and RuleML*. W3C Member Submission, 21, 79.

IBM. (1993). *IBM Dictionary of Computing*. New York: McGraw-Hill.

Isenor, A., & Lowry, R. K. (2005). *Final report of the ICES/IOC Study Group on the Development of Marine Data Exchange Systems using XML*. Academic Press.

Kimble, C., Grenier, C., & Goglio-Primard, K. (2010). Innovation and knowledge sharing across professional boundaries: Political interplay between boundary objects and brokers. *International Journal of Information Management*, *30*(5), 437–444. doi:10.1016/j.ijinfomgt.2010.02.002

Kuhn, T. S. (1962). *The structure of scientific revolutions*. Chicago, IL: University of Chicago Press.

Kumpati, M. (1985). Patent description. US Patent 4774661, filed November 19.

Latour, B., & Woolgar, S. (1979). *Laboratory life: The construction of scientific facts*. Princeton, NJ: Princeton University Press.

Lawlor, D. A., Davey Smith, G., & Ebrahim, S. (2004). Commentary: The hormone replace- ment-coronary heart disease conundrum: Is this the death of observational epidemiology? *International Journal of Epidemiology*, *33*(3), 464–467. doi:10.1093/ije/dyh124 PMID:15166201

Leadbetter, A., Cheatham, M., Shepherd, A., & Thomas, R. (2016). Linked Ocean Data 2.0. In P. Diviacco, A. Leadbetter, & H. Glaves (Eds.), *Oceanographic and Marine Cross-Domain Data Management for Sustainable Development*. Hershey, PA: IGI Global.

Leadbetter, A., Lowry, R., & Clements, O. (2014). Putting meaning into NETMAR – the open service network for marine environmental data. *International Journal of Digital Earth*, *7*(10), 811–828. doi:10.1080/17538947.2013.781243

Leadbetter, A. M., & Vodden, P. N. (2015). Semantic linking of complex properties, monitoring processes and facilities in web-based representations of the environment. *International Journal of Digital Earth*, 1-25.

Lebo, T., Sahoo, S., McGuinness, D., Belhajjame, K., Cheney, J., Corsar, D., ... & Zhao, J. (2013). *Provo: The provontology*. W3C Recommendation, 30.

Lezcano, L., Sicilia, M. A., & Rodríguez-Solano, C. (2011). Integrating reasoning and clinical archetypes using OWL ontologies and SWRL rules. *Journal of Biomedical Informatics*, *44*(2), 343–353. doi:10.1016/j.jbi.2010.11.005 PMID:21118725

Lowry, R., Rickards, L., & Brown, J. (2005). *Adding value to oceanographic data at the British Oceanographic Data Centre. In Ensuring Long-Term Preservation and Adding Value to Scientific and Technical Data (PV2005)* (pp. 21–23). Edinburgh, UK: Royal Society of Edinburgh.

McGuinness, D. (2003). Ontologies come of age. In D. Fensel, J. Hendler, H. Lieberman, & W. Wahlster (Eds.), *Spinning the Semantic Web: bringing the World Wide Web to its full potential* (pp. 171–196). Cambridge, MA: MIT.

Ma, X., Zheng, J. G., Goldstein, J. C., Zednik, S., Fu, L., Duggan, B., & Fox, P. et al. (2014). Ontology engineering in provenance enablement for the National Climate Assessment. *Environmental Modelling & Software*, *61*, 191–205. doi:10.1016/j.envsoft.2014.08.002

Mendelssohn, R., & Simons, R. A. (2008, December). ERDDAP-An Easier Way for Diverse Clients to Access Scientific Data From Diverse Sources. In *AGU Fall Meeting Abstracts* (Vol. 1, p. 9). Academic Press.

Miles, A., & Bechhofer, S. (2009). *SKOS simple knowledge organization system reference*. World Wide Web Consortium. Retrieved from http://www.w3.org/TR/2009/REC-skos-reference-20090818/

Morris, C. (1971). *Foundations of the theory of signs*. Chicago, IL: University of Chicago Press.

Murphy, K. (2012). The social pillar of sustainable development: a literature review and framework for policy analysis. *Sustainability: Science, Practice, & Policy, 8*(1), 15-29. Retrieved from http://www.google.it/archives/vol8iss1/1008-041.murphy.html

Navigli, R., & Velardi, P. (2008). From Glossaries to ontologies: extracting semantic structure from textual definitions. In P. Buitelaar & P. Cimiano (Eds.), *Ontology learning and population: bridging the gap between text and knowledge* (pp. 71–87). Amsterdam: IOS Press.

Peirce, C. S. (1958). *Collected Writings* (C. Hartshorne, P. Weiss, & A. W. Burks, Eds.). Cambridge, MA: Harvard University Press.

Polanyi, M. (1966). *The tacit dimension*. New York, NY: Anchor Day Books.

Popper, K. (1963). *Conjectures and refutations*. London, UK: Routledge.

Pshenichny, C. A. (2004). Classical logic and the problem of uncertainty. Geological Society, 239, 111-126.

Van Renssen, A. (2005). *Gellish, A generic extensible ontological language* (Doctoral dissertation). Delft: Delft University of Technology.

Scardamalia, M., & Bereiter, C. (2006). Knowledge building: Theory, pedagogy, and technology. In R. K. Sawyer (Ed.), *The Cambridge handbook of the learning sciences* (pp. 97–115). New York, NY: Cambridge University Press.

Searle, J. (1995). *The Construction of Social Reality*. New York: The Free Press.

Shadbolt, N., & Berners-Lee, T. (2008). Web science emerges. *Scientific American, 299*(4), 76–81. doi:10.1038/scientificamerican1008-76 PMID:18847088

Shannon, C. E., & Weaver, W. (1949). *A mathematical model of communication*. Chicago, IL: University of Illinois Press.

Shedroff, N. (1994). *Information interaction design: A unified field theory of design*. San Francisco, CA: Vivid Studios.

Star, S. L., & Griesemer, J. R. (1989). Institutional ecology, translations and boundary objects: Amateurs and professionals in Berkeley s museum of vertebrate zoology. *Social Studies of Science, 19*(4), 387–420. doi:10.1177/030631289019003001

UN resolution A/RES/70/1. (2015). *Transforming our world: the 2030 Agenda for Sustainable Development*. Retrieved from http://www.un.org/en/ga/search/view_doc.asp?symbol=A/RES/70/1 on 12/2/2016

UNESCO. (1987). GF3: A General Formatting System for geo-referenced data: Volume 2: technical description of the GF3 format and code tables. Paris: Intergovernmental Oceanographic Commission.

Van Saaze, V., & Dekker, A. (2013). Surprising usages of a documentation model: On the notion of boundary objects and beyond. *International Journal of Performance Arts & Digital Media*, *9*(1), 99–113. doi:10.1386/padm.9.1.99_1

Warner, A. (2002). *A Taxonomy Primer. School of Information*. University of Texas at Austin. Retrieved from http://www.ischool.utexas.edu/i385e/readings/Warner-aTaxonomyPrimer.html

Whitley, R. (2000). *The intellectual and social organization of the sciences*. Oxford, UK: Clarendon Press.

Woo, G. (1999). *The mathematics of natural catastrophes*. London, UK: Imperial College Press; doi:10.1142/9781860943867

World Economic and Social Survey 2013, Sustainable Development Challenges. (n.d.). Retrieved from https://sustainabledevelopment.un.org/content/documents/2843WESS2013.pdf

KEY TERMS AND DEFINITIONS

Archetype: A computable expression of a domain content model in the form of structured constraint statements, based on a reference (information) model. Archetypes are all expressed in the same formalism. In general, they are defined for wide re-use, however, they can be specialized to include local particularities. They can accommodate any number of natural languages and terminologies.

Boundary Object: An artefact that allow two different and incommensurable communities to nevertheless collaborate. They are weakly structured in common use, and become strongly structured in individual use.

Linked Data: Linked Data is a technique that uses the Word Wide Web to connect related data that was not previously linked, or uses the Web to lower the barriers to linking data which are currently linked using other methods. In practice, Linked Data often uses the Resource Description Framework model and Web addresses (as Uniform Resource Locators, or URLs) to achieve these linkages.

Ontology: In computer science, an ontology formally represents knowledge as a hierarchy of concepts within a domain, using a shared vocabulary to denote the types, properties and interrelationships of those concepts.

Semantic Web: The Semantic Web is the World Wide Web of data built on the Resource Description Framework model as a foundation for publishing and linking data online.

Tacit Knowledge: Knowledge that cannot be made explicit because obscured in socio-political phenomena or embedded in practices or even tools.

Theory Laden-ness: Refers to the fact that 7 experience is categorized and preconditioned by prior belief, and that the process of building knowledge always involves the use of concepts from some previous view. This holds both from a logical and a perceptive point of view.

ENDNOTE

1 Umberto Eco (1981) used the concept of abduction to describe Conan Doyle's Sherlock Holmes's way of reasoning.

APPENDIX: LISTING 1

```
archetype (adl_version=2.0.5)
  activity_of_observing.v.1.0.0
language
  original_language = <[iso_639-1::en]>
definition
  ACTIVITY_OF_OBSERVING[id1] matches {
        -- Maps to an O&M Observation or a Semantic Sensor Network Avtivity of
        -- Sensing
        RESULT_OF_OBSERVING[id2] matches{
          -- Maps to an O&M Result or a Semantic Sensor Network Observation
          result_value matches {|-inf..inf|}
            unit_of_measure matches {
                    DV_CODED_TEXT[id3] matches {
                            defining_code <[ac1]>
                    }
            }
        }
        PROPERTY_OBSERVED[id4] matches {
          -- Matches an O&M Observable Property or an ODO Parameter
          DV_CODED_TEXT matches {
            defining_code <[ac2]>
          }
          SENSING_INSTRUMENT[id5] matches {
                  DV_CODED_TEXT matches {
                    defining_code<[ac3]>
                  }
          }
        }
        CONTAINING_DATASET[id6] matches {http://?}
  }
terminology
  term_definitions = <
    ["en"] = <
      ["id1"] = <
        text = <"Activity of observing">;
        description =
                    <"The act of making an observation or measurement within the
                    environment">
      >
      ["id2"] = <
            text = <"Observation Result">
```

```
        description =
                  <"The result value and its associated unit of measure for an
                  observation event">
          >
          ["id3"] = <
            text = <"Unit of Measure">;
                description =
                  <"Measurement unit associated with result value of an
observation
                  event">
          >
          ["id4"] = <
            text = <"Observable Property">
                description =
                  <"The property which was observed to give rise to the
Observation
                  Result">
          >
          ["id5"] = <
            text = <"Instrument used to make the observation">
                description =
                  <"The sensing device by which an observation activity is
undertaken">
          >
          ["id6"] = <
            text = <"Containing dataset">
                description =
                  <"A link to the dataset definition which contains this
observation">
          >
        >
  >
  value_sets = <
    ["ac1"] = <
          id = <"ac1">
          members =
            [http://data.qudt.org/qudt/owl/1.0.0/unit.owl]
                -- uom to come from the QUDT ontology
        >
    ["ac2"] = <
      id = <"ac2">
          members =
            [http://vocab.nerc.ac.uk/standard_name/]
```

```
            -- using the Climate and Forecast conventions
  >
    ["ac3"] = <
      id = <"ac3">
      members =
        [http://vocab.nerc.ac.uk/collection/L22/]
            -- SeaVoX device catalogue
    >
>
```

Chapter 3
Developing a Common Global Framework for Marine Data Management

Helen M. Glaves
British Geological Survey, UK

ABSTRACT

The paradigm shift in marine research moving from the traditional discipline based methodology to a multidisciplinary, ecosystem level approach is being driven by changes in both the policies for the management and exploitation of the ocean, and the scientific method itself. The availability of large volumes of good quality data is fundamental to this increasingly holistic approach to ocean research but there are significant barriers to its re-use. The Ocean Data Interoperability Platform (ODIP) project has been funded in parallel by the European Commission, National Science Foundation in the USA and the Australian Government to promote the development of a common framework for marine data management that leverages the existing marine e-infrastructures which have been created in response to the need for greater sharing of marine data at a regional level.

INTRODUCTION

In recent years there has been a paradigm shift in marine research, moving from the traditional discipline based methodology employed at the national level by one or more organisations, to a multidisciplinary, ecosystem level approach conducted on an international scale.

This increasingly holistic approach to marine research is in part being driven by policy and legislation. The European Commission's Marine Knowledge 2020 initiative was launched in response to the need for a more coordinated approach to the management of marine data on a regional scale to support sustainable exploitation of the marine environment whilst still achieving the good environmental status of the seas (European Commission 2010). This initiative was aligned with the Marine Strategy Framework Directive that requires the European Member States to have coordinated marine monitoring programmes for the ongoing assessment of the environmental status of their marine waters (European Commission,

DOI: 10.4018/978-1-5225-0700-0.ch003

2008). It also emphasises the need for this marine monitoring data to be shared between organisations and across national boundaries in order to obtain wider assessments of the marine environment at a regional or sea basin level.

To address the requirements for increased availability of data for re-use that were highlighted in these EU policy documents the European Commission funded the European Marine Observation and Data Network (EMODNET). Its aim is to integrate the fragmented national and organisational marine data systems into a single coherent system that supports the discovery and access of data and data products for the European sea basins. The EMODNET initiative also forms a core element of the EU Blue Growth Strategy and, by further developing the existing framework, the European Commission aims to improve the availability of multidisciplinary basin level data and information to support policy driven marine research that forms part of this strategy.

The European Commission's Blue Growth strategy is promoting sustainable growth in the marine environment including the development of sea-basin strategies which foster cooperation between nations (European Commission, 2014). It recognises that national and regional data does not provide the information that is needed to fully understand the global ocean systems. The EU Blue Growth strategy therefore prioritises the need to address the gaps in knowledge and data about the state of the oceans, seabed resources, marine life and the potential risks to habitats and ecosystems.

The move towards basin level research in the north Atlantic has also been formalised through the Galway Statement on Atlantic Ocean Cooperation which launched the Transatlantic Ocean Research Alliance between Europe, the USA and Canada (European Commission, 2013). The goal of this agreement is to promote collaborative marine research in the Atlantic Ocean and the bordering sea areas. The Galway Statement identifies the Atlantic Ocean as a shared resource for the adjacent countries and recognises the importance of activities to coordinate data sharing and observing infrastructures to support a fundamental understanding of the Atlantic Ocean and adjacent bodies of water to support long-term sustainable management and use.

Similarly the Australian Government's position paper titled *Marine Nation 2025: Marine Science to Support Australia's Blue Economy* emphasised that "Australia's ongoing marine research success depends on improved management of national and global marine research data and information, and the engagement of the marine research community and marine research institutions in improving and supporting marine research data infrastructure" (Oceans Policy Science Advisory Group, 2013). This paper also highlighted the need for "a national marine science information management strategy to ensure that all data collected is stored and systematically described so that information is readily identifiable and accessible to stakeholders and users". The recommendations made in this discussion document gave rise to the *National Marine Science Plan 2015-2025* which emphasises the socio-economic benefit of the blue economy and the importance of sharing marine data to support multidisciplinary marine research (National Marine Science Committee, 2015). A key element for the delivery of this national strategy is recognised to be "greatly improved, openly available data as well as better understanding of the cumulative impacts of development, climate change and socio-economic factors on marine ecosystems".

As well as this policy driven shift to ecosystem level marine research there are also scientific and economic drivers for a basin level approach. Marine monitoring is essential for assessing the health of a particular ecosystem and for assessing the impacts of specific factors and activities on it. Data is fundamental to carrying out these marine ecosystem assessments and making forecasts of potential future changes to support sustainable use of the marine environment and the ecosystem services it delivers (UNEP-WCMC, 2011).

Marine ecosystem services are defined as the societal and economic benefits that are derived from the use and exploitation of the ocean, and they are increasingly becoming threatened by the more widespread and growing demands being made on the marine environment (Duffy, 2006). The potential implications of these increasing pressures on the marine ecosystem can be demonstrated through assessments of the impacts on ecosystem services at both the regional and national level. Organisations such as the International Council for the Exploration of the Sea (ICES) conduct quantitative evaluations and synthesis of information on physical, chemical, ecological, and human processes in the marine ecosystem to assess these impacts and develop integrated management plans for the sustainable use and protection of marine ecosystems. At the national level initiatives such as the Marine Ecosystems Research Programme (MERP) in the UK are seeking to address the current acknowledged gaps in marine ecosystem research by integrating existing data and targeted new data with current models and knowledge of ecosystem services. This will give an improved understanding of the whole UK marine ecosystem, how it responds to changes in pressures, and the consequences of those changes in terms of ecosystem services.

The accuracy and efficacy of any marine ecosystem assessment or forecasts of potential changes due to the impacts of the use and exploitation of the ocean are in part dependent on the availability of large volumes of high quality and fully documented data. The researcher conducting these studies must be able to discover, assess and access a range of multidisciplinary data that may only be available from several different distributed systems. It must be provided with adequate metadata and a range of additional supporting information that describes the data and its provenance to ensure that it is fit for purpose and allows the user to combine data from multiple sources and different domains for a specific application.

BARRIERS TO THE RE-USE OF MARINE DATA

The motivations and methods for capturing marine data have resulted in a high degree of heterogeneity. Data acquisition in the marine environment has often been undertaken to address a particular area of research or for a specific commercial application. It is also being collected by a large number of organisations and institutions around the world using a variety of sensors that are mounted on various platforms that range from traditional research vessels, moorings and buoys to modern satellites, gliders and even marine fauna. Each type of sensor can potentially deliver data in a number of different formats and this landscape is further complicated by data that are delivered in proprietary formats that are unique to a particular sensor or individual equipment manufacturer. The heterogeneity of the data, existing data management practices and organisational, national and regional policies on data sharing can all act as barriers to the future re-use of marine data.

Standardised Data Formats

A number of repositories and associated marine data infrastructures have attempted to reduce the heterogeneity of marine data by implementing the use of standardised formats for data discovery and delivery purposes. These standard formats provide the user with access to harmonised data from multiple distributed repositories that can be integrated for various applications including analytics, modelling and visualisation. The European SeaDataNet e-infrastructure has taken this approach by adopting standardised data transport formats and defining specific SeaDataNet profiles for certain data types (SeaDataNet, 2015). These selected standard formats include:

- **ODV4 ASCII data format:** the Ocean Data View (ODV) software and associated ODV data formats were created for the purposes of interactive analysis and visualization of oceanographic and other geo-referenced profile, time-series, trajectory and sequence data. The ODV data format provides dense storage and allows instant access to the data for arbitrary stations, even in very large data collections making it suitable for use on a range of platforms. The ODV "data collection" format has been optimized for variable length, irregularly spaced profile, trajectory, and time-series data and is compatible with several other commonly used international marine data formats. The most recent release of the ODV collection format (ODVCF6) is more powerful and flexible than previous versions and accepts data for virtually unlimited numbers of stations, samples and variables, and supports almost unlimited sets of metadata (Schlitzer, 2015)

  The ODVCF6 format also supports data quality flags for metadata and data variables selected from 15 of the most widely used oceanographic quality flagging schemes e.g. the International Oceanographic Data and Information Exchange (IODE) quality flag scheme (Intergovernmental Oceanographic Commission of UNESCO [IOC], 2013)

  The SeaDataNet ODV4 import format is a less flexible version of the ODV version 4 generic spreadsheet format that has been modified to include additional mandatory information required specifically for SeaDataNet applications. For example, to allow integration of data from multiple sources, each ODV file includes a mandatory SeaDataNet compliant semantic header that maps directly between the metadata and data column headings and the standardised SeaDataNet concepts. Selection of the ODV format as a standard data transport format for SeaDataNet has also provided additional support and a driver for the continued development of both the ODV software and formats.
- **Network Common Data Form (netCDF):** is a set of software libraries and machine-independent data formats that support the creation, access, and sharing of array-oriented scientific data (UNIDATA, 2015). It is also a community standard for sharing scientific data that is increasingly being used for oceanographic data.

  The netCDF Climate and Forecast (CF) conventions define metadata that provide definitive descriptions allowing data from different sources to be used in combination for specific applications. The CF conventions were originally developed by atmospheric modellers to ensure that data from different models were interoperable. They were subsequently developed further to include other Earth sciences disciplines and have been widely adopted by many oceanographic projects. The CF conventions define metadata that provide a definitive description of what the data represents, and the spatial and temporal properties of the data. This allows users of data from different sources to decide which quantities are comparable, and facilitates building of applications for extraction, re-gridding and display of the data (UNIDATA, 2015).

  The SeaDataNet profile of the NetCDF-CF standard for profiles, time series and trajectory data has been developed to reduce the level of flexibility of the generic standard and therefore enhance potential interoperability by increasing the number of mandatory fields, specifying mandatory data streams and introduction of additional attributes from other conventions. The SeaDataNet NetCDF – CF profile, based upon Version 1.6 of the CF Metadata Conventions, was proposed as an accepted data transport format to promote interoperability of marine data in Europe and beyond.
- **MEDATLAS format:** an auto-descriptive format used for vertical profiles, time-series data or trajectories. It was originally defined by IFREMER (France) in accordance with the recommendations of the international ICES/IOC Group of Experts on the Technical Aspects of Data Exchange

(GETADE) that includes in its terms of reference development of technical solutions for the management, exchange and easier integration of oceanographic data and information with data from other disciplines.

The SeaDataNet profile of the MEDATLAS format includes extensions that provide links between data and metadata in the SeaDataNet system, and also to standardised SeaDataNet semantic information such as detailed parameter descriptions that support interoperability across datasets from multiple sources. It was selected as an additional SeaDataNet format for use with vertical profiles or trajectory data (Schaap, 2017)

Spatial Coordinate Systems Including Datums

The format of the data is only one aspect of the challenges that are faced by the user seeking to integrate and re-use multidisciplinary marine data from different federated systems. Another source of inconsistency for users seeking to combine marine data from multiple distributed sources comes from the use of different spatial reference systems for unambiguously locating a point on the Earth's surface.

There are essentially two types of coordinate systems in use: a) global or geographic coordinate systems such as latitude and longitude that are measured from the centre of the Earth to a point on the surface using a 3-D grid applied to the globe, and b) projected coordinate systems which use a map projection (a mathematical transformation) to convert the Earth's curved surface into a 2-D Cartesian equal area grid with the origin at its centre. Geographic reference systems (GRS), defined using latitude, longitude and vertical height as specified in the ISO 19111:2007 standard for geographic information and spatial referencing by coordinates, are used extensively in the marine domain. However, it should be noted that there are some fundamental legacy issues associated with the use of this coordinate system which have direct impact on the potential re-use of oceanographic data. For many years latitude and longitude were recorded as degrees, minutes and seconds where a minute was defined as $1/60^{th}$ of a degree and similarly a second was $1/60^{th}$ of a minute but most, if not all, organisations have now moved to the decimal degrees system which express latitude and longitude geographic coordinates as decimal fractions. Nevertheless, there is still some inconsistency on how and where this had been applied such that there is still a significant amount of legacy marine data which includes spatial information that uses this earlier notation and therefore requires conversion to decimal degrees before it can be integrated with more modern data sets.

The most commonly used projected coordinate system in the marine domain is the Universal Transverse Mercator (UTM). The UTM coordinate system uses a 2-dimensional Cartesian co-ordinate system to locate a point on the surface of the Earth by dividing the Earth into sixty unique zones with each being equivalent to six-degrees of longitude. Within each zone a point is measured in metres as an easting and a northing coordinate. Users of marine data will also encounter the Military Grid Reference System (MGRS) used by NATO militaries for locating points on the earth and used extensively for marine data collected for military applications. The MGRS is derived from the Universal Transverse Mercator (UTM) grid system and the Universal polar stereographic (UPS) grid system, but uses a different labelling convention due to its greater level of precision.

In addition to these global spatial reference systems there are also a large number of 'local' geographic and projected coordinate systems that have been used for marine data acquisition. For example, in the United Kingdom both the projected British National Grid (BNG) system and the OSGB63 geographic coordinate system have been widely used for marine surveys in UK waters, and until more recently when WGS84 was adopted, the OSGB63 coordinate system was also used for the nautical charts produced by the UK Hydrographic Office (Iliffe, Ziebart, Turner, Oliveira & Adams, 2006)

To add an additional level of complexity, in order to unambiguously establish a location for an ocean observation using a coordinate system also requires a horizontal datum. Not to be confused with the vertical datum used as reference points for elevations of surfaces and features on the Earth surface, the horizontal datum defines the size and shape of the Earth (the ellipsoid), and the geometric relationship between it and a coordinate system grid. There are a large number of horizontal datums in existence, each based on specific ellipsoids, and varying from the global to national scale (Plag & Pearlman, 2009). The most commonly used global horizontal datum is the World Geodetic System 1984 (WGS84) which is based on the WGS84 ellipsoid but there are also a large number of continental scale datums (e.g. South Asia datum, based on the Modified Fischer 1960 ellipsoid) and national-scale datums (e.g. Ireland 1965, based on the Modified Airy ellipsoid) which were originally developed for mapping and navigation of specific geographic regions before the advent of the global systems (International Hydrographic Organisation [IHO], 2003).

The use of various coordinate reference systems, ellipsoids and datums makes integration and re-use of marine data potentially problematic often necessitating conversion to a more consistent global coordinate system before data sets from different sources can be used in combination. But this translation can be difficult when a specific point on the Earth can have very different coordinates depending on the datum used to make the measurement. The lack of adequate information regarding the specific coordinate reference system used for a dataset can also lead to significant errors in accurately locating it. For example, every coordinate system grid is aligned with an ellipsoid that approximates the shape of the Earth, therefore if one ellipsoid is replaced by another the corresponding coordinate grids also shift as in the case of the Australian Geodetic Datum (AGD) that is based on the Australian National Spheroid (ANS), which was superseded in 1994 by the Geocentric Datum of Australia (GDA) that uses the GRS80 ellipsoid which resulted in a shift of 200m for those coordinates using the AGD (Lowry, 2004)

Metadata Standards

Many of the issues associated with re-use of data in different formats, using different coordinate systems etc. can be avoided if the dataset is adequately documented. For the user to select, assess and make appropriate use of marine data necessitates that it is accompanied by detailed metadata and other supporting information (Leadbetter, Raymond, Chandler, Pikula, Pissierssens & Urban, 2013). A metadata standard has been defined as 'a requirement to establish a common understanding of the meaning or semantics of the data, to ensure correct and proper use and interpretation of the data by its owners and users'. There are currently a range of discovery metadata standards used for describing marine data. These standards include those that have wider applicability such as ISO 19115 which has been developed by the International Organisation for Standardisation (ISO) to define the requirements for describing geographic information and services and, although widely used for digital data, can also be used for other forms of geographical information such as charts and maps. The ISO 19115 metadata standard provides information about the identification, the extent, the quality, the spatial and temporal schema, spatial reference, and distribution of digital geographic data (International Organization for Standardization [ISO], 2009).

The Dublin Core metadata standard is also widely used for marine data. Maintained by the Dublin Core Metadata Initiative (DCMI), it defines a broad, generic and relatively small core element set that is used in conjunction with controlled vocabularies to describe a wide range of networked resources. The Dublin core standard has also been used as part of the Open Archives Initiative Protocol for Metadata Harvesting (OAI-PMH) which requires any implementation of the protocol to support representing metadata in the Dublin Core standard. Also commonly used for marine data but which has wider applicability is the ISO compliant Directory Interchange Format (DIF). It is a metadata standard for Earth science data that has eight core elements and developed for the creation of metadata records for NASA's Global Change Master Directory (GCMD) that holds more than 35,000 Earth science dataset and service descriptions relevant to global change and Earth science research (NASA, 2015). The Federal Geographic Data Committee (FGDC) Content Standard for Digital Geospatial Metadata (CSDGM) has also been widely adopted for use with marine data but in September 2010 the FGDC endorsed ISO 19115 and now encourages the federal agencies to adopt the recommended ISO metadata standard. The CSDGM standard is therefore superseded but still used by some repositories, particularly in the USA.

The use of these more widely applicable metadata standards supports cross-domain interoperability and facilitates multidisciplinary research by allowing researchers to discover and use data from more than one domain and also from a number of federated data systems across several geographic regions (Montgomery & Bermudez, 2010). But this is not the complete picture since, in addition to these wider metadata standards, there are also many instances of standards having been established within countries and regions that are in some cases incompatible with those devised in other areas making data discovery at the global level difficult.

Many European organisations and institutes responsible for collecting and managing marine data have aligned themselves with the Common Data Index (CDI) metadata standard that has been developed by SeaDataNet as part of a project to implement a pan-European marine data infrastructure and provide a data discovery and access service for harmonised oceanographic data (Boldrini, 2013a). The CDI metadata standard is based on the ISO 19115 standard and also the ISO 19139 standard that provides the associated XML implementation schema (Boldrini, 2013b; International Organization for Standardization [ISO], 2007). It has the core elements defined by the ISO 19115 standard but also includes a number of additional elements and extensions defined specifically for use with marine data (ISO, 2009). Although the Common Data Index metadata standard was originally developed for use in Europe on a regional level, it is also becoming the acknowledged standard for oceanographic data exchange on a global scale.

Metadata standards have also been created for use at the national level. The Marine Community Profile (MCP) metadata standard developed by the Australian Ocean Data Centre Joint Facility (AODCJF) is widely used by a range of organisations engaged in marine research across Australia and New Zealand (Marine Community Profile Governance Committee, 2013). The MCP standard is also a subset of the ISO 19115 international standard that includes all of the core metadata elements as well as defined supplementary elements and controlled vocabularies required to describe marine data resources. Later versions of the MCP (versions 1.3 and above) are based on the standard ISO 19139 XML implementation of ISO 19115.

While the standards described above are based on the relevant ISO standards they are a different application specific 'marine profiles' of the basic ISO19115 standard. This 'speciation' of discovery metadata profiles leads to incompatibilities between them which are difficult to resolve for the purposes of data re-use on a wider scale.

Controlled Vocabularies

Controlled vocabularies, which are essential for defining the allowable content for a particular metadata element, are utilised by the majority of marine data infrastructures at every level of granularity (Bermudez, Montgomery, Miller, Neiswender & Isenor, 2011). For example, in Europe the SeaDataNet marine data infrastructure has a number of vocabularies that have been developed specifically for this application which are maintained and made publicly available through the NERC Vocabulary Server (NVS 2.0). These vocabularies also have a strict governance structure in place to ensure their integrity. Similarly in Australia the Spatial Information Services Stack Vocabulary Service (SISSVoc) service provides a controlled vocabulary for the AODN data system while the Marine Metadata Interoperability (MMI) Ontology Registry and Repository (http://mmisw.org/) is widely used for the oceanographic domain in the USA. On an international scale ICES provides the RECO Vocabulary v2.0 which is used for all ICES databases and is made available for wider usage (International Council for the Exploration of the Sea [ICES], 2014).

The use of controlled vocabularies can also support interoperability between existing marine data infrastructures. This can be achieved either by the mapping between terms where there are comparable terms in each vocabulary or by the adoption of a single vocabulary which then provides the first tier of interoperability between existing data systems at the metadata level. The mapping between the vocabularies used within the different data systems requires the use of ontologies to assist with identifying similar terms. These ontologies provide a domain specific representation of knowledge that is documented using a standardized and structured syntax that describes the relationship between concepts. Metadata cross-walks also serve a similar purpose by providing the mappings between different metadata standards as human and computer readable documentation. These cross-walks are exceptionally useful for the integration of data from different disciplines and sources as they can be automated to allow the user to search across a number of metadata directories (Graybeal, Isenor, & Wang, 2009).

Data Access Policies

Facilitating technical interoperability between regional data systems is only one aspect of making data available for re-use. The availability of marine data is also determined by the data access policies that commonly vary between countries, regions and organisations. It is now widely acknowledged that any data that has been acquired using public money should be made openly available to users and in some regions this has been formalised by legislature, for example, the INSPIRE Directive, which forms part of European law, requires that those spatial datasets that come under one or more of the 34 environmental themes set out in the Directive Annexes are made available in a consistent formats to enable the sharing of environmental spatial information among public sector organisations and facilitate public access to spatial information across Europe (European Commission, 2007).

But the approach to making public sector data open and easily accessible is not globally consistent and not all marine data is captured by publicly funded bodies. A significant proportion of this data is captured for specific commercial purposes such as by the oil and gas industry or collected for the purposes of national maritime security and is therefore often not made freely available. For example, the UK Hydrographic Office is responsible for surveying the ocean floor for military purposes and much of this data is therefore restricted and not available for re-use.

ESTABLISHING A COMMON FRAMEWORK FOR MARINE DATA MANAGEMENT

The various barriers to the re-use of marine data have long been recognised and a number of initiatives have been established to develop marine data infrastructures at a regional level in an attempt to address these challenges. For example, the SeaDataNet Common Data Index (CDI) in Europe, the US National Oceanographic Data Centre (US NODC) in the USA and the Integrated Marine Observing System (IMOS) has each been developed to address specific regional requirements for access to a range of multidisciplinary marine data. However, a key issue associated with these regional marine data infrastructures is that their development has been driven by the requirements of their user community and guided by the funding agency. As a result these regional marine data infrastructures have been developed independently of those in other regions using different approaches, standards, technologies and best practice. To add a further layer of complexity to this landscape, there are also global marine data infrastructure initiatives e.g. the IOC - Intergovernmental Oceanographic Data Exchange (IODE), the Partnership for Observing the Global Oceans (POGO), as well as those with a wider remit for the delivery of environmental data from multiple domains e.g. GEOSS, COPERNICUS, FutureEarth etc. that are orientated to a different set of priorities.

Each of the regional data systems, although addressing the issues associated with the heterogeneity of data such as formats, coordinate systems etc. has developed and implemented its own solutions to these specific problems which are often incompatible with those developed elsewhere. But ecosystem level marine research requires a common framework for marine data management that supports the sharing of data across these regional and global data systems, and provides the user with access to the data available from these services via a single point of access. It must be based on the existing data systems and established by developing interoperability between them. However, achieving this objective potentially requires significant resources as well as a fundamental cultural change within the organisations and institutes responsible both for the management of these systems and the stewardship of the data.

Ocean Data Interoperability Platform (ODIP)

In October 2012 the European Commission, the National Science Foundation (NSF) in the USA and the Australian Government co-funded the Ocean Data Interoperability Platform (ODIP) project with the aim of promoting the development of a common global framework for marine data management. However, the scale of the effort required to develop this framework necessitated addressing selected key aspects of interoperability between the existing marine data systems (Glaves, Miller, Proctor & Schaap, 2013). To achieve these objectives ODIP brought together representatives from the regional marine data infrastructures in Europe, the USA and Australia along with other technical experts to build a community of practice that could evaluate the existing data systems and formulate technical solutions for establishing interoperability between them and also with the global multidisciplinary e-infrastructures (Figure 1). The ultimate aim of these activities was to create a network of integrated marine e-infrastructures whilst also developing an approach that could be implemented more widely (Glaves, Schaap, Arko & Proctor, 2014; Glaves, 2015).

Figure 1. The Ocean Data Interoperability Platform (ODIP) framework illustrating the regional marine data infrastructures and relationships with larger global data systems

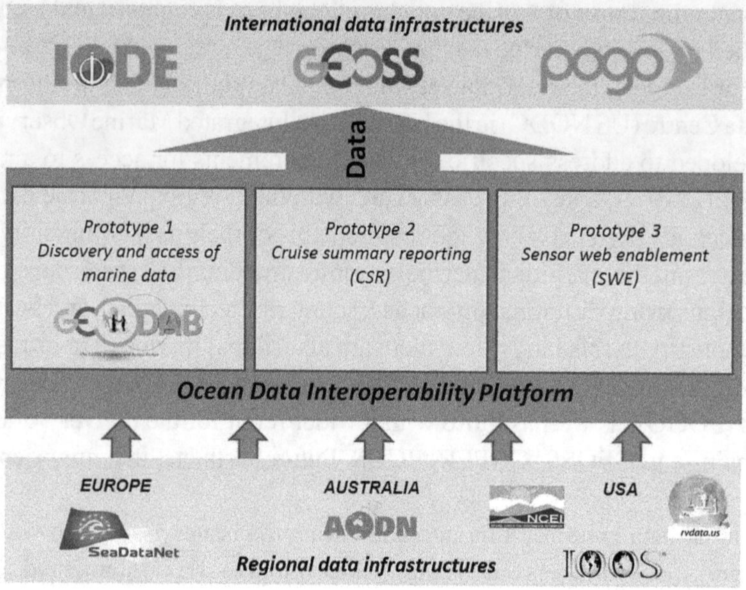

Prototype Interoperability Solutions

To begin the process of integration and also demonstrate its feasibility, the ODIP project developed a small number of prototype interoperability solutions. The formats of these prototypes were identified to demonstrate the methodology for integrating different aspects of the regional data systems into a coherent global framework. Each of the prototype development activities leveraged the on-going development activities of the existing regional data infrastructures with the objective of establishing interoperability between them and also with the relevant global systems such as the Global Earth Observation System of Systems (GEOSS) and the Partnership for Observation of the Global Oceans (POGO) portals.

Data Discovery and Access Services

The first of these ODIP prototype interoperability solutions has integrated existing key marine data discovery and access services in Europe, Australia and the USA to allow exchange of information about their respective data holdings, and also expose metadata for the data delivered by these regional data systems in the global IODE Ocean Data Portal and the GEOSS portal (Schaap, 2014). This prototype solution uses brokering technologies to establish interoperability between the regional data services and also with the selected global portals without making significant modifications to the existing services. To encourage wider adoption of this approach and to further develop the network of integrated regional data infrastructures necessitates that any implications of the adoption of the solution is minimised and that the benefits of this approach exceeds the inconvenience of any modifications that may be required. To develop this solution the Ocean Data Interoperability Platform project worked closely with the Global Earth Observation System of Systems (GEOSS) initiative that has developed the GEO Discovery and

Access Broker (GEO-DAB) for connecting heterogeneous/ distributed data resources that contribute to the GEOSS portal (Nativi, Craglia & Pearlman, 2012).

The GEO-DAB broker is part of the GEOSS Common Infrastructure (GCI) which consists of several software components and utilizes a suite of interoperability standards and best practice to interconnect the heterogeneous Earth Observation resources and information systems available within the framework of the Group on Earth Observations (GEO). The GCI is structured in such a way that it decouples the users of the distributed services from those providing them using the GEO-Discovery and Access Broker (DAB) framework, which provides functionality for discovery, access and semantic interoperability.

In contrast to the federation architecture that is often used for linking different information systems and relies on the adoption of a common data model and exchange protocol, the brokering approach is highly flexible providing mediation and transformation functionalities that build on existing data systems. The participating end systems retain a high degree of autonomy and are only required to document their existing interfaces and data models without any requirement to adopt a common model for data sharing (Nativi, Mazzetti, Santoro, Papeschi, Craglia, & Ochiai, 2015) The use of the brokering approach to establish interoperability across the regional marine data infrastructures and global data services represented in the Ocean Data Interoperability Platform project provided a flexible solution where the limited minimal overarching governance did not allow the enforcement of a solution that necessitated modifications to the existing systems or the adoption of a common model for information sharing as described by (Nativi et al., 2015).

Establishing interoperability between the regional data systems and with the global portals using the GEO-DAB brokering solution was conducted as a staged process which began by exposing data available through the European SeaDataNet Common Data Index (CDI) discovery and access service in the GEOSS portal and IODE Ocean Data Portal (ODP). The large number of datasets available in the SeaDataNet CDI data discovery and access service, which in October 2015 stands at in excess of 1.8 million CDI metadata records for individual datasets, necessitated that aggregations or 'collections' of metadata records 'granules' were created to reduce this number to a set of around 500 collections. This approach also employs a dynamic aggregation process which provides a current view of the available component datasets that incorporates any recent updates. Each of these collections also includes a web address (URL) to the regional data service which allows navigation to the original datasets ('granules') for viewing and/or download.

The collections, which in the case of the SeaDataNet service, were aggregated using specific parameters such as 'discipline' defined by the NVS 2.0 P08 vocabulary, or the data centre as defined using the European Directory of Marine Organisations (EDMO), are output as ISO 19139 compliant XML using a REST / CSW web service. This is then harvested by the GEO-DAB broker and converted to the GEO-DAB Generic Brokerage Reference Schema (shown in Figure 2) allowing the vocabulary terms from the original metadata record to be maintained alongside the ISO standard terms used by the GEO-DAB broker. The broker service then delivers the collections metadata as ISO 19139 XML in both an OAI-PMH service and a OGC Catalogue Service for the Web (CSW) Version 2.0.2 Service using the HTTP POST method and the 'GetRecords' operator. The CSW service is harvested and uploaded to the GEOSS portal while the OAI-PMH service and the Java-based Open Archives Initiative (jOAI) harvester are utilised for populating of the IODE Ocean Data Portal as illustrated in Figure 3.

The next stage of the process to integrate the regional data infrastructures and to expose the available data via the selected global ODP and GEOSS portals, has been to implement a similar approach for the data discovery and access services operated in the USA and Australia.

Figure 2. GEO-DAB generic brokerage reference schema: simplified conceptual model of GI-suite (Stefano Nativi; CNR-IIA)

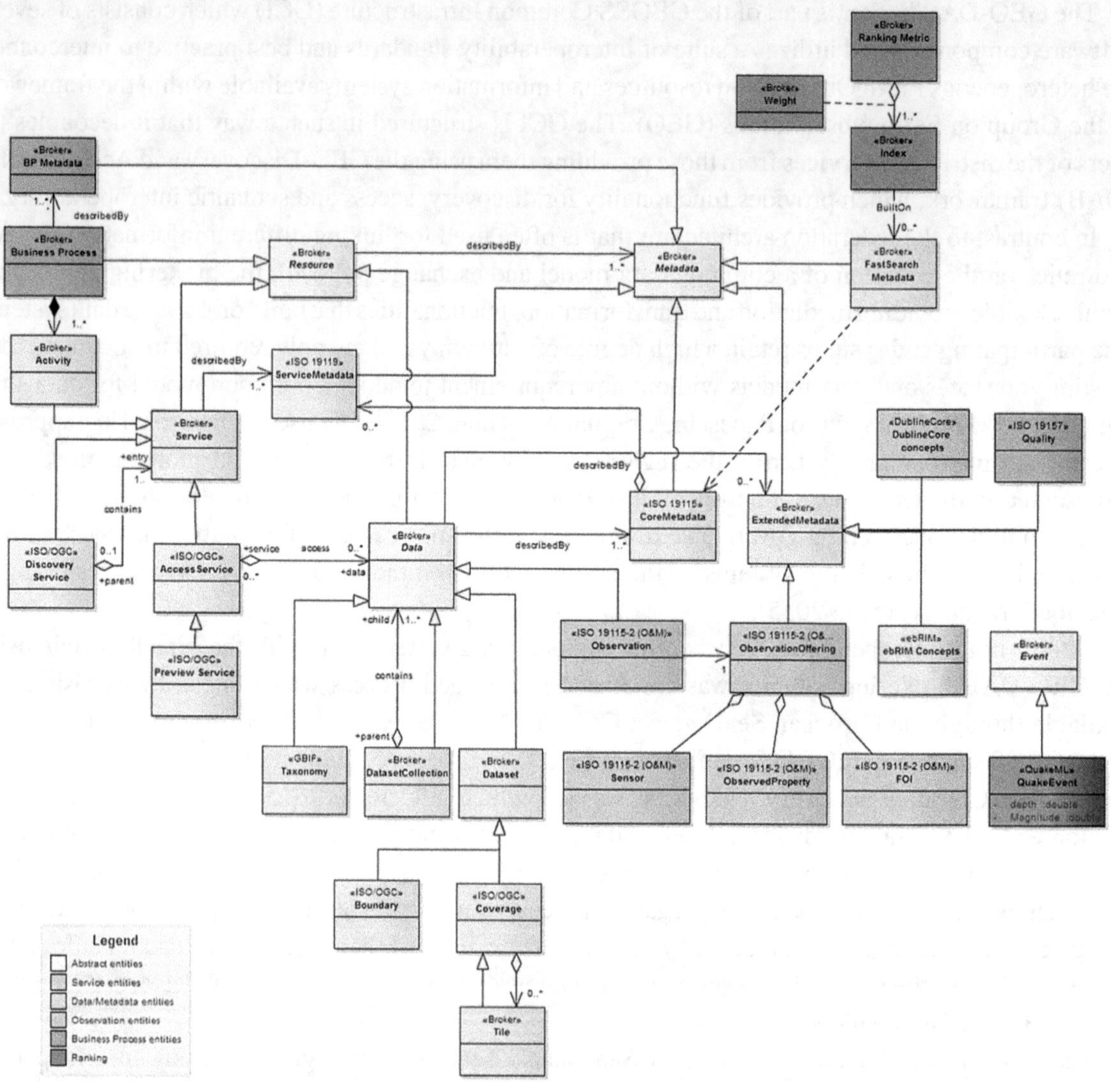

The National Centres for Environmental Information (formerly the National Oceanographic Data Centre or US NODC) which deals with in-situ, satellite and model data, was already providing collection level web services with CSW, OAI-PMH and OpenSearch endpoints for discovery. All metadata contains at least FTP and HTTP links to the data, and many also have Hyrax, THREDDS Data Server, and Live Access Server links. As of October 2015 the US NODC was delivering around 28,000 collections which represent aggregations of more than 1.3 million data set descriptions using ISO 19115 and ISO 19139 compliant metadata.

To integrate the US NODC with the other regional data infrastructures, the GEO-DAB brokerage service has made use of the available collections level CSW to harvest the metadata from the US NODC and converted the XML to the GEO-DAB Generic Brokerage Reference Schema (Figure 2) while making

Figure 3. SeaDataNet collections exposed in the IODE Ocean Data Portal

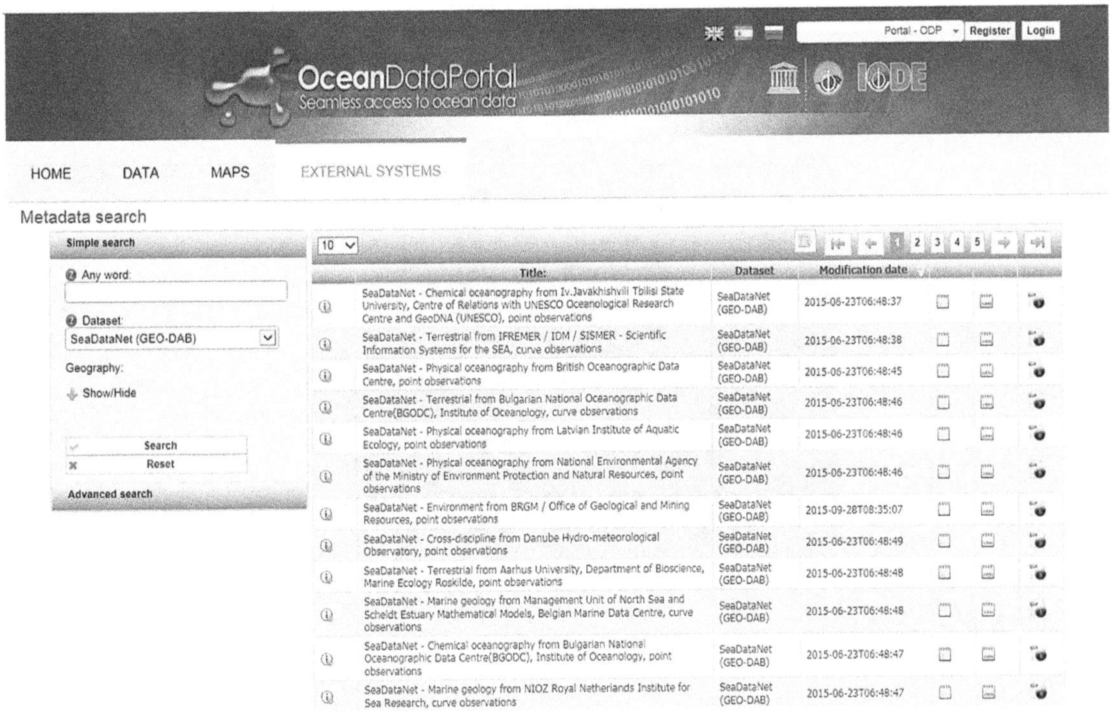

use of the US NODC vocabularies. The harvesting of the US NODC collections has been done using a similar approach to that taken for those from the SeaDataNet system with an OIA-PMH service used for delivery to the ODP and a CSW used to populate the GEOSS portal as shown in Figure 4.

The Australian Ocean Data Network (AODN) is a network of governmental and science organisations in Australia and New Zealand. The AODN123 ocean data portal which includes data from the six Commonwealth Agencies with responsibilities in the Australian marine jurisdiction, makes use of the Marine Community Profile (MCP) metadata profile that is based upon the ISO 19115 and ISO 19139 standards. It also makes use of GeoNetwork, a catalogue application for managing spatially referenced resources, for discovery (CSW) and mapping (WMS) services.

The AODN data discovery service originally operated at the granule level and has been upgraded to collections level but these are defined differently to those of SeaDataNet and the US NODC, and are aggregated according to the facility (similar to data originator), geometric object, and whether real-time or delayed mode of delivery. AODN provides CSW, OAI-PMH and OpenSearch endpoints for discovery of the data collections (around 110 in October 2015). Using a similar mechanism to that used for SeaDataNet and the US NODC, the GEO-DAB broker service now delivers the AODN collections as an OAI-PMH service and CSW but the Dublin Core vocabularies have been utilised. As a result the datasets available in the AODN regional data service can now potentially be exposed in the GEOSS portal and ODP.

The development of this prototype solution to integrate the regional data discovery and access services into GEOSS and IODE- Ocean Data Portal has successfully demonstrated how the brokering approach can be used to establish interoperability between the regional data infrastructures and the GEOSS portal and IODE-ODP. Collection level metadata from the three regional systems is now exposed in the global

Figure 4. NODC collections exposed in the GEOSS portal

portals making it possible for users to discover the available data through a single point of access and then use the links provided to return to the regional data systems to access more detailed information and download the available datasets.

The staged approach for integrating the regional marine data infrastructures into the global portals at the metadata level described here demonstrates how other data systems can potentially adopt a similar approach in order to become part of the larger federated network of marine data infrastructures which are also exposing data in global portals such as GEOSS and ODP.

Cruise Summary Reporting (CSR) Systems

A large number of research cruises are being conducted around the globe at any one time. There is also information available for a huge number of legacy cruises that have previously been carried out but this is largely fragmented and difficult to locate. In an era when funding for scientific research is becoming increasingly difficult to access there is a need to make the best use of existing resources to support the deployment of cruises. Initiatives such as EUROFLEETS aim to create a pan-European distributed research fleet infrastructure that has a common strategy and provides coordinated access to European marine research vessels and equipment. However, to support this activity and assist organisations with the planning of research cruises there is a need to provide better access to information about the legacy cruises that have been conducted. The Partnership for Observation of the Global Oceans (POGO), which is a forum to promote global oceanography and in particular the implementation of international and integrated global ocean observing systems, operates a catalogue of planned, current and past cruises in partnership with the Bundesamt für Seeschifffahrt und Hydrographie (BSH) in Germany, but incompatibilities between the regional cruise summary reporting systems and the POGO portal means that population of this centralised catalogue for some geographical regions is limited and in some areas completing lacking.

To address these inconsistencies ODIP has developed a mechanism to improve and streamline delivery of cruise summary reports (CSR) from the European, Australian and US systems to the POGO portal by promoting the use of common formats and controlled vocabularies.

The European Cruise Summary Report (CSR) system is an integral part of the SeaDataNet metadata infrastructure and at October 2015 contained entries for over 53,000 cruises that have been conducted by more than 2000 different research vessels since 1873. The Cruise Summary Reports system schema is ISO 19139 compliant in line with the INSPIRE directive and makes use of the standard SeaDataNet vocabularies which have been extended with several relevant attributes such as ports and platforms. The CSR also makes use of the ISO 19115 and 19115-2 content models for measurements and samples and also supports linking to track charts and station lists. The CSR web mapping service (WMS) displays cruise summary reports that have been supplied with the associated navigation tracks in a Geography Mark-up Language (GML) format.

CSR is also in use for the POGO research vessel portal which has a focus on ocean-going research vessels with a length of more than 60 metres and at around the same time as the development of the ODIP prototype solution began, a decision was taken that the global POGO portal would make use of the SeaDataNet Cruise Summary Report 3.0 schema. The adoption of the CSR 3.0 schema dictated that the other regional cruise summary reporting systems delivering cruise summary reports to the POGO portal should either adopt or achieve interoperability with this schema and the associated vocabularies. In addition GeoNetwork was selected for delivery of the CSR XML to the POGO portal. SeaDataNet already had a fully operational GeoNetwork instance for delivery of cruise summary reports to the POGO portal while the regional systems in the USA and Australia had their own systems for reporting details of research cruises.

The Rolling Deck to Repository (R2R) project is responsible for providing stewardship of underway marine sensor data collected by the United States academic research fleet. The R2R Cruise Catalog publishes cruise information in multiple formats that include ISO 19115 -2 metadata records; OGC complaint CSW; W3C Linked Open Data with a SPARQL endpoint and; DOI-DataCite Metadata records, all of which used controlled vocabularies and best practices where these are available. R2R also provides web services for cruise track lines (WFS delivering GML geometry) and controlled vocabularies (Simple Knowledge Organization System Reference (SKOS) compliant). The R2R metadata records are published using the NOAA Cruise Level Metadata Template that is based on ISO19115, ISO19115-2 and ISO19139 standards (NOAA, 2015). In November 2015 the R2R Cruise Catalog included information for 5831 cruises undertaken using 25 different vessels.

To move towards the delivery of cruise summary reports across the regional cruise summary report systems, the R2R initiative chose to adopt and implement the SeaDataNet CSR 3.0 schema which requires a mapping between the R2R vocabularies and those used by the SeaDataNet data infrastructure be created. This has led to the expansion of the SeaDataNet vocabularies and directories to include additional terms required for reporting US research cruises. The R2R Cruise Catalog has subsequently been upgraded to include these mappings which support interoperability with the SeaDataNet Cruise Summary Report system. To further support the development of interoperability between the regional cruise summary report systems R2R created a GeoNetWork search interface for the Cruise Catalog and API (CSW service) allowing the cruise metadata records to also be harvested and integrated into the POGO portal. As a result of this work cruise summary reports for US vessels held in the R2R Cruise Catalog can now also be accessed via the SeaDataNet Cruise Summary Report system and also discovered via the POGO portal.

Efforts to integrate the cruise summary reports for Australian cruises proved more challenging because this information is not held in a comparable system to those in Europe and the USA. Instead cruise information for Australian research vessels is stored in the CSIRO Marine and Atmospheric Research metadata system, MarLIN, along with metadata on datasets from the Marine National Facility (MNF) and other sources. The survey details, cruise plan, track and data are provided for each vessel and are relatively less well-structured when compared to the cruise summary reports available in the other regional systems and via the POGO portal. However, new Australian research vessels are coming on stream which will provide opportunities for implementing a more structured cruise reporting system and future collaborations with global initiative such as POGO will promote the adoption of this approach to allow details of Australian research cruises to be integrated into the global systems.

Further work will be undertaken to move towards a fully integrated cruise summary report system that can be adopted by other comparable systems elsewhere in the world. The aim is to support the development of a global system for capturing cruise summary reports that provides the user with the ability to discover all available cruises for an individual ecosystem or ocean basin. It will also support future cruise planning by providing a complete picture of all cruises that have been run in a specified area of interest irrespective of the nationality of the commissioning body.

Sensor Web Enablement

There are a wide range of ocean observing sensors deployed in the world's oceans at any given moment in time. The recent development in web technologies allows these sensors to be discoverable, accessible and useable via the world-wide web which has led to a corresponding rise in demand from users for (near) real-time access to data from them.

To support the development of interoperable 'sensor webs' the Open Geospatial Consortium (OGC) created a suite of specifications for standards and best practice for sensor web enablement (SWE) that include: a) Sensor Model Language (SensorML) that provides XML encodings for describing the sensors and measurement processes; b) Observations and Measurements (O&M) that defines general models and schemas for observations, and features involved in sampling when making those observations (also approved as ISO 19156:2011 standard); and c) Sensor Observation Service (SOS) specification for web service interfaces that allows querying of real-time observations and metadata from sensors. The data returned in response to a query provides a description of the senor itself encoded in SensorML and the measured value in O&M format.

Various initiatives around the world are progressing towards adopting and implementing these SWE standards for operational ocean observing systems. SensorML and O&M have been widely adopted to support more detailed description of the observations and the sensors used as part of the measuring process, and SOS web services are being used to provide direct access to the related data streams from real-time ocean monitoring networks and underway data from systems on board research vessels.

However, the OGC's sensor web enablement (SWE) standards are designed to be domain independent and deliberately provide a high degree of flexibility enabling implementation across the different disciplines and usage scenarios. There is therefore the potential for significant divergence between the various implementations of the SWE standards. To avoid potential issues with interoperability between the different initiatives there is a need for a common approach to how the SWE standards are applied.

Within the context of the Ocean Data Interoperability Platform (ODIP II) project efforts are being made to establish a coordinated approach to the development of common SWE standards that can be adopted not only within the marine domain but more widely where there is a requirement for real-time access to observing systems. ODIP is bringing together many of the initiatives in the marine domain that are seeking to implement sensor web enablement in order to identify the requirements for a common set of marine profiles of the OGC SWE standards that can be used as the basis for developments across multiple projects and organisations. For example, in Europe there are many projects that are addressing various aspects of SWE implementation including the EU-funded EUROFLEETS project that is working towards the use of SensorML for encoding underway ship summary reporting and also usage of SWE technologies for on-board data acquisition and shore to ship access. The SeaDataNet initiative is also defining SensorML profiles to describe instruments and sensors, and adapting the O&M data models for use with marine observation data. This implementation of SWE technologies will support delivery of data from the ocean observing systems to European data centres as well as integration with the SeaDataNet Common Data Index (CDI) data discovery and access service.

In addition to European efforts to adopt and implement the SWE standards there are also activities in the USA through the Integrated Ocean Observing System (US-IOOS) that are focussed on the application of the SWE standards for time series, profiles and trajectories data from buoys, tide gauges, ADCPs, gliders, and other common oceanographic data collection platforms. US-IOOS has also adopted the OGC SOS specification for web interfaces with the IOOS SOS v1.0 application profile having been fully developed that also conforms to the OGC SWE standards. US-IOOS currently has two SOS reference implementations that are based on the 52°North Sensor Observation Service (52North, 2014) and ncSOS services.

By bringing together the various initiatives seeking to adopt and implement the SWE standards for sensor observing systems with particular emphasis on, but not limited to, marine applications the Ocean Data Interoperability Platform project has played a key role in establishing a global "community of practice" with the aim of defining common SWE standards for the marine domain that can also potentially be adopted more widely.

CONCLUSION

The need for an ecosystem or sea-basin level approach to marine research has made the integration of the pre-existing marine data infrastructures into a larger coherent network of data services which allows the user to access multidisciplinary data from a range of distributed systems via a single point of access a necessity. The Ocean Data and Interoperability Platform (ODIP) project is an exemplar for an approach that leverages on existing development activities at a regional level to create an overarching common global framework for marine data management. This approach is one that is also entirely scalable and transferable to other domains.

Establishing interoperability across the small number of regional data infrastructures that have participated in an initiative such as ODIP and with the larger global GEOSS and ODP portals demonstrates how existing distributed data systems can become part of a larger framework. This makes marine data available to a wider range of multidisciplinary users who are able to use it to tackle the wider societal challenges that the global portals such as GEOSS have been established to address.

Marine data is often captured for a specific application such as resource assessment, or to investigate a particular area of interest. Due to the unpredictable and sometimes hostile nature of the marine environment, the capture of marine data is frequently challenging and requires a significant amount of resources to be invested. The data captured is also often unique providing an unrepeatable measurement of the state of the ocean at a given moment in time for a specific geographical location. Marine data is therefore irreplaceable and extremely valuable for future research. It should therefore be made as widely available for re-use as possible.

As described above, there are often significant barriers to the discovery and integration of marine data from distributed sources. By developing a better understanding of these issues and how they can potentially be addressed, the user seeking to make use of multidisciplinary marine data, some of which will be unfamiliar to them, for the purposes of ecosystem or ocean basin level can avoid serious inconsistencies in the results of their research due to making inappropriate use of data obtained from the diverse range of data services.

REFERENCES

Bermudez, L., Montgomery, E., Miller, S. P., Neiswender, C., & Isenor, A. (2011). *The Importance of Controlled Vocabularies*. The MMI Guides: Navigating the World of Marine Metadata. Retrieved from http://marinemetadata.org/guides/vocabs/vocimportance

Boldrini, E. (2013a). *SeaDataNet metadata profile of ISO 19115. Version 10.0.0*. Retrieved from http://www.seadatanet.org/content/download/20284/140589/file/CDI-profile-V10.0.pdf

Boldrini, E. (2013b). *SeaDataNet metadata profile of ISO 19115 –XML encoding. Version 10.0.0.* Retrieved from http://www.seadatanet.org/content/download/20285/140594/file/CDI-profile-XML-encoding-V10.0.pdf

Duffy, J. (2006). *Marine ecosystem services*. Retrieved from http://www.eoearth.org/view/article/154472

European Commission. (2007). *Directive 2007/2/EC of the European Parliament and of the Council of 14 March 2007 establishing an Infrastructure for Spatial Information in the European Community (INSPIRE)*. Retrieved from http://eur-lex.europa.eu/JOHtml.do?uri=OJ:L:2007:108:SOM:EN:HTML

European Commission. (2008). *Directive 2008/56/EC of the European Parliament and of the Council of 17 June 2008 establishing a framework for community action in the field of marine environmental policy (Marine Strategy Framework Directive)*. Retrieved from http://eur-lex.europa.eu/legal-content/EN/TXT/PDF/?uri=CELEX:32008L0056&from=EN

European Commission. (2010). *Communication from the European Commission to the European Parliament and the Council: Marine Knowledge 2020: marine data and observation for smart and sustainable growth COM/2010/0461*. Retrieved from http://eur-lex.europa.eu/legal-content/EN/TXT/PDF/?uri=CELEX:52010DC0461&from=EN

European Commission. (2013). *Galway Statement on Atlantic Ocean Cooperation: Launching a European Union – Canada – United States of America Research Alliance 24 May 2013*. Retrieved from https://ec.europa.eu/research/iscp/pdf/galway_statement_atlantic_ocean_ cooperation.pdf

European Commission. (2014). *Communication from the European Commission: Innovation in the Blue Economy: realising the potential of our seas and oceans for jobs and growth COM(2014) 254/2 (13/05/2014)*. Retrieved from http://eur-lex.europa.eu/legal-content/EN/TXT/PDF/ ?uri=CELEX:52014DC0254R(01)&from=EN

Glaves, H., Miller, S., Proctor, R., & Schaap, D. (2013). Mind the Gap: Furthering the development of an international collaboration in marine data management. *Geophysical Research Abstracts*, *15*, EGU2013–EGU12737.

Glaves, H., Schaap, D., Arko, R., & Proctor, R. (2014). Ocean Data Interoperability Platform (ODIP): Supporting the development of a common global framework for marine data management through international collaboration. *Geophysical Research Abstracts*, *16*, EGU2014–EGU14366.

(2015). Glaves (2015) Ocean Data Interoperability Platform (ODIP): Developing a common global framework for marine data management through international collaboration. *Geophysical Research Abstracts*, *17*, EGU2015–EGU12644.

Graybeal, J., Isenor, A., & Wang, H. (2009). Moving Between Standards (Crosswalking). In *The MMI Guides: Navigating the World of Marine Metadata*. Retrieved from http://marinemetadata.org/guides/mdatastandards/crosswalks

Iliffe, J.C., Ziebart, M., Turner, J., Oliveira, J., & Adams, R., (2006). The VORF Project - Joining up Land and Marine Data. *GIS Professional*, (13), 24-26.

Intergovernmental Oceanographic Commission of UNESCO. (2013). Ocean Data Standards, Vol.3: Recommendation for a Quality Flag Scheme for the Exchange of Oceanographic and Marine Meteorological Data. IOC Manuals and Guides, 54(3).

International Council for the Exploration of the Sea. (2014). Implementing the ICES strategic plan 2014–2018 Linking Science, Advice, Data and Information, and Secretariat. Author.

International Hydrographic Organisation. (2003). Users Handbook on Datum Transformations Involving WGS 84; 3rd Edition, July 2003 (Last correction August 2008) Special Publication No. 60. International Hydrographic Bureau.

International Organization for Standardization. (2007). *ISO 19139 Geographic information Metadata XML schema implementation. International Organization for Standardization*. Geneva: ISO.

International Organization for Standardization. (2009). *ISO 19115 Geographic Information - Metadata. International Organization for Standardization*. Geneva: ISO.

Leadbetter, A., Raymond, L., Chandler, C., Pikula, L., Pissierssens, P., & Urban, E. (2013). *Ocean Data Publication Cookbook, IOC Manuals and guides No. 64*. Retrieved from http://www.iode.org/index.php?option=com_oe&task=viewDocumentRecord&docID=10574

Lowry, J.B.C. (2004). *WGS - AGD - GDA: Selecting the correct datum, coordinate system and projection for north Australian applications*. Department of the Environment and Heritage, Internal Report 473.

Marine Community Profile Governance Committee. (2013). *Marine Community Profile Manual Release 1.4*. Retrieved from http://bluenet3.antcrc.utas.edu.au/mcp-doc/index.html

Montgomery, E., & Bermudez, L. (2010). Some Examples of Metadata. In *The MMI Guides: Navigating the World of Marine Metadata*. Retrieved from http://marinemetadata.org/guides/mdataintro/mdataexamples

NASA. (2015). *Directory Interchange Format (DIF) Writer's Guide, 2015. Global Change Master Directory*. National Aeronautics and Space Administration. Retrieved from http://gcmd.nasa.gov/add/difguide/

National Marine Science Committee. (2015). *National Marine Science Plan: Driving the development of Australia's blue economy*. Retrieved from http://frdc.com.au/environment/NMSC-WHITE/Documents/NMSP_TS_040116%20website%20update.pdf

Nativi, S., & Bigagli, L. (2009). Discovery, mediation, and access services for earth observation data. *IEEE J. Sel. Top. Appl. Earth Observ. Remote Sens.*, *2*(4), 233-240.

Nativi, S., Craglia, M., & Pearlman, J. (2012). The Brokering Approach for Multidisciplinary Interoperability: A Position Paper: National Research Council of Italy -Institute of Atmospheric Pollution Research(CNR-IIA). *International Journal of Spatial Data Infrastructures Research*, *7*, 1–15.

Nativi, S., Mazzetti, P., Santoro, M., Papeschi, F., Craglia, M., & Ochiai, O. (2015). Big Data challenges in building the Global Earth Observation System of Systems. *Environmental Modelling & Software*, *68*(June), 1–26.

NOAA. (2015). *NOAA Cruise Level Metadata Template*. Retrieved from https://geo-ide.noaa.gov/wiki/index.php? title=Cruise_Level_Metadata_Template

52. North. (2014). *Sensor Observation Service*. Retrieved from: http://52north.org/communities/sensorweb/sos/

Oceans Policy Science Advisory Group. (2013). *Marine Nation 2025: Marine Science to Support Australia's Blue Economy*. Retrieved from http://www.aims.gov.au/documents/30301/550211/Marine+Nation+2025_web.pdf/bd99cf13-84ae-4dbd-96ca-f1a330062cdf

Open Geospatial Consortium. (2011). *Observations and Measurements - XML Implementation.* OGC®project document: OGC 10-025r1. Retrieved from http://www.opengis.net/doc/IS/OMXML/2.0

Peter-Plag, H., & Pearlman, M. (Eds.). (2009). Global Geodetic Observing System: Meeting the Requirements of a Global Society on a Changing Planet in 2020. Springer Science & Business Media.

Schaap, D. (2014). ODIP - Ocean Data Interoperability Platform - developing interoperability pilot project 1. *Geophysical Research Abstracts, 16*, EGU2014–EGU2845.

Schaap, D. (2017). *SeaDataNet: Towards a Pan-European Infrastructure for Marine and Ocean Data Management. In Oceanographic and Marine Cross-Domain Data Management for Sustainable Development.* Hershey, PA: IGI Global.

Schlitzer, R. (2015). *Ocean Data View User's Guide. Version 4.7.4.* Retrieved from https://odv.awi.de/fileadmin/user_upload/odv/misc/odv4Guide.pdf

SeaDataNet. (2015). *SeaDataNet data file formats (D8.5).* Retrieved from http://www.seadatanet.org/content/download/16251/106283/file/SDN2_D85_WP8_Datafile_formats.pdf

UNEP-WCMC. (2011). Marine and coastal ecosystem services: Valuation methods and their application. UNEP-WCMC Biodiversity Series No. 33.

UNIDATA. (2015). *The NetCDF User's guide.* Boulder, CO: UCAR/Unidata. Retrieved from http://www.unidata.ucar. edu/ software/netcdf/docs/user_guide.html

KEY TERMS AND DEFINITIONS

Blue Economy: Economic development based around the marine environment that balances societal benefits against sustainable exploitation of the oceans.

Ecosystem Services: The collective benefits derived from an ecosystem by mankind that makes life both possible and fulfilling. These benefits are often subdivided into four categories which can be defined for the marine domain as: 'provisioning' that includes fisheries, fertilizer production; 'regulating' that supports the control of climatic effects and natural hazard protection; 'cultural' that provides recreational activities; and 'supporting' that underpins the other benefits e.g. nutrient and water cycling.

E-Infrastructure: Framework of interconnected distributed digital resources that includes, but is not limited to, hardware, software, data, services etc. that can be accessed and utilized irrespective of geographical location. Note that some authorities also include people and the organizational structures e.g. governance within their definition of an e-infrastructure.

Good Environmental Status (GES): A classification that arose from the European Commission's Marine Strategy Framework Directive (MSFD), it is defined as ecologically diverse oceans and seas that are clean, healthy and productive. This assessment is made via a suite of 11 qualitative descriptors that define how a marine environment that has achieved this status should appear.

Interoperability: The ability of a product or a system to interact with, or use parts of, another system or resource without any modification being necessary.

Quality Flag Scheme: A standardized classification system used to indicate the quality, reliability, any modifications and the level of validation for individual data values.

Research Cruise: A planned study of the marine environment involving a dedicated ocean-going vessel that is equipped to conduct research at sea. A research cruise conducts surveys of the marine environment that include data acquisition and/or collecting samples from the ocean or sea floor.

Chapter 4
Linked Ocean Data 2.0

Adam Leadbetter
Marine Institute, Ireland

Adam Shepherd
Woods Hole Oceanographic Institution, USA

Michelle Cheatham
Wright State University, USA

Rob Thomas
British Oceanographic Data Centre, National Oceanography Centre, UK

ABSTRACT

Within the theme of sustainable development, it is not desirable to either have data siloed in one location where it cannot be reused for purposes beyond which it was originally collected, or in a state where it cannot be integrated into a holistic view of the marine environment. As such, the links between datasets should be formally documented and exploited as best as possible. Given this, the use of Semantic Web technology and information modelling patterns are explored in this chapter with reference to the marine domain. Further, new strategies for adding semantic annotation to data in real-time are discussed and prototyped.

INTRODUCTION

Within the domain of marine data management, there is a history, beginning in the 1980s, of using controlled vocabularies to define the content of metadata fields and data file channels (UNESCO, 1987; Lowry, 2003; Merati & Burger, 2004; Lawrence et al., 2009; Schaap & Lowry, 2010). The availability of such vocabularies as Semantic Web resources through such platforms as the Marine Metadata Interoperability Ontology Registry and Repository (Graybeal et al., 2012) and the Natural Environment Research Council Vocabulary Server (Leadbetter et al., 2014) has led to the use of Linked Data (Berners-Lee, 2006) patterns to publish metadata and data concerning the marine domain. The Linked Ocean Data concept has been shown to facilitate distributed eScience through validation of chained web processing services and dynamic discovery and aggregation of datasets alongside increased impact of datasets through formalised links to the acquisition methodologies and associated publications interpreting the data (Leadbetter, 2015).

When considering sustainable development of the oceans, it is most important to be able to combine data from a range of producers in a single application. For example, the Marine Renewable Energy Portal

DOI: 10.4018/978-1-5225-0700-0.ch004

delivered by the Sustainable Energy Authority of Ireland and the Irish Marine Institute allows potential energy developers access to live dashboards of current and forecast wave, wind and tidal conditions using data from the Marine Institute, the Commissioner of Irish Lights and the United States' National Oceanic and Atmospheric Administration. However, these data are not connected in a formal Linked Data sense and moving in this direction would better allow users to incorporate other data sources, generate their own dashboards, and enquire of the data in new ways to generate new information and knowledge. Further, these principles can be extended to the creation of new decision support tools for planning operations on board research vessels, or at remote bases when considering the maintenance of missions of deployed Autonomous Underwater Vehicles.

In this chapter we will expand on the existing Linked Ocean Data research to include the emerging concepts of heterogeneous data integration through the use of ontology design patterns, and publishing Linked Data from sources of real-time observations. This will include:

- Revisiting the current state of the Linked Ocean Data cloud to show how it has expanded since it was first proposed in 2013
- Conceptualising the Linked Ocean Data cloud using advances in interchange visualisation techniques (Zeng et al., 2013; Krzywinski et al, 2011) to give new insights into the interconnectedness of the data and information represented within the cloud
- Examining how the emergence of the concept of ontology design patterns has been successfully applied in the ocean science domain
- Considering the relationships which are forming between Linked Data and Big Data and how these have begun to be explored in marine science
- Why the Linked Ocean Data paradigm is important for a sustainable approach to the exploration and development of the ocean

THE STATE OF THE LINKED OCEAN DATA CLOUD

The Linked Ocean Data concept was introduced by Leadbetter *et al.* (2013) and was refined by Leadbetter (2015). The initial Linked Ocean Data cloud consisted of 18 nodes, and by the final publication listed above a further two nodes were incorporated. However, the visualizations used in these descriptions of the Linked Ocean Data cloud have not shown any quantitative information concerning the linkages between the nodes of the cloud, solely the qualitative fact that nodes are linked. However, prior to the introduction of the Linked Ocean Data cloud, Leadbetter and Lowry (2012) experiment with visualisations of connections between controlled vocabularies using circos plots (Krzywinski 2009) which were originally designed for use in visualizing genomics data, but have since been repurposed to show customer flow in the motor industry, volume of courier shipments, database schemas, and presidential debates. The 2012 mappings from the NERC Vocabulary Server to other external controlled vocabulary services is shown in Figure 1. While providing some qualitative information about both the number and directions of the connections in the space occupied by the Linked Ocean Data cloud, this approach is not intuitively interpreted by those users who are unfamiliar with the circos concept. More recently, Zeng et al. (2013) have developed the circos plot concepts to show interchange patterns in the movements of passengers through public transport networks. From the 2015 state of the Linked Ocean Data cloud, it could be seen that there are several nodes where an interchange of data links may occur. Therefore an

Figure 1. External mappings from the NERC Vocabulary Server to other Semantic Web resources (1 o'clock to 7 o'clock) and to the NERC Vocabulary Server (other sectors) as of December 2012 (after Leadbetter & Lowry, 2012). Ribbon width is representative of the number of mappings, on a logarithmic scale.

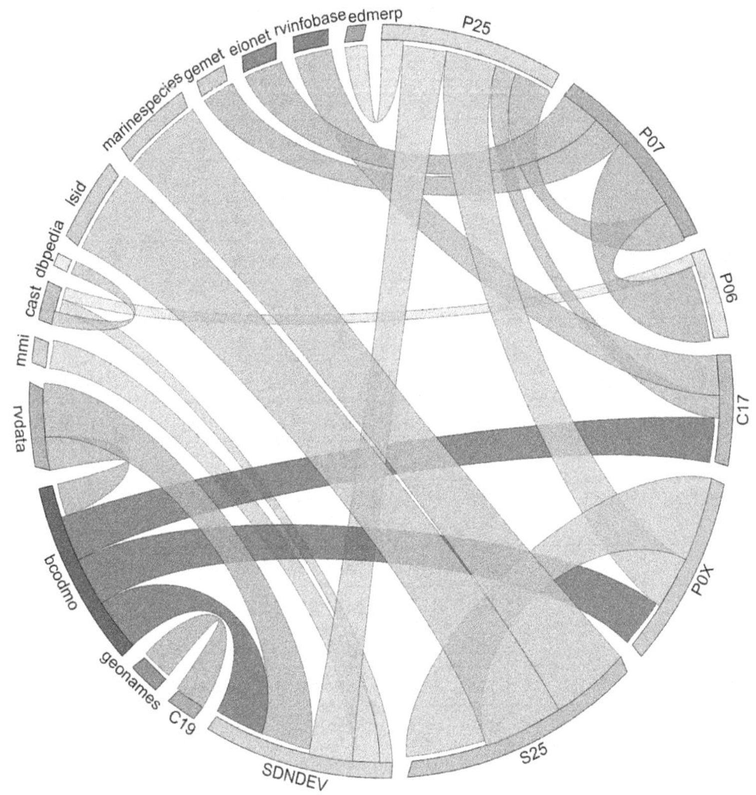

exploration of Zeng et al.'s interchange diagrams for the Linked Ocean Data space has been undertaken and is presented here.

The Vocabulary of Interlinked Datasets (VoID; Alexander & Hausenblas, 2009) provides a Semantic Web format for standardised descriptions of the connections from one Linked Data resource to another. The NERC Vocabulary Server (n.d.), which provides many controlled vocabularies used by Linked Ocean Data publishers, the Rolling Deck to Repository Project (n.d.) and the Biological and Chemical Data Management Office of the Woods Hole Oceanographic Institution (n.d.) all publish VoID documents making these data easy to collect and analyse. The state interlinkages between these Semantic Web resources as of 12th October 2015 is summarised in Table 1.

A further development of the circos visualisation by Krzywinski et al (2011) is the hive plot, which uses radial axes to collect related network nodes and the network edges are shown as arcs connecting these edges. This is useful, as the demonstration in Figure 2 shows that even with a small, but highly connected subset of the Linked Ocean Data cloud plotted in this way, there is still a difficulty in interpreting the information. As an example, the visualisation of the interchange paths between the Linked Ocean Data cloud nodes as shown in Figure 2 is also presented as a hive plot in Figure 3. One benefit of the hive plot approach over the other visualisation methods seen before is the hive panel view, which allows both the connections between the entire network of Linked Ocean Data Nodes and between given nodes in the system to be viewed easily in one plot group.

Table 1. The number of Linked Data relationships between nodes of the Linked Ocean Data cloud as of 20th October 2015, used to generate Figures 2 and 3

From Node	To Node	Number of Mappings	From Node	To Node	Number of Mappings
NVS-C19	geonames	126	BCODMO-Parameter	NVS-P03	4
NVS-L06	HeritageData	20	BCODMO-Parameter	NVS-P09	3
NVS-L06	MMISW-IOOS-Platform	26	BCODMO-Instrument	NVS-L05	93
NVS-L05	MMISW-TRDI-Glossary	2	BCODMO-Instrument	NVS-L22	33
NVS-L05	CAST	6	BCODMO-Instrument	NVS-C77	1
NVS-L22	MMISW-TRDI-Models	12	BCODMO-Platform	NVS-L06	168
NVS-C38	OrdSurv	205	R2R-Device	NVS-L05	90
NVS-P01	SISSVOC	58	R2R-Device	NVS-L22	24
NVS-P08	SISSVOC	1	R2R-Vessel	NVS-L06	1
NVS-P08	Programmes	10	R2R-Vessel	NVS-C19	42
NVS-P21	SISSVOC	3	R2R-Port	NVS-C38	258
NVS-P06	DBPEDIA	470	R2R-Holding	BCODMO-Deployment	362
NVS-P06	CAST	5	MMISW-CF	NVS-P07	2671
NVS-P25	CAST	1	MMISW-TRDI-Glossary	NVS-L05	26
NVS-S25	LSID	3050	MMISW-IOOS-Platform	NVS-L06	2
NVS-P07	EIONET	71	MMISW-TRDI-Models	NVS-L22	12
NVS-S27	CHEMDPlus	345	SISSVOC	QUDT	28
NVS-S27	OBO	269	SISSVOC	NVS-P01	84
BCODMO-Parameter	NVS-P01	101	SISSVOC	NVS-P02	22
BCODMO-Parameter	NVS-P02	5	SISSVOC	NVS-C38	1

Figure 2. A network interchange diagram (after Zeng et al., 2013) for four nodes of the Linked Ocean Data Cloud. Although there is now some clarity on the number and direction of mappings, the hub node of the NERC Vocabulary Server makes producing a coherent visualization of this type difficult

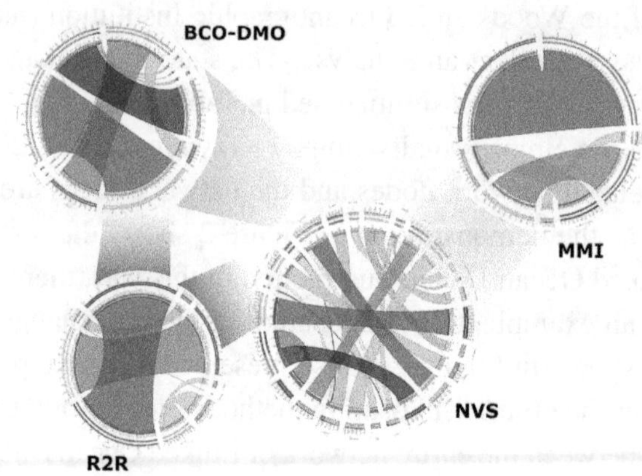

Figure 3. A hive panel showing the mappings between (a) all nodes of the Linked Ocean Data Cloud; and from (b) NERC Vocabulary Server; (c) the Biological and Chemical Oceanography Data Management Office (BCO-DMO); (d) Rolling Deck to Repository (R2R); (e) the Marine Metadata Interoperability (MMI) Ontology Register and Repository; and (f) CSIRO to other nodes on the Linked Ocean Data cloud. Clockwise from the top, the axes represent the NERC Vocabulary Server; NASA's Global Change Master Directory and Quantities, Units, Dimensions and Data Types; GeoNames; Heritage Data; MMI; Ordnance Survey; CSIRO; BBC; DBPEDIA; NERC's Chemical Analytical Services Thesaurus; Life Sciences ID; ChemDPlus; European Environment Agency; Open Biomedical Ontologies; R2R; BCO-DMO

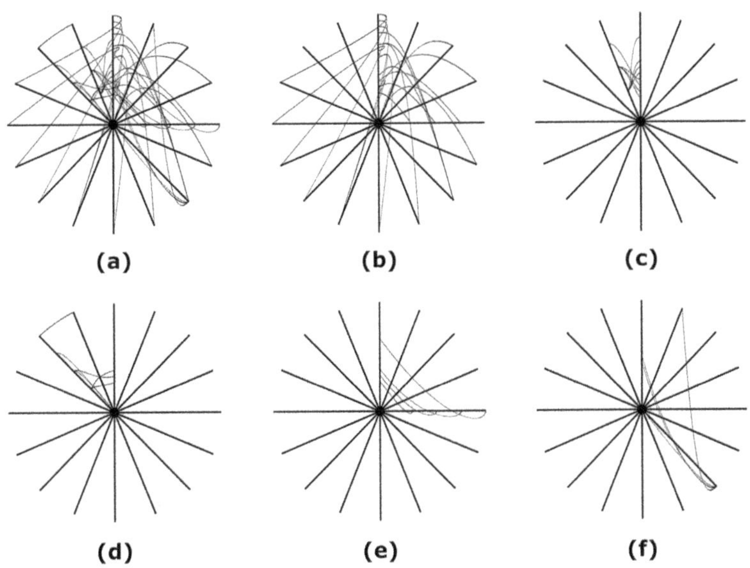

Integration of new data sources to the Linked Ocean Data cloud may be expedited if new data sources can be modelled against patterns which have emerged from a combined computer science and marine domain effort. The development of these ontology design patterns are described below.

ONTOLOGY DESIGN PATTERNS

Many different fields that are part craft and part science have found a need to encapsulate "best practices" for dealing with problems that recur frequently in those fields. For instance, in the 1970s an architect named Christopher Alexander developed a set of architectural patterns for different situations, such as a vestibule that is light and airy but not too hot when exposed to direct sunlight (Alexander, Ishikawa & Silverstein, 1977). Nearly twenty years later, a group of computer scientists known as the Gang of Four published a book of patterns related to needs that frequently arise during object oriented programming, such as the decorator pattern, which allows a programmer to change the behaviour of a particular object without affecting the behaviour of other objects of the same class (Gamma *et al.*, 1994). A decade later, the success of design patterns in architecture and computer science eventually led to them being applied to the growing field of data science, as ontology design patterns (ODPs) (Gangemi, 2005). An ODP is a reusable solution to a data modelling problem that commonly occurs across many different domains or within a wide variety of contexts within a single domain. For instance, concepts like Person or Event need to be represented in many different situations, while the concept of a Sample is key to many different geoscience datasets, from biology to chemistry to geology. When multiple scientific datasets need to be integrated, the join points tend to be precisely this type of core concept, which means that ODPs are very useful for data modelling and integration. This section will introduce some of the problems that arise when integrating scientific data, how ontology design patterns attempt to address these problems, and how they are being utilized in practice in the EarthCube GeoLink project.

Challenges of Scientific Data Integration

The complexity and scope of knowledge that man has gained about various aspects of the physical world has led to scientists to specialize further and further, to the point where some researchers now spend their entire careers studying the behaviour of one particular protein or a single species of plankton. This level of focus is unavoidable to some degree – no one person can be an expert in everything. However, it is sometimes easy to forget that the world around us is an interconnected system whose behaviour spans the artificial boundaries between traditional scientific disciplines, and often the greatest leaps forward in our understanding come at the intersection of different areas of study. These types of breakthroughs require the integration of data from many different scientific domains, and this integration must be done in such a way that the detail, uncertainty, and above all the context of the data are preserved.

The first challenge of scientific data integration is accessibility. Much of the data underpinning past and present scientific publications is not readily accessible – it exists only in isolated databases, as files on a grad student's computer, or in tables within PDF documents. The consequence of this is that it is often difficult to replicate published experimental results and to do new analyses on existing data. There is a monetary cost as well: if data is not stored and shared in an accessible manner then it must be collected independently by multiple researchers, using limited scientific funding that could better be spent elsewhere. Fortunately, many funding agencies have taken action on this issue and now require that data collected via funded programs be stored in official data repositories. This is a promising development, but many current data repositories do not make data integration easy. Traditional scientific data repositories are generally either relational databases or file servers containing spreadsheet, CSV, or irregularly structured text files. There can be various obstacles to retrieving this data, particularly due to a lack of consistency. For instance, some repositories might be accessible via websites or structured query mechanisms while others require a login and use of secure file transfer or copy protocols. Financial and legal concerns also inhibit data integration. Some data might be stored in proprietary database or file formats that require expensive software licenses to read, and licenses indicating what users are allowed to do with the data can be missing or restrictive, resulting in legal uncertainty.

One approach that has been proposed to address these accessibility issues is publishing data as "Linked Data." In a linked dataset, every entity is given a Uniform Resource Identifier (URI) that can be accessed via HTTP, similar to a standard web page. When this URI is dereferenced, there is structured data providing more information about the entity. This data is generally expressed using RDF, and it may include URIs to other, related, entities. Using these standards, it possible to make data available in a way that is both accessible and understandable.

An example to clarify this description seems warranted. Assume that Dr. Jane Doe, a scientist at State University, wants to publish a linked dataset containing information about the papers she has written, one of which is called "An Exploration of the Feasibility of Tenure." One way for Dr. Doe to do this is to acquire ownership of a domain name and assign URIs in that namespace to each of the entities in her dataset. For instance, if the domain name is profdoe.edu, then she might use the URI profdoe.edu/JaneDoe to represent herself and profdoe.edu/TenureFeasibilityExploration to represent the paper. Dr. Doe could then create files containing RDF statements about these entities and deploy them on a webserver. An RDF statement is a subject-predicate-object triple. For example, the following triple states that the paper's title is "An Exploration of the Feasibility of Tenure":

```
<profdoe.edu/TenureFeasibilityExploration>
<profdoe.edu/hasTitle>
"An Exploration of the Feasibility of Tenure"@en .
Similarly, this statement expresses that the paper was written by Dr. Doe:
<profdoe.edu/TenureFeasibilityExploration>
<profdoe.edu/hasAuthor>
<profdoe/JaneDoe> .
```

Using the linked data standards eliminates many of the problems described above, because the data is in a consistent format and accessible in a uniform manner rather than, for example, in proprietary databases behind firewalls. Problems related to licensing can be handled by linking to an appropriate license that has been encoded as linked data triples. For instance, the RDF triple below indicates that Professor Doe's linked dataset containing her publications (the dataset, not necessarily the publications themselves) is covered by version 3.0 of the creative commons "ShareAlike" license. Licensing information is important for those wishing to make use of the information available in the linked data cloud.

```
<profdoe.edu/publications.rdf>
cc:license
<http://creativecommons.org/licenses/by-sa/3.0/> .
```

Accessibility is only one requirement for semantic data integration, however. For data to be truly useful, scientists need to be able to interpret and use it after they acquire it. Doing this requires semantic context. In relational databases and spreadsheets, this is sometimes lacking because important information about what the various data fields mean and how they relate to one another is often implicit in the names of database tables and column headers, some of which are incomprehensible to anyone other than the dataset's creator. This problem is reduced to some degree if the linked data standards are used, because some of the meaning of these column names can be explicitly defined. For instance, the following RDF triples state that the "has Author" construct in our example relates a paper to a person and if Paper X "has Author" Person Y, then Person Y "wrote" Paper X.

```
<www.profdoe.edu/hasAuthor>
<rdfs:domain> <www.profdoe.edu/Paper>
<www.profdoe.edu/hasAuthor>
<rdfs:range> <www.profdoe.edu/Person>
<www.profdoe.edu/hasAuthor>
<owl:inverseOf> <www.profdoe.edu/wrote>
```

Using all of these RDF statements, a piece of software called a reasoner would be able to infer some semantic context for the data, such as that www.profdoe.edu/JaneDoe is a person and that she wrote the paper represented by www.profdoe.edu/TenureFeasibilityExploration, without the need for any natural language processing. These constraints enable a data provider to make the meaning of field names and relationships more precise, which in turn facilitates data integration, but there is still room for ambiguity. This potential for misunderstandings decreases as the links between datasets become more numerous, thereby further constraining potential interpretations of entities. For example, it is possible to express that the Jane Doe referred to using the URI www.profdoe.edu/JaneDoe is the same one that DBPedia, the linked data version of Wikipedia, refers to as dbpedia.org/page/Jane_Doe.

```
<www.profdoe.edu/JaneDoe>
<rdfs:seeAlso> <dbpedia.org/page/Jane_Doe>
```

Such links can be made at the schema level as well as the data (or instance) level. For instance, we can state that "hasAuthor" in this dataset is similar to "creator" in the Dublin Core terminology.

```
<www.profdoe.edu/hasAuthor>
<rdfs:seeAlso> <purl.org/dc/terms/creator>
```

While these examples are fictitious, Listing 1 contains a fully worked example RDF document describing a dataset published by the British Oceanographic Data Centre, and linked to many remote vocabulary resources. Leadbetter (2015) describes in detail the publication process for these RDF documents. Establishing these links can be very difficult, particularly if the datasets are large and complex, which is routinely the case in scientific domains. The fields of ontology alignment and co-reference resolution seek to develop tools and techniques to facilitate the identification of links between datasets (Euzenat & Shvaiko, 2007), however scientific datasets are particularly challenging for several reasons. Perhaps most obviously, such datasets can be extremely large. Consider climate data collected on a 1° grid – this results in over a petabyte of data, more than enough to swamp most existing data integration techniques. Additionally, scientific datasets generally have a spatiotemporal aspect, but current alignment algorithms struggle with finding relationships across this type of data because of the variety of ways to express it. For example, spatial regions can be represented by geopolitical entities (whose borders change over time), by the names of nearby points of interest, or by polygons whose points are given via latitude and longitude. Similarly, issues pertaining to measurement resolution, time zones, the international dateline, etc. can confuse the comparison of timestamps of data observations. Furthermore, scientific datasets frequently involve data of very different modalities, from audio recordings of dolphin calls to radar images of storms to spectroscopy of cellular organisms. Such data is also obviously collected at widely differing scales, from micrometres to kilometres. And oftentimes the data that needs to be integrated is from domains with only a small degree of semantic overlap, as is the case with, for example, one dataset containing information about NSF project awards and another with the salinity values for ocean water collected during oceanographic cruises, several of which were funded by NSF.

At this point we can clearly see that successfully integrating scientific datasets holds the promise of major advances in our knowledge of the world around us, but achieving this goal is likely to be exceedingly challenging. What is the best way to start? One proposal that is gaining traction is to focus on the key concepts that recur frequently across many subdomains. This idea of focusing on the few similari-

Listing 1. A fully worked example RDF document, as published by the British Oceanographic Data Centre within their Published Data Library. The RDF document describes the dataset, its authors, and how the dataset should be cited within the scientific literature.

```
@prefix dc1: <http://purl.org/dc/elements/1.1/> .
@prefix ns1: <http://www.opengis.net/> .

<https://www.bodc.ac.uk/data/published_data_library/catalogue/10.5285/0c9b8ad1-
ba02-0c0e-e053-6c86abc03d26/> dc1:Period "start=1915-01-01; end=2014-12-31;
scheme=W3C-DTF"@en;
    dc1:bibliographicCitation """
                Haigh I.D.; Wadey M.P.; Gallop S.L.; Loehr H.; Nicholls R.J.;
Horsburgh K.J.; Brown J.; Bradshaw E. (2015). A database of 100 years (1915-
2014) of coastal flooding in the UK. British Oceanographic Data Centre - Natu-
ral Environment Research Council, UK. doi:10/zcm."""@en;
    dc1:contributor "Bradshaw E. "@en,
        "Brown J. "@en,
        "Gallop S.L. "@en,
        "Horsburgh K.J. "@en,
        "Loehr H. "@en,
        "Nicholls R.J. "@en,
        "Wadey M.P. "@en;
    dc1:coverage <http://vocab.nerc.ac.uk/collection/C19/current/1/>,
        <http://vocab.nerc.ac.uk/collection/C19/current/1_2/>,
        <http://vocab.nerc.ac.uk/collection/C19/current/1_4/>,
        <http://vocab.nerc.ac.uk/collection/C19/current/1_5/>,
        <http://vocab.nerc.ac.uk/collection/C19/current/1_7/>;
    dc1:creator "Haigh I.D. "@en;
    dc1:description "This database, and the accompanying website called 'Surge-
Watch' (http://surgewatch.stg.rlp.io), provides a systematic UK-wide record of
high sea level and coastal flood events over the last 100 years (1915-2014).
Derived using records from the National Tide Gauge Network, a dataset of ex-
ceedence probabilities from the Environment Agency and meteorological fields
from the 20th Century Reanalysis, the database captures information of 96
storm events that generated the highest sea levels around the UK since 1915.
For each event, the database contains information about: (1) the storm that
generated that event; (2) the sea levels recorded around the UK during the
event; and (3) the occurrence and severity of coastal flooding as consequence
of the event. The data are presented to be easily assessable and understand-
able to a wide range of interested parties.  The database contains 100 files;
four CSV files and 96 PDF files. Two CSV files contain the meteorological and
sea level data for each of the 96 events. A third file contains the list of
the top 20 largest skew surges at each of the 40 study tide gauge site. In
the file containing the sea level and skew surge data, the tide gauge sites
```

continued on following page

Listing 1. Continued

```
are numbered 1 to 40. A fourth accompanying CSV file lists, for reference, the
site name and location (longitude and latitude). A description of the param-
eters in each of the four CSV files is given in the table below. There are
also 96 separate PDF files containing the event commentaries. For each event
these contain a concise narrative of the meteorological and sea level condi-
tions experienced during the event, and a succinct description of the evidence
available in support of coastal flooding, with a brief account of the recorded
consequences to people and property. In addition, these contain graphical rep-
resentation of the storm track and mean sea level pressure and wind fields at
the time of maximum high water, the return period and skew surge magnitudes at
sites around the UK, and a table of the date and time, offset return period,
water level, predicted tide and skew surge for each site where the 1 in 5 year
threshold was reached or exceeded for each event.  A detailed description of
how the database was created is given in Haigh et al. (2015).  Coastal flooding
caused by extreme sea levels can be devastating, with long-lasting and diverse
consequences. The UK has a long history of severe coastal flooding. The recent
2013-14 winter in particular, produced a sequence of some of the worst coastal
flooding the UK has experienced in the last 100 years. At present 2.5 mil-
lion properties and £150 billion of assets are potentially exposed to coastal
flooding. Yet despite these concerns, there is no formal, national framework
in the UK to record flood severity and consequences and thus benefit an under-
standing of coastal flooding mechanisms and consequences. Without a systematic
record of flood events, assessment of coastal flooding around the UK coast
is limited.   The database was created at the School of Ocean and Earth Sci-
ence, National Oceanography Centre, University of Southampton with help from
the Faculty of Engineering and the Environment, University of Southampton, the
National Oceanography Centre and the British Oceanographic Data Centre. Col-
lation of the database and the development of the website was funded through
a Natural Environment Research Council (NERC) impact acceleration grant. The
database contributes to the objectives of UK Engineering and Physical Sciences
Research Council (EPSRC) consortium project FLOOD Memory (EP/K013513/1)."@en;
    dc1:format <http://vocab.nerc.ac.uk/collection/M01/current/DEL/>,
        <http://vocab.nerc.ac.uk/collection/M01/current/DOC/>;
    dc1:identifier <http://dx.doi.org/10.5285/0c9b8ad1-ba02-0c0e-e053-
6c86abc03d26>,
                <http://dx.doi.org/10/zcm>;
    dc1:language <urn:ietf:params:language:en-GB>;
    dc1:publisher <http://www.bodc.ac.uk/>;
    dc1:relation <https://www.bodc.ac.uk/data/information_and_inventories/ed-
med/report/6120/>;
    dc1:subject <http://www.isotc211.org/2005/resources/Codelist/gmxCodelists.
xml#MD_TopicCategoryCode_elevation,>,  <http://www.isotc211.org/2005/resourc-
es/Codelist/gmxCodelists.xml#MD_TopicCategoryCode_oceans>;
```

continued on following page

Listing 1. Continued

```
    dcl:title "A database of 100 years (1915-2014) of coastal flooding in the
UK."@en;
    ns1:gmlid <http://www.isotc211.org/2005/resources/Codelist/gmxCodelists.
xml#MD_TopicCategoryCode_elevation,>,     <http://www.isotc211.org/2005/re-
sources/Codelist/gmxCodelists.xml#MD_TopicCategoryCode_oceans>;
```

ties that exist amidst the many differences inherent in the datasets is at the core of the ontology design pattern approach to data modelling and integration.

Fundamentals of ODPs

While an ODP differs in important ways from an ontology, the basic components of both are the same: classes, instances, and properties. A class represents a grouping of objects with similar characteristics. Classes are often arranged in a hierarchy using subclass relationships. For instance, Submarine may be a subclass of Vessel. An instance (or individual, the terms are often used interchangeably) is a particular object. An instance has a type that is some class within the ontology. For example, the Anorep 1 is an instance of type Submarine. This is somewhat analogous to classes and instances of those classes in object-oriented programming languages, such as Java. Relationships between instances, such as captainOf and hasName, are called properties. All properties are directed binary relations that map an instance with a type from the domain to something in the range. Properties that map an instance to another instance (e.g. captainOf, which may map an instance of type Person to an instance of type Vessel) are object properties, whereas properties that map an instance to a literal value (e.g. hasName, which maps an instance of type Person to a string value) are datatype properties. Common data types include integers, doubles, strings, and dateTime. Both object properties and data properties must involve an instance. A third type of property, called an annotation property, can be used to describe relationships between any types of entities (i.e. instances, classes or other properties). All of this information: classes, properties, and any restrictions on them, such as cardinality, disjointness, etc., are called the schema, or T-box (for terminology), of the ontology. Conversely, the instance data, or A-box (for assertions), contains assertions about individuals using data from the T-box. Both T-box and A-box statements are generally expressed using the Web Ontology Language (OWL). A more formal and extensive treatment of these topics can be found in Hitzler, Krotzsch, and Rudolph (2011).

While ODPs are made up of the same components as full ontologies they differ in important ways, the biggest of which is that an ODP focuses on only one generic notion. The OWL axiomatization of the ODP is carefully formulated such that no overly strong (i.e., application-specific) ontological commitment is made by the pattern, in the same way that an architectural pattern for a vestibule avoids making constraints on the rest of the building. In comparison to a monolithic upper ontology, an ODP can be seen as a snippet that defines only one particular notion without the excessive semantic constraints that an upper ontology may entail. An analogy may be helpful here. An ODP differs from an ontology in the same way that a paragraph is differs from a project report. The key difference is that a good paragraph (and a good ODP) very tightly describes a single idea. You likely can re-use a single paragraph you write about some aspect of your work in many different project reports, but it's unlikely that you can re-use a full project report in many different situations, simply because the complete report says too much to be fully reusable in different situations. This is true even if the project report is extremely well written, by people who are established experts in their field. However, if one writes a project report on the same general subject as a different report that she wrote previously, she may very well cite that other report in the current one, referring to it when it makes sense. This is the relationship between data integrated by stitching together ODPs, and existing full domain ontologies – it is generally not possible to use the full domain ontologies directly, but one can, and should, refer to them by creating the appropriate links.

Figure 4 shows the core of an ODP for representing an oceanographic cruise, which is defined as an expedition undertaken by a vessel in the ocean or other navigable water body in order to conduct oceanographic research activities. Cruises hold a critical role in the ocean sciences because most field observations, data acquisition, and scientific experiments can only be accomplished when researchers are on a scientific expedition. From a data integration perspective, the notion of a cruise is important because it acts as a type of "glue" connecting all of the data and research results resulting from activities carried out during an oceanographic study.

ODPs can leverage one another through axioms that formalize horizontal links between them. This allows applications to formally reason over the data and draw new inferences. For example, a key component of a cruise is the path the vessel takes on its voyage. An ODP to represent a trajectory already existed (Hu, *et al.* 2013). When the cruise ODP was created, axioms to connect the relevant concepts within that pattern to the aforementioned trajectory pattern were specified. For example, axiom 1 shows that a cruise has exactly one trajectory and is undertaken by exactly one vessel.

$$\text{Cruise} \sqsubseteq (=1\ \text{hasTrajectory.Trajectory}) \sqcap (=1\ \text{isUndertakenBy.Vessel}) \quad (1)$$

ODPs can also extend one another in order to represent different levels of abstraction. For instance, an oceanographic cruise is a specialization of a more general event pattern. This is shown intuitively in Figure 5 and specified formally in axioms 2 through 6a,b.

$$\text{Cruise} \sqsubseteq \text{Event} \quad (2)$$

Figure 4. An oceanographic cruise ODP

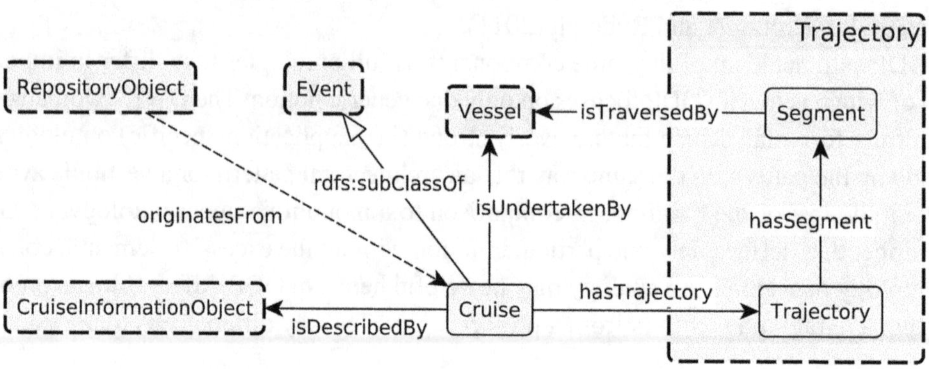

$$\text{CruiseRoleType} \sqsubseteq \text{RoleType} \tag{3}$$

$$\text{CruiseRoleType(x) for every role type x defined for the pattern} \tag{4}$$

$$\text{RCruise} \circ \text{owl:topObjectProperty} \circ \text{RCruiseRoleType} \sqsubseteq \text{providesRoleType} \tag{5}$$

$$\text{Cruise} \equiv \exists\text{RCruise.Self}, \text{CruiseRoleType} \equiv \exists\text{RCruiseRoleType.Self (6a,b)}$$

ODPs are useful for more than just data integration. Another important benefit is their usability by domain scientists unfamiliar with knowledge representation and reasoning technologies. Because ODPs represent key concepts in a domain that recur frequently across many datasets, they can serve as semantic "anchors" that allow someone looking at a dataset for the first time to get conceptually oriented. The ability of ODPs to specialize other patterns in order to represent data at a variety of abstraction levels also allows individuals with different levels of understanding of the domain, such as students, to work with the data at a level that matches their expertise.

GeoLink: ODPs in Action

As discussed previously, understanding past and present data related to the world around us to the point where it is possible to actually predict aspects of its future state is extremely challenging. The earth is a complex and interconnected system that no one scientist, research group, or even field of study can hope to understand. Instead, geoscientists of all different stripes must work together to make progress: geologists, meteorologists, climatologists, ecologists, archaeologists, and so on. The National Science Foundation (NSF) has recognized this need, and in 2011 it launched the EarthCube initiative. The goal of this effort is to galvanize a community-driven approach to collaboration across traditional geoscience domains. Rather than impose standards and infrastructure from the top down, the NSF is providing

Figure 5. The cruise ODP as a specialization of the event ODP

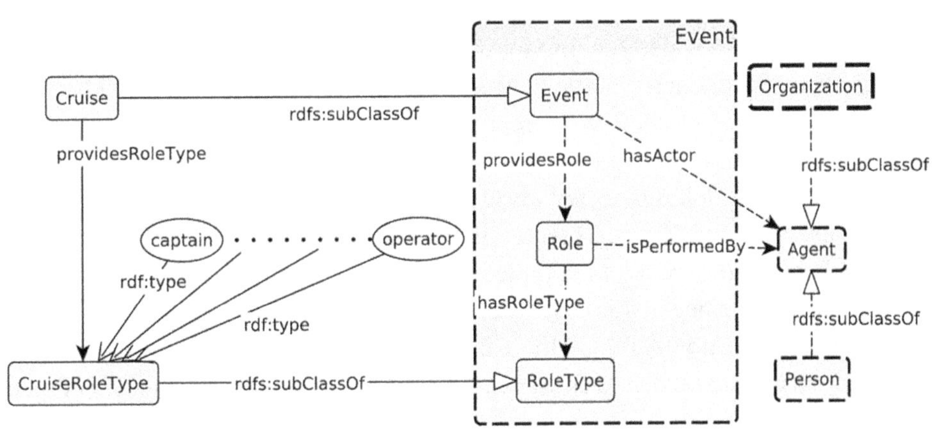

funding to various interdisciplinary teams to explore alternatives, identify the best ones, and disseminate those findings across the community. While this is obviously not something that can be done quickly or easily, the NSF has committed to seeing the effort through until at least 2022.

One of the projects funded under the EarthCube initiative is "GeoLink – Leveraging Semantics and Linked Data for Data Sharing and Discovery in the Geosciences". The goal of GeoLink is to lower barriers to cross-repository data discovery and access, while respecting and preserving repository autonomy and heterogeneity, by applying ODPs to link the repositories' holdings. Spanning ocean, earth and polar sciences, GeoLink's data providers are Rolling Deck to Repository (R2R), the Biological and Chemical Oceanographic Data Management Office (BCO-DMO), Integrated Earth Data Applications (IEDA), the Long-Term Ecological Research Network (LTER), DataONE, and the International Ocean Discovery Program (IODP). If a researcher could query across all of these repositories, tasks such as the following become feasible:

- "Find abundance data of *Euphausia pacifica* sampled with the MOCNESS instrument."
- "Find all NSF awards and publications related to the location called 'Station Papa'."
- "Find data of all measurements of salinity taken at 'Station Papa' over the last 40 years."
- "Find people who collected biodiversity data along the Gulf of Mexico."

These tasks may seem simple on first glance, but consider a couple of features of the queries necessary to accomplish them. First, each of these queries involves multiple facets of the data, including measurements, instruments, scientists, research themes, and time and place. Put in the language of the database community, there are many 'joins' involved, and these joins are over very different data types. Second, while many of these facets appear in multiple datasets, their role, prominence, and meaning within those datasets varies greatly. This is because each repository has a unique point of view on the domain. For instance, some repositories may see a notion such as 'instrument type' as central to their holdings, some as data that is potentially interesting but not required, and others may see instrument type as a detail that is unimportant to the community they serve. Performing queries involving facets that are ascribed such different roles in different datasets is a difficult challenge, and the main goal of the GeoLink effort.

As a sample case study, let us consider a vessel operator that wants to determine their organization's impact on research related to a particular type of phytoplankton. In this case, the organization may want to issue a query such as "Find me all fluorescence data collected during research cruises on vessels operated by Organization A." Such a query involves data stored in at least two repositories: R2R and BCO-DMO. R2R's data schema revolves around the concept of a research cruise – what vessel sailed, the track of the cruise, who provided funding for the cruise, etc. Information about the data the vessel's instrumentation collected is included in the R2R repository, but it is not the central focus. Conversely, BCO-DMO is centred on the data gathered by researchers and the instruments used to collect it. Because R2R gets its data from the vessel's operating organization, when a group of researchers brings their own instrumentation aboard a vessel, it does not always show up in the R2R repository. The result is that, while the same cruises are typically represented in both R2R and BCO-DMO, these are two separate representations each with a different, but complementary, focus.

Completing the task described in this case study requires two subtasks: (1) finding all of the research cruises operated by Organization A (using the R2R repository) and all the cruises in BCO-DMO that involved the collection of data related to fluorescence and then (2) finding the intersection of these two sets. Prior to GeoLink, this was surprisingly difficult. One challenge is that a 'cruise' in R2R is not called a 'cruise' in BCO-DMO; rather it is a 'deployment' with a 'platform' of type 'vessel'. A deployment is defined as the activity of data collection made possible by some platform, such as a vessel, laboratory, aircraft, or satellite. The platform is the base from which instruments are deployed to collect data. The implication is that we must query R2R and BCO-DMO differently for the same concepts, due to slight variations in their vocabularies. Once we have the cruises from R2R involving Organization A and the deployments of type vessel with datasets involving fluorescence from BCO-DMO, we need to find the overlap between these two sets. This leads us to our second challenge: determining when a particular Cruise X in R2R was the same as Cruise Y in BCO-DMO. Prior to GeoLink, this was done via ad-hoc scripts and manual input from the owners of both ontologies.

This use case illustrates that even relatively simple queries require an alignment of both the vocabularies and the instances in the relevant repositories. Arriving at the realization that a cruise in R2R is equivalent to a deployment on a platform of type vessel in BCO-DMO required the expertise and in-depth collaboration of the repository custodians. The same is true for error-free confirmation that Cruise X in R2R is the same as Cruise Y in BCO-DMO. This is not scalable, particularly when queries require data in more than two repositories.

GeoLink address this issue through the use of ontology design patterns as a mediating semantic layer focused on the areas of conceptual overlap between ontologies in the domain. Once the vocabularies of individual ontologies are aligned to these patterns (subtask 1 from the use case), automated co-reference resolution becomes much more accurate (subtask 2), due to the larger volume of data available about each item. For instance, if we know from R2R that a person with the name Michael Smith served as Chief Scientist on Cruise X which was funded by Organization Y, and from BCO-DMO that Cruise X studied ocean salinity, and from a third repository that a paper with "ocean salinity" as a keyword that acknowledges Organization Y as a funding source was written by M. Smith, it is highly likely that this author is the same Michael Smith mentioned in R2R. In this way, the GeoLink project is showing that ontology design patterns facilitate the alignment of vocabularies and instance data across repositories, even when the repositories' views of the domain vary widely.

One potential criticism of GeoLink is that the ODP-based approach "reinvents the wheel" rather than leveraging the many carefully developed domain ontologies that are relevant, such as OBOE (for ecological observational data) (Madin, *et al.* 2007), OBO Foundry (a collection of open biomedical ontologies) (Smith, *et al.* 2007), O&M (for observations and measurements) (Probst, 2006), and ENVO (for environments) (Buttigieg, *et al.* 2013). The GeoLink approach is not meant to imply a negative view on the utility or correctness of those ontologies, but rather to merely acknowledge that wide-ranging domain ontologies, while useful in many situations, are quite difficult to use when integrating large and heterogeneous datasets such as those found in the geosciences. For instance in the OBOE ontology, if an observation occurred within a certain context (i.e. environment), then the entity that was observed is considered to also be in that context. This might not be a safe assumption for some datasets.

Of course, many individuals and applications would greatly benefit from being able to query the GeoLink data using terms found in established domain ontologies. For instance, existing applications often automatically query for instances of foaf:Person when looking for the individuals described in a dataset. GeoLink can respond appropriately to such queries by establishing the relevant links between its internal schema entities, such as gl:Person, and related concepts in external ontologies like foaf:Person. Care must be taken when encoding these relationships, however.

One of the powerful aspects of linked data and ontologies is the ability to make inferences based on the available facts. For instance, assume that the knowledge base contains the fact that any Chief Scientist must have a university degree in a scientific discipline, and that Jane Doe was the Chief Scientist on Cruise123. Even if there is no information about Jane Doe's degree in the knowledge base, it can be inferred that she must have one. Therefore a query to the knowledge base for any people with a scientific degree would include Jane Doe among the results. Inferencing can also help to recognize when data has been entered incorrectly. One of the axioms in the Cruise ODP is that a cruise involves exactly one vessel. If a single cruise has two vessels associated with it, those vessels must either actually be the same ship or there is a mistake in the data.

Inferences are made through a piece of software called a reasoner, and they represent one of the biggest benefits of the semantic web – a knowledge base that you can't reason over is essentially just an "RDFized" database, to which the word "semantic" cannot be reasonably applied. However, if the relationships between the GeoLink design patterns and domain ontologies are specified using strong logical commitments such as owl:equivalentClass or rdf:subClassOf, then reasoners may behave in undesired ways when they are applied to the dataset. To see this, consider that a Program in GeoLink can be considered similar to a Project in the project management ontology PROMONT (Abels, *et al.* 2007). Furthermore, the GeoLink entities Person and PrincipleInvestigatorRole are similar to PROMONT's Employee and Task. One way to represent these relationships is through the following triples:

<gl:Program> <owl:equivalentClass> <PROMONT:Project> .
<gl:Person> <owl:equivalentClass> <PROMONT:Employee> .
<gl:PrincipleInvestigatorRole> <owl:equivalentClass> <PROMONT:Task> .

Assume that Jane Doe is the Principal Investigator on both Program123 and Program456. The start and end dates of these programs are such that there is some overlap. Everything may work fine at first, but PROMONT is designed for precise project management applications. What if the developers of PROMONT decide to add an axiom to their ontology indicating that it is not possible for an employee to be working on two tasks simultaneously? An action beyond GeoLink's control has now caused a logical inconsistency in the knowledge base, and a reasoner will fail. In order to avoid problems such as this, GeoLink plans to establish links to domain ontologies using the gl:matches predicate, as shown in the example below.

<gl:Program> <gl:matches> <PROMONT:Project> .

The gl:matches entity is defined very broadly to mean that two entities linked with this predicate are "sufficiently similar that they can be used interchangeably in some information retrieval applications." Note that this is the same definition as the one for skos:closeMatch (Alexander, 1977). The difference is that, unlike skos:closeMatch, the gl:matches property is not defined to be associative.

With this setup, users and applications can query GeoLink for instances of PROMONT:Project by finding all instances of any GeoLink class that gl:matches PROMONT:Project. This level of indirection makes it clear to the consumer of the data that it may not be appropriate to use the GeoLink instances in all situations in which a PROMONT:Project is expected. This approach allows GeoLink to be queried using terminology familiar to domain experts while still supporting reasoning.

Unlike traditional mission-driven science funding initiatives, the National Science Foundation funds hypothesis-driven research. This difference produces research quality data with varying types of heterogeneity in size and scope. Because of this variability, integrating datasets across NSF data repositories is a difficult task to achieve and scale. For the data repositories participating in the GeoLink project, Ontology Design Patterns, as a mediating semantic layer, have helped to integrate these heterogeneous data holding for producing new information and knowledge. Now that this is possible, with improved accuracy of the asserted shared connections and intersection points, this new knowledge can be put in the hands of researchers and policy makers affecting decisions related to the Earth's ecosystems.

STREAMING LINKED DATA

It has now been seen that Linked Ocean Data can be created within design patterns which give a structure to data made available online. Traditionally, oceanographic data has been published in a *post-facto* delayed mode which improves the base archives used to develop climatologies, extend time series, and contribute to studies of climate change, among other activities (JCOMM, 2009). However, there has been a more recent push to deliver marine science data in real-time (Fredericks, 2015). Operational data aggregation and assembly from distributed data sources will be essential to the ability to adequately describe and predict the physical, chemical and biological state of the ocean. These activities demand a trustworthy and consistent quality description for every observation distributed as part of the global ocean observing system.

At the same time, the "Big Data" paradigm has been increasingly prevalent in the global consciousness. Big Data recognises that data are being generated at an increased volume, at a greater speed, covering a greater range of parameters than ever before. Activities such as the International Argo Programme and the Everyone's Gliding Observatories have added to the instrumentation of the ocean which has vastly increased the volume of data being captured, and the platforms deployed by these programmes often report in real-time increasing the data velocity. Novel sensors, such as the biogeochemical sensors developed within the SenseOCEAN project, are increasing the variety of the data these platforms collect meaning the marine science community is beginning to move into a Big Ocean Data paradigm. This paradigm allows the use of emerging software architecture models for the streaming of data to be leveraged within the ocean sciences domain. Big Data is not solely linked to the volume of data being processed, but also has characteristics of: "velocity", i.e. Big Data is often available in real-time; and "variety", i.e. Big Data is often complex with a number of dimensions (De Mauro, Greco, and Grimaldi 2015). Other characteristics which are often applied to the concept of Big Data are its "veracity" which indicates that the quality of the source data must be considered and is vital to its effective processing; and the "complexity" of the data as the sources may be many and varied such that the data must be linked, connected, and correlated so users can grasp the information the data is supposed to convey (Hilbert 2013). Bearing in mind both the "velocity" and the "complexity" aspects of Big Data, it can be seen that there is value in the exploration of streaming Linked Data from oceanographic instrumentation.

The Header-Dictionary-Triples (RDF-HDT) data structure and serialization format offers a compact structure for storing triples and a binary serialization of the RDF model which allows RDF datasets to be compressed while maintaining search and browse capabilities (Fernández, Martínez-Prieto, Gutiérrez, Polleres, & Arias 2013). Initially, this would seem to be the approach to follow for streaming Linked Ocean Data from instruments to data centres and then to the World Wide Web. However, the experiences of the SenseOCEAN project have shown that there is a practical issue with attempting to deploy any extra software layers, such as the RDF-HDT layer, on to instruments to be deployed remotely in the marine environment as the bandwidth between sensor, logger and transmitter is highly limited - particularly between sensor and logger. Also the deployment of extra software for processing on autonomous platforms, such as Argo floats and Autonomous Underwater Vehicles which are where much of the extra volume of observational data in the marine sciences are collected, significantly increases the battery usage on those platforms and therefore reduces the length of time for which they can be deployed. In order to build Linked Ocean Data into a Big Ocean Data paradigm, an alternative solution must therefore be sought.

As the demand for data to be delivered in real-time from a range of internet applications, new architectures for software processing such as the *lambda*- and *kappa*-architectures, have been developed. In *lambda*-architecture, an immutable sequence of records is captured and fed into a batch system and a stream processing system in parallel (Figure 6). The transformation logic is, however, implemented twice, once in the batch system and once in the stream processing system. Results from both systems are stitched together at query time to produce a complete answer (Marz and Warren, 2015). However, particularly in ocean science scenarios, the desired workflow is to process the data in some rapid manner as close to the time of collection as possible using the *a priori* knowledge of the dataset, and then re-process once the *post hoc* knowledge base is increased. This fits well within the basis of the *kappa*-architecture, which was proposed by Kreps (2014) as an alternative to the *lambda*-architecture (Figure 7). The lambda architecture concentrates on the ability to reprocess the full data stream at a later date through storing the full data message queue in a message queue which allows for multiple subscribers (for example, Apache Kafka). When reprocessing is required, new processing job code is introduced to the stream processing system which can be run against the entire message queue to generate an *n+1* version of the output.

Figure 6. The lambda-architecture. Example stream processing systems in the second stage include the Apache projects Hadoop, Samza and Storm.

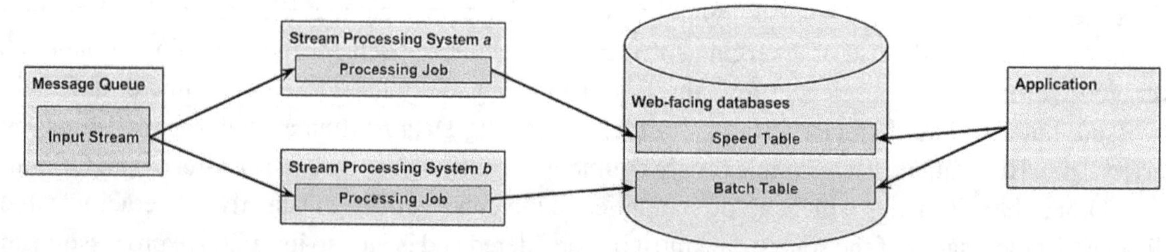

Figure 7. The kappa-architecture

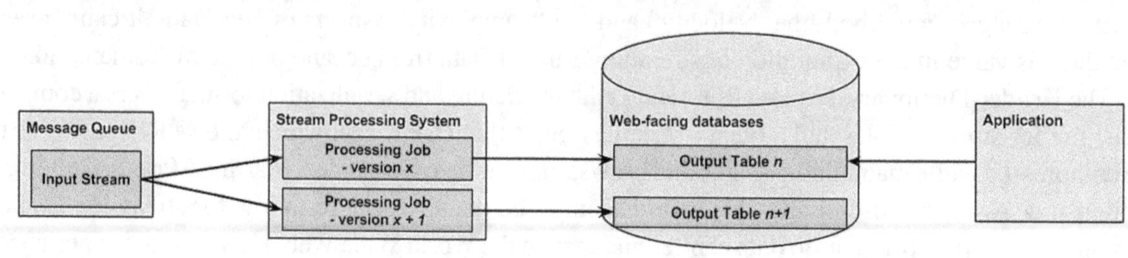

In terms of streaming Linked Data within an architecture like this, a possible approach is outlined below. The initial message received from the instrumentation is placed into a message queue, a processing logic node converts this into a full RDF document and posts this to a new message queue. The RDF document is read from its message queue and parsed into its component triples, each of which are added to the next message queue. A final processing node reads this message queue and passes the triples to a SPARQL endpoint using an UPDATE command. This combination of messaging queues and nodes containing processing logic which subscribe to these queues is known as a "topology". Specifically, a topology is a graph of computation in which the nodes represent individual functions enacted on data and the edges denote the pathways by which data may move between nodes. A demonstration topology has been created using Apache Kafka as the messaging queue and the R statistical programming language as the stream processing system. R has been demonstrated as an appropriate platform for processing Semantic Web data streams by Willighagen (2014), and the support for interfacing with Kafka from R is very good, using a simple Application Programming Interface to push messages to and read messages from a queue. The topology is outlined in Figure 8 and detailed in Listing 2.

Figure 8. An example Big Ocean Data topology for streaming Linked Ocean Data

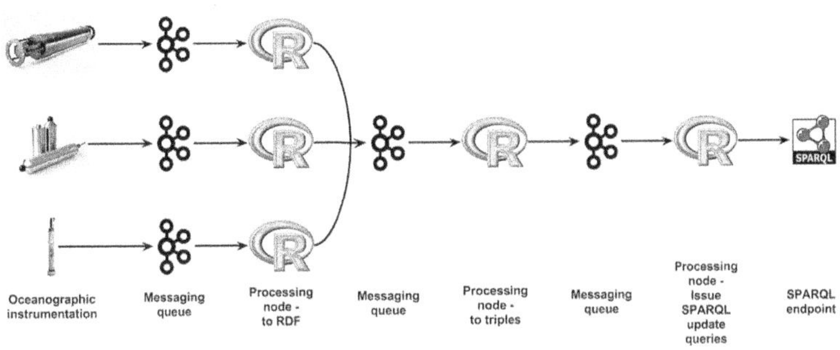

Listing 2. An R statistical programming language encoding of the topology in Figure 8

```
library(rkafka)
library(elastic)
library(rrdf)
library(jsonlite)
message2kafka <- function(kafkaIP, kafkaTopic, message) {
  if (missing(kafkaIP)) {
    stop('Missing input argument: kafkaIP')
  } else if (missing(kafkaTopic)) {
    stop('Missing input argument: kafkaTopic')
  } else if (missing(message)) {
    stop('Missing input argument: message')
  } else {
    messageProducer <- rkafka.createProducer(kafkaIP)
    rkafka.send(messageProducer, kafkaTopic, kafkaIP, message)
    rkafka.closeProducer(messageProducer)
```

continued on following page

Listing 2. Continued

```
  }
}
kafka2r <-function(zookeeperIP, kafkaTopic) {
  rdfConsumer <- rkafka.createConsumer(zookeeperIP,
                                      kafkaTopic, consumerTimeoutMs = "10000")
  msg <- rkafka.read(rdfConsumer)
  rkafka.closeConsumer(rdfConsumer)
  msg
}
rdfmultiplex <- function(zookeeperIP, kafkaTopic) {
  tripleInAsJSON <- ".."
  while(tripleInAsJSON != "") {
    tripleInAsJSON  <-  kafka2r(zookeeperIP, kafkaTopic)  # Try Catch Loop needed here
    tripleIn <- fromJSON(tripleInAsJSON)
    print(tripleIn)
  }
}
rdfxmlfile2kafka <- function(fileName, kafkaIP, kafkaTopic) {
  if (file.exists(fileName)) {
    rawRdfXml  <-  readChar(fileName, file.info(fileName)$size)
    message2kafka(kafkaIP, kafkaTopic, rawRdfXml)
  } else {
    stop('Specified input file does not exist')
  }
}
rdf2triples <-function(formatRDF, zookeeperIP,
                       kafkaInTopic, kafkaIP, kafkaOutTopic) {
  rdfXmlStr <- kafka2r(zookeeperIP, kafkaInTopic)
  tripleStore <- new.rdf(ontology = FALSE)
  fromString.rdf(toString(rdfXmlStr), formatRDF, appendTo = tripleStore)
  triples <- sparql.rdf(tripleStore, "select * where {?a ?b ?c}")
  for (i in 1:nrow(triples)) {
    jsonTriples <- (paste0('{"s": "', unname(triples[i, 1]), '", "p": "',
                 unname(triples[i, 2]), '", "o": "',
                 unname(triples[i, 3]), '"}'))
    message2kafka(kafkaIP,kafkaOutTopic,jsonTriples)
  }
}
# read an XML file into Kafka
rdfxmlfile2kafka('~/rdf/oa_overlay.xml', 'localhost:9092', 'rawrdfxml')
# Read the RDF/XML from Kafka and create a series of RDF triples
rdf2triples('RDF/XML', 'localhost:2181', 'rawrdfxml', 'localhost:9092', 'rdftriples')
# Multiplex the RDF Triples
rdfmultiplex('localhost:2181','rdftriples')
```

The topology shown in Figure 8 takes raw oceanographic instrument output, normally as text over a serial port, as its inputs (to the far left of the image within Figure 8). A script, for example written in Python or R, listens to the port over which the instrument is streaming its outputs and these are pushed to the Kafka messaging queue. The R topology reads the raw data from the message queues for each instrument and converts the full reading to an RDF XML document which is pushed onto a combined message queue for each instrument. The R topology then reads each of these documents and breaks them down into the component triples, each of which are passed on to a new message queue. The final node of the R topology takes the triples in turn and issues SPARQL update queries to push them on to a SPARQL endpoint. This topology has been successfully tested on local desktop machines in batch mode in order to prove the validity of the concept as far as the input to the SPARQL endpoint, but it remains to be proven in a production environment outside of a batch processing mode. However, the careful preparation of the data into a standardised structure for use in data intensive science supports the concept of High Performance Data as proposed by Evans *et al.* (2015).

One such standardised structure is the Observations & Measurements (O&M) model (Cox, 2006) which is both an International Organization for Standardization (ISO) and Open Geospatial Consortium (OGC) standard which defines a conceptual schema encoding for observations, and for features involved in sampling when making observations. The OGC make specific use of O&M in their Sensor Web Enablement standards suite, where it provides the response model for the Sensor Observation Service. In this case, Sensor Web Enablement and Sensor Observation Services are of particular interest as they are intended, respectively, to make sensors discoverable, accessible and usable and to query both real-time and time-series sensor data via the Web. However, one drawback has been the reliance on the service response documents being delivered via eXstensible Markup Language (XML) documents, which are heavy-weight for use in real-time systems and are not favoured by web developers looking to build responsive user interfaces. One option is to use JavaScript Object Notation as an alternative to XML as it is a lighter-weight document meaning quicker response times from applications and easier development of interfaces. As Tim Bray noted in 2013 "… these days, if you want to interchange tuples or tables of tuples or numbers and strings, you have JSON. If you want to do nontrivial publishing automation, use XML. If you want to interchange smart bitmaps of page images, there's PDF. I personally think we're probably done with inventing low-level textual interchange formats" (Bray, 2013).

One barrier to this has been the lack of either a formal schema for JSON to allow the validation of a JSON document and an Observations & Measurements encoding for JSON. The first issue has been addressed by the emergence and coming to maturity of JSON Schema (Galiegue, Zyp, & Court, 2013) specifies a JSON-based format to define the structure of JSON data for validation, documentation, and interaction control. The second barrier is overcome by the publication of a JSON Schema for O&M (OM-JSON; Cox & Taylor, 2015). This allows for a JSON object to be created containing the full O&M data model, and to be validated using the appropriate JSON Schema. This is of particular interest in the Linked Ocean Data space, as OM-JSON specifies that such properties as: the Observed Property; the Procedure; the Feature of Interest; and the Units of Measure are all defined by HTTP URIs. Therefore OM-JSON is immediately Linked Data at one level or another, all the more so if common controlled vocabularies, such as the NERC Vocabulary Server, are used to provide the URIs to such resources (see Listing 3).

The example OM-JSON encoding shown in Listing 3 is from an experimental Sensor Observation Service instance which takes JSON feeds from instruments deployed within the marine environment, and maps these data onto the O&M data model (Irish Marine Institute, n.d.). A further step towards

Listing 3. An example of schema compliant OM-JSON created from a lightweight, experimental Sensor Observation Service.


```
{
  "foi": {
    "href": "http://example.marine.ie/feature/galwayBayCableObservatory"
  },
  "id": "http://example.marine.ie/ctd/Idronaut/3137/b457-a9c72a392d90",
  "member": [ {
    "id": "Temp",
    "observedProperty": {
      "href": "http://vocab.nerc.ac.uk/collection/P01/current/TEMPPR01/"
    },
    "procedure": {
      "href": "http://vocab.nerc.ac.uk/collection/L22/current/TOOL0861/"
    },
    "result": {
      "uom": "http://vocab.nerc.ac.uk/collection/P06/current/UPAA/",
      "value": 13.988
    },
    "resultTime": "2015-10-12T15:33:44Z",
    "type": "Measurement"
  }, {
    "id": "Sal",
    "observedProperty": {
      "href": "http://vocab.nerc.ac.uk/collection/P01/current/PSALCU01/"
    },
    "procedure": {
      "href": "http://vocab.nerc.ac.uk/collection/L22/current/TOOL0861/"
    },
    "result": {
      "uom": "http://vocab.nerc.ac.uk/collection/P06/current/UUUU/",
      "value": 34.775
    },
    "resultTime": "2015-10-12T15:33:44Z",
    "type": "Measurement"
  }, {
    "id": "Press",
    "observedProperty": {
      "href": "http://vocab.nerc.ac.uk/collection/P07/current/CFSN0330/"
    },
    "procedure": {
      "href": "http://vocab.nerc.ac.uk/collection/L22/current/TOOL0861/"
    },
```


continued on following page

Listing 3. Continued


```
    "result": {
      "uom": "http://vocab.nerc.ac.uk/collection/P06/current/UPDB/",
      "value": 27.7
    },
    "resultTime": "2015-10-12T15:33:44Z",
    "type": "Measurement"
  }, {
    "id": "SoundV",
    "observedProperty": {
      "href": "http://vocab.nerc.ac.uk/collection/P01/current/SVELCV01/"
    },
    "procedure": {
      "href": "http://vocab.nerc.ac.uk/collection/L22/current/TOOL0861/"
    },
    "result": {
      "uom": "http://vocab.nerc.ac.uk/collection/P06/current/UVAA/",
      "value": 1503.6205
    },
    "resultTime": "2015-10-12T15:33:44Z",
    "type": "Measurement"
  }, {
    "id": "Press",
    "observedProperty": {
      "href": "http://vocab.nerc.ac.uk/collection/P01/current/CNDCST01/"
    },
    "procedure": {
      "href": "http://vocab.nerc.ac.uk/collection/L22/current/TOOL0861/"
    },
    "result": {
      "uom": "http://vocab.nerc.ac.uk/collection/P06/current/UECA/",
      "value": 41.694
    },
    "resultTime": "2015-10-12T15:33:44Z",
    "type": "Measurement"
  } ],
  "phenomenonTime": {
    "instant": "2015-10-12T15:33:44Z"
  }
}
```

making such an OM-JSON file 5-star Linked Data would be to overlay JSON-LD patterns on the OM-JSON. JSON-LD is a method of transporting Linked Data using JSON. JSON-LD is designed around the concept of a "context" to provide additional mappings from JSON to an RDF model. The context links object properties in a JSON document to concepts in an ontology. In order to map the JSON-LD syntax to RDF, JSON-LD allows values to be coerced to a specified type or to be tagged with a language. A context can be embedded directly in a JSON-LD document or put into a separate file and referenced from different documents (from traditional JSON documents via an HTTP Link header). As an ontology for O&M already exists in OWL (Cox, 2013) it is possible to do this for OM-JSON by inserting URIs from the O&M ontology into the OM-JSON context. One strategy for creating an OM-JSON context is given in Listing 4. In order to show the knowledge encapsulated by combing Listing 3 and Listing 4, Listing 5 shows the RDF graph inferred through overlaying the JSON-LD context of Listing 4 on to the OM-JSON of Listing 3.

FUTURE RESEARCH DIRECTIONS

The solutions outlined above to "Born Semantic" data are as yet only prototypes and have neither been used in a production environment nor, in keeping with the Big Data paradigm, have they been shown to scale. The next steps in their development should be to remove the reliance on an intermediate platform

Listing 4. A possible strategy for creating a truly Linked Data Sensor Observation Service involves the output in Listing 3 overlain with a JSON-LD Context document of the type shown here.

```
"@context": {
    "observedProperty": "http://def.seegrid.csiro.au/isotc211/iso19156/2011/
observation#observedProperty",
    "href": "@id",
    "foi": "http://def.seegrid.csiro.au/isotc211/iso19156/2011/
observation#featureOfInterest",
    "id": "@id",
    "phenomenonTime": "http://def.seegrid.csiro.au/isotc211/iso19156/2011/
observation#phenomenonTime",
    "instant": "@value",
    "member": "http://def.seegrid.csiro.au/isotc211/iso19156/2011/observation#
relatedObservation",
    "procedure": "http://def.seegrid.csiro.au/isotc211/iso19156/2011/
observation#Process",
    "resultTime": "http://def.seegrid.csiro.au/isotc211/iso19156/2011/
observation#resultTime",
    "result": "http://def.seegrid.csiro.au/isotc211/iso19156/2011/
observation#result",
    "value": "@value"
  }
```

Listing 5. The RDF graph inferred by overlaying the OM-JSON of Listing 3 with the JSON-LD context of Listing 4.

```
<http://example.marine.ie/ctd/Idronaut/3137/b457-a9c72a392d90> <http://def.
seegrid.csiro.au/isotc211/iso19156/2011/observation#featureOfInterest> <http://
example.marine.ie/feature/galwayBayCableObservatory> ;
        <http://def.seegrid.csiro.au/isotc211/iso19156/2011/
observation#phenomenonTime> "2015-10-12T15:33:44Z" ;
        <http://def.seegrid.csiro.au/isotc211/iso19156/2011/
observation#relatedObservation> < http://example.marine.ie/ctd/Idronaut/3137/
b457-a9c72a392d90/Press>, < http://example.marine.ie/ctd/Idronaut/3137/
b457-a9c72a392d90/Sal>, < http://example.marine.ie/ctd/Idronaut/3137/b457-
a9c72a392d90/SoundV>, < http://example.marine.ie/ctd/Idronaut/3137/b457-
a9c72a392d90/Temp> .

< http://example.marine.ie/ctd/Idronaut/3137/b457-a9c72a392d90/Press> <http://
def.seegrid.csiro.au/isotc211/iso19156/2011/observation#Process> <http://vo-
cab.nerc.ac.uk/collection/L22/current/TOOL0861/> ;
        <http://def.seegrid.csiro.au/isotc211/iso19156/2011/
observation#observedProperty> <http://vocab.nerc.ac.uk/collection/P01/current/
CNDCST01/>, <http://vocab.nerc.ac.uk/collection/P07/current/CFSN0330/> ;
        <http://def.seegrid.csiro.au/isotc211/iso19156/2011/
observation#result> "2.77E1"^^<http://www.w3.org/2001/XMLSchema#double>,
"4.1694E1"^^<http://www.w3.org/2001/XMLSchema#double> ;
        <http://def.seegrid.csiro.au/isotc211/iso19156/2011/
observation#resultTime> "2015-10-12T15:33:44Z" .

< http://example.marine.ie/ctd/Idronaut/3137/b457-a9c72a392d90/Sal> <http://
def.seegrid.csiro.au/isotc211/iso19156/2011/observation#Process> <http://vo-
cab.nerc.ac.uk/collection/L22/current/TOOL0861/> ;
        <http://def.seegrid.csiro.au/isotc211/iso19156/2011/
observation#observedProperty> <http://vocab.nerc.ac.uk/collection/P01/current/
PSALCU01/> ;
        <http://def.seegrid.csiro.au/isotc211/iso19156/2011/
observation#result> "3.4775E1"^^<http://www.w3.org/2001/XMLSchema#double> ;
        <http://def.seegrid.csiro.au/isotc211/iso19156/2011/
observation#resultTime> "2015-10-12T15:33:44Z" .

< http://example.marine.ie/ctd/Idronaut/3137/b457-a9c72a392d90/SoundV> <http://
def.seegrid.csiro.au/isotc211/iso19156/2011/observation#Process> <http://vo-
cab.nerc.ac.uk/collection/L22/current/TOOL0861/> ;
        <http://def.seegrid.csiro.au/isotc211/iso19156/2011/
observation#observedProperty> <http://vocab.nerc.ac.uk/collection/P01/current/
SVELCV01/> ;
```

continued on following page

Listing 5. Continued

```
        <http://def.seegrid.csiro.au/isotc211/iso19156/2011/
observation#result> "1.5036205E3"^^<http://www.w3.org/2001/XMLSchema#double> ;
        <http://def.seegrid.csiro.au/isotc211/iso19156/2011/
observation#resultTime> "2015-10-12T15:33:44Z" .

< http://example.marine.ie/ctd/Idronaut/3137/b457-a9c72a392d90/Temp> <http://
def.seegrid.csiro.au/isotc211/iso19156/2011/observation#Process> <http://vo-
cab.nerc.ac.uk/collection/L22/current/TOOL0861/> ;
        <http://def.seegrid.csiro.au/isotc211/iso19156/2011/
observation#observedProperty> <http://vocab.nerc.ac.uk/collection/P01/current/
TEMPPR01/> ;
        <http://def.seegrid.csiro.au/isotc211/iso19156/2011/
observation#result> "1.3988E1"^^<http://www.w3.org/2001/XMLSchema#double> ;
        <http://def.seegrid.csiro.au/isotc211/iso19156/2011/
observation#resultTime> "2015-10-12T15:33:44Z" .
```

such as R for the encoding of the nodes of the topology, and the code should be moved out to either a more formal programming language (such as Java or Go) or to a dedicated stream processing platform such either Spark or Storm from the Apache Software Foundation.

Similarly, the High Performance Datasets described above are of low-volume, but only of a medium complexity. As the creation of terabyte scale ocean datasets becomes much more prevalent, the question of converting 'Big Data' sets that comprise thousands of individual heterogeneous files (e.g., bathymetry data sets) into 'High Performance Data' (HPD) sets that can be accessed in High Performance Computing environments becomes much more relevant. This is a known problem where both climate and oceans modellers want access to national scale, calibrated higher resolution bathymetry data sets, as reported by the Australian National University to the September 2015 Ocean Data Interoperability Platform workshop. A related Big Data issue relevant to sustainable development of the coastline is the merging the high resolution LiDAR data sets (in LAS formats) with shallow water bathymetry (in CARIS, ASCII, ESRI Grid, or possibly a well-managed netCDF flavour) to create high resolution coastal elevation data sets for accurate tsunami and storm surge modelling. These remain open research issues.

Further, the extension of Observations and Measurements through application schema is a well-known through domain specialization (Cox, 2013). However, in the past this has mainly been through an informal use of common feature-type catalogues; sensor registers; parameter dictionaries; and result formats for a given domain. As shown by Diviacco and Leadbetter (2016, see chapter X this volume) this does not lead to a sustainable development paradigm. However, the lightweight formalisation of the JSON schema approach as used in OM-JSON is already being extended by the Ocean Acidification community in a more formalised way, and future research will determine if this presents a better route for the extension of these core data models.

CONCLUSION

In this chapter, we have seen how the concept of Linked Ocean Data is exploiting the potential of third generation World Wide Web technology, specifically the Semantic Web and Linked Data, to provide interconnections between datasets published online. For sustainable development of the marine environment, where a holistic view of the marine system is required, Linked Data approaches allow for the easier integration of new data sources, particularly where relevant Ontology Design Patterns have been published and Linked Data can be created to those patterns. Finally, the emergence of Big Data streaming systems with small chunks of processing logic placed between fully re-analysable message queues presents a solution to the question of data being "Born Semantic" or "Born Linked" allowing new observations of the marine environment to be connected to the Linked Ocean Data cloud from the moment they are captured. Through the emergence of OM-JSON, the Born Semantic data can also be compliant with standards for accessing sensor observations through Web interfaces.

REFERENCES

Alexander, C., Ishikawa, S., & Silverstein, M. (1977). *A pattern language: towns, buildings, construction* (Vol. 2). Oxford, UK: Oxford University Press.

Alexander, K., & Hausenblas, M. (2009). Describing linked datasets-on the design and usage of void, the vocabulary of interlinked datasets. In *Linked Data on the Web Workshop (LDOW 09), in conjunction with 18th International World Wide Web Conference (WWW 09).*

Berners-Lee, T. (2006). *Linked Data - Design Issues.* World Wide Web Consortium. Retrieved May 29, 2015 from http://www.w3.org/DesignIssues/LinkedData.html

Bray, T. (2013). *XML's 15th Birthday.* Retrieved February 2, 2016 from http://www.tbray.org/ongoing/When/201x/2013/02/10/XML-at-15

Cox, S. (2006). *Observations and measurements. Open Geospatial Consortium Best Practices Document.* Open Geospatial Consortium.

Cox, S. (2013, October). An explicit OWL representation of ISO/OGC Observations and Measurements. In SSN@ ISWC (pp. 1-18).

Cox, S. (2013, December). *Observations to information.* Abstract IN42A-01 presented at 2013, Fall Meeting, AGU, San Francisco, CA.

Cox, S., & Taylor, P. (2015). OM-JSON - a JSON implementation of O&M. Presentation to the Open Geospatial Consortium Technical and Planning Committee Sensor Web Enablement Domain Working Group, Nottingham, UK.

De Mauro, A., Greco, M., & Grimaldi, M. (2015, February). What is big data? A consensual definition and a review of key research topics. In AIP Conference Proceedings (Vol. 1644, pp. 97-104).

Diviacco, P., & Leadbetter, A. (2016). Balancing Formalization and Representation in cross-domain data management for sustainable development. In P. Diviacco, A. Leadbetter, & H. Glaves (Eds.), *Oceanographic and Marine Cross-Domain Data Management for Sustainable Development*. Hershey, PA: IGI Global.

Euzenat, J., & Shvaiko, P. (2007). *Ontology matching*. Heidelberg, Germany: Springer.

Evans, B., Wyborn, L., Pugh, T., Allen, C., Antony, J., Gohar, K., . . . Bell, G. (2015). The NCI High Performance Computing and High Performance Data Platform to Support the Analysis of Petascale Environmental Data Collections. In Environmental Software Systems. Infrastructures, Services and Applications (pp. 569-577). Springer International Publishing. doi:10.1007/978-3-319-15994-2_58

Fernández, J. D., Martínez-Prieto, M. A., Gutiérrez, C., Polleres, A., & Arias, M. (2013). Binary RDF representation for publication and exchange (HDT). *Web Semantics: Science, Services, and Agents on the World Wide Web, 19*, 22–41. doi:10.1016/j.websem.2013.01.002

Fredericks, J. (2015). Persistence of knowledge across layered architectures. In P. Diviacco, P. Fox, C. Pshenichny, & A. Leadbetter (Eds.), *Collaborative Knowledge in Scientific Research Networks*. Hershey, PA: IGI Global. doi:10.4018/978-1-4666-6567-5.ch013

Galiegue, F., Zyp, K., & Court, G. (2013). *JSON Schema: core definitions and terminology*. Retrieved October 12, 2015 from http://tools.ietf.org/html/draft-zyp-json-schema-04

Gamma, E., Helm, R., Johnson, R., & Vlissides, J. (1994). *Design patterns: elements of reusable object-oriented software*. New York, NY: Pearson Education.

Gangemi, A. (2005). Ontology design patterns for semantic web content. In *The Semantic Web–ISWC 2005* (pp. 262–276). Berlin: Springer Berlin Heidelberg. doi:10.1007/11574620_21

Graybeal, J., Isenor, A., & Rueda, C. (2012). Semantic mediation of vocabularies for ocean observing systems. *Computers & Geosciences, 40*, 120–131. doi:10.1016/j.cageo.2011.08.002

Guizzardi, G. (2006, July). The role of foundational ontologies for conceptual modeling and domain ontology representation. In *Databases and Information Systems, 2006 7th International Baltic Conference on* (pp. 17-25). IEEE. doi:10.1109/DBIS.2006.1678468

Hilbert, M. (2013). *Big data for development: From information-to knowledge societies*. Available at SSRN 2205145.

Hitzler, P., Krotzsch, M., & Rudolph, S. (2009). *Foundations of semantic web technologies*. CRC Press.

Hu, Y., Janowicz, K., Carral, D., Scheider, S., Kuhn, W., Berg-Cross, G., & Kolas, D. (2013). A geo-ontology design pattern for semantic trajectories. In *Spatial Information Theory* (pp. 438–456). Springer International Publishing.

Irish Marine Institute. (n.d.). *Sensor Observation Service*. Retrieved from: https://github.com/IrishMarineInstitute/sensor-observation-service

JCOMM. (2009). *Cookbook for Submitting Data in Real Time and In Delayed Mode.* Joint World Meteorological Organisation-Intergovernmental Oceanographic Commision Technical Commission for Oceanography and Marine Meteorology. Retrieved May 29, 2015 from http://www.jcomm.info/index.php?option=com_content&view=article&id=37

Kreps, J. (2014) *Questioning the lambda architecture.* O'Reilly Radar. Retrieved August 28, 2015 from http://radar.oreilly.com/2014/07/questioning-the-lambda-architecture.html

Krzywinski, M., Birol, I., Jones, S. J., & Marra, M. A. (2012). Hive plots—rational approach to visualizing networks. *Briefings in Bioinformatics, 13*(5), 627–644. doi:10.1093/bib/bbr069 PMID:22155641

Krzywinski, M., Schein, J., Birol, I., Connors, J., Gascoyne, R., Horsman, D., & Marra, M. A. et al. (2009). Circos: An information aesthetic for comparative genomics. *Genome Research, 19*(9), 1639–1645. doi:10.1101/gr.092759.109 PMID:19541911

Lawrence, B., Lowry, R., Miller, P., Snaith, H., & Woolf, A. (2009). Information in environmental data grids. *Philosophical Transactions of the Royal Society of London A: Mathematical, Physical and Engineering Sciences, 367*(1890), 1003–1014. doi:10.1098/rsta.2008.0237

Leadbetter, A. (2015). Linked ocean data. In T. Narock & P. Fox (Eds.), The Semantic Web in Earth and Space Science: Current Status and Future Directions. Amsterdam: IOS Press.

Leadbetter, A., Arko, B., Chandler, C., Shepherd, A., & Lowry, R. (2013). Linked Data: An Oceanographic Perspective. *Journal of Ocean Technology, 8*(3), 7–12.

Leadbetter, A., & Lowry, R. (2012). *The NERC Vocabulary Server: Version 2.0.* Abstract IN51D-1709 presented at 2012, Fall Meeting, AGU, San Francisco, CA.

Leadbetter, A., Lowry, R., & Clements, D. (2014). Putting meaning into NETMAR - the open service network for marine environmental science. *International Journal of Digital Earth, 7*(10), 811–828. doi:10.1080/17538947.2013.781243

Lowry, R. (2003). *EnParDis - Enabling parameter discovery.* Paper presented at 2003 Global Organisation for Earth System Science Portals. Central Laboratory of the Research Councils, Daresbury Laboratory.

Marz, N., & Warren, J. (2015). *Big Data: Principles and best practices of scalable realtime data systems.* Manning Publications Co.

Merati, N., & Burger, E. (2004). *Implementing Marine XML for NOAA Observing Data.* Paper presented at 2004 User Conference, Esri, Redlands, CA.

Rolling Deck. (n.d.). *Dataset Description for Rolling Deck to Repository (R2R) Linked Data.* Retrieved from: http://linked.rvdata.us/.well-known/void

Schaap, D., & Lowry, R. (2010). SeaDataNet - Pan-European infrastructure for marine and ocean data management: Unified access to distributed data sets. *International Journal of Digital Earth, 3*(sup1Supplement 1), 50–69. doi:10.1080/17538941003660974

UNESCO. (1987). GF3: A General Formatting System for geo-referenced data. Volume 2: technical description of the GF3 format and code tables. Paris: Intergovernmental Oceanographic Commission.

Vocab. (n.d.). Retrieved from: http://vocab.nerc.ac.uk/.well-known/void

Willighagen, E. (2014). Accessing biological data in R with semantic web technologies (No. e185v3). *PeerJ PrePrints*. Retrieved from https://dx.doi.org/10.7287/peerj.preprints.185v3

Woods Hole. (n.d.). Retrieved from: http://www.bco-dmo.org/.well-known/void

Zeng, W., Fu, C.-W., Müller Arisona, S., & Qu, H. (2013). Visualizing Interchange Patterns in Massive Movement Data. *Computer Graphics Forum*, *32*(3), 271–280. doi:10.1111/cgf.12114

ADDITIONAL READING

Bizer, C., Heath, T., & Berners-Lee, T. (2009). Linked data-the story so far. *Semantic Services, Interoperability and Web Applications: Emerging Concepts*, 205-227.

Kleppmann, M. (2015). *Apache Kafka, Samza, and the UNIX Philosophy of Distributed Data*. Retrieved October 12, 2015 from http://www.confluent.io/blog/apache-kafka-samza-and-the-unix-philosophy-of-distributed-data

Morrison, J. P. (2010). *Flow-Based Programming: A new approach to application development*. CreateSpace.

Narock, T., & Fox, P. (Eds.). (2015). *The Semantic Web in Earth and Space Science. Current Status and Future Directions*. IOS Press.

KEY TERMS AND DEFINITIONS

Big Data: Big Data can be simply defined as "having more data than I had yesterday, and not knowing what to do with it". Key aspects of Big Data are the data volume; the speed of data production and transfer (its velocity); and the wide variety of data types introduced.

Big Ocean Data: An application of the ideas of Big Data specifically to the marine science domain.

Controlled Vocabulary: A controlled vocabulary provides a way to organise knowledge for subsequent retrieval. Vocabularies are used in subject indexing schemes, subject headings, and their content can be organised hierarchically to create thesauri, taxonomies and other forms of knowledge organization systems. Controlled vocabulary schemes mandate the use of predefined, authorised terms that have been preselected by the individual or group governing the vocabulary, in contrast to natural language vocabularies where no such restriction is put in place.

Linked Data: Linked Data is a technique that uses the Word Wide Web to connect related data that was not previously linked, or uses the Web to lower the barriers to linking data which are currently linked using other methods. In practice, Linked Data often uses the Resource Description Framework model and Web addresses (as Uniform Resource Locators, or URLs) to achieve these linkages.

Linked Ocean Data: A subset of Linked Data, with a specific focus on the marine science domain.

Ontology: In computer science, an ontology formally represents knowledge as a hierarchy of concepts within a domain, using a shared vocabulary to denote the types, properties and interrelationships of those concepts.

Ontology Design Pattern: An Ontology Design Pattern is a reusable solution to a data modelling problem that commonly occurs across many different domains or within a wide variety of contexts within a single domain.

Resource Description Framework (RDF): RDF is a standard model for the exchange of data over the World Wide Web. The RDF data model is based upon the idea of making statements about resources in the form of subject-predicate-object expressions, known as triples. A classic example of a subject-predicate-object triple is "sky"-"has colour"-"blue".

Semantic Web: The Semantic Web is the World Wide Web of data built on the Resource Description Framework model as a foundation for publishing and linking data online.

Streaming Data: Streaming data is an analytic computing paradigm that is focused on speed of throughput of data.

Topology: In the context of processing streams of data, a topology is a graph of computation. Each node in the graph contains the processing logic and the graph's edges (or connections between the nodes) indicate the pathways of data between the nodes.

Chapter 5
Documenting Provenance for Reproducible Marine Ecosystem Assessment in Open Science

Xiaogang Ma
Rensselaer Polytechnic Institute, & University of Idaho, USA

Stace E. Beaulieu
Woods Hole Oceanographic Institution, USA

Linyun Fu
Rensselaer Polytechnic Institute, USA

Peter Fox
Rensselaer Polytechnic Institute, USA

Massimo Di Stefano
Rensselaer Polytechnic Institute, USA & University of New Hampshire, USA

Patrick West
Rensselaer Polytechnic Institute, USA

ABSTRACT

Open Science not only means the openness of various resources involved in a scientific study but also the connections among those resources that demonstrate the origin, or provenance, of a scientific finding or derived dataset. In this chapter, the authors used the PROV Ontology, a community standard for representing and exchanging machine-readable provenance information in the Semantic Web, and extended it for capturing provenance in the IPython Notebook, a software platform that enables transparent workflows. The developed work was used in conjunction with scientists' workflows in the Ecosystem Assessment Program of the U.S. NOAA Northeast Fisheries Science Center. This work provides a pathway towards formal, well-annotated provenance in an electronic notebook. Not only will the use of such technologies and standards facilitate the verifiability and reproducibility of ecosystem assessments, their use will also provide solid support for Open Science at the interface of science and ecosystem management for sustainable marine ecosystems.

INTRODUCTION

Open Science has been receiving significant attention in recent years (Nosek et al., 2015). The Open Science movement includes open access to publications (Harnad & Brody, 2004), open data (Uhlir & Schröder, 2007; Glaves, H., 2017, chapter 3 this book), open source software and web services (Hey &

DOI: 10.4018/978-1-5225-0700-0.ch005

Payne, 2015), open collection of physical samples used in research (Lehnert, Vinayagamoorthy, Djapic, & Klump, 2006), as well as Science 2.0 (Shneiderman, 2008) that uses social media to exchange ideas and facilitate collaborations among scientists. Reproducibility is one of the key points that Open Science addresses (Open Science Collaboration, 2013; Yaffe, 2015). Efficient organization of the data, software programs, and samples used in a publication allows the authors to replicate the research. Making those resources open will further allow readers of that publication to verify the reported findings and, possibly, to reuse the resources in new studies. Besides reproducibility of research, the Open Science movement is also important in "actionable science" and connecting science to policy and management decisions (Ma, Fox, Tilmes, Jacobs, & Waple, 2014a; Palmer, 2012; Reichman, Jones, & Schildhauer, 2011).

Making publications, data, code, and samples open does not mean sharing them as fragmented pieces. In the era of the World Wide Web, various resources shared online can be linked in innovative ways (Emile-Geay & Eshleman, 2013). The work of provenance provides both methodological and technological approaches to associate those resources of Open Science on the Web and make the lineage and workflow of research traceable. The literal meaning of provenance is "the origin of something". In scientific works, documenting provenance on the Web involves linking and presenting the 'graph' structure of a network of research activities, people, and organizations involved in the production of scientific findings. This network includes the supporting observations, datasets, models, and methods used to generate those findings. On the Web, provenance documentation can be achieved through four primary steps: categorization of resources, assigning a unique identifier to each resource, annotation of those resources, and linking among them (Ma et al., 2014a).

In this chapter, the authors present work on provenance documentation and transparent workflows in assessment reports for Large Marine Ecosystems (LMEs) (See Appendix A for a list of acronyms). The LME concept was developed by the U.S. National Oceanic and Atmospheric Administration (NOAA), in consultation with international partners, for conservation and management of living marine resources (Sherman, 1991). LMEs are large areas of ocean about 200,000 km^2 or greater, adjacent to the continents where primary productivity is generally higher than in open ocean areas (NOAA, 2011). To understand and compare the relative impacts of fishing efforts, habitat degradation, climate change, and natural cycles on marine ecosystems, many different types of data must be collected, integrated, and interpreted. Data collected for monitoring ecosystem indicators in an Integrated Ecosystem Assessment (Levin, Fogarty, Murawski, & Fluharty, 2009) may be as diverse as satellite-derived sea surface temperature, counts of zooplankton species from net tows, and landings data from commercial fisheries. Provenance documentation is increasingly recognized as important for data contributed to community repositories and archives, inclusive of NOAA data collections and products derived from these data (NOAA 2011).

As part of the efforts to employ cyberinfrastructure data technologies to facilitate quantitative analysis and synthesis in LME assessments, the work of provenance documentation presented in this chapter benefited significantly from Semantic Web technologies (Leadbetter, Cheatham, Sepherd and Thomas, 2017, chapter 4 this book). A key step in this work is to have a machine readable conceptual model (i.e., an ontology) for objects and relationships in the provenance for datasets and other products in the assessment of LMEs. A motivation for enabling machine-readable provenance for NOAA's assessment of LMEs is the U.S. Executive Order "Making Open and Machine Readable the New Default for Government Information" (Obama, 2013). In the remainder of this chapter, Section 2 provides more details about provenance representation and documentation in the Semantic Web, as well as an interactive environment - the IPython Notebook - for data analysis for LME assessment. Section 3 describes the ontology developed to capture provenance in the IPython Notebook. Section 4 presents how this ontology can be implemented, using a few case studies from an Ecosystem Status Report for the U.S. Northeast Shelf LME. Section 5 compares the approach in this paper with related studies and discusses directions for future work. Finally, Section 6 concludes this chapter.

PROVENANCE DOCUMENTATION AND TRANSPARENT WORKFLOW

The Semantic Web

This study was conducted in the environment of the Semantic Web, which extends the existing World Wide Web by adding machine-readable meanings (semantics) to content on the Web (Berners-Lee, Hendler, & Lassila, 2001). In the conventional Web, or the Web of Documents, markup languages such as the HyperText Markup Language and identifiers such as the Uniform Resource Locator enable the connections among various resources. However, such markup and connections are just at the textual level, for which humans can read and understand, but it is not possible for machines to interpret the meaning of the text due to the absence of a structure for concept definition. The Semantic Web, or the Web of Data, overcomes this gap by treating resources as objects and adding meanings to those objects and the relations among them. Given a subject of study, the objects and connections under that subject and their definitions are stored in a well-organized structure called ontology. By authoring ontologies in a standardized language encoding, each resource on the Web can be categorized within a context, annotated and connected to other resources. Moreover, the meanings of those categories, annotations and connections can be read by machines and utilized for various purposes.

In computer science, ontology is defined as "… the formal specification of a shared conceptualization of a domain" (Gruber, 1995), in which a domain may be related to tasks or to a scientific discipline or community such as oceanography, or fisheries science. In a broad perspective, there could be several types of ontologies in science according to the richness and formalization of the modeling and encoding. Aspects such as glossary, taxonomy, thesaurus, conceptual schema, and formal description with logic constraints, form a so-called *ontology spectrum* (McGuinness, 2003). Ontologies can also be classified according to their level of dependence on a specific task or point of view (Guarino, 1997), such as a top-level ontology, a domain or task ontology, and an application ontology.

The World Wide Web Consortium (W3C) develops standards for both the conventional Web and the Semantic Web. Presently, most ontologies are encoded in the Resource Description Framework (RDF), which is a fundamental format in the Semantic Web. There are also standard vocabularies such as the RDF Schema (RDFS) (W3C, 2014a) and the Web Ontology Language (OWL) that provide building blocks for constructing ontologies.

Normally, an ontology encoded in RDF includes a set of defined classes (formalization of the objects) and properties (often, but not always, a formalization of the relations). There are two types of properties: the object property and the data type property. The former is for the relationship between individuals and the latter links individuals to literal values, for example strings and date/time. RDF uses a triple structure of <subject, predicate, object> in its encoding structure. For example, the triple <prov:Person rdfs:subClassOf prov:Agent> asserts that 'prov:Person' is a subclass of 'prov:Agent'. The 'prov' and 'rdfs' in those triples are namespace prefixes, which represent the namespaces 'http://www.w3.org/ns/prov#' of the W3C PROV Ontology (PROV-O) (Lebo, Sahoo, & McGuinness, 2013) and 'http://www.w3.org/2000/01/rdf-schema#' of the RDF Schema. The namespace prefixes specify where a resource comes from, i.e. its authority. Through the use of namespaces, each element in an RDF triple has a unique Uniform Resource Identifier (URI). For example, the URI of 'prov:Person' is 'http://www.w3.org/ns/prov#Person'. An ontology in RDF can be queried by several languages, such as the SPARQL Protocol and RDF Query Language (SPARQL)[1]. The capability to extract parts of ontologies, using the triple structure as a basis, allows the discovery of global knowledge from the graph. For example, find all relations between Object A and Object B, or find all things that Object A (e.g. prov: Person) is related to; all relations, all objects. Lastly, Ontologies, RDF, SPARQL and other building blocks provide the fundamentals for Linked Data (Bizer et al., 2009). Linked Data represents a less formal encoding of a Web of Data, and is especially flexible and useful for graphs of multiple domains, or contexts. Berners-Lee (2006) proposed the five star scheme for the publishing of Linked Data, summarized as:

1. Make data available on the Web;
2. Use a machine readable format;
3. Use a non-proprietary format;
4. Use open standards from W3C (RDF and SPARQL); and
5. Link data to other people's data.

A goal of this work was to provide standardized annotations (metadata) for data products, so that a human or a machine could trace back to the source observational data and models used to reach a scientific finding. The Semantic Web provides an environment for publishing key provenance encodings and in turn using them. For a persistent knowledge base a "triple-store", or an "RDF database" is a common implementation to enable machine accessibility; instead of following links in a Web of Data. An RDF database is a type of database that stores and retrieves RDF information. As noted above, the 'subject-predicate-object' data structure of triples make the RDF database comparable to a graph database (Angles & Gutierrez, 2008). Both the RDF database and the graph database can store and query data in 'graph' structures, and the RDF database still has its own features, such as reasoning based on existing triples.

Provenance Representation and Documentation in the Semantic Web

The Semantic Web provides an environment for standardized provenance documentation. As mentioned above, the literal meaning of provenance is the origin of something. In a general sense, conventional work on metadata, i.e., data that describes data, can also be considered as relevant to the work on provenance. Nevertheless, provenance documentation provides a key context for such metadata, i.e. it extends the scope, because metadata conventionally focuses more on the description or annotation of certain objects, whereas provenance documentation covers not only the annotations of objects but also the relations among them. For example, the metadata of a dataset may include an item called 'Data Collector', and the record of it will normally be a literal value, such as the name of a person. Therein the dataset is the object, and the metadata items are annotations in literal values. If provenance documentation is used for the same example, then there will be two objects, the dataset and the person, and the 'Data Collector' will be the relation between them. There will also be annotations describing the person and possibly connections linking this person to other objects, such as their role, organization and so on.

Prior to PROV-O there were several notable works on Web-based provenance models and ontologies. Earlier work include the Proof Markup Language (PML) (Pinheiro da Silva, McGuinness, & Fikes, 2006), the KARMA framework (Simmhan, Plale, & Gannon, 2006), the PrOM framework (Sahoo, Barga, Sheth,

Thirunarayan, & Hitzler, 2009), the Open Provenance Model (OPM) (Moreau et al., 2011), and more. The culminating work resulted in PROV-O (Lebo et al., 2013). PROV-O arose as a consensus of a broader community. Thus, it shares a few key concepts with PML and OPM (Table 1). However it is also more comprehensive in that it includes detailed classes and properties not provided in PML or OPM. Work on PML, OPM and PROV-O were all largely initiated in the Semantic Web community. However, there is work in other fields, for example, in the field of Geographic Information Science, the ISO 19115 and ISO 19115-2 standards (International Organization for Standardization, 2003, 2009) provide a specific lineage and citation model for describing the provenance of geospatial data. There are also applications of these models in the Semantic Web (Yue, Gong, & Di, 2010). Table 1 lists a few key concepts from ISO 19115 and compares with PROV-O, PML and OPM.

These provenance models and their ontologic encodings provide ways to record structured, machine readable and meaningful provenance information (Moreau, 2011). In the scientific work of analyzing a set of observations or experiments, they can be used to describe the various datasets, platforms, instruments, methods, and outputs in a workflow, as well as the people and institutions involved and the experiments and steps conducted. For example, the three top classes of PROV-O (i.e. Entity, Agent and Activity) and the interrelationships among them offer a concise yet powerful structure that covers almost all the elements in such a workflow (Figure 1). PROV-O also provides classes and properties either at a more detailed level or with domain-specific meanings to meet the needs of provenance documentation. Provenance information encoded in such structures enables scientists to publish results together with provenance information, and also allows readers to trace the provenance information to see how a result or finding was generated. Using PROV-O, the provenance information encoded in RDF format

Table 1. Similar classes and properties in PROV-O, PML, OPM, and ISO 19115 Lineage and Citation model

	PROV-O	PML	OPM	ISO 19115
Classes	prov:Entity	pmlj:Nodeset (+ pmlp:Information)	opmv:Artifact	LI_Source
	prov:Activity	pmlj:InferenceStep (+ pmlp:InferenceRule)	opmv:Process	LI_ProcessStep
	prov:Agent	pmlp:InferenceEngine	opmv:Agent	CI_ResponsibleParty
	prov:Role	Not established	opmo:Role	CI_RoleCode
Properties	prov:used	pmlj:hasAntecedentList (+ pmlj:NodeSetList)	opmv:used	source
	prov:wasGeneratedBy	pmlj:isConsequentOf	opmv:wasGeneratedBy	sourceStep
	prov:wasAssociatedWith	pmlp:hasInferenceEngine	opmv:wasControlledBy	processor
	prov:wasAttributedTo	Not established	Not established	Not established
	prov:wasDerivedFrom	Not established	opmv:wasDerivedFrom	sourceCitation
	prov:wasInformedBy	Not established	opmv:wasTriggeredBy	Not established
	prov:actedOnBehalfOf	Not established	Not established	Not established
	prov:startedAtTime	Not established	opmv:wasStartedAt	Not established
	prov:endedAtTime	Not established	opmv:wasEndedAt	dateTime

(from Ma et al., 2014b)

Figure 1. The three top classes in W3C PROV-O
(from Lebo et al., 2013)

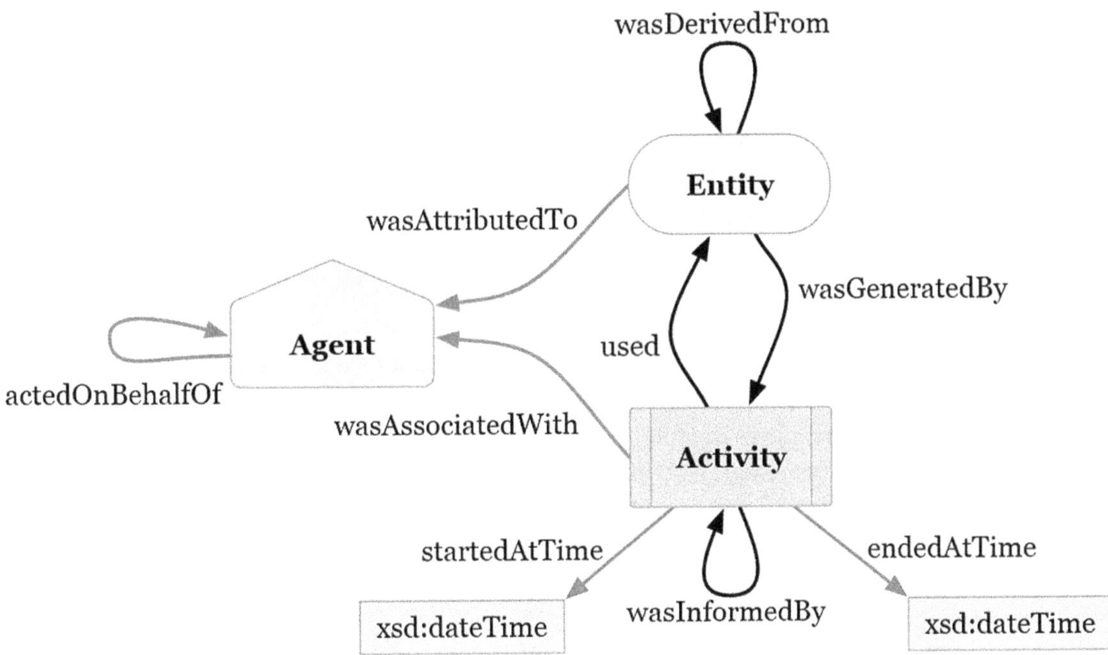

is exchangeable over the Web with applications using RDF. These applications can retrieve provenance that can be used with other RDF data on various topics, and especially, be analyzed by tools that process provenance information.

PROV-O was approved by W3C as a recommendation (a standard in Web terminology) in 2013. The work presented here was one of the first studies to test the use of PROV-O in a domain context. To make the provenance information captured for LME assessments interoperable with other provenance information in the larger Semantic Web, the authors used and extended PROV-O for the assessment specific aspects of provenance representation and documentation. The PROV-O classes and properties provide an abstract overview about a workflow and the various resources and people involved. In the assessment domain, i.e. quantitative analysis and synthesis in an LME assessment, concepts with detailed disciplinary meanings are needed, such that the captured provenance information is more readable and understandable to researchers in that field of study. In order to achieve that, those concepts can be derived from the terminology used by the researchers in their studies. In the next section, an examination of workflows used for LME assessment provides details on domain specific concepts.

Interactive and Transparent Workflows of Data Analysis

One of the primary objectives of this work is to document transparent workflows in LME assessment. A workflow is understood as a sequence of steps for achieving a specific objective. Resources may be used in those steps and outputs generated, and each step may be associated with a person or specific software. To realize a transparent workflow, not only the datasets, methods and outputs of various steps should be documented, but also the context information, such as when and where an experiment was conducted, the software programs used, the original intention of a certain method, and more (Goecks, Nekrutenko, Taylor, & Team, 2010). In addition, to fully realize transparent workflow, comprehensive and integrative provenance information is needed in order to link the disparate elements in that workflow (Howe et al., 2008).

The particular software environment examined in this study for interactive workflows of data analysis was the IPython Notebook, developed in 2011 by a group led by Fernando Pérez, University of California, Berkeley and Brian Granger at California Polytechnic State University. Based on Interactive Python, a rich architecture for interactive computing (Rossant, 2013), the IPython Notebook provides a browser-based front end, in which users can combine code, its execution, rich text, mathematics, plots and rich media in a single user interface. IPython Notebook gives scientists the ability to work on diverse and heterogeneous data and information sources, that are realized in interactive command windows. This enables users to write and run code, access datasets, import data analysis packages and view outputs such as tables or plots. It also provides an effective way to share the source code used to generate data products and allows other people to execute, reuse, and modify the code to build their own research or to collaborate with other researchers. Such features make IPython Notebooks a powerful platform to support transparent workflow and reproducible research (Shen, 2014), since all inputs and outputs are recorded in a notebook document that can be shared on the Web. Pérez, Granger and their colleagues have been moving the IPython Notebook into a new project called Jupyter, which aims to support more programming languages, such as Julia, R and Scala.

In this study, the authors focused on the metadata associated with the workflow embedded in an IPython Notebook as well as domain-specific concepts in Integrated Ecosystem Assessments. The work described here is an extension of the ECO-OP project (ECO-OP is an abbreviation for ECOsystem interoperability; TWC, 2014), a collaboration between computer scientists, oceanographers, and fisheries scientists. Earlier work in the ECO-OP project identified the IPython Notebook as a technology that could be adopted by NOAA scientists to share workflows for data products in an Ecosystem Status Report (Beaulieu et al., 2013; Di Stefano et al., 2014). This work on standardized provenance metadata is timely, as the ISO 19115 standard in Table 1 was recommended relatively recently by the NOAA Data Documentation Procedural Directive (NOAA 2011). A key feature of the IPython Notebook is that the metadata can be generated simultaneously during the access and processing of data, which provide rich resource information for provenance documentation. An ontology is also needed, as a data model, to transform those metadata into machine readable provenance information, and in turn to be published in the Semantic Web (or deposited in a triple-store).

THE ECO-OP PROVENANCE CAPTURE ONTOLOGY

Each instance of a workflow created in the IPython Notebook environment is called an IPython Notebook. Within the user interface, such a notebook appears as an ordered list of cells for resource input, processing and output, including dataset(s), text, code, computational libraries, tables, diagrams as well as other media formats (Figure 2). Each IPython Notebook's file format is a JSON (JavaScript Object Notation) file with an .ipynb extension. If an IPython Notebook is open, it has exactly one interactive session connected to an IPython kernel, which interprets and runs the code in the session and returns the results. There is a growing list of kernels for different programming languages (IPython, 2016).

Figure 2. Components in the user interface of an IPython Notebook. Cells with Python or other supported language are numbered sequentially and the activities of some Cells are indicated with inputs (left) and outputs (right). The example uses a screen grab of the Ecosystem Assessment Program Climate Forcing IPython Notebook at http://nbviewer.ipython.org/gist/anonymous/6153235.

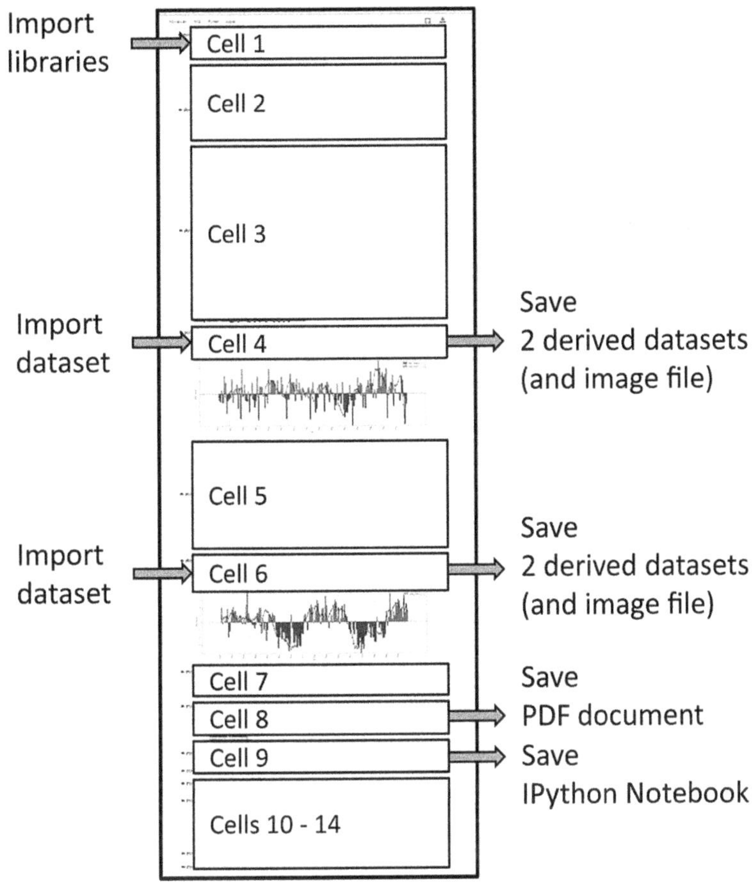

Given the structure and characteristics of the IPython Notebook, the authors designed an ontology, called the ECO-OP Provenance Capture Ontology (prov-ecoop) for representing provenance information in the workflow of an IPython Notebook (Figure 3). This ontology is general to the use of IPython Notebook for data processing and documentation; its specific application to marine ecosystem assessment is presented in Section 4. The two diagrams (a) and (b) in Figure 3 are linked by two classes: ecoop:IPythonNotebookRun and ecoop:CodeCellRun (left in Figure 3 (a) and upper right in Figure 3 (b)). ecoop:IPythonNotebookRun is regarded as an activity of running an IPython notebook, in which there could be several sub-activities of running code cells in the notebook. Figure 3 (a) presents classes about components in the environment of the IPython Notebook and the relationships between those classes. In contrast, Figure 3 (b) offers more details about the inputs and outputs in the running of a code cell, such as libraries and datasets can be used as input to generate images, arrays, scripts or new datasets. For the running of a whole notebook, the output can be included in a document formatted in LaTeX, HTML or PDF. Components in the document, such as figures, tables and listings, are made by using outputs such as images, arrays, and scripts generated from the running of code cells. Most object properties in Figure 3 are presented in pairs, called inverse properties, to realize the forward and backward connections.

To enable the interoperability of the developed ontology with other ontologies and applications, a few classes and properties from PROV-O and the DCMI Terms (Dublin Core, 2012) were used directly in the prov-ecoop ontology, such as those shown in Figure 3. Additional ontologies used in prov-ecoop

Figure 3. Key components of the prov-ecoop ontology for representing provenance information of the workflow in an IPython Notebook represented in a Concept Map format. Rounded rectangles contain the namespace and name of the class (i.e. object) and labelled directional arrows connecting classes are the named and typed properties between them (i.e. relations). Top: (a) Components relevant to the environment of IPython Notebook; bottom: (b) Components relevant to the inputs and outputs in an IPython notebook. The data type properties of each class are eliminated here to save space.
The prov-ecoop ontology file (in Turtle serialization; W3C, 2014c) is accessible at: https://github.com/tetherless-world/ecoop/blob/master/prov/ecoopProv.ttl.

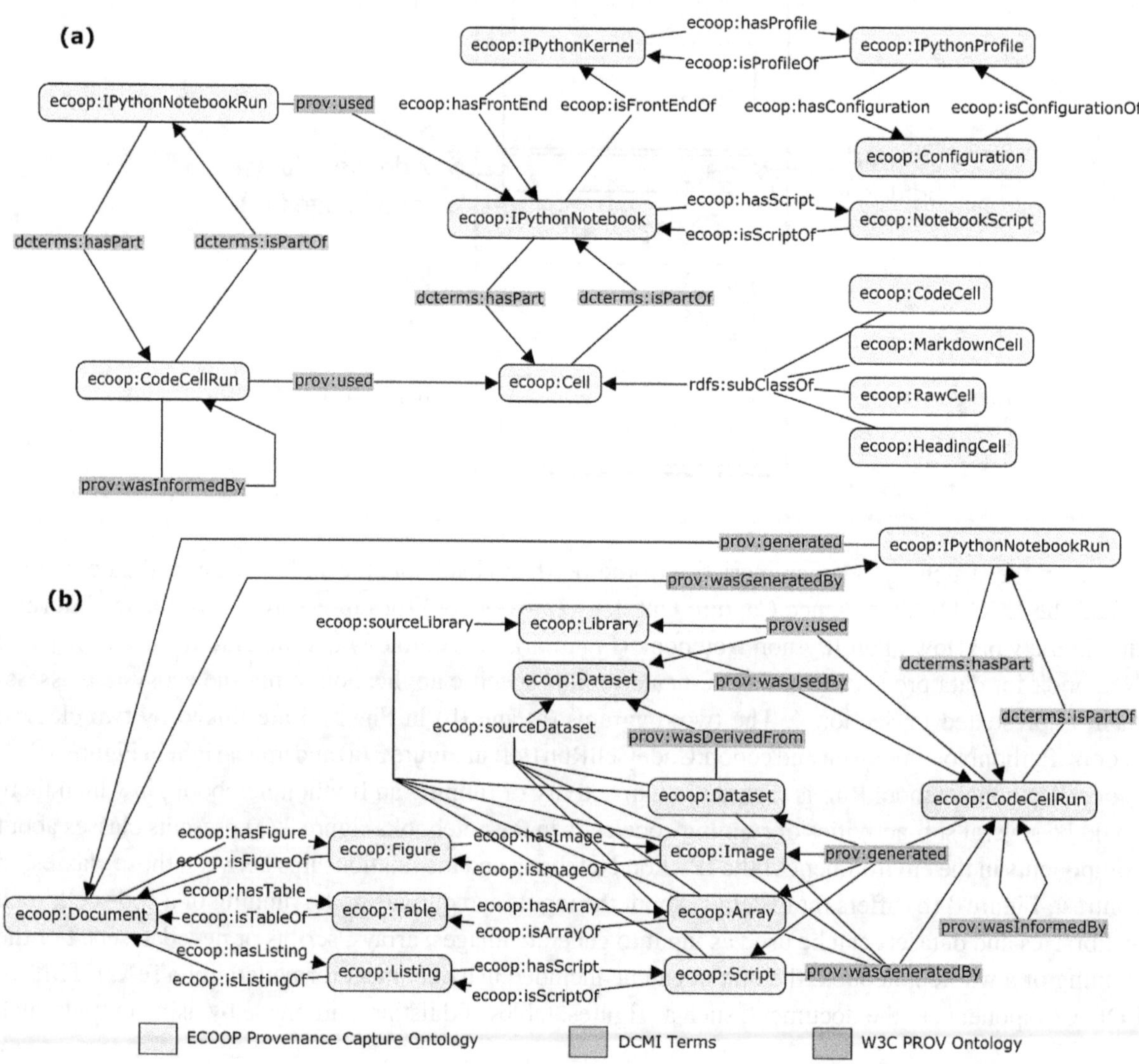

Figure 4. Relationships of some classes in the prov-ecoop ontology to classes in the PROV-O ontology

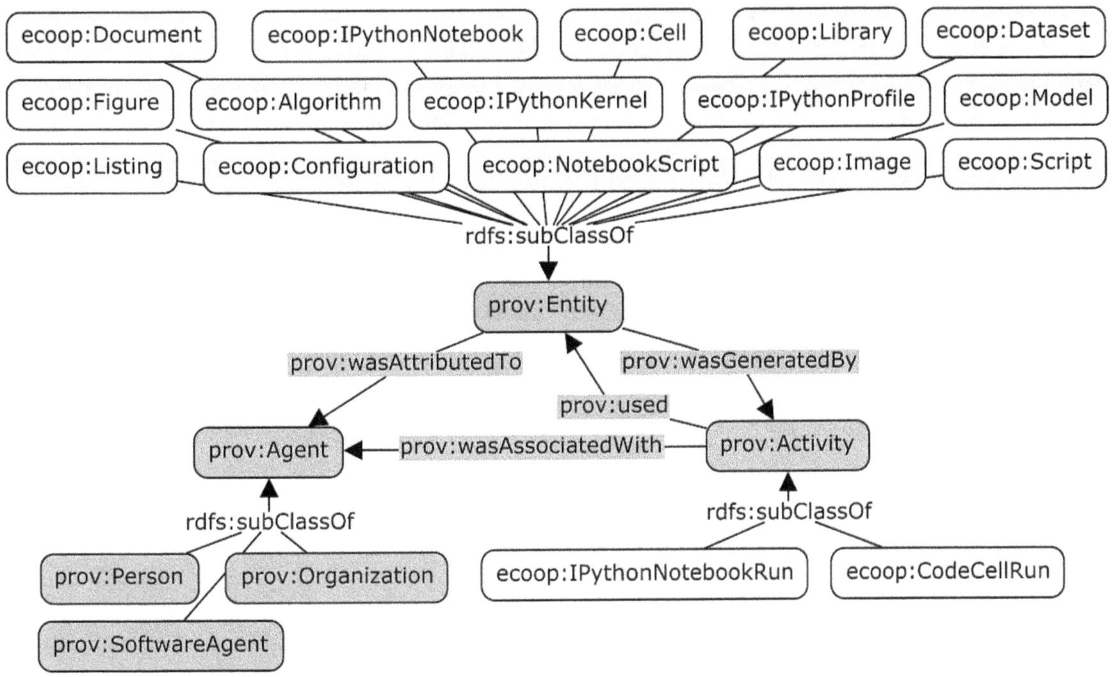

include Data Catalog Vocabulary (DCAT) (W3C, 2014b), Friend Of A Friend vocabulary (FOAF) (2014), Simple Knowledge Organization System (SKOS) (W3C, 2004), and Quantities, Units, Dimensions and Types(QUDT, 2014). Moreover, a few classes and properties in the prov-ecoop ontology were related to those in PROV-O by using the assertions of subclass or subproperty, such as those shown in Figure 4.

The prov-ecoop ontology, as an application ontology, represents key concepts and relationships in the IPython Notebook. Components in the ontology are also consistent with those in PROV-O. Using the developed ontology, functionalities are developed to collect the provenance information generated at each run of the workflow in an IPython Notebook. The captured provenance information is stored in the JSON-LD format (JSON for Linking Data, n.d.), which is compatible with the Semantic Web.

CASE STUDIES

Case studies for the use of the prov-ecoop ontology were examined in conjunction with scientists' workflows in the Ecosystem Assessment Program (EcoAP) of the U.S. NOAA Northeast Fisheries Science Center (NEFSC). EcoAP scientists assess conditions, threats, and impacts and develop models to predict responses in the U.S. Northeast Shelf LME (see an example; NEFSC, 2008) as one part of overall Integrated Ecosystem Assessments (Levin et al. 2009; Levin et al. 2014). The data and information products generated by the environmental and fisheries scientists in EcoAP are important for implementing ecosystem-based management in this LME, a guiding principle of U.S. National Ocean Policy.

The three case studies, described below in increasing complexity, were developed from a use case for the biannual Ecosystem Status Report (ESR) for the U.S. Northeast Shelf LME. The ESR has synthesized data products ranging from climate indicators, to regional maps of primary production, to fisheries indicators for the LME (Ecosystem Assessment Program, 2009; Ecosystem Assessment Program, 2012). A specific stated goal of the ESR use case was to make the generation of the report reproducible. Ultimately, the technologies adopted for the ESR use case included the IPython Notebook for interactive and transparent workflows of data analysis and provenance documentation (Beaulieu et al., 2013; Di Stefano et al., 2014). For the purposes of this study, the authors addressed use cases specific to adopting the prov-ecoop ontology and hence the PROV-O standard to production of the ESR. Thus, the goal is to show how a prototype system is being developed in which the user (the scientist who is generating a dataset for an ecosystem status report) is also generating machine-readable provenance information that can be published in the Semantic Web. One of the claims being tested is that capturing standardized information during execution of the workflow is more efficient than providing the scientist with a metadata template afterwards. This claim shares common thoughts with research in the Rolling Deck to Repository (R2R) program (Arko et al. 2013; Leadbetter 2015) and the Eurofleets project (Diviacco et al. 2015).

Climate Forcing Chapter in the Ecosystem Status Report

The first case study applies the prov-ecoop ontology to a chapter titled "Climate Forcing" in the 2009 ESR (Ecosystem Assessment Program, 2009). A portion of this chapter, which includes derived data products for several climate indicators, was successfully reproduced in an IPython Notebook (Beaulieu et al., 2013; Di Stefano et al., 2014). This case study was examined by iterating a use case with the goal to capture semantic provenance for entities produced by a NEFSC scientist (the principal actor) who ran the "ESR_Climate_forcing_notebook" (Figure 2). The text below describes how to represent and implement provenance for the derived data products, the chapter document, and the IPython Notebook itself.

To summarize the use case:

A researcher is using an IPython Notebook to produce a document relevant to climate forcing. The document includes a figure called 'Figure 1: North Atlantic Oscillation,' in which there is an image that is based on derived datasets and is generated by scripts in a software platform, and similarly a 'Figure 2: Atlantic Multidecadal Oscillation.' The IPython Notebook (semi-)automatically assigns an URI to each entity involved in the workflow of creating the document.

A few instance objects were recognized in the use case description, which can be re-stated using classes from the prov-ecoop ontology:

The EcoAP scientist (prov:Agent, foaf:Agent), being part of the EcoAP group (prov:Group, foaf:Group) acting on behalf (prov:actedOnBehalfOf) of NEFSC (prov:Organization, foaf:Organization), ran (ecoop:IPythonNotebookRun) an ESR_Climate_forcing_notebook (ecoop:IPythonNotebook) to generate (prov:generated) a PDF document for the Climate Forcing chapter (ecoop:Document). The document which was attributed to (prov:wasAttributedTo) the EcoAP scientist includes (ecoop:hasFigure) a set of figures (ecoop:Figure) with (ecoop:hasImage) images (ecoop:Image) that were created by using (prov:sourceDataset) some datasets (ecoop:Dataset). Those datasets were attributed to (prov:wasAttributedTo) the same EcoAP scientist who ran the ESR_Climate_forcing_notebook and were derived from (prov:wasDerivedFrom) imported source datasets (ecoop:Dataset).

Figure 5. An RDF graph for key objects in the ESR climate forcing use case. The integers shown for Cells in this diagram correspond to Cells in the Notebook depicted in Figure 2. Hard sided rectangles contain instances of classes (also known as individuals), i.e. specific values. Color shading is used to highlight which ontology classes in PROV-O the instances are related to (see upper left). Note the use of ecoop:, prov:, foaf: and dcterms: namespaces for the property relations (labelled arrows).

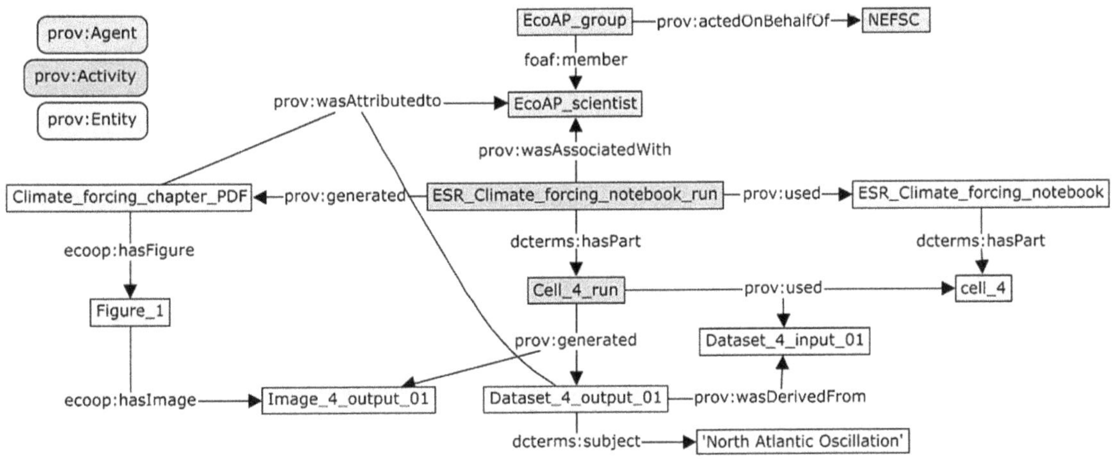

Those recognized objects and their relationships are depicted in Figure 5 using classes and properties from prov-ecoop and PROV-O.

In Figure 5, the nodes in yellow color show the key entities involved in this use case. Taking the dataset 'Dataset_4_output_01' as an example, it is an instance of the class ecoop:Dataset. The dataset was generated from the running of Cell 4, which imported a source dataset, 'Dataset_4_input_01'. An image 'Image_4_output_01' was also generated and saved from the running of Cell 4 and was used in a figure in the "Climate Forcing" chapter. In addition to those relationships among objects, it is also useful to manually assign some keywords to the dataset 'Dataset_4_output_01' to provide domain-specific meaning to the provenance. In Figure 5, the label 'North Atlantic Oscillation' (GCMD, 2016a) that tags 'Dataset_4_output_01' is an SKOS Concept from the NASA Global Change Master Directory (GCMD) (NASA, 2016) Science Keywords.

Fisheries Indicator and Regional Map of Primary Production in the Ecosystem Status Report

The other two case studies, a fisheries indicator and a regional map of primary production, are from chapters on "Anthropogenic Factors" and "Primary and Secondary Production," respectively, in the 2011 ESR (Ecosystem Assessment Program, 2012). These two case studies are presented together here because they both are related to cross-domain datasets and require more computer technologies for data acquisition and processing, which made them more complex to be implemented in the IPython Notebook. These two case studies were not fully implemented in the IPython Notebook; at the time of this project, kernels were not available for all of the programming languages used. These case studies are typical examples of the work for provenance documentation and tracing, because provenance and context of studies underpinned by multiple resources and stakeholders have received increasing attention in the community of Earth science informatics (Ramdeen & Hills, 2013). For each case study an image of the synthesized data product will be shown below, together with a sketch or description of a scientist's workflow for generating the image. The implementation in IPython Notebook or Jupyter Notebook and manual annotation to domain-specific vocabularies will also be proposed and discussed below. Assignment to concepts from the following well-recognized vocabularies were considered: NASA GCMD as above, AGROVOC (Rajbhandari & Keizer, 2012), the SeaDataNet Parameter Discovery Vocabulary (Leadbetter, Lowry, & Clements, 2012), and the U.S. Data Categories for Marine Planning (Lightsom, Cicchetti, & Wahle, 2015).

The fisheries indicator that was examined was the mean trophic level of the catch for the Northeast Shelf LME during 1960-2010 (Figure 6 (a)). In practice, this indicator is assembled by two EcoAP scientists, a natural scientist and a social scientist, using datasets contributed by four other agents. Figure 6 (b) shows a diagram developed from a sketch of the EcoAP scientists' workflow for generating the plots in Figure 6 (a). More specifically, Figure 6 (b) is a proposed implementation in IPython Notebook or Jupyter Notebook, which would be possible using the kernel for the R programming language. The large rounded rectangle in Figure 6 (b) indicates an IPython Notebook entity, and there are several Cell entities and the corresponding activities of running those Cells. The workflow is initiated at left with five source dataset entities. There are two software platform entities in addition to the IPython Notebook and the activities running them and generating a few intermediate derived datasets in the middle. Ultimately a synthesized dataset is saved as the result of running the notebook (lower right).

For the fisheries indicator, the proposed implementation in IPython Notebook would allow for automated acquisition of provenance metadata linking the final derived dataset to the multiple source datasets, the code used in data processing, and the responsible agents. For additional semantic provenance that would enhance the discoverability and understanding of the datasets used in the workflow, manual assignment of concepts from GCMD might include "fisheries," (GCMD, 2016b) "trophic dynamics," (GCMD, 2016c) or "biospheric indicator" (GCMD, 2016d) Annotation of concepts from AGROVOC might include "multispecies fisheries" (Skosmos, 2014) or "trophic levels," (Skosmos, 2015) and SeaDataNet concepts might include "Fish and shellfish catch statistics" (NERC, 2010). The terms in the U.S. Data Categories for Marine Planning are not yet available in RDF; once semantically enabled, Data Content Subjects in the hierarchy "Ocean Uses > Harvesting Living Resources >" may be considered, but the Data Content Type would be "Distributions" for source datasets and "Assessments" for the final derived dataset.

The case study for the regional map of primary production is truly cross-disciplinary in that the final derived dataset involves the application of a primary production model to satellite ocean color and temperature datasets. This case study is more complex in terms of the number of inputs and number of processing steps in the EcoAP scientist's workflow. It uses more classes of the prov-ecoop ontology, and the source datasets would be assigned to very different concepts than the final derived dataset. The image shown in Figure 7 is not the synthesized data product in the 2011 ESR (Figure 4 in Ecosystem Assessment Program (2012)), but is shown to demonstrate that many layers of spatial data may be represented in a map generated in an IPython Notebook. At the time of writing this manuscript, it was not possible to reproduce the EcoAP scientist's workflow in IPython Notebook because of specific software used for data processing. However, as of fall 2015, there is a Jupyter Notebook kernel for the Interactive Data Language (IDL, 2016), and thus, the workflow for the primary production map now is likely possible within Jupyter.

Figure 6. Domain scientists' workflow for using multiple source datasets to generate the fisheries indicator plots. (a) Mean trophic level of the catch by Ecological Production Unit for the U.S. Northeast Shelf LME (from Figure 8.2 in Ecosystem Assessment Program (2012)); (b) Proposed implementation of the workflow using IPython Notebook to generate the result in (a). Types of entities depicted: 1 - IPython Notebook; 2 - Cell; 3 - Dataset, inclusive of source, intermediate, and final derived (lower right) datasets; 4 - script written in other programming language (R) that was split into five Cells grouped together as a collection entity (W3C, 2013); 5 - other software environments in which data processing was performed. Four different agents are identified as contributing source datasets. Types of activities shown: 1 - CellRun; 2 - other activities performed in other software environments.

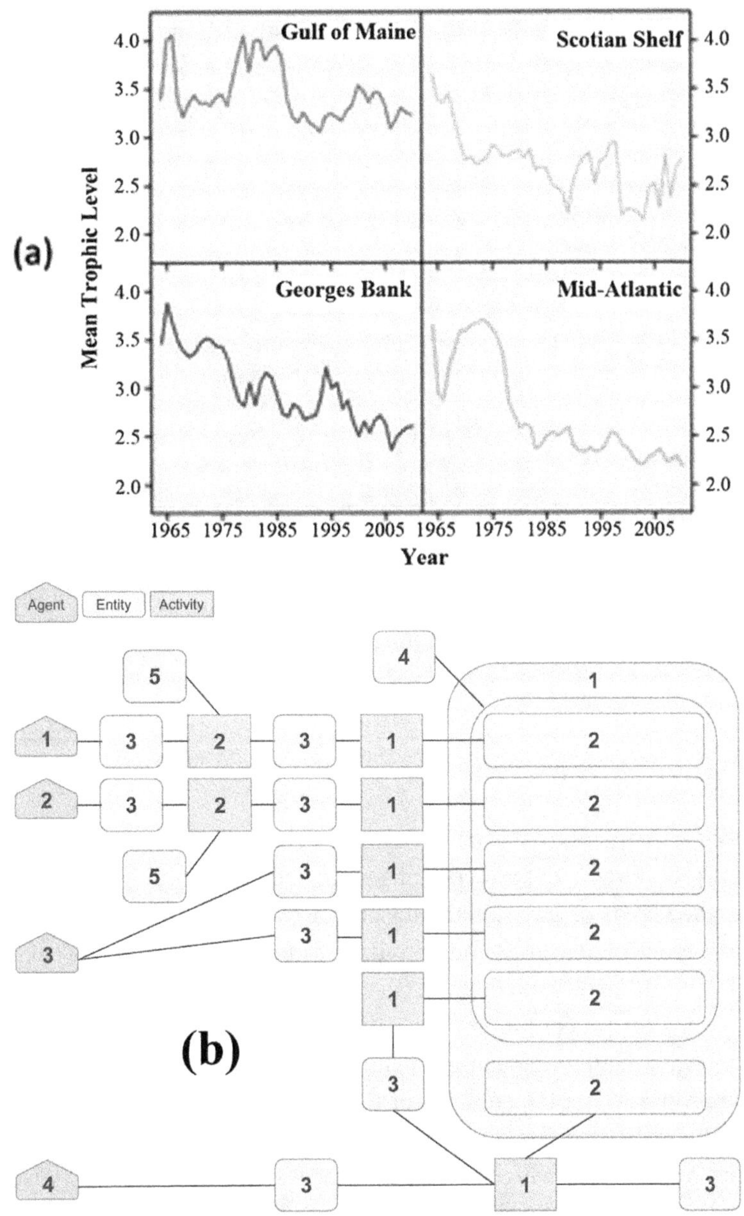

Figure 7. A test image in IPython Notebook showing a map of chlorophyll-a concentration (Chl-a) derived from satellite ocean color data, overlaid with boundaries for the NES LME Ecological Production units and USGS terrestrial water features (from Di Stefano, Fox, Beaulieu, & Maffei, 2012). Chl-a is an intermediate dataset used in the algorithm for the primary production model.

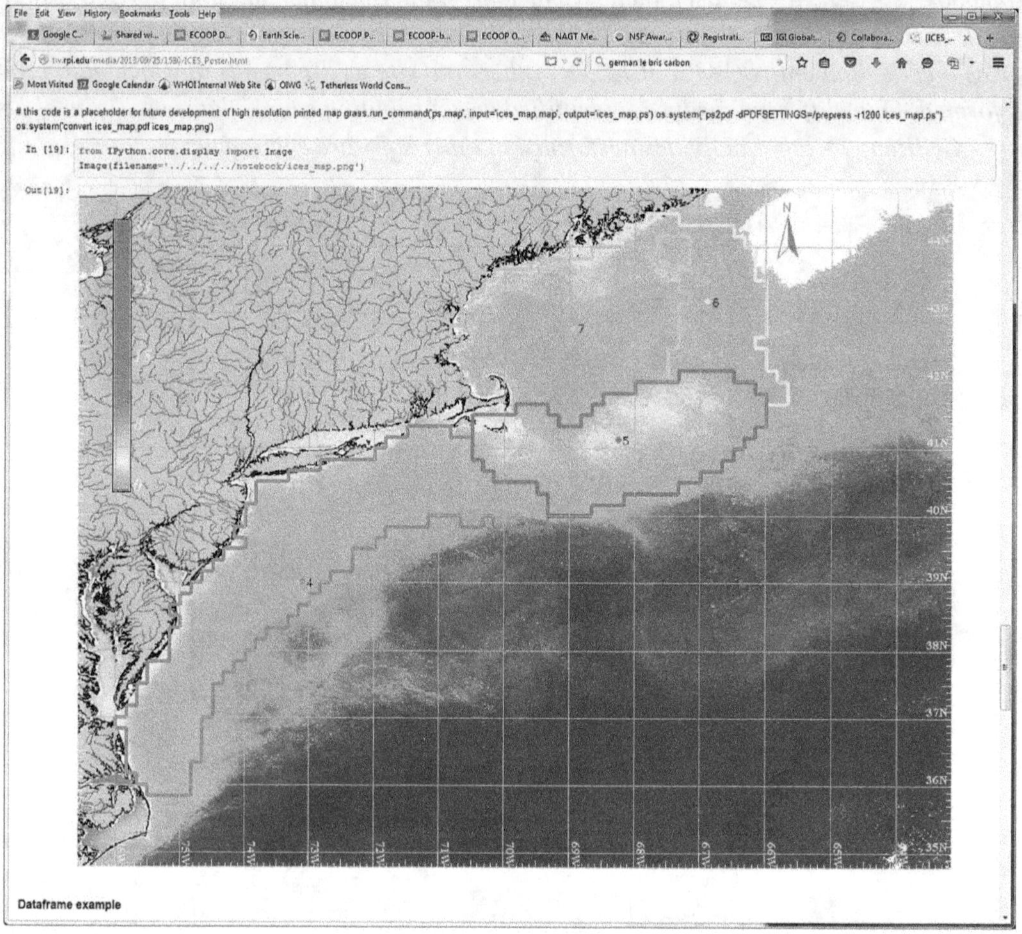

The authors worked with the EcoAP scientist to sketch the workflow for the regional map of primary production. The workflow starts with two types of source datasets - satellite ocean color and satellite sea surface temperature (thousands of individual datasets for each type) - contributed by two agents. Data processing involves 12 types of activities and 13 types of intermediate datasets (matched to NASA data levels 1a through 4), and ultimately results in a final derived dataset and image entity for contribution to the ESR. Presently the data processing involves two main software environments, SeaDAS and IDL. Installation instructions are available for implementing SeaDAS tools in a Python virtual environment (SeaDas, 2016). The proposed implementation of the workflow in a Jupyter Notebook would involve Cells specific to activities utilizing SeaDAS tools to process satellite ocean color datasets, followed by splitting an IDL script into Cells at which additional processing steps would be discriminated. It would be important to associate one of the CellRun activities to an algorithm (ecoop:Algorithm) for a specific primary production model (an instance of ecoop:Model). Because this workflow has thousands of source datasets, and the provenance chain from source datasets to final derived dataset involves so many steps, we would recommend visualizing the trace in software such as PROV-O-Viz (Hoekstra & Groth, 2015). For additional semantics capture for the provenance, manual assignment of concepts from GCMD might include "ocean color," (GCMD, 2016e) "sea surface temperature,"(GCMD, 2016f) or "primary production" (GCMD, 2016g). SeaDataNet annotations might include "Ocean colour and earth-leaving visible waveband spectral radiation," (NERC, 2006) "Skin temperature of the water column," (NERC, 2004) or "Primary production in the water column" (NERC, 2003). Although AGROVOC and U.S. Data Categories for Marine Planning provide appropriate matches of terms to the final derived dataset, i.e., "primary production" and "biological production" respectively, currently neither has good matches for source datasets.

DISCUSSION

The approach presented in this paper provides an innovative way to record provenance of scientific findings, allowing the tracing from derived data products back to their original sources, the methods used, the processing conducted, and the people who participated in the scientific research. The work presented here for the IPython Notebook may also be coupled with broader methodologies and architectures for semantic provenance documentation. For example, the Provenance Management System (PROMS) in Car, Stenson, & Hartcher (2014) and the semantic workflow system WINGS in Gil (2014) both can be used in further research. The entities generated by the workflow, such as derived datasets, may also be annotated with concepts from controlled vocabularies represented as Linked Data on the Web (Gallant, Schmidt, & Car, 2015). For example, this work on LMEs was related to a number of Linked Data vocabularies (i.e., NASA GCMD, AGROVOC, and SeaDataNet) and the U.S. Data Categories for Marine Planning. This work was guided by case studies in the field of marine ecosystems, but the methods and technologies for provenance documentation and tracing can be applied to other disciplines as well after minor adaptations (Reichman et al., 2011).

In this study the enhanced IPython Notebook is intended to support science collaborations for Integrated Ecosystem Assessments based on information science and semantic technologies (Fox, 2014). Best practices for semantic representation and annotation of provenance are emerging, and the authors provided some examples with the case studies above. Although functionalities to collect provenance generated at each run of an enhanced IPython Notebook may be technically feasible (Di Stefano et al. 2014), we recognize that there are challenges in helping the end user scientists adopt these technologies. Projects such as ECO-OP, in which computer scientists engaged directly with fisheries scientists, and broader community efforts such as the Geoscience Papers of the Future Initiative will aid the adoption of these and other Open Science technologies by domain scientists (Gil et al. 2015).

Enhanced semantics for provenance is likely to be important for "actionable science" (Palmer, 2012), such as the science reported in the third U.S. National Climate Assessment and the Global Change Information System (GCIS) which also used PROV-O for provenance documentation through the GCIS ontology (Ma et al., 2014b). The prov-ecoop ontology represents key concepts in a workflow that generates derived datasets or scientific findings, and the GCIS ontology represents key concepts in a scientific report and their relationships with the underpinning resources. The former in combination with the latter could provide a knowledge base that extends from the desktop of a scientist using IPython Notebook to the "key messages" or "traceable accounts" in GCIS. Another example is eReefs, which conducted an integrated use of the PROMS provenance service along with vocabulary services for a multi-agency assessment of the Great Barrier Reef in Australia (Car, 2013). Semantic provenance would aid the integration and understanding of domain-specific indicators, such as indicators of marine ecosystems, into more comprehensive reports on the state of the environment. For example, in 2011 the U.S. Government Accountability Office conducted a study on key indicators of economy, society, and environment (inclusive of ecosystems) and proposed that a comprehensive indicator system be developed "for integrating and disseminating this information to better inform the nation about complex challenges" (U.S. Government Accountability Office, 2011).

The knowledge network of LMEs is necessarily complex. The overarching goal of sustainable governance of LMEs is based on the best science, data, and information, oftentimes represented as composite indicators (Levin et al., 2014; Diviacco and Leadbetter, 2017, chapter 2 this book). The graph structure in the RDF database could be inclusive of a larger knowledge base comprising multiple reports and the work of many scientists. Tracing synthesized data products such as ecosystem indicators back to their original data sources is important for verification and understanding by diverse stakeholders, and can be done by using PROV-O and linking to domain-specific ontologies such as prov-ecoop to give meaning to the source datasets and derived products. As recognized by Reichman et al. (2011): "Provenance is especially important to support scientific results used in policy and management decisions," and it is important that "...the data, analytical processes, and results become part of a knowledge base supporting evidence-based science, to better inform decision-making in conservation and resource management."

The case studies were presented for one government agency assessing an LME that straddles international boundaries. Documenting provenance and reproducible workflows would assist comparison with other stakeholders' assessment of the same and other LMEs and the ocean as a whole. The past decade has seen a growing number of regional and global ocean assessments that employ indicators and other composite data products, such as the Ocean Health Index (Halpern et al., 2012; Halpern et al., 2014), the International Programme on the State of the Ocean (2016), and the United Nations World Ocean Assessment (2013). There is great motivation to bring standards and best practices of provenance documentation to the desktops of the scientists delivering, visualizing, and interpreting derived datasets for assessment reports. Enabling scientists to automatically generate provenance metadata during their data processing workflows is an incentive because it is tedious and inefficient to document provenance after the fact.

In the field of oceanography, international standards have already been used to promote data interoperability, such as the Observations and Measurements (O&M) (OGC, 2016) and SensorML (OGC, 2016) developed by the Open Geospatial Consortium (OGC). Applications of these can be seen in national or international data management initiatives such as NOAA's Integrated Ocean Observing System in USA (2008), the Infrastructure for Spatial Information in the European Community (2013). OGC also organized the Ocean Science Interoperability Experiment (OGC, 2016). While a primary purpose of those standards is to enable online exchange of structured datasets using standardized schemas, there can also be further works to leverage the standards in the Semantic Web. Recently, Cox (2016) discussed ontologies for observations and sampling features in O&M. He also mapped a few classes from O&M and the Semantic Sensor Network Ontology (SSNO, 2009) to PROV-O, and discussed the differences between the meanings of a few classes. For example, in O&M, the Observation class is about an activity, in which there is a process (using sensor, instrument, algorithm) generating a result. In contrast, the Observation class in SSNO is about the output or record of an observation event.

The presented work is methodologically consistent with existing works on provenance in data preservation, curation and reuse (Ding, Michaelis, McCusker, & McGuinness, 2011; Mayernik et al., 2013;

Missier Sahoo, Zhao, Goble, & Sheth, 2010) and open software (De Nies et al., 2013; Ratnakar & Gil, 2015), though developed more specifically for the IPython Notebook environment and focusing more on the provenance of workflow. Davidson & Freire (2008) discussed four open problems for provenance in the context of workflows: (1) information management infrastructure, (2) provenance analytics and visualization, (3) interoperability, and (4) connecting database and workflow provenance. The work presented in this chapter partly address questions (3) and (4), and the Semantic Web environment addresses question (1). Future work can be conducted to address question (2). There is already existing work that can be used or adapted for these directions, such as the PROV-O-Viz (Hoekstra & Groth, 2015). Other studies in computer sciences are also examining the use of the PROV provenance standard with electronic notebooks such as the IPython Notebook (Wibisono et al., 2015). Similar, recent work on provenance capture in an online interactive workflow include the noWorkflow system built by Pimentel, Braganholo, Murta, & Freire (2015) that can automatically collect structured provenance for Python scripts in IPython Notebook, and analyze the provenance information, reason about, and debug the developed work. Narock and Yoon (2015) used a service provenance ontology to capture the execution details of workflow constituent web services. Realized in conjunction with a multi-agent system, their work is able to automate provenance aggregation and collation, and scale up to large number of workflows. The environment of the Semantic Web also enables the integration of knowledge for reliable use of technologies (Blythe & Dadi, 2012; Leadbetter, Lowry, & Clements, 2014). Together with well-documented provenance information, the work of data discovery, access and processing will be greatly facilitated.

CONCLUSION

Open Science is already changing the way people perform their science. Strong emphasis on the sharing of integrated ecosystem assessments among scientists, managers, stakeholders, and the public, inevitably relies upon oceanographic and marine sciences. Besides highlighting the need for accessibility to various resources involved in scientific research, this chapter addresses directly the provenance of figures, output datasets or indicators, and research findings that are prominent in modern assessment reports. To realize the objective of provenance documentation: verifiability, transparency and reproducibility (to name a few), a Semantic Web community standard for provenance representation and documentation, the PROV Ontology, was extended for use in capturing provenance in reports that are part of Large Marine Ecosystem assessments. The provenance representation is embedded in IPython Notebook as the platform for heterogenous science workflows. The developed representation: an ontology, was tested using several case studies in the preparation of materials for the Ecosystem Status Report for the U.S. Northeast Shelf Large Marine Ecosystem. The work indicated that the Semantic Web approach provides a robust and rigorous environment for integrated data and information management and exchange, that in the future is expected to become fundamental to the conduct of Open Science. Well-organized provenance information, through well-structured semantics, will facilitate the reproducibility of Large Marine Ecosystem assessments.

ACKNOWLEDGMENT

The authors thank M. Fogarty, K. Hyde, S. Lucey, G. DePiper, and other members of the Ecosystem Assessment Program at the NOAA Northeast Fisheries Science Center for case studies, context, access to data, testing, and evaluation of the notebooks presented herein. In particular, this work would not have been possible without the significant effort and collegiality of Jon Hare (NOAA NEFSC), and the authors extend special thanks to him. We also acknowledge A. Maffei for comments on the manuscript. Support for this research was provided by NSF #0955649 with additional support to SB by a USGS Cooperative Agreement and the Investment in Science Fund at WHOI.

REFERENCES

Angles, R., & Gutierrez, C. (2008). Survey of graph database models. *ACM Computing Surveys*, *40*(1), 1–39. doi:10.1145/1322432.1322433

Arko, R., Chandler, C., Stocks, K., Smith, S., Clark, P., Shepherd, A., . . . Beaulieu, S. (2013). Rolling Deck to Repository (R2R): Collaborative Development of Linked Data for Oceanographic Research. *Geophysical Research Abstracts, 15.*

Beaulieu, S., Fox, P., Maffei, A., Hare, J., Fogarty, M., & Di Stefano, M. (2013). *Traceability, reproducibility, and scalability in Integrated Ecosystem Assessments*. Oral presentation at ESIP Federation Summer Meeting 2013.

Berners-Lee, T. (2006). *Linked Data*. Accessible at: http://www.w3.org/DesignIssues/LinkedData.html

Berners-Lee, T., Hendler, J., & Lassila, O. (2001). The Semantic Web. *Scientific American*, *284*(5), 34–43. doi:10.1038/scientificamerican0501-34 PMID:11396337

Bizer, C., Heath, T., & Berners-Lee, T. (2009). Linked Data - The story so far. *International Journal on Semantic Web and Information Systems*, *5*(3), 1–22. doi:10.4018/jswis.2009081901

Blythe, J. N., & Dadi, U. (2012). Knowledge integration as a method to develop capacity for evaluating technical information on biodiversity and ocean currents for integrated coastal management. *Environmental Science & Policy*, *19-20*, 49–58. doi:10.1016/j.envsci.2012.01.007

Car, N. J. (2013). The eReefs Information Architecture. In J. Piantadosi, R. S. Anderssen, & J. Boland (Eds.), Proceedings of the 20th International Congress on Modelling and Simulation (pp. 831–837). Adelaide, Australia. Accessible at http://www.mssanz.org.au/modsim2013/C7/car2.pdf

Car, N. J., Stenson, M. P., & Hartcher, M. (2014). A Provenance Methodology And Architecture For Scientific Projects Containing Automated And Manual Processes. In *Proceedings of the 11th Conference on Hydroinformatics*. Accessible at: http://academicworks.cuny.edu/cc_conf_hic/57

Cox, S. J. D. (2016). (in press). Ontology for observations and sampling features, with alignments to existing models. *Semantic Web*. doi:10.3233/SW-160214

Davidson, S. B., & Freire, J. (2008). Provenance and scientific workflows: challenges and opportunities. In *Proceedings of the 2008 ACM SIGMOD international conference on Management of data*. doi:10.1145/1376616.1376772

De Nies, T., Magliacane, S., Verborgh, R., Coppens, S., Groth, P., Mannens, E., & Van de Walle, R. (2013). Git2PROV: exposing version control system content as W3C PROV. In *Posters & Demonstrations Track within the 12th International Semantic Web Conference* (ISWC-2013).

Di Stefano, M., Fox, P., Beaulieu, S., & Maffei, A. (2012). *The integrated ecosystems assessment initiative—enabling the assessment of impacts on large marine ecosystems: informatics to the forefront of science-based decision support*. Poster Presentation at 2012 ICES Annual Science Conference, Bergen, Norway. Accessible at: http://tw.rpi.edu/media/2012/10/08/cd52/ICES_2012.pdf

Di Stefano, M., Fu, L., West, P., Ma, X., Beaulieu, S., & Fox, P. (2014). *Development of cyberinfrastructure to facilitate collaboration and knowledge sharing for marine Integrated Ecosystem Assessments*. Poster Presentation at ESIP Summer Meeting 2014, Frisco, CO. Accessible at: http://tw.rpi.edu/web/doc/ECOOP-ESIPSummer2014-Poster

Ding, L., Michaelis, J., McCusker, J., & McGuinness, D. L. (2011). Linked provenance data: A semantic web-based approach to interoperable workflow traces. *Future Generation Computer Systems*, *27*(6), 797–805. doi:10.1016/j.future.2010.10.011

Diviacco, P., De Cauwer, K., Leadbetter, A., Sorribas, J., Stojanov, Y., Busato, A., & Cova, A. (2015). Bridging semantically different paradigms in the field of marine acquisition event logging. *Earth Science Informatics*, *8*(1), 135–146. doi:10.1007/s12145-014-0192-0

Diviacco, P., & Leadbetter, A. (2017). Balancing Formalization and Representation in Cross-Domain Data Management for Sustainable Development. In P. Diviacco, A. Leadbetter, & H. Glaves (Eds.), *Oceanographic and Marine Cross-Domain Data Management for Sustainable Development*. Hershey, PA: IGI Global.

Dublin Core. (2012). *DCMI Metadata Terms*. Retrieved from: http://dublincore.org/documents/dcmi-terms/

Ecosystem Assessment Program. (2009). *Ecosystem Assessment Report for the Northeast U.S.* Continental *Shelf Large Marine Ecosystem*. U.S. Department of Commerce, Northeast Fisheries Science Center Reference Document 09-11. Available online at http://www.nefsc.noaa.gov/publications/crd/crd0911/

Ecosystem Assessment Program. (2012). *Ecosystem Status Report for the Northeast Shelf Large Marine Ecosystem - 2011*. U.S. Department of Commerce, Northeast Fisheries Science Center Reference Document 12-07. Available online at http://nefsc.noaa.gov/publications/crd/crd1207/

Emile-Geay, J., & Eshleman, J. A. (2013). Toward a semantic web of paleoclimatology. *Geochemistry Geophysics Geosystems*, *14*(2), 457–469. doi:10.1002/ggge.20067

FOAF. (2014). *FOAF Vocabulary Specification 0.99*. Retrieved from: http://xmlns.com/foaf/spec/

Fox, P. (2014). *Facilitating Next Generation Science Collaboration: Marine Ecosystems Status Reports and Assessments*. Oral presentation at 2014 IMBER Open Science Conference, Bergen, Norway. Accessible at: http://tw.rpi.edu/web/doc/ECOOP_IMBER_D2_Fox_20140624

Gallant, S. N., Schmidt, R. K., & Car, N. J. (2015). Implementing a glossary and vocabulary service in an interdisciplinary environmental assessment for decision makers. In R. Denzer, M. Argent, G. Schimak, & J. Hrebícek (Eds.), Environmental Software Systems. Infrastructures, Services and Applications. Springer. doi:10.1007/978-3-319-15994-2_38

GCMD. (2016a). Retrieved from: http://gcmdservices.gsfc.nasa.gov/kms/concept/c5e1c055-768e-4aa3-a0a1-3adfda8ecdca?format=rdf

GCMD. (2016b). Retrieved from: http://gcmdservices.gsfc.nasa.gov/kms/concept/c7112a64-be39-414a-9125-f63ab44ecb5b?format=rdf

GCMD. (2016c). Retrieved from: http://gcmdservices.gsfc.nasa.gov/kms/concept/bd46a0bf-5c06-48af-a6c9-022417b1fffd?format=rdf

GCMD. (2016d). Retrieved from: http://gcmdservices.gsfc.nasa.gov/kms/concept/76943142-e5a9-4ecf-b496-050dd3d97101?format=rdf

GCMD. (2016e). Retrieved from: http://gcmdservices.gsfc.nasa.gov/kms/concept/78f5a84f-1b5b-44a9-97e7-4a1996cd2e36?format=rdf

GCMD. (2016f). Retrieved from: http://gcmdservices.gsfc.nasa.gov/kms/concept/bd24a9a9-7d52-4c29-b2a0-6cefd216ae78?format=rdf

GCMD. (2016g). Retrieved from: http://gcmdservices.gsfc.nasa.gov/kms/concept/ecd03762-df34-49b7-91f2-d8a51acd270e?format=rdf

Gil, Y. (2014). Intelligent Workflow Systems and Provenance-Aware Software. In *Proceedings of the Seventh International Congress on Environmental Modeling and Software*.

Gil, Y., Michel, F., Ratnakar, V., Hauder, M., Duffy, C., Dugan, H., & Hanson, P. (2015) A task-centered framework for computationally-grounded science collaborations. In *Proceedings of the Eleventh IEEE International Conference on e-Science*. doi:10.1109/eScience.2015.76

Glaves, H. (2017). Developing a Common Global Framework for Marine Data Management. In P. Diviacco, A. Leadbetter, & H. Glaves (Eds.), *Oceanographic and Marine Cross-Domain Data Management for Sustainable Development*. Hershey, PA: IGI Global.

Goecks, J., Nekrutenko, A., Taylor, J., & Team, T. G. (2010). Galaxy: A comprehensive approach for supporting accessible, reproducible, and transparent computational research in the life sciences. *Genome Biology*, *11*(8), R86. doi:10.1186/gb-2010-11-8-r86 PMID:20738864

Gruber, T. R. (1995). Toward principles for the design of ontologies used for knowledge sharing. *International Journal of Human-Computer Studies*, *43*(5-6), 907–928. doi:10.1006/ijhc.1995.1081

Guarino, N. (1997). Understanding, building and using ontologies. *International Journal of Human-Computer Studies*, *46*(2-3), 293–310. doi:10.1006/ijhc.1996.0091

Halpern, B. S., Longo, C., Hardy, D., McLeod, K. L., Samhouri, J. F., Katona, S. K., & Zeller, D. et al. (2012). An index to assess the health and benefits of the global ocean. *Nature*, *488*(7413), 615–620. doi:10.1038/nature11397 PMID:22895186

Halpern, B. S., Longo, C., Scarborough, C., Hardy, D., Best, B. D., Doney, S. C., & Samhouri, J. F. et al. (2014). Assessing the Health of the U.S. West Coast with a Regional-Scale Application of the Ocean Health Index. *PLoS ONE*, *9*(6), e98995. doi:10.1371/journal.pone.0098995 PMID:24941007

Harnad, S., & Brody, T. (2004). Comparing the impact of open access (OA) vs. non-OA articles in the same journals. *D-Lib Magazine*, *10*(6). doi:10.1045/june2004-harnad

Hey, T., & Payne, M. C. (2015). Open science decoded. *Nature Physics*, *11*(5), 367–369. doi:10.1038/nphys3313

Hoekstra, R., & Groth, R. (2015). PROV-O-Viz - Understanding the role of activities in provenance. In B. Ludäscher & B. Plale (Eds.), *Provenance and Annotation of Data and Processes* (pp. 215–220). Springer. doi:10.1007/978-3-319-16462-5_18

Howe, B., Lawson, P., Bellinger, R., Anderson, E., Santos, E., Freire, J., & Silva, C. et al. (2008). End-to-End eScience: Integrating Workflow, Query, Visualization, and Provenance at an Ocean Observatory. In *Proceedings of the IEEE Fourth International Conference on eScience*. doi:10.1109/eScience.2008.67

IDL. (2016). Retrieved from: http://www.exelisvis.com/IDL

Infrastructure for Spatial Information in the European Community. (2013). Retrieved from: http://inspire.ec.europa.eu/index.cfm/pageid/241/documentid/3107

Integrated Ocean Observing System (IooS) Implements OGC Standards. (2008). *Another U.S. Government Agency Chooses OGC Standards*. Retrieved from: http://xml.coverpages.org/IOOS-OGC.html

International Organization for Standardization (ISO). (2003). *ISO 19115: Geographic Information — Metadata*. Geneva: ISO.

International Programme on the State of the Ocean. (2016). Retrieved from: http://www.stateoftheocean.org

IPython. (2016). *IPython.Jupyter kernals*. Retrieved from: https://github.com/ipython/ipython/wiki/IPython-kernels-for-other-languages

Tetherless World Constellation (TWC). (2014). *ECO-OP*. Retrieved from: http://tw.rpi.edu/web/project/ECOOP

ISO. (2009).*ISO 19115-2: Geographic Information — Metadata — Part 2: Extensions for Imagery and Gridded Data*. ISO.

JSON for Linking Data. (n.d.). Retrieved from: http://json-ld.org/

Leadbetter, A., Cheatham, M., Shepherd, A., & Thomas, R. (2017). Linked Ocean Data 2.0. In P. Diviacco, A. Leadbetter, & H. Glaves (Eds.), *Oceanographic and Marine Cross-Domain Data Management for Sustainable Development*. Hershey, PA: IGI Global.

Leadbetter, A., Lowry, R., & Clements, D. O. (2012). The NERC Vocabulary Server: Version 2.0. In *Geophysical Research Abstracts 14,Proceedings EGU General Assembly 2012*.

Leadbetter, A. M. (2015). Linked Ocean Data. In T. Narock & P. Fox (Eds.), *The Semantic Web in Earth and Space Science: Current Status and Future Directions* (pp. 11–31). Berlin: IOS Press.

Leadbetter, A. M., Lowry, R. K., & Clements, D. O. (2014). Putting meaning into NETMAR - the open service network for marine environmental data. *International Journal of Digital Earth*, *7*(10), 811–828. doi:10.1080/17538947.2013.781243

Lebo, T., Sahoo, S., & McGuinness, D. (2013). *PROV-O: The PROV Ontology*. Accessible at: http://www.w3.org/TR/prov-o/

Lehnert, K. A., Vinayagamoorthy, S., Djapic, B., & Klump, J. (2006). *The digital sample: metadata, unique identification, and links to data and publications*. American Geophysical Union Fall Meeting 2006, San Francisco, CA.

Levin, P. S., Fogarty, M. J., Murawski, S. A., & Fluharty, D. (2009). Integrated Ecosystem Assessments: Developing the Scientific Basis for Ecosystem-Based Management of the Ocean. *PLoS Biology*, *7*(1), e1000014. doi:10.1371/journal.pbio.1000014 PMID:19166267

Levin, P. S., Kelble, C. R., Shuford, R. L., Ainsworth, C., deReynier, Y., Dunsmore, R., & Werner, F. et al. (2014). Guidance for implementation of integrated ecosystem assessments: A US perspective. *ICES Journal of Marine Science*, *71*(5), 1198–1204. doi:10.1093/icesjms/fst112

Lightsom, F. L., Cicchetti, G., & Wahle, C. M. (2015). *Data Categories for Marine Planning*. USGS Open-File Report 2015-1046. USGS. doi:10.3133/ofr20151046

Ma, X., Fox, P., Tilmes, C., Jacobs, K., & Waple, A. (2014a). Capturing provenance of global change information. *Nature Climate Change*, *4*(6), 409–413. doi:10.1038/nclimate2141

Ma, X., Zheng, J. G., Goldstein, J., Zednik, S., Fu, L., Duggan, B., & Fox, P. et al. (2014b). Ontology engineering in provenance enablement for the National Climate Assessment. *Environmental Modelling & Software*, *61*, 191–205. doi:10.1016/j.envsoft.2014.08.002

Mayernik, M., DiLauro, T., Duerr, R., Metsger, E., Thessen, A. E., & Choudhury, G. S. (2013). Data conservancy provenance, context, and lineage services: Key components for data preservation and curation. *Data Science Journal*, *12*(0), 158–171. doi:10.2481/dsj.12-039

McGuinness, D. L. (2003). Ontologies come of age. In D. Fensel, J. Hendler, H. Lieberman, & W. Wahlster (Eds.), *Spinning the Semantic Web: Bringing the World Wide Web to Its Full Potential* (pp. 171–196). Cambridge, MA: MIT Press.

Missier, P., Sahoo, S., Zhao, J., Goble, C., & Sheth, A. (2010). Janus: From Workflows to Semantic Provenance and Linked Open Data. In D. McGuinness, J. Michaelis, & L. Moreau (Eds.), *Provenance and Annotation of Data and Processes* (pp. 129–141). Berlin: Springer. doi:10.1007/978-3-642-17819-1_16

Moreau, L. (2011). Provenance-based reproducibility in the Semantic Web. *Web Semantics: Science, Services, and Agents on the World Wide Web*, *9*(2), 202–221. doi:10.1016/j.websem.2011.03.001

Moreau, L., Clifford, B., Freire, J., Futrelle, J., Gil, Y., Groth, P., & Van den Bussche, J. et al. (2011). The Open Provenance Model core specification (v1.1). *Future Generation Computer Systems*, *27*(6), 743–756.

Narock, T. W., & Yoon, V. (2015). An Agent-Based Approach for Capturing and Linking Provenance in Geoscience Workflows. *Computers & Geosciences*, *79*, 58–68. doi:10.1016/j.cageo.2015.03.004

NASA. (2016). *Global Change Master Directory*. Retrieved from: http://gcmd.nasa.gov/

NEFSC. (2008). *Examples of Management Jurisdictions of the Northeast Shelf Ecosystem*. Retrieved from: http://www.nefsc.noaa.gov/publications/tm/tm208/images/f7.gif

NERC. (2003). Retrieved from: http://vocab.nerc.ac.uk/collection/P02/current/PPRD/

NERC. (2004). Retrieved from: http://vocab.nerc.ac.uk/collection/P02/current/PSST/

NERC. (2006). Retrieved from: http://vocab.nerc.ac.uk/collection/P02/current/R410/

NERC. (2010). Retrieved from: http://vocab.nerc.ac.uk/collection/P02/current/FCST/

NOAA Environmental Data Management Committee. (2011). *NOAA Data Documentation Procedural Directive Version 1.0*. Available online: https://nosc.noaa.gov/EDMC/PD.DD.php

Nosek, B. A., Alter, G., Banks, G. C., Borsboom, D., Bowman, S. D., Breckler, S. J., & Yarkoni, T. et al. (2015). Promoti.ng an open research culture. *Science*, *348*(6242), 1422–1425. doi:10.1126/science.aab2374 PMID:26113702

Obama, B. (2013). *Making Open and Machine Readable the New Default for Government Information*. US Government Executive Order, The White House. 2013. Accessible at: http://www.whitehouse.gov/the-press-office/2013/05/09/executive-order-making-open-and-machine-readable-new-default-government

OGC. (2016). *Ocean Science Interoperability Experiment*. Retrieved from: http://www.opengeospatial.org/projects/initiatives/oceansie

Open Science Collaboration. (2012). An open, large-scale, collaborative effort to estimate the reproducibility of psychological science. *Perspectives on Psychological Science*, *7*(6), 657–660. doi:10.1177/1745691612462588 PMID:26168127

Palmer, M. A. (2012). Socioenvironmental Sustainability and Actionable Science. *Bioscience*, *62*(1), 5–6. doi:10.1525/bio.2012.62.1.2

Pimentel, J. F. N., Braganholo, V., Murta, L., & Freire, J. (2015). Collecting and analyzing provenance on interactive notebooks: when IPython meets noWorkflow. In *Proceedings of the 7th USENIX Conference on Theory and Practice of Provenance (TaPP'15)*. USENIX Association.

Pinheiro da Silva, P., McGuinness, D. L., & Fikes, R. (2006). A proof markup language for Semantic Web services. *Information Systems*, *31*(4-5), 381–395. doi:10.1016/j.is.2005.02.003

QUDT. (2014). *QUDT – Quanitites, Units, Dimensions and Data Types Ontologies*. Retrieved from: http://www.qudt.org/

Rajbhandari, S., & Keizer, J. (2012). The AGROVOC Concept Scheme – A Walkthrough. *Journal of Integrative Agriculture*, *11*(5), 694–699. doi:10.1016/S2095-3119(12)60058-6

Ramdeen, S., & Hills, D. J. (2013). *ESIP's Emerging Provenance and Context Content Standard Use Cases: Developing Examples and Models for Data Stewardship*. American Geophysical Union Fall Meeting 2013, San Francisco, CA.

Ratnakar, V., & Gil, Y. (2015). *OntoSoft ontology*. Accessible at: http://ontosoft.org/ontology/software/

Reichman, O. J., Jones, M. B., & Schildhauer, M. P. (2011). Science Challenges and Opportunities of Open Data in Ecology. *Science*, *331*(6018), 703–705. doi:10.1126/science.1197962 PMID:21311007

Rossant, C. (2013). *Learning IPython for Interactive Computing and Data Visualization*. Birmingham, UK: Packt Publishing Ltd.

Sahoo, S. S., Barga, R., Sheth, A. P., Thirunarayan, K., & Hitzler, P. (2009). *PrOM: A Semantic Web Framework for Provenance Management in Science*. Kno.e.sis Center Technical Report KNOESIS-TR-2009. Accessible at: http://corescholar.libraries.wright.edu/knoesis/445

SeaDas. (2016). Retrieved from: http://seadas.gsfc.nasa.gov

Semantic Sensor Network Ontology (SSNO). (2009). Retrieved from: https://www.w3.org/2005/Incubator/ssn/ssnx/ssn

Shen, H. (2014). Interactive notebooks: Sharing the code. *Nature*, *515*(7525), 151–152. doi:10.1038/515151a PMID:25373681

Sherman, K. (1991). The Large Marine Ecosystem Concept: Research and Management Strategy for Living Marine Resources. *Ecological Applications*, *1*(4), 350–360. doi:10.2307/1941896

Shneiderman, B. (2008). Science 2.0. *Science*, *319*(5868), 1349–1350. doi:10.1126/science.1153539 PMID:18323442

Simmhan, Y. L., Plale, B., & Gannon, D. (2006). A framework for collecting provenance in data-centric scientific workflows. In *Proceedings of ICWS'06- IEEE International Conference on Web Services*. doi:10.1109/ICWS.2006.5

Skosmos. (2014). *Multispecies Fisheries*. Retrieved from: http://aims.fao.org/skosmos/agrovoc/en/page/c_4987

Skosmos. (2015). *Trophic Levels*. Retrieved from: http://aims.fao.org/skosmos/agrovoc/en/page/c_7969

Uhlir, P. F., & Schröder, P. (2007). Open data for global science. *Data Science Journal*, *6*, OD36–OD53. doi:10.2481/dsj.6.OD36

United Nations World Ocean Assessment. (2013). Retrieved from: http://www.worldoceanassessment.org/

U.S. Government Accountability Office. (2011). *Key Indicator Systems: Experiences of Other National and Subnational Systems Offer Insights for the United States*. Report to Congressional Addressees GAO-11-396. Available online: http://www.gao.gov/products/GAO-11-396

W3C. (2004). *SKOS Simple Knowledge Organization System*. Retrieved from: http://www.w3.org/2004/02/skos/

W3C. (2013). *PROV-O: The Prov Ontology*. Retrieved from: http://www.w3.org/TR/prov-o/#Collection

W3C. (2014a). *RDF Schema 1.1. W3C Recommendation*. Retrieved from: http://www.w3.org/TR/rdf-schema/

W3C. (2014b). *Data Catalog Vocabulary (DCAT)*. Retrieved from: http://www.w3.org/TR/vocab-dcat/

W3C. (2014c). *RDF 1.1 Turtle*. Retrieved from: http://www.w3.org/TR/turtle/

Wibisono, A., Bloem, P., de Vries, G. K. D., Groth, P., Belloum, A., & Bubak, M. (2015). Generating Scientific Documentation for Computational Experiments Using Provenance. In B. Ludäscher & B. Plale (Eds.), *Provenance and Annotation of Data and Processes* (pp. 168–179). Springer. doi:10.1007/978-3-319-16462-5_13

Yaffe, M. B. (2015). Reproducibility in science. *Science Signaling*, *8*(371), eg5–eg5. doi:10.1126/scisignal.aaa5764 PMID:25852185

Yue, P., Gong, J., & Di, L. (2010). Augmenting geospatial data provenance through metadata tracking in geospatial service chaining. *Computers & Geosciences*, *36*(3), 270–281. doi:10.1016/j.cageo.2009.09.002

KEY TERMS AND DEFINITIONS

Integrated Ecosystem Assessment (IEA): An assessment of a coupled natural-human ecosystem in relation to management objectives, involving integration and analysis of information on natural and socioeconomic factors. The concept of IEA for marine ecosystems was proposed by Phillip S. Levin and colleagues in the late 2000s.

Large Marine Ecosystem (LME): A large area of ocean about 200,000 km^2 or greater, adjacent to the continents where primary productivity is generally higher than in open ocean areas. The U.S. NOAA recognizes 64 LMEs globally.

Ontology: In computer science, an ontology is the formal specification of a shared conceptualization of a domain.

Open Science: The movement to make the process of scientific research transparent and accessible. It covers various topics such as Open Access, Open Data, Open Source (of software program), Open Collection (of samples), Transparent Workflow, and more.

Provenance: The origin of something. In scientific works, provenance documentation means linking a range of observations and model outputs, research activities, people and organizations involved in the production of a scientific finding with the supporting data sets and methods used to generate it.

Reproducibility: The capacity of a scientific work to be duplicated with consistent results.

Semantic Web: An extension to the existing World Wide Web by adding machine-readable meanings to contents on the Web. It is called a 'Web of Data' while the classic Web is called a 'Web of Documents'.

Workflow: A sequence of steps or processes through which a work passes from initiation to completion.

APPENDIX

Table 2. Acronyms used in this chapter

Acronym	Full Name
DCAT	Data Catalog Vocabulary
DCMI	Dublin Core Metadata Initiative
EcoAP	NEFSC Ecosystem Assessment Program
ECO-OP	ECOsystem interOPerability
ESR	Ecosystem Status Report
FOAF	Friend Of A Friend vocabulary
GCIS	Global Change Information System
GCMD	NASA Global Change Master Directory
IDL	Interactive Data Language
ISO	International Organization for Standardization
JSON	JavaScript Object Notation
JSON-LD	JSON for Linking Data
LME	Large Marine Ecosystem
NEFSC	U.S. NOAA Northeast Fisheries Science Center
NOAA	U.S. National Oceanic and Atmospheric Administration
O&M	OGC Observations and Measurements
OGC	Open Geospatial Consortium
OPM	Open Provenance Model
OWL	Web Ontology Language
PML	Proof Markup Language
PROMS	Provenance Management System
prov-ecoop	ECO-OP Provenance Capture Ontology
PROV-O	W3C PROV Ontology
QUDT	Quantities, Units, Dimensions and Types
R2R	Rolling Deck to Repository
RDF	Resource Description Framework
RDFS	RDF Schema
SKOS	Simple Knowledge Organization System
SPARQL	SPARQL Protocol and RDF Query Language
URI	Uniform Resource Identifier
W3C	World Wide Web Consortium

Chapter 6
Semantic Search Engine for Data Management and Sustainable Development:
Marine Planning Service Platform

Giuseppe Manzella
ETT SpA, Italy

Roberto Bartolini
CNR ILC, Italy

Franco Bustaffa
Delta Progetti, Italy

Paolo D'Angelo
ETT SpA, Italy

Maurizio De Mattei
Delta Progetti, Italy

Francesca Frontini
CNR ILC, Italy

Maurizio Maltese
Delta Progetti, Italy

Daniele Medone
Delta Progetti, Italy

Monica Monachini
CNR ILC, Italy

Antonio Novellino
ETT SpA, Italy

Andrea Spada
SyO srl, Italy

ABSTRACT

This chapter presents a computer platform supporting a Marine Information and Knowledge System based on a repository that gathers, classify and structures marine scientific literature and data, guaranteeing their accessibility by means of standard protocols. This requires the access to quality controlled data and to information that is provided in grey literature and/or in relevant scientific literature. There exist efforts to develop search engines to find author's contributions to scientific literature or publications. This implies the use of persistent identifiers. However very few efforts are dedicated to link publications to data that was used, or cited in them or that can be of importance for the published studies. Full-text technologies are often unsuccessful since they assume the presence of specific keywords in the text; to fix this problem, it is suggested to use different semantic technologies for retrieving the text and data and thus getting much more complying results.

DOI: 10.4018/978-1-5225-0700-0.ch006

INTRODUCTION

The value of marine environmental data is very high for the cost inherent to data collection, as well as for the knowledge that can be extracted from them. It is estimated that the EU Member States spend approximately € 1.85 billion a year on marine research (European Commission, 2010). About half is on infrastructures for facilitating observation. This includes ships, underwater observatories, floating buoys, drifting devices, remotely operated or autonomous underwater vehicles, as well as many other platforms, all equipped with a range of sensors and analytical capabilities. Unfortunately, a significant amount of data has been lost and is being lost for many reasons.

The NOAA National Data Buoy Centre reported that about 15-20% of data are lost for vandalism to buoys. However, the major quantity of data have been lost for the changes in recording technologies, that have affected the data integrity and rescue. Before electronic computers came into general use, oceanographic data were recorded in manuscripts, data reports, and card index files. With the advent of electronic data storage, oceanographic observations were increasingly recorded on magnetic media such as tapes and disks. Unfortunately, all these media are subject to degradation over time with subsequent loss of unique data. This has occurred in some cases, but unfortunately technology turnover is not the only reason for data loss. There are still a lot of researchers that are not ‘publishing’ their data in data centre systems, and with the result that data is lost when the researcher retires. Unfortunately large amounts of research funds are spent every year, while already existing data remain underutilised.

It has been underlined that data from the marine environment is a valuable asset; use and re-use can address threats to the marine environment, and can be used for the development of policies and legislation to protect vulnerable areas of our coasts and oceans, in understanding trends and in forecasting future changes. More in general, better quality and more easily accessible marine data can support the ‘blue growth’ or, in other words, the further sustainable economic development.

Data is analysed, synthesised, interpreted and transformed into information, and, as a final step, can produce knowledge. The outcome of this process is published as a scientific article. The intangible value of the data has pushed public authorities and organizations to encourage free and open access to data. In 2003 the ‘Berlin Declaration on Access to Knowledge in the Sciences and Humanities’ was published in order to “promote the Internet as a functional instrument for a global scientific knowledge base and human reflection and to specify measures which research policy makers, research institutions, funding agencies, libraries, archives and museums need to consider.” (UNESCO, 2013). To make data usable in a tangible way it is necessary to accompany data with documentation, i.e. protocols, reports, grey literature published papers.

There are many reasons limiting the open and free access to data and documents, among which the Intellectual Property Rights and Copyrights. The idea of universal access to research, education, and culture is made possible by the Internet, but existing legal and social systems don’t always allow that idea to be realized. To achieve the vision of universal access the Creative Commons (Clarke, 2001) is trying to create a balance between the reality of the Internet and the reality of copyright laws (Clarke, 2005).

The Marine Planning and Service Platform (MAPS) project started from a schematic depiction of the flow from research to library resources that is interlinking documentation and their underlying data. Within the project ‘documentation’ intend protocols and reports, as well as grey literature and papers published in conventional scientific journals. MAPS has developed a web search engine where the information retrieval is obtained from metadata and full text indexing and the information allow to select the underlying data from a database. With the existing information systems it is easy to connect the data with publications, and provide information on technologies used for data acquisition, laboratory analysis tools, protocols (tools and technologies being, in a wider sense, part of those cultural artefacts developed in a certain historical period by scientists.

BACKGROUND

The accessibility to information, management of knowledge and its dissemination to the public, has a long history. However, there were, and still there are, many weak points in the transmission of knowledge among scientific communities and from scientific communities to the public.

In 1945 Wannever pointed out: 'Professionally our methods of transmitting and reviewing the results of research are generations old and by now are totally inadequate for their purpose. If the aggregate time spent in writing scholarly works and in reading them could be evaluated, the ratio between these amounts of time might well be startling. ... Mendel's concept of the laws of genetics was lost to the world for a generation because his publication did not reach the few who were capable of grasping and extending it; and this sort of catastrophe is undoubtedly being repeated all about us, as truly significant attainments become lost in the mass of the inconsequential.' The main statement of Wannever was essentially containing the idea behind the MAPS project: 'A record if it is to be useful to science, must be continuously extended, it must be stored, and above all it must be consulted.'

Many authors have underlined that knowledge can be tacit or explicit (e.g. Polanyi, 1966, Brown & Duguid 1998). The tacit knowledge is sometimes referred as 'know-how', while the explicit knowledge is sometimes referred as know-what (Brown & Duguid 1998). The former refers to codified knowledge, such as that found in documents, while the latter refers mainly to personal/experience-based knowledge.

In the MAPS concepts, tacit knowledge is part of the cultural environment of a person and the belonging to a community of practice. Although it is difficult to transfer to another person the tacit knowledge by means of (e.g.) writing it down, it is possible to built a knowledge management system were different types of knowledge can be discussed in some way. The personal contacts, people interactions and social media are means to transfer knowledge. Manzella and Manzella (2015) have carried out an experiment in a classroom with students having different cultural background (biologist, engineers, physicist, geologist), each having initial different personal beliefs deriving from their specific studies. They have been guided towards publications and to work with data, giving their own interpretation. Discussions on the different ideas allowed clarification of issues, reduction of uncertainties and shared understanding. However, it is a belief of the corresponding author of this chapter, that the tacit knowledge cannot be entirely transmitted. Any other person will elaborate the experiences of others and, after a knowledge building process, acquire intimately concepts, but in different forms.

The technological developments linked to computers and internet have offered the necessary tools for recording data, publications and any cultural artefacts and make them available to any user. The interlink between documentation and data needs two main components: an information system managing data and metadata and a semantic search system that extract structured information from a text. MAPS started from the consideration that a further development of a knowledge building system can support the social dimension of the sustainable development, providing tools to 'raise the level of understanding and commitment to action on the part of individuals, ... organisations ...' (WCED, 1987).

Information Retrieval and Semantic Search

Information retrieval is the activity of obtaining user needed information from a collection of information resources. The idea of using computers to search for relevant pieces of information has practically started from the end of the second world war. In 1992 there was the first Text Retrieval Conference that catalyzed research on methods that scale to huge corpora. The introduction of web search engines has boosted the need for very large scale retrieval systems even further. A web search engine is a software system that is designed to search for information on the World Wide Web. The information may be a mix of web pages, images, and other types of files. Some search engines also mine data available in databases or open directories. Unlike web directories, which are maintained only by human editors, search engines also maintain real-time information by running an algorithm on a web crawler. The currently popular search systems, including those implemented by common search engines, are mainly based on matching of strings: the words of the query are treated as a set of key terms whose presence or absence is sought in previously indexed texts (e.g. Brin, 1998; Kogut & Holmes, 2001).

Semantic search systems are slowly spreading amongst the generalist search engines. They consider context of search, location, intent, variation of words, synonyms, generalized and specialized queries, concept matching and natural language queries to provide relevant search results. In general, a preliminary analysis of the texts in question by means of existing technologies relying on lexical and conceptual generic resources is always necessary for the realisation of a statistical semantic search system in a specific domain. Later on, the results of this analysis – possibly manually corrected - are used for extracting new knowledge through specific terminology that will enrich the lexical resources: in this way these resources will act as a training set for the linguistic analysis systems. In the environmental science, a semantic search system was developed in the EU – FP7 project KYOTO (Knowledge Yielding Ontologies for Transition-based Organization), had the goal of providing a system for semantic search in order to allow expert users to model and improve their domain ontology with terms and concepts automatically extracted (Vossen et al, 2008).

Text Mining

The term Text Mining (TM) are all those semantic technologies that aim to extract structured information from unstructured data (Rajman & Besançon, 1998).

The main applications of Text Mining and Information Retrieval technologies are the search systems that allow to retrieve texts or portions of text contained in a document base through a string of natural language search, carried out by the user with a simple query.

Research systems commonly in use, including those implemented by common search engines, are based primarily on matching strings: the words of the query are treated as a set of key terms whose presence or absence is sought in the texts previously indexed.

There are several statistical algorithms to determine the similarity of the query text and one contained in the documents (Lott, 2012). It is important to note, however, that such algorithms do not make a semantic analysis of the content of the query and even the content of the documents under consideration. Such systems therefore have many problems given the well-known ambiguity of natural language.

For this reason, the research in this field is pushed toward systems that are able to understand and structure, at least in part, semantic information contained in the text. In semantic search, words are not regarded as mere strings, but as a unit with meaning that can be reported by the context in which they are located; this allows to identify the concepts and semantic relationships that are expressed in the words of the query and go and see if these concepts or relationships (or concepts / relationships like) are present in the texts in question. Simultaneously, the system allows to ignore and discard only superficially different concepts that are represented by the same strings. To operate the system must be able to make the semantic disambiguation of queries. In this context it is essential to have access to knowledge bases that identify related concepts.

If more simple research systems semantics are limited to recognize and manage homonymy and synonymy, the most sophisticated one want to extract complex levels of information, such as (i.e.) 'events'. For the extraction of such levels of information is necessary to recover the semantic relationships between terms, from grammatical relations, the events, the participants in the action and their semantic roles.

Information System

During the last decades important marine data infrastructures have been developed at international level. Good examples are provided by the IOC-UNESCO IODE Ocean Data Portal (Belov & Mikhailov, 2012), the Australian Integrated Marine Observing System (Proctor et al, 2010), the European SeaDataNet (Maillard et al, 2007), the US NOAA National Oceanographic Data Centre (Boyer et al., 2013) and many others.

These information systems have the primary goal to provide transparent access to marine data sets and data products. They have developed standardized distributed systems for managing large and diverse datasets (i.e. temperature, salinity current, sea level, and many other chemical, physical and biological properties) collected by different sensors on different platforms. ISO and OGC compliant discovery services and metadata models have been adopted in order to assure interoperability.

To make scientific web portals effective elements of knowledge management, the information systems should be linked to the semantic search systems (and vice versa) by providing the necessary information on geographical and temporal coverage, parameters. This is the goal of MAPS project.

SPECIFIC ONTOLOGY BASED SEMANTIC SEARCH ENGINE

Behind the MAPS project, there is the idea of implementing a 'knowledge management' system as a basic element for the wider objective of participation to societal advances in science, technology and sustainable development. The knowledge management system is based on two main components:

1. knowledge building environment - a continuous learning practice that allows the presentation of theories, the comparison with observations, understanding and resolution of controversies
2. service and planning platform - an open and free access to data/products and information related to the data acquisition (including environmental conditions), instrumentation, protocols for in situ and ex situ practices used for quality assessment and quality control.

The knowledge management model proposed is represented in Figure 1, where the knowledge building element is strongly linked to the information system, that is providing access to data and products, and must include also links to documents.

Figure 1. MAPS concept showing the interlink between documentation and data

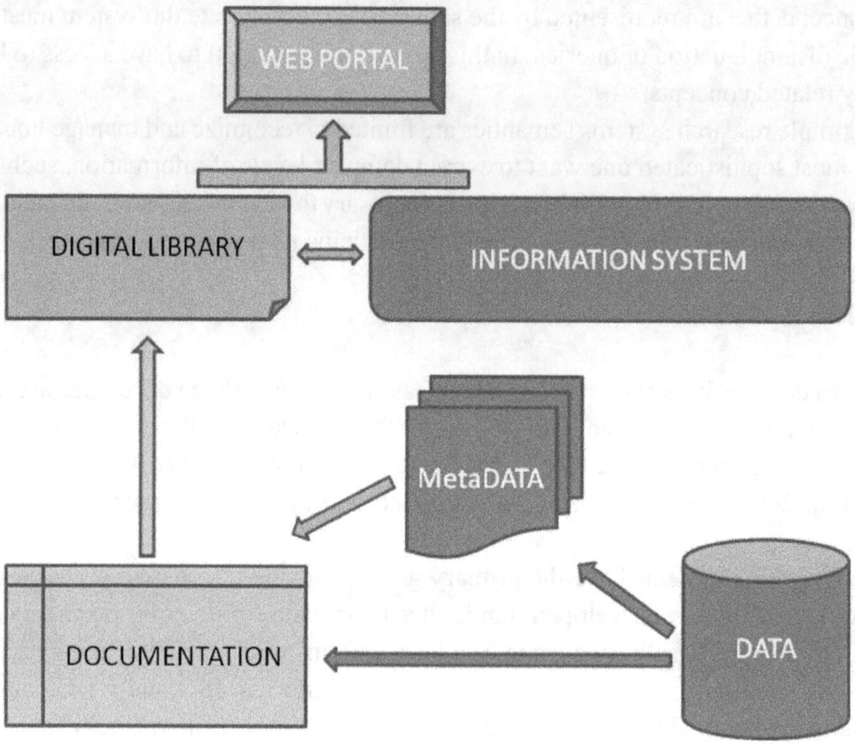

A portal managing at the same time both data and documentation allows any user to find and assess scientifically credible information about the (e.g.) the marine environment.

In general, the realization of a semantic search system in a specific domain requires a first analysis of texts based on technologies that rely on conceptual generic lexical resources. This first analysis is used to extract new knowledge through specific terminology: in this way there will be an adjustment of the systems of linguistic analysis to lexical resources. This 'linguistic annotation' is added to the text information that aim to make explicit the implicit structured information in the document. The annotation can be done manually, but generally for large amounts of texts exist linguistic analysis tools that automatically write down the various levels that are often dependent on each other, as some are preparatory to the identification of others.

The tools of linguistic analysis can be 'language specific', if only one language is used, or language independent, if more than one languages are used. Typically the tools that work on more than one language extract information of the highest level and lean annotation lower levels (for example morphosyntactical). Without going into detail, the automatic language analysis tools fall into two broad categories:

1. Tools that implement in their code explicit rules ("rules");

2. Statistical tools.

Statistical tools that implement algorithms in a supervised machine learning (with a corpus of training annotated by hand) or unsupervised (inductively) can "learn" to recognize the regularities in the text. For a state of the art tools on linguistic see among others Parra-Arnau et al 2010. For the importance of record in semantic search see also Uren et al., 2006.

The different levels of linguistic annotations are:

- **Language Detection:** The process of association of a generic text to the language in which it was written. Typically it is based on the calculation of the number of stop words (prepositions, article, words from classes closed) within the text.
- **Tokenization:** The process of dividing a text in minimum units of analysis called tokens, eg words, punctuation, dates, initials etc. The process of Tokenization can be very complex for some languages and relatively simpler for others, for example in Italian, thanks to the presence of spaces and stop symbols, the process is quite easy.
- **Lemmatisation:** The process of reducing the inflected form of a word to its canonical form (unmarked), called lemma. The standardized form is the gateway to the language vocabulary. The presence of linguistic ambiguity means that the association is typically one-to-many.
- **Analysis Morphology and/or Part-of-Speech Tagging:** This consists in associating with each word of the text to its Part-of-Speech and possibly other morphological features such as the kind, number, mode and time for verbs. If first phase lemmatisation the task is reduced to morphological disambiguation. Generally the approach to this task and statistical or methods are used supervised learning machine learning which allow accuracies around 95%.
- **Shallow Parsing:** The subdivision of a text phrases for analyzing morphosyntactic, or in blocks nominal, prepositional, verbal etc. (If these elements are flat, without presenting a branched structure, non-recursive, they are called chunks) and represent a linguistic unit in large granularity: for example a nominal chunk consists of the name preceded by 'any article, prepositional chunks are introduced by preposition more head carrier of lexical meaning.
- **Deep Parsing:** With this kind of deep analysis means the task of identifying dependencies between the functional portions of text, such as dependencies "be subject" or "be" a verb, than the phase of surface analysis that implicitly identified only relationships between words.
- **Named Entity Recognition and Classification (NERC):** Consists in 'assign a semantic category in a sequence of words close, perceived as a concrete entity par. The state of the NERC currently allows the identification and classification of high-level categories such as people, locations, organizations, and artifacts.
- **Coreference Recognition (COREF):** binds to the Named Entity Recognition (NER) since it allows to recognize portions of text that coreference to the same entity, for example pronouns or other; in this way it is possible to recover property or event in which such entity is also involved because it is not named in full.
- **Word Sense Disambiguation (WSD):** This consists in 'assignment of a proper contextual meaning (sense) to a particular word in the text or speech. Many international research groups are working to WSD, using a variety of approaches. However, to date, there is no record of accurate systems of WSD with broad coverage [28]. Currently the accuracy achieved for English texts with trained systems on a limited number of words is 60-70% (due to the large effort required for the

manual annotation of examples). The WSD is still one of the most important open problems in NLP.

- **Semantic Role Labeling (SRL):** The assignment of the appropriate relationship between a predicate and its syntactic constituents. Typical topics include semantic agent, patient, instrument, subject added, complementary rental, time, mode, etc. cause. Recognize and label semantic arguments is a fundamental task to answer the following questions: "Who", "When", "What", "Where", "Why", as they are fundamental issues to be solved in the areas of Information Extraction (IE), Question Answering (QA) and Text Summarization (TS).

Currently, there are a number of tools for the analysis of natural language, but given the fragmentation of the methodologies and technologies is impossible to refer to a single source for a complete list; as regards the English and other European languages:

- **TreeTagger:** http://www.cis.uni-muenchen.de/~schmid/tools/TreeTagger/_
- **Freeling:** http://nlp.lsi.upc.edu/freeling/_
- **openNLP:** https://opennlp.apache.org/index.html_
- **Stanford CoreNLP:** http://nlp.stanford.edu/software/corenlp.shtm_
- **Morph-it:** http://dev.sslmit.unibo.it/linguistics/morph-it.php_

To evaluate a tool, the scientific community provides 'gold standard' (benchmarks), manually annotated corpora to calculate precision / recall and accuracy. In addition to accuracy it is also important usability of the instrument, the possibility of access to the code to change the rules or to improve the training with new data. For this reason, the tools to be favored are those distributed as open source.

It is important to note that many of the tools available freely online cannot be used for commercial purposes, but only for research purposes.

In MAPS have been used the tool suite developed for the project (opener) Open Polarity Enhanced Name Entity Recognition. For a description of the project and tools made from Apache OpenNLP, whose license also allows commercial use, see the section on the relevant projects.

SPECIFIC DATA ACCESS INFRASTRUCTURE

Any observing or modeling system needs a robust data and information system, capable of combining data and knowledge gathered over time with new observations and modelling results to provide a range of integrated, interdisciplinary datasets, indicators, visualizations, scenarios, and other information products. The present efforts on data preservation and information systems include multiple sources of information, involve multiple stakeholders, support effective decisions at global to local scales, provide full and open access to data (Diviacco & Leadbetter, 2017, chapter 2 this book).

The information system is the real interface for most users, providing the data, information and products required to support research and address societal decisions in diverse areas such as climate studies and adaptation, disaster warning and mitigation, and ecosystem-based management. Beyond the observation elements there are many modelling, data assimilation and synthesis activities that provide added value to observations and meet specific user requirements for information.

Providing open access to data and related products reduces duplication among the user community and promotes data standards and broad accessibility This supports the principle of "measure once/use many times.

The Physics portal of the European Marine Observation and Data Network (EMODnet) has been used to provide both real time and historical data, using the experience done within the SeadataNet network of National Oceanographic Data Centers. EMODnet Physics provides a single point of free and open access to marine real time and archived data on physical conditions of all European Seas as monitored by fixed platforms, ferry box, ARGO, gliders, etc.

The main elements derived from the SeaDataNet 'system' are the vocabularies and the Common Data Index (CDI), that can be considered a de-facto standard for marine metadata in Europe. The CDI format is a marine profile of the ISO19139 metadata content standard compliant with INSPIRE Directive Implementing Rules.

In MAPS, the information system has been a scientific field of study addressing both operational activities and the distribution and use of information to knowledge communities. This means that the Information system in MAPS case has been referred also to the interaction that can occur across organizational boundaries. In other words, the information system must include important elements of information management as well as knowledge management together. In practice, the system must be able to (e.g. Ward & Peppard, 2002):

- Manage and maintain data to be interpreted in order to render information
- The information has to be understood in order to emerge as knowledge
- Knowledge must be used by a wide variety of stakeholders, including decision makers
- Stakeholders use of knowledge is expected to deliver meaningful results.

It has been underlined by some authors that this approach (called DIKAR: Data, Information, Knowledge, Action and Result) is aligning technology and organisational strategies, and it can be seen as a strategic change in information management systems. The recognition that information management is an investment that must deliver meaningful results is important to all modern organisations that depend on information and good knowledge for their success.

SOLUTIONS AND RECOMMENDATIONS

In this paragraph are presented some important projects in which semantic search technologies and / or textual analysis have been developed for the extraction and retrieval of information from texts domain specific.

KYOTO

The project Yielding Kyoto Knowledge Ontologies for Transition-based Organization has aimed to provide a system for searching deep semantic content, and that would allow experienced users to model and improve its domain ontology with terms and concepts automatically extracted. As domain environment for Kyoto was chosen environmental sector. It has gone from a knowledge base consisting of predefined generic ontology (SUMO for example) connected to the WordNet of each language treated by the project;

It is then used this lexical basis of conceptual knowledge both to allow the semantic analysis and research on a number of documents to extract new candidate information from such documents; eventually the system enables experienced users to evaluate and integrate the concepts and terms candidates in their knowledge base, improving the analysis of new documents.

- **Link Project Site:** http://kyoto-project.eu/xmlgroup.iit.cnr.it/kyoto/index.html_
- **Link Tools:** http://kyoto-project.eu/xmlgroup.iit.cnr.it/kyoto/index6b4e.html_
- **Link Deliverables:** http://kyoto-project.eu/xmlgroup.iit.cnr.it/kyoto/index4160.html_

GLOSS

The project GLOSS Multi Lingual Information Access to Multi Media Contents (MLIA2MMC) is an emerging technology and enabling that permit new forms of multilingual access to information, by combining the latest developments in Text Mining, Knowledge Engineering (IC), Language Processing Natural (TAL) and Semantic Interpretation (IS). The ability of these tools is to provide targeted and effective access to content: location of information in an easy to read grid enriched with functional claims (who does what), time (when), geographic (where). This resilience has become a key feature for the search engines of the future. The overall objective of the project GLOSS - derived from Kyoto - but addressed the geographical domain, is focused on graphical visualization tools. GLOSS The project was designed to carry out a research of industrial type starting with the current state of the art technology MLIA2MMC, thereby allowing the construction of an integrated environment for the consultation, the production and sharing of knowledge.

- **Link with Project:** http://weblab.iit.cnr.it/gloss/_

PANACEA

The project PANACEA Platform for Automatic, Normalized Annotation and Cost- Effective Acquisition of Language Resources for Human Language Technologies (STREP – Specific Targeted Research Project - under EU-FP7) has developed a factory of linguistic resources (both data and algorithms) that automates all the steps involved in acquisition, production, maintenance and upgrading of resources required by the machine translation. The factory also includes other language technologies such as those inherent alignment of parallel corpora, the production of bilingual dictionary and the production of lexicons very informative. The reference domain is environmental / legal.

- **Link Project Site:** http://www.panacea-lr.eu/_
- **Link Tools:** http://www.panacea-lr.eu/en/info-for-professionals/the-platform/_
- **Link Deliverables:** http://www.panacea-lr.eu/en/project/work-packages/_

Opener

The project Opener Open Polarity Enhanced Named Entity Recognition is funded by the European Commission under the 7th Framework Programme. The main goal of opener and provide a set of tools is immediately usable by researchers from small and medium enterprise to perform activities of natural language processing, free and easy to integrate into their workflow. Opener also wants to be able to detect and disambiguate entities mentioned and perform sentiment analysis and opinion on the texts detection, so you can capture information and steal the sentiment and market views about the products reviewed on the web.

- **Link Project Site:** www.opener-projcct.org_
- **Link Tools:** http://www.opener-project.eu/webservices_
- **Link Deliverables:** http://www.opener-project.eu/documentation/_
- **Link Extraction Multiwords:** https://github.com/opener-project/multiword-generation_

BootStrep

The project BooStrep Bootstrapping of Ontologies and Terminologies Strategic Research Project was funded within the 6th Framework Programme of the EC. The project - aimed at the biological realm - combined in a common framework - terminology resources and existing databases, and has implemented a system of analysis of texts for the acquisition of new terms, concepts and relationships, or a semi-automatic process of construction ontologies in biology.

- **Link Project Site:** http://www.bootstrep.org/_
- **Link Tools:** http://www.bootstrep.org/resources.html_
- **Link to the Result of the Project:** http://www.ebi.ac.uk/Rebholz-srv/BioLexicon/biolexicon.html_

Information System

The MAPS information system was developed in a previous regional project (MARINE) and is constituted by a fully distributed system where data and metadata are collected and maintained by different project partners. The system is INSPIRE compliant and is using the following elements from the European project SeaDataNet:

- Common Data Index (CDI) indexing the data and thus being the element of the catalogue, allowing search, localization, selection, semantic interoperability, asset management and availability; CDI is containing also the links to documentation;
- BODC vocabularies;
- NetCDF and Ocean Data View formats

The services provided by the information system are Catalogue Service for data search, Web Map Service for consultation, Web Feature Service for downloading, Coordinate Transformation and Service Chain to claim other services (see for the general concepts Glaves, 2017, chapter 3 this book).

Each node of the federated information system is structured as follows:

- *PlatformList.xml* file is giving the list of platforms that have collected data made available though ftp protocol
- *PlatformsCDI.xml* file is giving information on each platform listed in *PlatformList.xml*

- A directory *latestData* is containing the data collected during the last 30 days; one file is containing daily data from one platform; daily subdirectory are continuously formed;
- A directory *monthlyData* is archiving the data older that 30 days; it is organised as the *latestData* directory;
- A directory *biblioData* is archiving documentation on data acquisition, protocols, reports; it is organised in subdirectories for the different platforms.

PlatformList.xml files have KML format and allow the easy management of geospatial data.

The MARINE information system used in MAPS is schematized in Figure 2.

The Common Data Index

The use of the Common Data Index (CDI) XML is of strategic importance in the project. It has been developed within the European projects SeaSearch, SeaDataNet and SeaDataNet2. However, some additional elements have been implemented in MAPS in order make it more effective in combination with a search engine.

From the original CDI (n.d.), MAPS has used those elements that allow the selection of important information, such as the parameter name (e.g.)

```
<gmd:keyword>
<sdn:SDN_ParameterDiscoveryCode codeSpace="SeaDataNet" codeListValue="TEMP"
codeList="http://.../schemas/SeaDataNet/Codelist/sdnCodelists.xml#SDN_Paramet
erDiscoveryCode">Temperature of the water column</sdn:SDN_ParameterDiscovery-
Code>
</gmd:keyword>
```

```
<gmd:keyword>
<sdn:SDN_DeviceCategoryCode codeSpace="SeaDataNet" codeListValue="13"
codeList="http://.../schemas/SeaDataNet/Codelist/sdnCodelists.xml#SDN_DeviceCa
tegoryCode">bathythermographs</sdn:SDN_DeviceCategoryCode>
</gmd:keyword>
```

Some elements were better defined, such as the information on publications

```
<gmd:CI_Citation> <gmd:title>
<gco:CharacterString>Improved near real-time data management procedures for
the Mediterranean ocean Forecasting System-Voluntary Observing Ship program</
gco:CharacterString>
</gmd:title>
<gmd:date> <gmd:CI_Date> <gmd:date>
<gco:Date>2003</gco:Date>
</gmd:date>
<gmd:dateType>
```

Figure 2. The federated information systems used in MAPS project

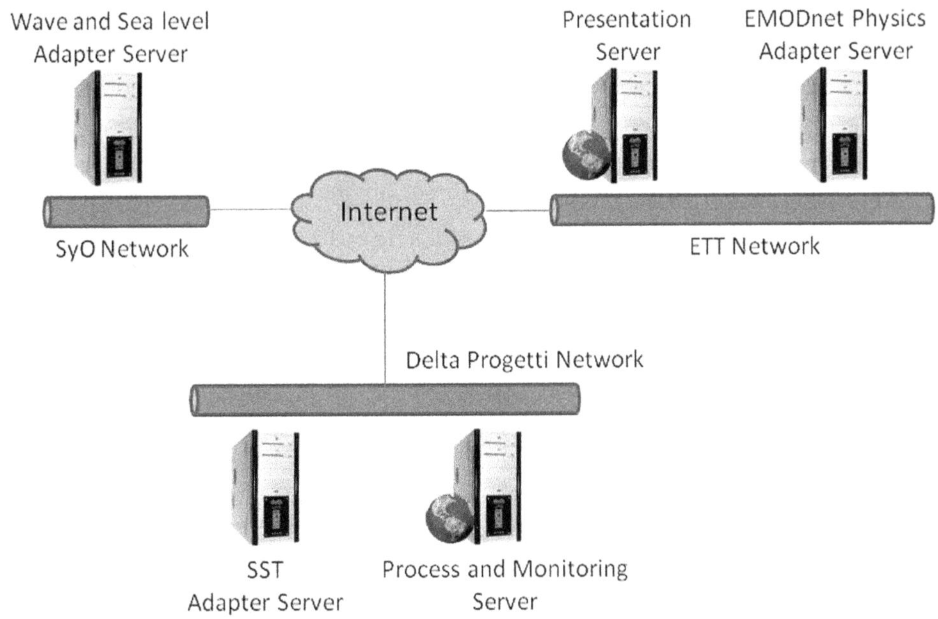


```
<gmd:CI_DateTypeCode codeList="http://.../schemas/SeaDataNet/Codelist/gmx-
Codelists.xml#CI_DateTypeCode" codeListValue="publication" codeSpace="ISOTC211
/19115">publication</gmd:CI_DateTypeCode>
</gmd:dateType> </gmd:CI_Date> </gmd:date>
<gmd:citedResponsibleParty> <gmd:CI_ResponsibleParty> <gmd:individualName>
<gco:CharacterString>G. M. R. Manzella</gco:CharacterString>
</gmd:individualName>
<gmd:organisationName>
<gco:CharacterString>ENEA</gco:CharacterString>
</gmd:organisationName>
…
<gmd:role> <gmd:series> <gmd:CI_Series>
<gmd:name> <gco:CharacterString>Annales Geophysicae</gco:CharacterString> </
gmd:name>
<gmd:issueIdentification> <gco:CharacterString>1</gco:CharacterString> </gmd:i
ssueIdentification>
```

```
</gmd:CI_Series> </gmd:series>
<sdn:onlineResource> <gmd:CI_OnlineResource> <gmd:linkage>
<gmd:URL>http://www.ann-geophys.net/21/49/2003/angeo-21-49-2003.html</gmd:URL>
</gmd:linkage> </gmd:CI_OnlineResource> </sdn:onlineResource> </gmd:CI_Ci-
tation> </sdn:additionalDocumentation> </gmd:MD_DataIdentification> </
gmd:identificationInfo>
```


For the digital identifier (DOI) the NOAA (http://www.ngdc.noaa.gov/docucomp/page) elements have been used


```
<gmd:identifier> <gmd:MD_Identifier> <gmd:code>
<gmx:Anchor xlink:href="http://dx.doi.org/10.7289/V5154F01" xlink:title="DOI"
link:actuate="onRequest">doi:10.7289/V5154F01</gmx:Anchor>
</gmd:code> </gmd:MD_Identifier> </gmd:identifier>
```


As well as for right constraints


```
<gmd:resourceConstraints xlink:title="NGDC Data Citation Statement">
<gmd:MD_LegalConstraints> <gmd:useLimitation>
<gmd:MD_RestrictionCode codeList="http://.../schema/resources/Codelist/gmx-
Codelists.xml#MD_RestrictionCode"
codeListValue="otherRestrictions">otherRestrictions</gmd:MD_RestrictionCode>
</gmd:useLimitation>
<gmd:otherConstraints>
<gco:CharacterString>Cite as: Manzella, G. M. R., Scoccimarro, E., Pinardi,
N., and Tonani, M.: Improved near real-time data management procedures for the
Mediterranean ocean Forecasting System-Voluntary Observing Ship program, Ann.
Geophys., 21, 49-62. doi:10.5194/angeo-21-49-2003, 2003</gco:CharacterString>
</gmd:otherConstraints> </gmd:MD_LegalConstraints> </gmd:resourceConstraints>
```


In this way, the Common Data Index has all the elements selected in MAPS for the semantic search.

Digital Library

The library is containing the documentation and the search engine that is composed by:

- **Syntactic Parser:** This module is responsible for the extraction of "rich words" from the text: the whole document gets parsed to extract the words which are more meaningful for the main argument of the document, and applies the extraction in the form of N-grams (mono-grams, bi-grams, tri-grams).
- **MAPS Database:** This module is a simple database which contains all the N-grams used by MAPS (physical parameters from SeaDataNet vocabularies) to define our marine "ontology".
- **Term Extractor:** This module performs the most important task of filtering the N-grams extracted from the text by the parser with the provided oceanographic terminology. It checks N-grams

supplied by the syntactic parser and then matches them with the terms stored in the MAPS database. Found matches are returned back to the parser with flexed form appearing in the source text.

- **A "Relaxed" Extractor:** This option can be activated when the search engine is launched. It was introduced to give the user a chance to increase the ontology with new N-grams combining existing mono-grams and bi-grams in the database with rich-words found within the source text.

The innovation of a semantic engine lies in the fact that the process is not just about the retrieval of already known documents by means of a simple term query but rather the retrieval of a population of documents whose existence was unknown. The system answers by showing a screenshot of results ordered according to the following criteria:

- **Relevance:** of the document with respect to the concept that is searched
- **Date:** of publication of the paper
- **Source:** data provider as defined in the SeaDataNet Common Data Index
- **Matrix:** environmental matrices as defined in the oceanographic field
- **Geographic area:** area specified in the text
- **Clustering:** the process of organizing objects into groups whose members are similar

The clustering returns as the output the related documents. For each document the MAPS visualization provides:

- Title, author, source/provider of data, web address
- Tagging of key terms or concepts
- Summary of the document
- Visualization of the whole document

Library Model

MAPS can be applied to documents in English and in Italian. The design of the semantic engine architecture is based on the integration of already existing linguistic/semantic modules emphasizing the adaptation to the lexical and terminological document bases of the specific marine domain. A general source of inspiration for MAPS's terminology base is WordNet (Fellbaum, 1998), a lexical database that groups English words into sets of synonyms, providing short definitions and recording a number of relations among these synonym sets or their members. WordNet can thus be seen as a combination of dictionary and thesaurus. The WordNet model was applied to Italian by Roventini et al (2000), with the creation of ItalWordNet, and used in several projects to enhance semantic search in documents; Marinelli and Roventini (2006) enhanced ItalWordNet with maritime terms.

Search Engine

The MAPS system was conceived so as to receive objects of various nature (text and data) that users need to consult via a web interface. MAPS objects are also documents that need to be discovered in their content, as well as by means of their metadata. Figure 3 represents the search architecture adopted

Figure 3. The semantic search system

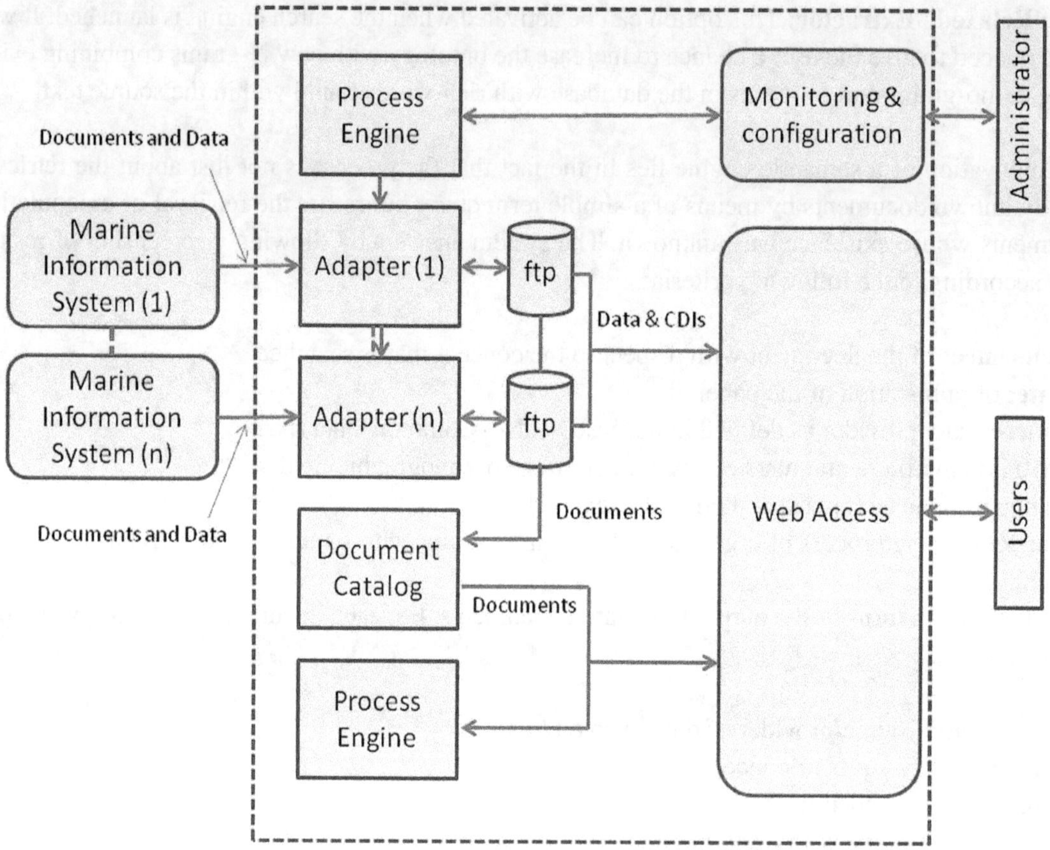

for MAPS. The documents are indexed and made searchable by the Semantic Engine and then queried from the web interface by users together with other types of objects (data).

In particular, the Search Engine uses semantic-conceptual technologies in order to extract key concepts from unstructured text such as technical documents (reports and grey literature) and scientific papers. Any text has to be made indexable and searchable by the end user in the same way as the structured data (such as oceanographic observations and metadata) are.

In order to achieve this purpose a semantic system has been implemented using natural language processing techniques and language engineering tools. They allow the correct indexing of documents and the correct processing of natural language queries in the human-machine search interface. This process is completed by incrementing the terminological knowledge base of domain concepts with new elements found in the texts.

As shown in Figure 4, the Semantic system contains a Natural Language Processing (NLP) pipeline that analyses texts thanks to NLP tools and pre-existing terminological databases, both generic and domain specific. The result of the NLP pipeline is an annotated text that allows for terminological extraction of relevant concepts. Such terms are then used for indexing, but can also enrich a domain lexicon Data Base that can be later re-fed into the NLP pipeline.

Figure 4. The semantic engine.

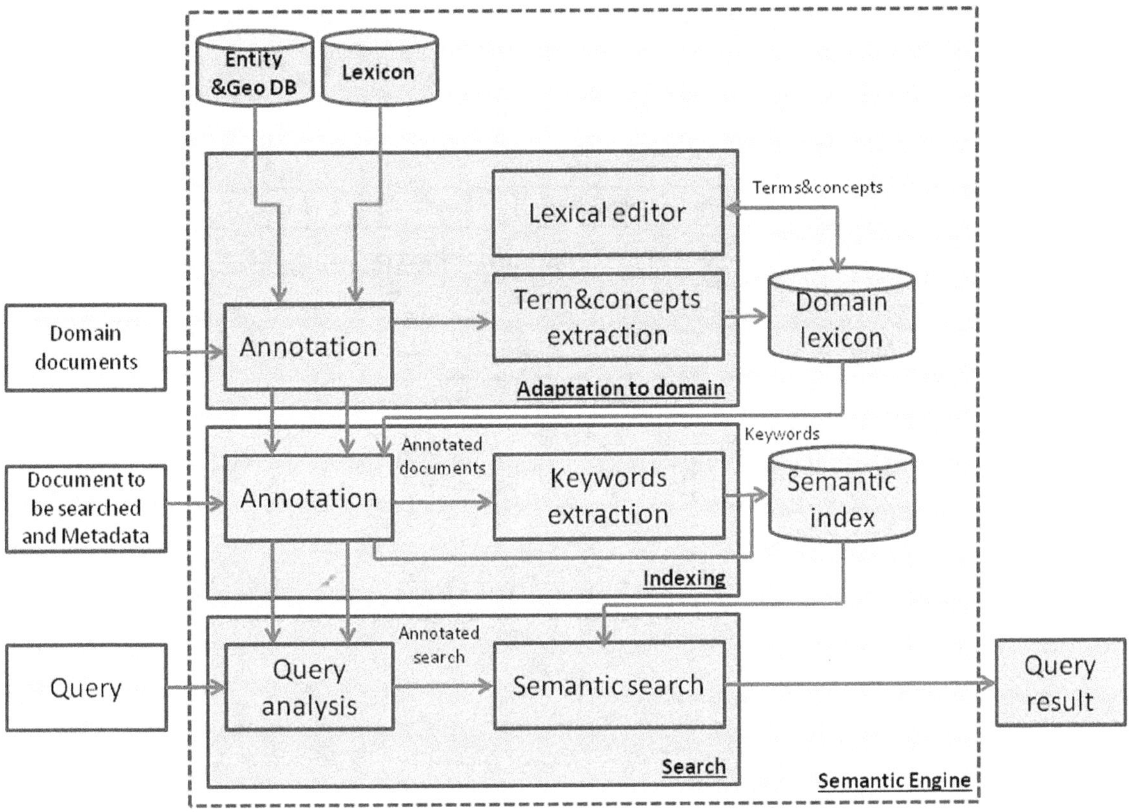

The NLP and indexing processes are asynchronous and are activated whenever a new set of texts enters the document base.

The query analysis pipeline is instead a synchronous process, that is called whenever the search engine is interrogated by a user with a query in natural language. For this reason the query analysis system is built in a simplified way that guarantees to obtain similar results to the more complex NLP and indexing pipeline (so as to ensure the extraction of relevant texts), but with better performances in terms of response time - at least for such smaller snippets of texts that are likely to be entered by humans in a query. The logical steps for NLP and term extraction are shown in Figure 5.

The first step is the transformation of the original document (often in pdf format) in plain utf-8 format text. This is done with an open source command-line utility (pdftotext) converting PDF files to plain text files - i.e. extracting text data from PDF-encapsulated files.

Subsequently, since documents may be in English or Italian, a language detector is used in order to call for the correct language specific NLP pipeline. This is done by means of the Google language detector cld2 (https://code.google.com/p/cld2/). Simplified ad-hoc versions for the two query pipelines have been implemented, using frequent terms in both languages. In practice, the logical steps for the NLP pipelines are the same for both languages, but some pre-existing tools are used whenever possible:

Figure 5. Steps of natural language processing and term extraction.

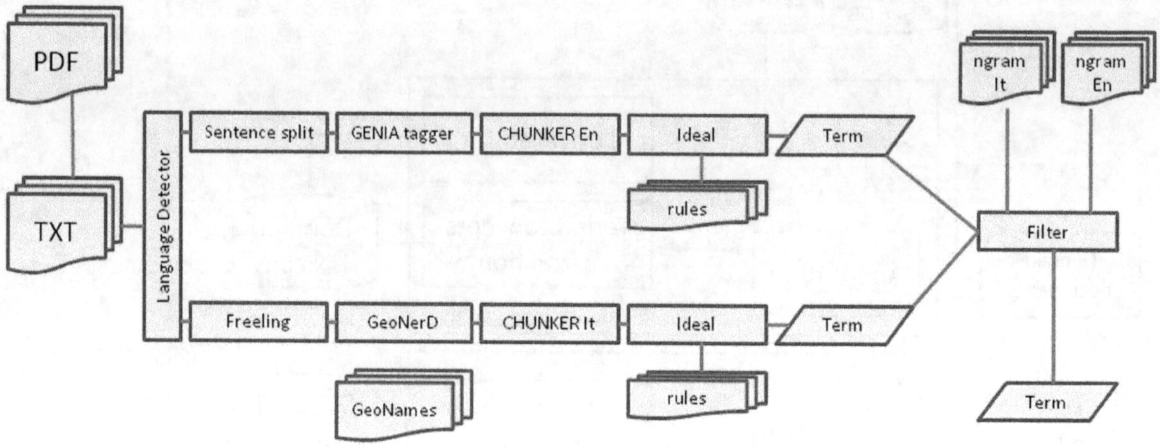

- sentence splitting, dividing the text in sentences
- word tokenization, splitting sentences into words
- lemmatization and morphological analysis (part of speech tagging)
- toponym detection, identifying geographic names
- basic syntactic analysis (chunking) dividing the sentence into non recursive constituents

More specifically the component used for document in English are:

- For sentence splitting the SentenSentenceDetector of the Apache OpenNLP suite was used (https://opennlp.apache.org/)
- For tokenization, lemmatization and morphological analysis the Genia Tagger was adapted (http://www.nactem.ac.uk/tsujii/GENIA/tagger/)
- For toponym detection and identification of geographic names GeoNerD (http://sourceforge.net/u/geonerd/profile/) ad hoc develop method using a knowledge base (Geonames) and a rule based algorithm

- For syntactic analysis Chunker Eng, ad hoc developed module using a rule based algorithm (e.g. Wang, 2008)

The component used for document in Italian are:

- Freeling, an open source language analysis tool suite available for several languages among which Italian (http://nlp.lsi.upc.edu/freeling/).
- GeoNerD: ad hoc develop method using a knowledge base (Geonames) and a rule based algorithm.
- Chunker Ita, ad hoc developed module using a rule based algorithm (Lenci et al., 2003).

Terminology Extraction

The NLP pipeline produces an intermediate annotated document, which is preliminary to terminology extraction; the latter in turn is necessary in order to be able to correctly index the document in the document base to be later semantically searched. The terminology extraction tool, Ideal, takes the chunked text as input, containing all required morphosyntactic information. Ideal is a rule based system and specific rules were developed for Italian and English. This tool was developed specifically for MAPS, being the most important part of the NLP pipeline. The rules, written in an ad hoc language (Bartolini et al, 2004), are designed to extract all simple and complex noun phrases in the text. The extractor works on the chunked text searching for patterns such as:

- nominal phrase;
- nominal phrase followed by one or more adjectival or prepositional phrases

All possible intermediary marches are produced, and they are taken to represent concepts of different degrees of specificity. Thus the algorithm not only returns candidate terms, but sub-trees of candidate terms linked by hierarchical relations. For instance a document containing the text mean sea salinity will produce the following "family" candidate terms:

- salinity (more generic)
- mean salinity (less generic)
- mean sea salinity (specific)

In particular each complex term is analysed into a lemmatized head term (salinity) and all the possible specifiers, with the frequency with which they occur in the text.

In order to guarantee that such terms are domain relevant terms a filtering operation is then performed. To this purpose a list of concepts is stored in the terminological base of the semantic engine, drawn from SeaDataNet vocabularies (see Buck & Lowry, 2017, chapter 1 this book). Whenever a set of related terms is extracted, it is compared to the list of terms in the terminological base. If one of the related terms or a sub component thereof is also present in the terminology base, the whole family of related terms is validated.

Filtering can be obtained by:

- **Exact Matching:** only the terms corresponding to those found in the terminological base are used for indexing;
- **Relaxed Match:** one of the related terms is also present in the TB, the whole family of related terms is validated.

So in the previous example, if salinity is present in the domain term base, in the case of exact matching only salinity is used for indexing the document, whereas in the case of relaxed matching salinity, mean salinity and mean sea salinity are used for indexing.

While the exact matching method may be used to increase precision in the semantic search, the second option is more interesting, as it allows for the discovery of new complex domain terms that are not already in the domain terminology. For instance from given the domain term data, it can be derived satellite data, wavemetrics data, oceanographic data; from wave it can be derived mean wave height, wave height variation, ...

The NLP and terminology extraction pipeline has a set of terms in a standardised json format as output, which is then used as input for the indexing function in the Apache Lucene Core.

Similarly, the query analysis pipeline produces the same output that can be used to search the existing index for matching terms.

The Service Platform

During the last stages of the MAPS project, a trial activity was carried out in order to assess the capabilities of the developed search engine. The trial activities were broken down in two phases:

- The first one was responsible of assessing whether stakeholders requirements were correctly implemented in the pilot system;
- The second phase had the objective to evaluate whether the search engine is capable of retrieving relevant documents.

The assessment of requirements is usually carried out by planning and executing a number of tests in order to verify the behaviour of the system in given operational conditions. This approach is usually adopted during user acceptance project phases, where stated requirements are verified. However, the trial was finalized more on evaluating the quality attributes of the implementation than verifying that every function was properly executing. As such, it was decided to evaluate the functionality of the pilot system by measuring the quality attributes defined by a software quality model versus stakeholders' requirements.

A software quality model is a collection of quality attributes that a piece of software should have. The quality of the developed software is then assessed by evaluating whether or not it possesses all the required attributes. Attributes are generally structured in two or more hierarchical levels so that it is easier to figure them out and verify the software against them.

There are several quality models available. The first software quality models were developed by McCall (1977) and Bohem (1978). More recently, ISO has developed a number of standards defining software quality models, the most known of which is the ISO/IEC 9126 standard. This model has been used for assessing the search engine.

The ISO 9126 quality model was based on the McCall and Boehm models. The model has two main parts consisting of:

- Internal and external quality attributes
- Quality in use attributes.

Internal quality attributes are referred to the system properties that can be evaluated without executing, while external quality attributes refer to the system properties that can be assessed by observing it during execution and are experienced by users when the system is in operation.

The quality in use attributes are referred to usage of the product and regards its effectiveness productivity, security and satisfaction.

The ISO/IEC 9126 quality model is a two level model as it consists of a set of characteristics (or attributes) and sub-characteristics. For instance, the internal and external quality characteristics are: Functionality, Reliability, Usability, Effiency, Maintenability and Portability; the sub-characteristics of Functionality are Suitability, Accurancy, Interoperability, Security and Compliance.

The ISO/IEC 9126 software quality model has been applied as follows. Initially a number of characteristics and associated sub-characteristics of interest for the project was identified. Subsequently requirements were grouped in 4 classes and then evaluated by determining whether each requirement implementation has or has not each sub-characteristic. Finally, a rating, in the range 0 to 10, has been computed for each requirement group and sub-characteristic by evaluating the ratio (expressed as a fraction of 10) between the number of the requirements in the group satisfying the sub-characteristic and the total number of requirements. The results of this evaluation are reported in Table 1.

The total rating achieved is 8.4 out of 10, which is a remarkably good result. However, several areas have to be improved, especially is the presentation and usability extents.

In the second phase of the trial, the search engine was experimented to assess how much it was capable of retrieving relevant documents against the submitted queries.

Information Retrieval has elaborated several techniques to evaluate the performance of retrieval systems. Most of the methods use a defined collection of documents and a collection of queries for those documents; by comparing the query results provided by a group of experts with those returned by the search engine, it is possible to evaluate the performance of the engine. There are several techniques for comparing experts and search engine results, such as Precision and Recall, Accuracy, f-measure, ranked results, and so on. The major difficulty with these methods is linked with the need of experts and the fact that this may result in a subjective evaluation.

To overcome this difficulty, benchmarking techniques has been proposed, such as TREC (Text Retrieval Conference) and others. Basically, benchmarking techniques compare search engines using a well-defined collection of documents and queries. These techniques are not subjective as the involvement of experts is not necessary; however the collection of documents and queries are fixed and cross-domain in order to compare effectively different search engines.

Another set of techniques for evaluating the effectiveness of information retrial systems is based on evaluating ranked retrieval results, where importance is placed not only on obtaining the maximum number of relevant documents but also on returning relevant documents higher in the ranked list. A commonly used method to rank outputs is to compute precision at various level of recall (see below for a definition of precision and recall).

Table 1. Identification of sub-characteristics in MAPS documents (see text above for details)

Requirements	Adequacy	Accuracy	Compatibility	Learnability	Operability	Attractivity	Effectiveness	Productivity	Fulfillment
Formulation search									
Natural language search	■	■		■	■	■	■		■
Key concepts search	■	■	■		■	■	■	■	
Possibility of using logical operators	■	■						■	
Formulate queries in Italian	■	■	■	■	■	■	■	■	■
Formulate queries in English	■	■	■	■	■	■	■	■	■
Narrowing searches based on platforms, instrument, environmental matrix, geographic area	■	■	■	■	■	■	■	■	■
Rating	**10**	**10**	**7**	**7**	**8**	**8**	**8**	**8**	**7**
Presentation of search results									
Sort the results by relevance, date of publication, matrix environment, geography	■	■		■	■	■	■	■	■
Of each document display: title, author, source and internet link	■	■	■	■	■	■	■	■	
Document visualization	■	■	■	■	■		■	■	■
Rating	**10**	**10**	**7**	**10**	**10**	**7**	**10**	**10**	**7**
Refining search results									
Refining search on the base of file type (doc, pdf, etc)	■	■	■	■	■		■	■	
Refining search on the base of the author	■	■	■	■	■	■	■	■	■
Repeating one previous search			■	■		■	■	■	■
Rating	**7**	**7**	**10**	**10**	**7**	**7**	**10**	**10**	**7**
Averaged rating	**8.9**	**8.9**	**7.8**	**8.9**	**8.3**	**7.2**	**9.4**	**9.4**	**6.7**
Mean total rating					**8.4**				

Other measures have also been conceived to evaluate different information retrieval problems. For example, to measure the success of search tasks where just one relevant document is required, measures such as mean reciprocal rank (MRR – e.g. Chapelle et al., 2009), can be used.

Several techniques are then available for measuring the effectiveness of search engines (or information retrieval systems) and, in practice it is important to select an evaluation measure that is suitable for the given task.

We decided to adopt the Precision/Recall method for its simplicity and because benchmarking techniques, which involve the use of fixed document collections of several different domains, were not feasible as the MAPS search engine was specifically customized for the marine domain.

Precision and recall are defined as follows:

$$precision = \frac{\left|\left\{relevant\ documents\right\} \cap \left\{retrieved\ documents\right\}\right|}{\left|\left\{retrieved\ documents\right\}\right|}$$

$$recall = \frac{\left|\left\{relevant\ documents\right\} \cap \left\{retrieved\ documents\right\}\right|}{\left|\left\{relevant\ documents\right\}\right|}$$

using a mathematical notation

$$p = \frac{n_R^R}{n^R}\ r = \frac{n_R^R}{n_R}$$

where n^R is the number of document retrieved by the search, n_R is the number of relevant documents in the documents collection and n_R^R is the number of relevant documents retrieved by the search.

For computing precision and recall, a collection of 15 documents and an assortment of 9 queries (Q1, …, Q9 in Table 2) were defined; for each document-query pair, a number of experts determined the corresponding relevance (n_R in Table 2). Then the 15 documents have been loaded in an empty installation of the MAPS pilot system and the 9 queries have been executed, annotating the number of retrieved documents (n^R in Table 2) and the number of relevant documents (n_R^R in Table 2) retrieved by each query. Finally, overall precision and recall values have been computed by averaging the values resulting from each query. The results obtained are reported in Table 2.

By averaging the precision and the recall computed for each query, we obtained that about 80% of relevant documents were retrieved while 75% of the retrieved documents were relevant. Which is a promising result.

Further analysis shows that the behavior of the system is largely dependent on the marine domain customization (that we implemented by using a domain ontology). For instance, by restricting the ontology to the terms of the P02 BODC vocabulary, the system performs better with documents describing marine data acquisition procedures and protocols. Including additional concepts to the ontology, the system starts to perform also with other types of marine documents, such as articles and studies.

Table 2. Precision and Recall values evaluated in the trial (see text above for details)

	Q1	Q2	Q3	Q4	Q5	Q6	Q7	Q8	Q9
# of Relevant docs (n_R)	2	2	4	2	2	2	2	4	1
# of Retrieved & Relevant docs (n_R^R)	2	2	2	0	0	2	2	3	1
# of Retrieved docs (n^R)	2	4	2	0	0	2	2	3	1
Precision ($p = \frac{n_R^R}{n^R}$)	1.00	1.00	0.50	0.00	0.00	1.00	1.00	0.75	1.00
Recall ($r = \frac{n_R^R}{n_R}$)	1.00	1.00	0.50	0.00	0.00	1.00	1.00	0.75	1.00
Precision (Average)	0.75								
Recall (Average)	0.81								

FUTURE RESEARCH DIRECTIONS

It is well-know that document indexing techniques are not sufficient to satisfy user information needs that go beyond the limits of a simple term matching search. Therefore the MAPS search engine is enriched with semantic technologies aimed at providing an accurate representation of the content of vast repositories of unstructured documents for semantic indexing purposes. Today language technologies make it possible for scientists and developers to produce software applications capable of revealing the semantic properties of textual elements and associating them with conceptual structures. Search functions coupled with semantic-conceptual technologies are able to interpret the underlying search criteria and thus better identify the data and the corresponding documents. This enables an effective and selective access to information even in the presence of significant collections of data.

CONCLUSION

The MAPS project, supported by regional funding POR-FESR for industrial development of enterprises associated to the Liguria Cluster of Marine Technologies, is part of a long term activity aiming at building a computer platform for supporting a Marine Information and Knowledge System. This integrated MAPS platform offers the advantage to have all content in one place and allows linking different information that exists, thus helping data management activities.

This study has attempted to provide semantic annotations to grey literature documents of the oceanography domain. The initial experiment has revealed that available methods are capable of assisting the process of semantic annotation with promising results. The incorporation of ontologies and knowledge resources in a rule-based information extraction technique promises to enable rich semantic indexing of grey literature documents.

Flexibility is a necessary requirement for any new use that the system intends to embrace while additional efforts are required for further exploitation of the technique and adoption of formal evaluation methods for assessing the performance of the method with measurable criteria.

REFERENCES

Bartolini, R., Lenci, A., Montemagni, S., Pirrelli, V., & Claudia Soria, C. (2004). Semantic mark-up of Italian legal texts through NLP-based techniques. In *Proceedings of LREC 2004.*

Belov, S., & Mikhailov, N. (2012). *Technical Workshop on the IODE Ocean Data Portal.* Retrieved from ftp://ftpdownload:ftpdownload@odp.meteo.ru/5-ODP_V2.pdf

Boehm, B., Brown, J. R., Kaspar, H., Lipow, M., McLeod, G., & Merritt, M. (1978). Characteristics of Software Quality. North Holland.

Bosma, W. E., Vossen, P., Soroa, A., Rigau, G., Tesconi, M., Marchetti, A., & Aliprandi, C. et al. (2009). Kaf: a generic semantic annotation format. In *Proceedings of the GL2009 Workshop on Semantic Annotation.*

Boyer, T. P., Antonov, J. I., Baranova, O. K., Coleman, C., Garcia, H. E., Grodsky, A., . . . Zweng, M. M. (2013). World Ocean Database 2013. NOAA Atlas NESDIS 72.

Brin, S. (1998). Extracting Patterns and Relations from the World Wide Web. In *Proceedings of the WebDB Workshop at 6th International Conference on Extending Database Technology.*

Brown, J. S., & Duguid, P. (1998). Organising knowledge. *California Management Review, 40*(3), 90–111. doi:10.2307/41165945

Buck, J., & Lowry, R. (2017). Oceanographic Data Management; Quills and Free Text to the digital age and 'Big Data. In P. Diviacco, A. Leadbetter, & H. Glaves (Eds.), *Oceanographic and Marine Cross-Domain Data Management for Sustainable Development.* Hershey, PA: IGI Global.

Chapelle, O., Metlzer, D., Zhang, Y., & Grinspan, P. (2009). Expected Reciprocal Rank for Graded Relevance. In *Proceedings of the Conference on Information and Knowledge Management.*

Clarke, R. (2005). A proposal for an open content licence for research paper (Pr)ePrints. *First Monday, 10*(8), 1–11. doi:10.5210/fm.v10i8.1262

Clough, P., & Sanderson, M. (2013). Evaluating the performance of information retrieval systems using test collections. *Information Research, 18*(2). Retrieved from http://www.informationr.net/ir/18-2/paper582.html

Commons, C. (2001). *"Some Rights Reserved": Building a Layer of Reasonable Copyright.* Retrieved 2005-09-14 from the World Wide Web: http://creativecommons.org

Diviacco, P., & Leadbetter, A. (2017). Balancing Formalization and Representation in Cross-Domain Data Management for Sustainable Development. In P. Diviacco, A. Leadbetter, & H. Glaves (Eds.), *Oceanographic and Marine Cross-Domain Data Management for Sustainable Development*. Hershey, PA: IGI Global.

European Commission. (2010). *European Marine Observation and Data Network: Impact Assessment*. SEC(2010) 999 Final.

European Commission. (2012). *Green Paper Marine Knowledge 2020*. COM(2012)473Final. Retrieved from http://ec.europa.eu/maritimeaffairs/documentation/publications/documents/marine-knowledge-2020-green-paper_en.pdf

Fellbaum, C. (1998). *WordNet: An Electronic Lexical Database*. MIT Press.

Glaves, H. (2017). Developing a Common Global Framework for Marine Data Management. In P. Diviacco, A. Leadbetter, & H. Glaves (Eds.), *Oceanographic and Marine Cross-Domain Data Management for Sustainable Development*. Hershey, PA: IGI Global.

Kogut, P., & Holmes, W. (2001). *AeroDAML: Applying Information Extraction to Generate DAML Annotations from Web Pages*. In First International Conference on Knowledge Capture (K-CAP 2001). Workshop on Knowledge Markup and Semantic Annotation. Retrieved from http://citeseerx.ist.psu.edu/viewdoc/download?doi=10.1.1.21.8180&rep=rep1&type=pdf

Lenci, A., Montemagni, S., & Pirrelli, V. (2003). CHUNK-IT. An Italian Shallow Parser for Robust Syntactic Annotation. *Linguistica Computazionale*, *16-17*, 353–386.

Lott, B. (2012). *Survey of Keyword Extraction Techniques*. UNM Education.

Maillard, C., Lowry, R. K., Maudire, G., & Schaap, D. (2007). SeaDataNet: Development of a Pan-European Infrastructure for Ocean and Marine Data Management. *Oceans*, *2007*. doi:10.1109/OCEANSE.2007.4302435

Manzella, G. & Manzella, A. (2015). Apeiron: engage students in earth and ocean sciences. *International Journal of Knowledge Society Research, 6*(4), 100-111. DOI: .10.4018/IJKSR.2015100107

Marinelli, R., & Roventini, A. (2006). The Italian maritime lexicon and the ItalWordNet semantic database. In E. M. Bermudez & L. R. Miyares (Eds.), *Linguistic in the twenty first century* (pp. 173–181). Cambridge Scholar Press.

McCall, J.A., Richards, P.K. & Walters, G.F. (1977). *Factors in software quality.* National Technical Information Services (NTIS), RADC-TR-77-369.

Parra-Arnau, J., Rebollo-Monedero, D., & Forné, J. (2010). *A Privacy-Preserving Architecture for the Semantic Web based on Tag Suppression.* CON-SOLIDER INGENIO 2010 Project. Retrieved from https://pdfs.semanticscholar.org/949a/1cc930d4050a8b2166e58a76a8e244e17cdb.pdf

Polanyi, M. (1966). *The Tacit Dimension.* Chicago: University of Chicago Press.

Powers, D. M. W. (2011). Evaluation: From Precision, Recall and F-Measure to ROC, Informedness, Markedness & Correlation. *Journal of Machine Learning Technologies*, *2*(1), 37–63.

Proctor, R., Roberts, K., & Ward, B. J. (2010). A data delivery system for IMOS, the Australian Integrated Marine Observing System. *Advances in Geosciences*, *28*, 11–16. doi:10.5194/adgeo-28-11-2010

Rajman, M., & Besançon, R. (1998). Text mining: natural language techniques and text mining applications. In Data Mining and Reverse Engineering. Springer US.

Roventini, A., Alonge, A., Calzolari, N., Magnini, B., & Bertagna, F. (2000). ItalWordNet: a Large Semantic Database for Italian. In *Proceedings of the 2nd International Conference on Language Resources and Evaluation (LREC 2000)*. The European Language Resources Association (ELRA).

SeaDataNet. (n.d.). Retrieved from: www.seadatanet.org

UNESCO. (2013). *Berlin Declaration*. Retrieved from http://www.unesco.org/new/en/social-and-human-sciences/themes/physical-education-and-sport/mineps-2013/declaration/

Uren, V., Cimiano, P., Iria, J., Handschuh, S., Vargas-Vera, M., Motta, E., & Ciravegna, F. (2006). Semantic annotation for knowledge management: Requirements and a survey of the state of the art. *Web Semantics: Science, Services, and Agents on the World Wide Web*, *4*(1), 14–28. doi:10.1016/j.websem.2005.10.002

Vossen, P., Agirre, E., Calzolari, N., Fellbaum, C., Hsieh, S., Huang, C., ... van Gent, J. (2008). KYOTO: A System for Mining, Structuring, and Distributing Knowledge Across Languages and Cultures. In *Proceedings of the Fourth International Global Word Net Conference - GWC 2008*. doi:10.4018/978-1-61520-883-8.ch012

Wang, Y. (2008). *Software Engineering Foundations, a software science perspective*. Auerbach Publications.

Wannever, B. (1945). As we may think. *The Atlantic Monthly*. Retrieved from http://www.theatlantic.com/magazine/archive/1945/07/as-we-may-think/303881/

Ward, J., & Peppard, J. (2002). *Strategic Planning for Information Systems* (3rd ed.). Chichester, UK: Wiley.

WCED. (1987). *Report of the World Commission on Environment and Development: Our Common Future*. Retrieved from http://www.un-documents.net/our-common-future.pdf

KEY TERMS AND DEFINITIONS

Data: Re-interpretable representation of information in a formalized manner suitable for communication, interpretation, or processing [ISO/IEC 2382-1:1993].

Information: Knowledge concerning objects, such as facts, events, things, processes, or ideas, including concepts, that within a certain context has a particular meaning [ISO/IEC 2382- 1:1993].

Information System: Any organized system for the collection, organization, storage and communication of information.

Information Retrieval: The activity of obtaining information resources relevant to an information need from a collection of documents.

Lemmatisation: The process of grouping together the different inflected forms of a word so they can be analysed as a single item.

Part of Speech: A category of words which have similar grammatical properties. Commonly listed English parts of speech are *noun, verb, adjective, adverb, pronoun, preposition, conjunction, interjection.*

Search Engine Indexing: Collects, parses, and stores data to facilitate fast and accurate information retrieval.

Tokenization: The process of breaking a stream of text up into words, phrases, symbols, or other meaningful elements called tokens. The list of tokens becomes input for further processing such as parsing or text mining.

Chapter 7
SeaDataNet:
Towards a Pan-European Infrastructure for Marine and Ocean Data Management

Dick Schaap
Marine Information Service (MARIS), The Netherlands

ABSTRACT

SeaDataNet is actively operating and further developing a Pan-European infrastructure for managing, indexing and providing access to ocean and marine data sets and data products, acquired from research cruises and other observational activities in European marine waters and global oceans. It is undertaken by the National Oceanographic Data Centres (NODCs), and marine information services of major research institutes, from 35 coastal states bordering the European seas. Its data centres have the essential capabilities and facilities for data quality control, long term stewardship, retrieval and distribution. Development and provision of standards, services and tools for marine and ocean data management has proven instrumental for wider uptake. This is further encouraged by cooperation and synergy with EU marine research projects, and by adoption of the SeaDataNet infrastructure in the EMODnet (European Marine European Marine Observation and Data Network) development. EMODnet is an initiative in the framework of the EU Marine Directive.

INTRODUCTION

The initial steps towards developing a pan-European infrastructure for marine data management were undertaken with support of the European Commission through the EU SEA-SEARCH project which ran from 2002 to 2005. This was followed in 2006 by the SeaDataNet project (2006 – 2011) and the subsequent SeaDataNet II project (2011 – 2015). Through these projects a consortium of European oceanographic data centres have been actively developing the SeaDataNet pan-European infrastructure for managing, indexing and providing access to ocean and marine data sets and data products that have been acquired from research cruises and other observational activities in European seas and global oceans. The SeaDataNet consortium consists of National Oceanographic Data Centres (NODCs), and the marine information services of major research institutes, from 35 coastal states bordering the European

DOI: 10.4018/978-1-5225-0700-0.ch007

seas, and also includes expert modelling centres, IT experts, and the international Intergovernmental Oceanographic Commission (IOC) of UNESCO, International Council for the Exploration of the Sea (ICES) and EU Joint Research Centre (EU-JRC) in its network. The data centres involved in SeaDataNet are highly skilled having been actively engaged in data management for many decades; they also have the essential capabilities and facilities for quality control, long-term stewardship, retrieval and distribution of marine data.

Preparations are now underway with the European Commission for another phase of the project that will address new challenges such as big data, new types of instrumentation and data, new IT standards, and cloud infrastructures. The objective of this chapter is to give an overview of the current approach and architecture of the SeaDataNet infrastructure including its component services, how it has expanded over time to a network of more than 100 connected data centres, and its role as a key element of the technologies behind a number of the European Marine Observation and Data Network (EMODnet) portals.

BACKGROUND

Oceanographic data is highly diverse and includes a wide range of measurements and variables that have been generated by a broad, multidisciplinary spectrum of projects and programs. It has been collected by hundreds of research institutes, governmental organizations and private companies using various heterogeneous observing sensors which are installed on a variety of platforms such as research vessels, submarines, aircraft, moorings, satellites, buoys etc. to measure a range of physical, chemical, biological, geological and geophysical parameters. As well as the data captured directly from observing the marine environment, secondary data about the marine domain results from the analysis of water and sediment samples collected in the oceans.

The nature of the marine environment means that data acquisition is undertaken at very considerable cost and the resulting observations are often unique and irreplaceable. This data is also fundamental to all aspects of marine research, assessment and planning, and to make best use of it necessitates a robust operational infrastructure, based on European and internationally agreed standards covering data quality, and long-term stewardship as well as technical and semantic aspects of interoperability.

A number of directives and communications from the European Commission including the Marine Strategy Framework Directive (MSFD, 2008) highlight the need for a coordinated approach to marine monitoring in European waters to support sustainable exploitation while maintaining the good environmental status (GES) of the seas. The MSFD has also given rise to the European Marine Observation and Data Network (EMODnet) that aims to integrate national and regional marine data infrastructures to allow stakeholders to access data and data products for the European seas. The EU communication "Marine Knowledge 2020" (European Commission 2010) further emphasizes the need for a systematic approach to support the development of integrated European marine data management.

The SeaDataNet data infrastructure has become a core element of this coordinated approach to marine data management in Europe. It has been developed through a series of projects with many of the SeaDataNet data services and directories having been created within the framework of its precursors, the EU-funded SEA-SEARCH project that was established in response to the lack of awareness of available marine data and information, and the EU-funded SeaDataNet project that was undertaken to overcome the fragmented delivery of data and information. The main objective of the latter project was to provide those stakeholders seeking to make use of marine data with a centralized overview and access to the available resources. Since these initial projects the SeaDataNet e-infrastructure has been developed further through a number of EU-funded projects that have supported the continuation of this pan-European marine data infrastructure.

Challenges

In its communication to the European Parliament and the Council entitled "Marine Knowledge 2020 - marine data and observation for smart and sustainable growth" (European Commission 2012) the EU states that "the creation of marine knowledge begins with observation of the seas and oceans". An impact assessment estimates that the collection of marine data by public institutions in EU Member States costs more than €1 billion annually, with a further €0.4 billion for marine-related satellite data. It also states that these data are largely collected "with a specific purpose in mind – for instance to exploit marine resources, to ensure safe navigation, to monitor compliance with regulations or to test a scientific hypothesis".

A related public consultation concluded that users continued to find it hard to discover what data already exists. The opportunity to develop innovative new products and services based on these data are therefore potentially lost due to restrictions on access, use and re-use or the pricing policy of some data providers. In addition the fragmented standards, different formats and nomenclature, lack of information on precision and accuracy, and insufficient temporal or spatial resolution act as further barriers to data re-use. Glaves, H. (2017, chapter 3 this book)

SOLUTIONS AND RECOMMENDATIONS

The SeaDataNet data infrastructure has been developed to address some of these issues by providing a coordinated approach to marine data management in Europe. It comprises a network of more than 100 interconnected data centres from across Europe that are responsible for the stewardship of marine data, coupled with the central SeaDataNet data discovery and access portal (http://www.seadatanet.org). The portal provides users with a unified and transparent overview of the metadata for the individual datasets and controlled access to the large collections of data sets, managed by the interconnected data centres.

The SeaDataNet portal infrastructure comprises the following middleware services:

- **Common Data Index (CDI) Discovery and Access Service:** delivering metadata directories and user interfaces that allow potential users to locate, assess and access the datasets available from the participating data centres.
- **Vocabulary Services:** used for populating the CDI metadata fields to support interoperability
- **Security Services:** supporting authentication, authorization and accounting
- **Delivery Services:** allows users to request and download available data sets and products
- **Viewing Services:** map interface that allows visualization of the metadata for the datasets held by the data centres
- **Product Viewing Services:** visualization tools for generic and standard map products
- **Monitoring Services:** providing usage and performance statistics for the SeaDataNet system and transaction management for data requests from users.
- **Maintenance Services:** entry and updating of metadata for the available datasets by data centres

The SeaDataNet architecture has been designed to provide a coherent system of the various services. A range of technical components have been defined for the purposes of implementation that make use of recent web technologies and also comprise Java components, to provide multi-platform support and syntactic interoperability.

Directories and Catalogues

SeaDataNet currently maintains and operates several directories as part of the overall e-infrastructure that provide information relating to organizations with interests in the marine domain and their engagement in marine research projects, management of large datasets, and data acquisition both by research vessel and through various monitoring programmes in the European seas and global oceans. Each of these directories is based on the ISO 19115 content model where applicable and has been harmonized using common vocabularies and mutually comparable XML schemas for exchange (XSD) between the national and pan-European directories wherever possible. These directories include:

- **European Directory of Marine Organisations (EDMO):** contains up-to-date addresses and activity profiles of research institutes, data holding centres, monitoring agencies, governmental and private organisations that are in some way engaged in oceanographic and marine research activities, data and information management and/or data acquisition activities. EDMO currently lists and describes more than 3000 organisations (March 2016)
- **European Directory of Marine Environmental Data (EDMED):** a comprehensive reference to the marine data sets and collections held within European research laboratories providing marine scientists, engineers and policy makers with a simple mechanism for their identification. It covers a wide range of disciplines including marine meteorology; physical, chemical and biological oceanography; sedimentology; marine biology and fisheries; environmental quality; coastal and estuarine studies; marine geology and geophysics; etc. Data sets are described in EDMED irrespective of their format (e.g. digital databases or files, analogue records, paper charts, hard-copy tabulations, photographs and videos, geological samples, biological specimens etc.). EDMED describes currently more than 3500 data sets, held by more than 700 data holding centres across Europe.
- **European Directory of Marine Environmental Research Projects (EDMERP):** provides information on marine research projects for a wide range of disciplines including marine meteorology; physical, chemical and biological oceanography; sedimentology; marine biology and fisheries; environmental quality; coastal and estuarine studies; marine geology and geophysics etc. The primary objective of EDMERP is to support users in identifying research activities and connecting them to the related research managers and organisations across Europe. EDMERP currently describes more than 2500 research projects.
- **European Directory of the Ocean Observing Systems (EDIOS)**: gives an overview of the ocean measuring and monitoring systems operated by European countries. The directory includes discovery information on location, measured parameters, data availability, responsible institutes and links to data-holding agencies plus additional technical information on instruments such as sampling frequency etc. EDIOS can also be used to deliver an overview of marine monitoring programmes that are relevant for the implementation of the Marine Strategy Framework Directive

(MSFD 2008). EDIOS currently includes more than 10,000 entries for European marine monitoring systems.

- **Cruise Summary Reports (CSR):** a directory of the cruises or field experiments conducted at sea. The cruise summary report (CSR) is submitted to the relevant National Oceanographic Data Centre (NODC) and provides a first level inventory of the measurements and samples collected during the cruise or field experiment. Formerly known as the Report of Observations/Samples collected by Oceanographic Programmes (ROSCOPS), the Cruise Summary Reports directory currently covers cruises conducted since 1873 by more than 2000 different research vessels and includes records for a total of more than 45,000 cruises.

Common Standards

SeaDataNet has developed and maintains a set of common standards for the marine domain by adopting and adapting relevant ISO and OGC standards. These standards, which have been created in collaboration with European and international experts within the frameworks of the IOC-IODE (Intergovernmental Oceanographic Commission's International Oceanographic Data and Information Exchange programme), ICES (International Council for Exploitation of the Seas) and ODIP (Ocean Data Interoperability Platform), include:

- **Common Data Index (CDI):** a profile of the ISO 19115 standard for describing metadata for marine data sets (Boldrini 2013a; Boldrini 2013b). It has subsequently been upgraded to the ISO19139 profile for compliance with the European INSPIRE directive that aims to create a European Union (EU) spatial data infrastructure that will enable the sharing of environmental spatial information among public sector organisations and better facilitate public access to spatial information across Europe (EU 2007).
- **Cruise Summary Reports (CSR):** marine metadata profiles derived from the ISO 19115-2 standard which also include XML schemas and related schematron rules (Boldrini 2013c; Boldrini 2013d). The updated ISO 19139 schema that forms part of the EU INSPIRE Directive Implementing Rules is also fully mature and has been widely adopted.
- **Controlled Vocabularies**: the common vocabularies established by SeaDataNet consist of more than 60 lists that include more than 160,000 standardised terms that cover a broad spectrum of disciplines relevant to both the oceanographic and the wider community.

The use of common vocabularies is an important prerequisite towards consistency and semantic interoperability; their use resolves the problems of ambiguity associated with data mark-up and also enables records to be interpreted by computers. It also opens up the data sets to the possibilities of computer aided manipulation, distribution and long term re-use. Common vocabularies require both content governance (concept population and semantic descriptions) and technical governance (content storage and distribution) services to ensure consistency. Technical governance for SeaDataNet vocabularies is based on the NERC DataGrid (NDG) Vocabulary Server Web Service API while the content governance is administered by a combined SeaDataNet and MarineXML Vocabulary Content Governance Group (SeaVoX), and moderated by the British Oceanographic Data Centre (BODC).

In order to support the evolving requirements of the user community it was necessary to upgrade the previous NERC Vocabulary Service (NVS). An enhanced instance of the service, NVS 2.0, was developed that included the following improvements:

- a move to the accepted version of the World Wide Web Consortium's (W3C) Simple Knowledge Organization System (SKOS) specification for encoding the data dictionaries and taxonomies served through the NERC Vocabulary Server (NVS) (Buck and Lowry (2017, chapter 1 this book)
- the ability to serve multilingual titles and definitions for resources
- the provision for mappings to external resources enabling the results of ontology extension to be delivered.

The updated NVS 2.0 has been operational and stable since early 2013. All SeaDataNet resources (CDI, EDMERP, CSR, EDMED, and EDIOS), formats (ODV, NetCDF), tools (MIKADO, NEMO, ODV) and services now use NVS 2.0 which supports improved version management and unique URNs for concepts.

- **Standard Data Exchange Formats:** the SeaDataNet infrastructure provides access to the available data sets via dedicated download services. To support delivery of this data to users a suite of common data transport formats have been implemented that are compatible with the other SeaDataNet standards in use, such as the common vocabularies and the quality flags, and also with the SeaDataNet data analysis and presentation tools: Ocean Data View (ODV) and Data-Interpolating Variational Analysis (DIVA). (Barth, et al. (2017, chapter 15 this book)
  The standardised data transport formats implemented by SeaDataNet are:
  - SeaDataNet ODV4: an ASCII output format used for profiles, time series and trajectory data.
  - SeaDataNet NetCDF with CF compliance: used for profiles, time series and trajectories,
  - SeaDataNet MEDATLAS: an additional optional format that can also be used for vertical profiles and trajectories
  - NetCDF with CF compliance: suitable for 3D observation data such as ADCP
- **Standardised Quality Assurance (QA) and Quality Control (QC) Procedures**: the quality of data is dependent on a number of factors including the type of sensor, the calibration of the instrument, methodologies for obtaining measurements etc. SeaDataNet has implemented common quality control procedures for the interconnected data centres (SeaDataNet 2010) "to ensure the data consistency within a single data set and within a collection of data sets and to ensure that the quality and errors of the data are apparent to the user who has sufficient information to assess its suitability for a task." (IOC/CEC Manual, 1993) The SeaDataNet standards, which include those for quality control and quality flags, have been developed in cooperation with international organisations, for example, IOC-IODE and ICES, and other relevant projects and programmes such as WOCE, JGOFS, MyOcean etc. in order to derive a common set of recommended standards for the quality control of various marine data types.

Software Tools

In addition to the implementation of these standards and services, SeaDataNet has also developed a set of dedicated software tools that are available for use by the data centres and can also be freely downloaded by users via the SeaDataNet portal. The SeaDataNet software tools are provided for a range of applications including metadata generation, statistical analysis, grid interpolation, data visualization etc. The tools available for download and use include:

- **MIKADO**: a UNIX or Windows based software tool, written in Java, and used to create XML metadata entries for the SeaDataNet directories EDMED, CSR, EDMERP, CDI and EDIOS. These XML entries can either be created manually or automatically generated through direct interfacing with local databases. MIKADO uses the most recent versions of the SeaDataNet XML schemas in combination with the common vocabularies (NVS 2.0), the EDMO and EDMERP directories and the new INSPIRE compliant CDI and CSR schemas and XML formats.
- **Med2MedSDN**: for the purposes of standardisation and international exchange, specific extensions to the MEDATLAS format have been defined for SeaDataNet. These extensions provide a linkage between the data and metadata in the SeaDataNet system, and also to standardised SeaDataNet semantic information such as detailed parameter descriptions. The Med2MedSDN software is used to reformat MEDATLAS file(s) without the SeaDataNet extensions to the SeaDataNet MEDATLAS format. Users can convert individual MEDATLAS files or a directory containing multiple MEDATLAS files.
- **NEMO:** data centres may need to pre-process their data holdings into the standard data transport formats required for delivery via the SeaDataNet infrastructure. The NEMO software tool is provided to support conversion of data in any ASCII format to the SeaDataNet ODV and MEDATLAS ASCII formats or to the SeaDataNet NetCDF (CF) format.
- **Download Manager (DM):** The SeaDataNet data discovery and access service supports automatic processing of data set requests submitted to the participating data centres as far as possible. To implement this functionality the data centres require a local installation of the Download Manager (DM) software tool that handles all communications between the local data centre system and the CDI Request Status Manager (RSM) that oversees the processing and administration of all requests for access to data sets submitted by the Common Data Index discovery and access service. The RSM also allows users to monitor the progress of their requests for data following submission to the responsible data centre via the SeaDataNet portal.
- **Ocean Data View (ODV):** a software package used for the interactive exploration, analysis and visualization of oceanographic and other geo-referenced profile, time-series, trajectory or sequence data. The software provides functionality for displaying raw data points or gridded fields based on the original data. It has two fast weighted-averaging gridding algorithms built in and also allows integration of the advanced DIVA software tool. Data from the ARGO, World Ocean Circulation Experiment (WOCE), World Ocean Database, World Ocean Atlas, SeaDataNet, and MEDAR/MEDATLAS projects can be directly imported into ODV. It also supports the netCDF format allowing exploration of various compliant datasets.
- **Data-Interpolating Variational Analysis (DIVA):** software tool developed for gridding in-situ data. The analysis is performed using a finite element method allowing for a spatial variable resolution and a good representation of the coastline and isobaths. The finite-element grid of DIVA

also has the advantage of being able to adequately resolve complex coastlines such as those found around some European seas.(Barth, et al. (2017, chapter 15 this book)

SeaDataNet Architecture

The SeaDataNet architecture has been designed as a multidisciplinary system. It is able to support a wide variety of data types and serve a number of different communities. SeaDataNet is also actively sharing its technologies and expertise, expanding its approach to other disciplines, and establishing collaborative relationships with other well established infrastructures in the marine domain. As a result SeaDataNet has taken an active role in a number of related European marine data management projects including Geo-Seas, Upgrade Black Sea Scene, EuroFleets, Jerico, Jerico Next, CLIPC, Micro-B3, AtlantOS to name but a few.

SeaDataNet has also established a close cooperation with EuroGOOS, an international non-profit organisation committed to European ocean observing and operating within the context of the Global Ocean Observing System of the Intergovernmental Oceanographic Commission of UNESCO (IOC GOOS); and also with the Copernicus Marine Environment Monitoring Service where SeaDataNet supports improvements in the service for handling historical/reprocessed data.

Common Data Index (CDI) Data Discovery and Access Service

The core element of the SeaDataNet architecture is the Common Data Index (CDI) data discovery and access service that provides on-line access to a vast array of marine datasets managed by distributed data centres across Europe. The CDI discovery and access service provides users with a highly detailed overview of the geographical coverage and associated metadata data for the available data sets. The CDI service provides a single 'shopping basket' type interface that allows users to select datasets of interest and submit requests for access to the individual data centres. The user can then monitor the progress of the request via the Request Status Manager (RSM) and, on completion, download datasets in the SeaDataNet standard formats via the portal.

The CDI data discovery and access service has two user interfaces:

- **CDI Quick Search interface:** provides users with a faceted search mechanism. The results are delivered to the user as an overview of the relevant CDI entries both on a map and filtered by the user selected metadata fields. The user can then refine the search to reduce the number of results using additional search terms to drill down further into the records returned by the search.
- **CDI Extended Search interface:** search mechanism using a combination of several criteria that can be defined by the user as either free text or terms selected from pull down menus. The user can further refine the query criteria to filter the results returned by the query.

As a result of SeaDataNet participation in a range of European projects the number of interconnected data centres has risen to more than 100, providing metadata for and access to in excess of 1.8 million datasets that originate from over 500 organisations across Europe. This data has been acquired in European waters and the global oceans and covers physical, geological, chemical, biological and geophysical parameters (Figure 1).

Figure 1. Overview of datasets available in the Common Data Index (October 2015)

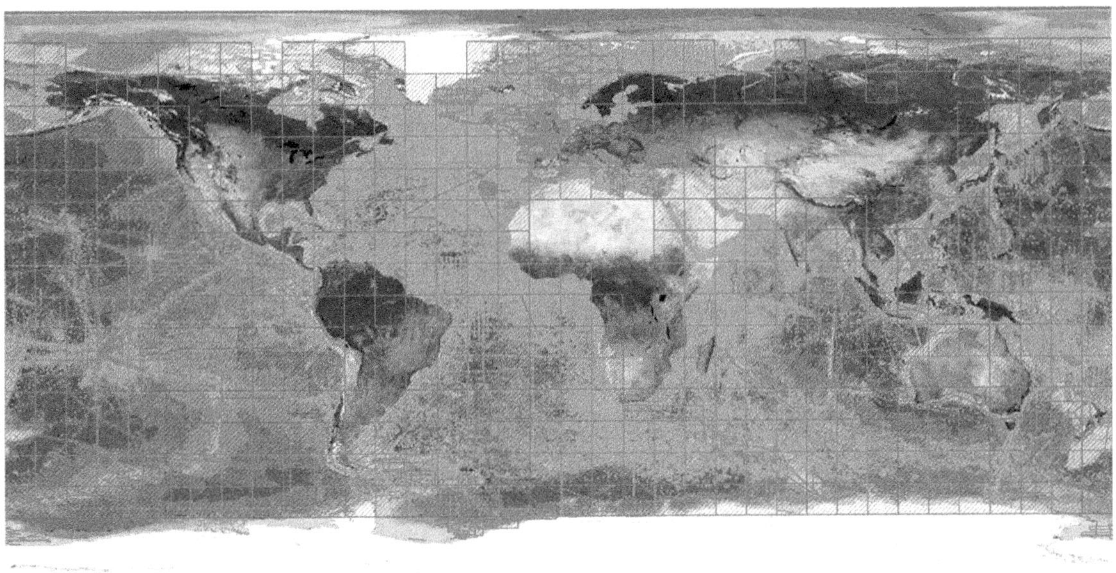

How Does it Work?

Figure 2 illustrates the process for discovering and accessing marine datasets in the Common Data Index (CDI). Both of the CDI user interfaces allow users to define a set of search criteria which filter the results of the query that are then returned both as a list of available datasets and displayed as geographical locations on the map-based interface. Users can then display the full metadata record for a specific dataset by selecting it from the list of query results. The extended metadata record provides details of the what, where, when, how, and who for the data set and also includes full details of the data access restrictions that apply.

Users are free to query and browse the CDI metadata directory anonymously but, in order to submit requests for data access, users must be registered in the SeaDataNet central user directory and have agreed to the terms of the SeaDataNet User Licence. This registration process provides users with access to the

Figure 2. 'Shopping basket' process for requesting access to data sets for downloading from distributed data centres

full functionality of the SeaDataNet discovery and access service including use of the shopping basket mechanism to order multiple datasets from different data centres simultaneously.

All requests for data access are forwarded automatically from the SeaDataNet portal to the relevant data centres. The user placing the request also receives a confirmation by e-mail that includes details of how to monitor the progress of their order. The process of submitting user requests to access datasets held by the distributed data centres is controlled by the Request Status Manager (RSM) service. It is also used for registered users to check the status of their requests and download the data sets once the order has been processed by the data centre. The connected data centres can also use the RSM to manually intervene where negotiation with the user is required before access to a specific dataset can be granted and to monitor usage statistics for their data holdings.

As well as user services the SeaDataNet CDI also features machine-to-machine interfaces which are used internally to combine specific data sets for the purposes of generating harmonised and validated data collections and data products. These machine-to-machine services include a CDI robot harvester which uses the CDI shopping mechanism to discover and retrieve specific data sets from distributed data centres and maintains a central data buffer for a given set of criteria. These facilities are used for generating SeaDataNet products such as a European Temperature and Salinity Climatology and EMODnet products such as Interpolated Chemistry maps for European seas.

SeaDataNet and the European Marine Observation and Data Network (EMODnet)

SeaDataNet directly supports the goals of the European Marine Observation and Data Network (EMODnet) by providing key technical elements for the discipline specific portals e.g. bathmetry, chemistry. The EMODnet initiative was created in response to European Commission's Marine Strategy Framework Directive (MSFD, 2008) that aims to establish a more coordinated approach to marine monitoring in European waters. It has been set-up as a network of existing and developing European observation systems, linked by a data management structure covering all European coastal waters, shelf seas and surrounding ocean basins. It will facilitate long-term and sustainable access to the high-quality data necessary to understand the biological, chemical and physical behaviour of seas and oceans. Key elements for harmonisation and interoperability are establishing and adopting common standards and protocols for quality control, metadata and data formats, vocabularies and technical protocols such as those developed and implemented by SeaDataNet. EMODnet is coordinated at EU level with the other European directives and large-scale framework programmes on European and global scales (GMES and GEOSS), that urge access to, and exchange of, environmental data and information.

SeaDataNet supports the EMODnet objectives both by providing the required operational infrastructure and further developing its services and products in response to the evolving requirements for implementation of European policy for the sustainable exploitation of the seas.

The EMODnet bathymetry and chemistry portals make use of the SeaDataNet standards and basic infrastructure. As a result a number of additional European marine data centres have become connected to SeaDataNet infrastructure and are delivering their metadata and data via the CDI discovery and access services.

The SeaDataNet infrastructure has played a major role in the development and implementation of EMODnet. This is illustrated through its contribution to the development of the discipline specific portals that make up the EMODnet framework.

EMODnet Chemistry

The EMODNet Chemistry 'lot' focuses on assembling measurement data, metadata and data products, for groups of chemical variables in the marine environment. It also aims to facilitate data access and re-use, and provide dynamic visualization services for aggregated and validated data collections. The focus for EMODnet Chemistry is on the following European seas: Baltic Sea, North Sea area (including Norwegian Sea and Celtic Sea), Atlantic Sea area (including Atlantic Coast and Macaronesia), Black Sea, and Mediterranean Sea. The data types of interest include observations for nutrients, chlorophyll, oxygen and contaminants in three matrices (water, seabed and biota).

Methodology

EMODnet Chemistry has aligned it activities with two key European directives:

- **INSPIRE (2007):** standards and implementing rules for a harmonized and interoperable EU Geographic Information Infrastructure for Geographic Data, Metadata and Services;
- **MSFD (2008):** for monitoring seas and oceans at National, Regional and EU level. The reporting (WISE Marine) of the environmental status is based on: Descriptors, Criteria and Indicators with the objective of the definition and achievement of a Good Environmental Status for the Euroepan seas.

EMODnet Chemistry has adopted the SeaDataNet Common Data Index (CDI) infrastructure with its network of distributed National Oceanographic Data Centres (NODCs) and marine data centres to provide an integrated and harmonised overview and access to marine chemical data. It relies on the network of connected data centres that are in charge of collecting, quality checking, standardizing and sharing oceanographic data in each of the countries around the European seas. The NODCs are in permanent contact with the data originators within their country to facilitate the submission of new datasets generated during research cruises or monitoring activities to the European decentralized infrastructure. EMODnet Chemistry follows the approach of adopting and adapting the SeaDataNet infrastructure and standards for chemical data, using:

- SeaDataNet CDI metadata format and XML schemas, based on the ISO 19115 and 19139 standards for data indexing
- SeaDataNet common vocabularies covering a broad spectrum of disciplines to mark-up both metadata and data
- Standard data transport formats ODV ASCII, MEDATLAS and NetCDF (CF) for delivery of data
- Common quality control methods, standards and quality flag scale
- Web services for connecting various modules in the architecture
- OGC services (WPS, WMS, WFS) for viewing and downloading services for data products or subsets via the OPeNDAP protocol.

Data Validation Loop

A particular challenge when merging heterogeneous data coming from different sources, as in the case of the SeaDataNet infrastructure, is the quality of the data which also has a strong influence on any resulting product that is generated. In an attempt to address this issue two parallel quality management workflows are operated:

- A data quality loop, performed "ex-post" at a regional scale on the local data buffers;
- Collection of quality information "ex-ante", to enrich metadata with information related to the source laboratory's analysis (based on ISO/IEC 17025/2005).

The data quality loop includes an initial set of checks that are carried out by the originators prior to the inclusion of the data in the infrastructure. The NODCs then undertake to further enrich the data with related metadata (e.g. who, when, where and how measurements were taken) in order to ensure the long term re-use of the harvested information, and to introduce a set of basin scale quality control checks to flag any outliers in the data. These checks include:

- **Position:** Latitude and longitude must be valid i.e. latitude in range -90 to 90 and longitude in the range -180 to 180, and position must not be on land
- **Date:** Date and time of an observation has to be valid, i.e. year with 4 digits, month between 1 and 12, day in the range expected for the month, hour between 0 and 23, minute between 0 and 59
- **Missing Value:** Where a data value is missing a pre-defined number will represent the absent data
- **Range:** Tests that the observed parameter values are within the expected extremes encountered in the ocean and/or in specific regions)
- **Spike:** Tests differences between sequential measurements, where one measurement differs significantly from those adjacent to it)
- **Pre-Existing Statistics**: Tests that values are within ranges defined by statistics for specific areas/ regions with reference to a specific climatology, as e.g. Levitus or Medar vertical climatologies)
- **Visualization and/or Manual Checking**: a visual screening with the support of plots to highlight outliers and unexpected trends in the data caused, for example, by use of wrong measurement unit, are valuable for certain data types

Once the data has been checked, flagged and the metadata completed, they are then added to the SeaDataNet CDI discovery and access service. As a result of this activity in EMODnet Chemistry more than 600,000 additional entries were added to the CDI which can be discovered and accessed both through the SeaDataNet portal and the dedicated EMODnet Chemistry portal (http://www.EMODnet-chemistry.eu).

SeaDataNet is a distributed infrastructure with the data managed by the interconnected European data centres. However for the purposes of EMODnet Chemistry the metadata and data sets need to be aggregated at the regional level for generating data products that can support the implementation of the Marine Strategy Framework Directive (MSFD) by contributing to the determination of MSFD indicators, assessing environmental conditions and impacts for each of the regional European seas. To facilitate this process, the robot harvester system was developed as additional service within the SeaDataNet architecture. It is configured with space, time and parameter limits to collect relevant metadata and data from the distributed data centres using machine-to-machine services. This data is then held in five distinct buffers, one for each of the European regional seas, which are then prepared and transferred to the regional coordinators who perform the aggregation and regional quality control following the SeaDataNet common guidelines and using the Ocean Data View (ODV) software to create the harmonised regional

EMODnet Chemistry data collections. For the purposes of quality control regular reports are sent to the data collators who identify any errors or anomalies in the master copy of the data and correct them. This feedback loop guarantees the data quality and ensures that users have access to the highest quality version of the datasets.

From Data to Products and from Products to Knowledge

One challenge for EMODnet Chemistry is to collect, check, standardize and make available the fragmented and heterogeneous marine chemistry data, and to build knowledge from the assembled information. This is done by preparing a set of aggregated data products that highlight specific aspects of the data. As a result, a set of products based on the regional quality checked, aggregated and harmonised data collections are generated that include:

- Interpolated maps generated with a 10-year moving window over the last 55 years allowing rapid visualisation of temporal trends; they are produced on an annual and seasonal scale and for selected nutrients in the water column according to data availability using the Data-Interpolating Variational Analysis (DIVA) software tool. (Barth, et al. (2017, chapter 15 this book)
- Dynamic station distribution maps, time series plots and vertical profile plots are created by users through customized selection of parameter, depth and time range (using standard OGC web processing services (WPS) integrated into the viewing service).

As shown in Figures 3, 4 and 5, these aggregated data products can be visualised using the OceanBrowser tool, a web service based on Open Geospatial Consortium (OGC) standards originally developed for SeaDataNet, it has been adapted for the requirements of EMODNet Chemistry.

Figure 3. OceanBrowser interface in the EMODnet Chemistry portal showing visualisation of time series data for concentrations of phosphates by season over 10 years in the Baltic Sea

Figure 4. OceanBrowser interface in the EMODnet Chemistry portal showing the positions and density of the chemical observations included in the aggregated, validated and harmonised data collection

Figure 5. OceanBrowser interface in the EMODnet Chemistry portal showing a visualisation of the data at selected locations in the dynamic plots created using WPS services

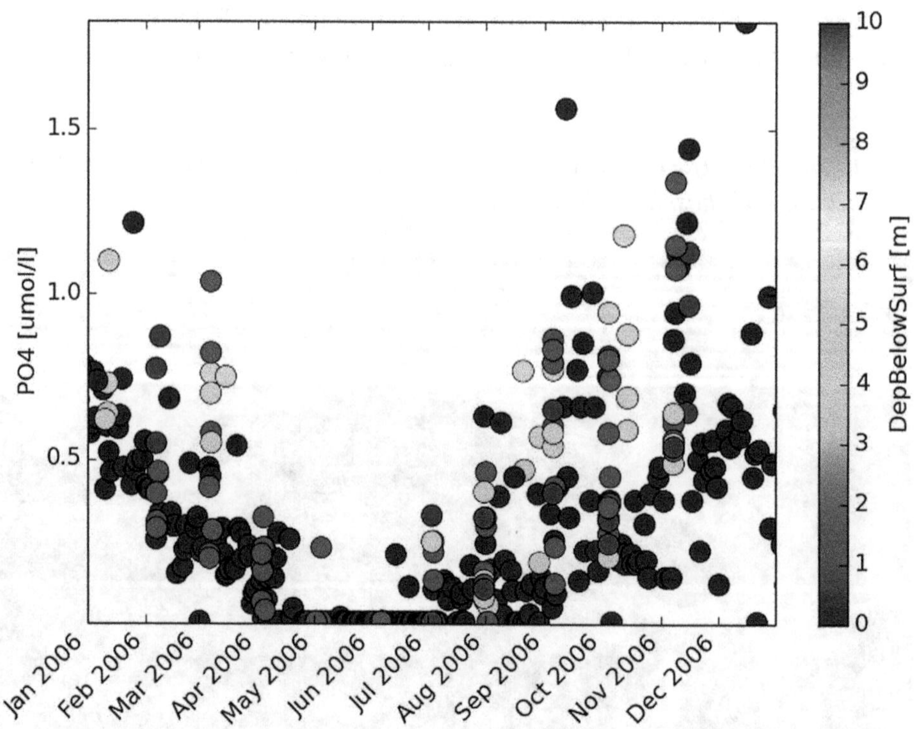

Due to the relatively low density of the observations for chemical parameters in comparison to those for other physical parameters it is also necessary to clearly show the relationship between the derived interpolated data products and underlying observations. For this reason each of the interpolated DIVA maps is also displayed with a layer that shows the underlying observations that have been used as the basis for the map, and which are also directly linked to the CDI data discovery and access service.

The EMODNet Chemistry tools and services provide a rapid overview of the chemical measurements covered by the MSFD Descriptors 5 (Eutrophication), 8 (Contaminants and pollution effects) and 9 (Contaminants in fish and seafood) that are used to assess the Good Environmental Status (GES) of the European seas. The customised tools, provided as part of the EMODnet viewing and downloading services, allow the extraction of the knowledge and information necessary to support the aims and objectives of the European Marine Strategy Framework Directive relating to the sustainable use of marine resources.

EMODnet Bathymetry

The EMODnet Bathymetry portal (http://www.EMODnet-bathymetry.eu) provides a further exemplar for the adoption of the SeaDataNet approach. The objective of the EMODnet Bathymetry pilot project is the discovery and access of bathymetric data both as raw survey data and as composite DTMs that are held by an increasing number of data providers in the governmental and research sectors. The portal also provides a service for viewing and downloading a harmonised Digital Terrain Model (DTM) for the European sea regions that has been generated by the EMODnet Bathymetry partnership using data collated from multiple sources.

Methodology

EMODnet Bathymetry has adopted the SeaDataNet Common Data Index (CDI) infrastructure for the purposes of collating and delivering an integrated and harmonised overview of bathymetric survey data sets that are being managed by national hydrographic services, marine research institutes and industry in Europe. The EMODNet Bathymetry consortium consists of representatives from these organisations as well as other potential data providers that are being encouraged to make with their bathymetric data available for wider use. As a result of this initiative there are approximately 14,000 bathymetric survey datasets from 27 data centers available via EMODnet Bathymetry CDI discovery and access service that cover 14 countries and originate from 167 institutes (March 2016). In addition a number of data providers have supplied composite data products (DTMs or clouds of soundings from multiple surveys) as an alternative to the bathymetric data sets which are delivered via the EMODnet Sextant Catalogue service; an INSPIRE compliant web service using the marine profile of the ISO19139 standard and the European Directory of Marine Organisations (EDMO), the European Directory of Marine Environmental Research Projects (EDMERP) and the SeaDataNet Common Vocabularies to ensure consistency with the survey data delivered via EMODnet Bathymetry CDI discovery and access service.

The EMODnet Bathymetry initiative has also generated a digital terrain model (DTM) for all of the European sea basins incorporating around 7000 of the available bathymetry data sets and 30 composite DTMs. Any gaps in the DTM coverage have been completed using either the latest General Bathymetric Chart of the Oceans (GEBCO 2014 30" gridded data) DTM or data from the Baltic Sea Bathymetry Database operated by the Baltic Sea Hydrographic Commission.

Figure 6. Overview of bathymetric survey data sets available via the EMODnet Bathymetry CDI discovery and access service

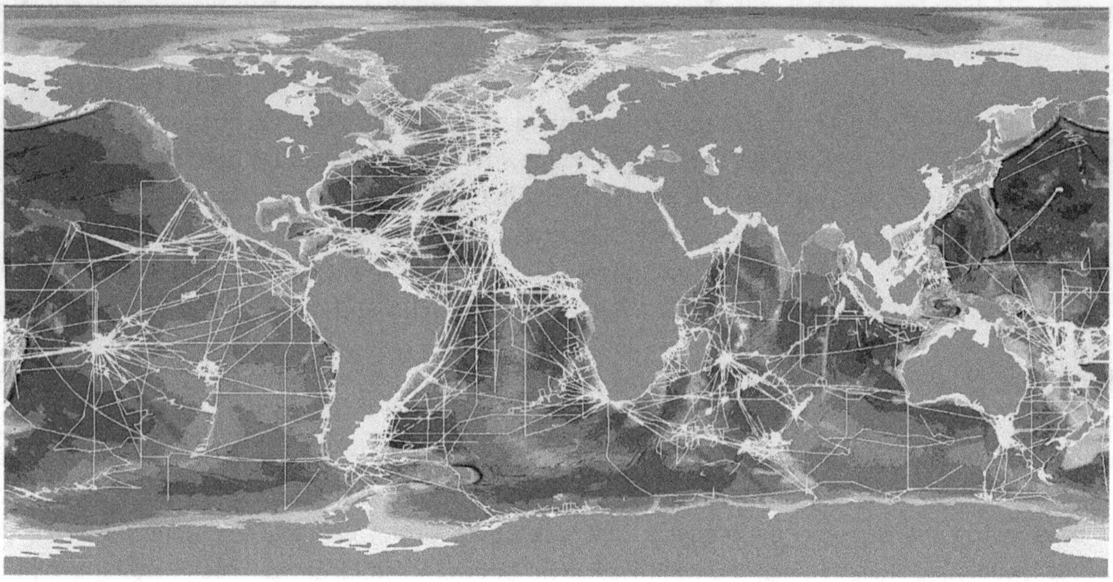

The methodology for generating the EMODnet bathymetric DTM for the seas of Europe involves a number of actors including local data providers, regional coordinators and those with overall responsibility for integrating the various components of the DTM to produce the single unified data product. The data providers contribute either bathymetric survey data or composite DTMs that have been created following the guidelines in the EMODnet Bathymetry processing manual (Vossen 2011) and preferably using the standard Global Oceanographic Bathymetry Explorer (GLOBE) software.

GLOBE is recommended for the production of the digital terrain models due to the in-built functionality that has been incorporated, and subsequently updated and enhanced, according to the specific requirements of EMODnet Bathymetry. For example, a spline interpolation method and additional functionality to allow re-projection of the raw datasets before adding to the regional EMODnet DTMs had been added in recent years which has greatly reduced gridding artefacts in the resulting regional DTMs.

The GLOBE software can also handle a range of data types e.g. multibeam surveys, seismic surveys, images etc., and in multiple formats including NetCDF (CF), standard grids and sensor acquisition formats which can be used to create post-processed products compatible with WMS, GoogleEarth etc. It also provides support for seafloor mapping (elevation, backscatter, gridded, 4D data, water column echoes, rasters and grids) and includes a 3D viewer that is based on the of NASA WorldWind technologies. This range of functionality has made the GLOBE software the preferred solution for the majority of the EMODnet partners but for a small number an alternative in-house solution such as CARIS tools was selected for creation of the component DTMs.

Once the local data centres have delivered their bathymetric data or regional DTMs the nominated regional coordinators select, compile and process those datasets that will generate the best possible digital terrain model (DTM) for their sea region also using the GLOBE software. These regional digital terrain models will then be integrated into the harmonised EMODnet DTM with final quality control/assurance tests carried out to ensure that the model exposed via the EMODnet Bathymetry Viewing and

Download service is of the highest possible quality and that the CDI and Sextant references present in the DTM cells are valid. These cross-references allow users to interactively identify the survey data sets and composite DTM products have been used to generate the values in each grid cell of the resulting EMODnet DTM.

The EMODnet Bathymetry digital terrain model (DTM) is based on a latitude/longitude horizontal coordinate system and a WGS84 datum at a resolution of 0.125 arc-minutes. Each grid cell has a minimum, maximum and average water depth over the cell, plus ancillary information such as statistics for the average water depth (standard deviation, number of interpolations, number of elementary surfaces), smoothed average water depth determined by a spline function and offset of smoothed water depth related to the average. In addition each grid cell will contain an identifier indicating the source material used for the computation of the depth combined with the ID of the data provider as shown in the European Directory of Marine Organisations (EDMO). In those cells where more than one dataset is referenced the source dataset whose soundings are the most frequently represented in the cell will be identified and associated to the description of the survey dataset in the CDI Data Discovery and Access service. For those instances where a composite DTM has been used to populate cells without soundings a reference to the associated resource in the Sextant Catalogue service is provided. In those cases where there is no other data a reference to the GEBCO 2014 grid is provided (Figure 7).

The most recent version of the EMODnet Bathymetry DTM, illustrated in Figure 8, was released and made available in the Bathymetry Viewing and Downloading Service during September 2015. The EMODnet DTM will be upgraded further in the future to incorporate additional survey data and improvement of the digital bathymetry.

The EMODnet Bathymetry portal (http://www.EMODnet-bathymetry.eu) provides access to interconnected services that allow users to identify and request access to the bathymetric survey data sets (via the CDI service) and the composite DTM data products (via Sextant Catalogue) for the European seas.

Figure 7. Source references layer with indication of which data sources have been used for the resulting DTM

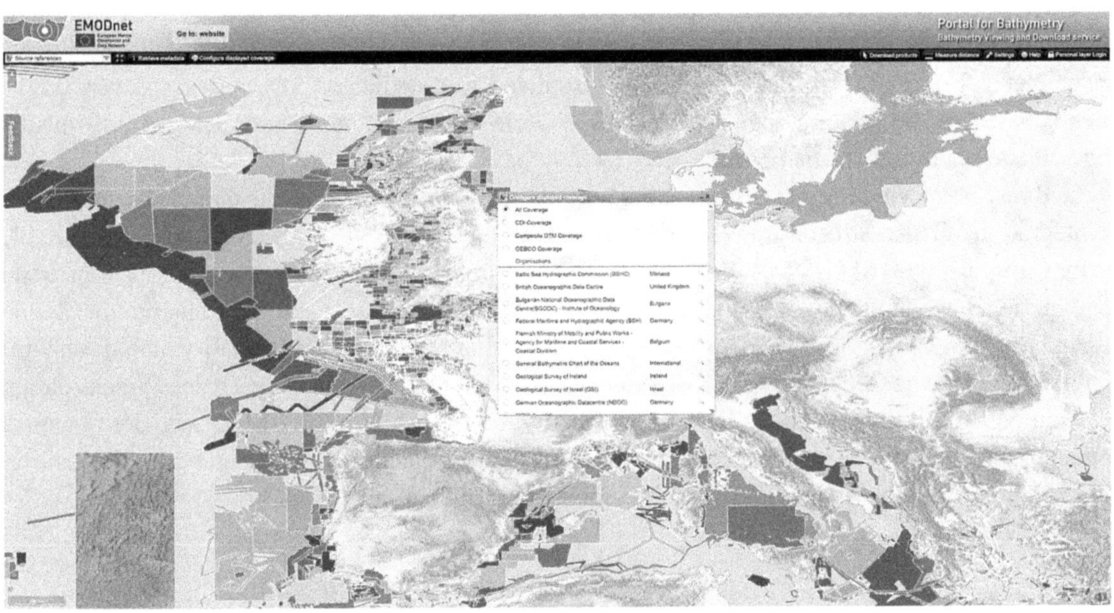

Figure 8. Harmonised bathymetry for all European seas – release September 2015

The associated Bathymetry Viewing and Downloading service which is based on a PostgreSQL 9.3 RDBMS and the latest version of GeoServer in combination with OpenLayers 3 for developing the user interface, incorporates extensive functionality and GIS layers (see Table 1) that allow users to view, interrogate and download the latest version of the EMODnet DTM and also view and download the EMODnet digital bathymetry.

The Bathymetry Viewing and Downloading service functionality includes:

- Gridded water depth on a DTM grid of 1/8 by 1/8 arc minute of longitude and latitude (approximately 230 by 230 metres resolution)
- option to view depth parameters for individual DTM cells and references to source data
- option to download DTM in 16 tiles in various formats including: ESRI ASCII grid, XYZ, EMODnet CSV, NetCDF (CF), GeoTiff and Secure Digital (SD)

The NetCDF (CF) DTM files can also be visualised in a dedicated 3D viewer software package which is based on the existing open source NASA World Wind JSK application and is freely available for download via the EMODnet portal.

The Bathymetry Viewing and Downloading Service has also been upgraded to include a seamless land-sea integration of Open Street Map WMS and the land topography WMS with a resolution comparable to that for the updated DTM (1/8 by 1/8 arc minutes). The addition of these map services has improved the user experience when viewing the updated EMODnet DTM by giving a smooth transition between the land and sea, and no visible grid edges in the model. Further additional functionality includes a wrecks layer derived from the OceanWise WMS service that uses the UK Hydrographic Office Wrecks database as its source. An undersea features layer is provided via the WMS from GEBCO who are responsible for maintaining and delivering a gazetteer of the names, generic feature type and geographic position of features on the seafloor via its Sub-Committee on Undersea Features (SCUFN).

Table 1. Layers available the in EMODnet Bathymetry Viewing and Downloading service

Layer	Description
Geographic grid	Geographic grid lines (latitude and longitude graticule)
Coastlines	Provided by OpenStreetMap 2014
Sea names	Maintained and provided by SeaDataNet
Depth contours	Generalised bathymetric contour lines (50, 100, 200, 500, 1000, 2000, 5000 meter intervals)
GEBCO Undersea Features	GEBCO Sub-Committee on Undersea Feature Names (SCUFN) maintains and provides a gazetteer of the names, generic feature type and geographic position of features on the sea floor.
Wrecks	Wrecks from the UKHO database provided by OceanWise.
Source references	Coverage of individual bathymetric surveys and composite DTMs that contribute to the EMODnet Bathymetry DTM.
Survey tracks/polygons	Tracks/polygons of all bathymetric surveys which are included in the CDI Data Discovery and Access service.
Land geography and topography	Maintained and provided by OpenStreetMap.
Coastal areas with high resolution bathymetry	Demonstrator for integrating high resolution coastal sections
Mean depth in multi-colour style (no land data)	Mean depth based on source resolution of 1/8 arc minute (~250 meter). Ocean overlay without land cover in multi-colour style.
Mean depth in rainbow colour ramp (no land data)	Mean depth based on source resolution of 1/8 arc minute (~250 meter). Ocean overlay without land cover in rainbow (multi-colour) ramp style.
Mean depth full coverage	Mean depth based on source resolution of 1/8 arc minute (~250 meter). Full coverage ocean and land DTM in traditional atlas style colours.
GEBCO bathymetry base map	Backdrop base map showing the Earth based on satellite imagery (Blue Marble) and GEBCO 2014 bathymetry (General Bathymetric Chart of the Oceans) with source resolution of ½ arc minute ((~1000 meter).

Figure 9. DTM loaded into EMODnet Bathymetry 3D viewer

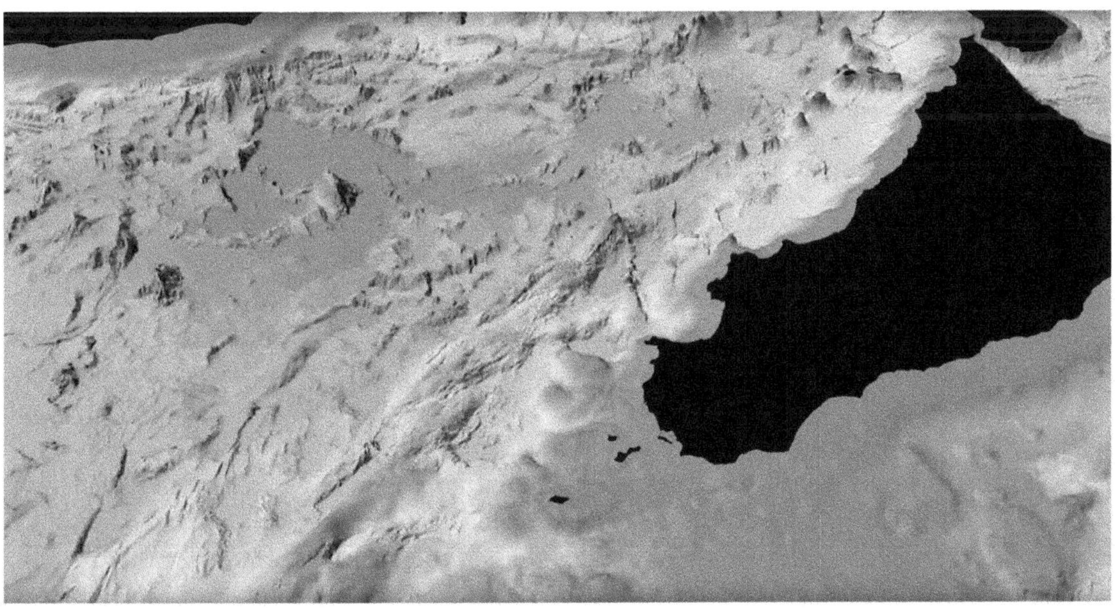

EMODnet Bathymetric Grids: Incorporation into GEBCO

The General Bathymetric Chart of the Oceans (GEBCO), through its Sub-Committee on Regional Undersea Mapping (SCRUM), aims to work with regional marine mapping initiatives such as EMODnet, and capitalise on the regional expertise to build authoritative regional and global bathymetric models. As part of a bilateral cooperation GEBCO integrates the updated releases of the EMODnet DTM into its GEBCO grid at ½ minute resolution while EMODnet uses the GEBCO grid to fill gaps in the coverage of the DTM. This synergy and two-way exchange between EMODnet and GEBCO will ultimately lead to better products for both initiatives and also minimise the issues where EMODnet grids based upon survey data are joined or adjacent to those created by GEBCO.

FUTURE DIRECTIONS

The continued cooperation and EU-funded support for the consecutive SeaDataNet and SeaDataNet II projects (funded as a Research Infrastructure in the European Commission's FP6 and FP7 framework programmes), and the implementation of the SeaDataNet infrastructure and standards in the development of the European Marine Observation and Data Network (EMODnet) have contributed significantly to the long-term accessibility of heterogeneous marine observations. An infrastructure of more than 100 interconnected data centres from 35 countries around the European seas is providing access to in excess of 1.8 million datasets, a range of pan-European metadata and data services, and the SeaDataNet standards which have also been widely adopted throughout the marine community.

Nevertheless, the SeaDataNet initiative continues to face new challenges due to the highly diverse nature of marine data, the number of existing European data centres dealing with marine data that have yet to become part of the federated SeaDataNet e-infrastructure, and the drive for a common global approach to marine data management being promoted by international initiatives. The specific challenges that need to be addressed to ensure the continued sustainability of the SeaDataNet data infrastructure include:

- Optimization and streamlining of data transmission directly from the ocean observing systems to the data centres using new protocols such as the Sensor Web Enablement (SWE) suite of standards, both for research vessels and automatic observatories. (Buck and Lowry (2017, chapter 1 this book; Oggioni et al. (2017, chapter 9 this book; Diviacco et al. (2017, chapter 8 this book). The objectives are: 1) making data from ocean sensors available as soon as possible, 2) guarantee that all metadata related to the observation are delivered with the data and are well preserved; and 3) directly involve platform/network operators in the data management by providing them with services for synthesis of observation sampling, feedback and alert systems for data availability and quality, and long term storage facilities (e.g. cloud technologies). To ensure consistency and alignment these activities will be conducted in partnership with other related projects and programmes including those at the international level that are addressing similar issues.
- Optimization of the distributed SeaDataNet data infrastructure to make it more reliable, robust and easier for users by, for example, reducing the latency of data publication and providing access to very comprehensive data sets. The system will also be enhanced with additional monitoring services, secured exchanges over open networks and integrated user management using identity

federation services such as that provided by GÉANT (A European initiative delivering advanced networks and associated e-infrastructure services for education and research).

- To expand the range of services of the SeaDataNet infrastructure by specifying, developing and deploying advanced e-services as part of a Virtual Research Environment (VRE) to facilitate individual and collaborative research using, handling, curating, quality controlling, transforming and processing marine and ocean data into value-added analyses, harmonised data collections, and data products which can be integrated, visualised and published using OGC and high level visualisation services.
- Enhancement of data quality control to support delivery of large and coherent aggregated data sets
- Strengthening of links between scientific publications and the related datasets through the use of digital object identifiers (DOI) for both observations and generated data products.
- Building links between data management, data publishing and related data citation services to encourage researchers to deposit their data sets in recognised data centres thus improving the availability and access to marine data.
- Reinforcing and promoting mechanisms for online data ingestion from various sources, including autonomous observng platforms, by the data centres in order to stimulate more contributions of data from third parties;
- Further expansion of the SeaDataNet network to other European countries especially those in the Mediterranean and Black Sea areas not currently connected to the e-infrastructure.
- Promoting and reinforcing interoperability on a global scale through international initiatives such as the Ocean Data Interoperability Platform (Glaves, H. (2017, chapter 3 this book) and the International Oceanographic Data and Information Exchange (IODE) of the Intergovernmental Oceanographic Commission (IOC).
- Supporting and contributing to international earth observation initiatives such as the Global Earth Observation System of Systems (GEOSS).

CONCLUSION

SeaDataNet is actively operating a highly successful pan-European e-infrastructure for managing, indexing and providing access to marine data sets and data products acquired from research cruises and other ocean observing activities in the European seas and global oceans. This network of interconnected data centres continues to expand through cooperation between the SeaDataNet consortium and other related EU-funded projects, and also by the active involvement of SeaDataNet in the development of the European Marine Observing and Data Network (EMODnet). As a result there is steady increase in the number of data sets and metadata references being added to the SeaDataNet services and therefore more widely available for re-use.

The standards developed by SeaDataNet are also being widely adopted both within Europe and globally. These standards have been implemented by some of the key initiatives in the marine domain such as EuroGOOS and Copernicus Marine Environment Monitoring Service (CMEMS) as well as by the many individual organisations that are involved in marine data collection and management. SeaDataNet is also providing standards, technologies and operational services for several large scale EU-funded projects such as the EMODnet Bathymetry and Chemistry 'lots'. The adoption of the SeaDataNet standards and services by various European initiatives supporting access to data for marine research will promote the development of an overarching e-infrastructure that facilities the discovery and access of marine data on as large a scale as possible, and will in turn support the generation of Europe-wide data products which are fit for the purposes of various user groups.

REFERENCES

Barth, A., Watelet, S., Troupin, C., Alvera-Azcarate, A., & Beckers, J. M. (2017). Analysis of Ocean In Situ Observations and Web-Based Visualization. In *Oceanographic and Marine Cross-Domain Data Management for Sustainable Development*. Hershey, PA: IGI Global.

Boldrini, E. (2013a). *SeaDataNet metadata profile of ISO 19115*. Retrieved from http://www.seadatanet.org/content/download/20284/140589/file/CDI-profile-V10.0.pdf

Boldrini, E. (2013b). *SeaDataNet metadata profile of ISO 19115 –XML encoding*. Retrieved from http://www.seadatanet.org/content/download/20285/140594/file/CDI-profile-XML-encoding-V10.0.pdf

Boldrini, E. (2013c). *SeaDataNet CSR metadata profile*. Retrieved from http://www.seadatanet.org/content/download/20282/140579/file/CSR-profile-V3.0.pdf

Boldrini. (2013d). *SeaDataNet CSR metadata profile of ISO 19115-2 – XML encoding*. Academic Press.

Buck, J., & Lowry, R. (2017). Oceanographic Data Management; Quills and Free Text to the digital age and 'Big Data. In *Oceanographic and Marine Cross-Domain Data Management for Sustainable Development*. Hershey, PA: IGI Global.

Diviacco, P., Sorribas, J., De Cauwer, K., Sinquin, J. M., Casas, R., Busato, A., & Scory, S. et al. (2017). Repositioning data management near data acquisition. In *Oceanographic and Marine Cross-Domain Data Management for Sustainable Development*. Hershey, PA: IGI Global.

European Commission. (2007). *Directive 2007/2/EC of the European Parliament and of the Council of 14 March 2007 establishing an Infrastructure for Spatial Information in the European Community (INSPIRE)*. Available at: http://eur-lex.europa.eu/JOHtml.do?uri=OJ:L:2007:108:SOM:EN:HTML

European Commission. (2008). *Directive 2008/56/EC of the European Parliament and of the Council of 17 June 2008 establishing a framework for community action in the field of marine environmental policy (Marine Strategy Framework Directive)*. Available at: http://eur-lex.europa.eu/legal-content/EN/TXT/PDF/?uri=CELEX:32008L0056&from=EN

European Commission. (2010). *Communication from the European Commission to the European Parliament and the Council: Marine Knowledge 2020: Marine data and observation for smart and sustainable growth COM/2010/0461*. Author.

European Commission. (2014). *Communication from the European Commission: Innovation in the Blue Economy: Realising the potential of our seas and oceans for jobs and growth COM(2014) 254/2 (13/05/2014)*. Author.

Glaves, H. (2017). Developing a common global framework for marine data management. In *Oceanographic and Marine Cross-Domain Data Management for Sustainable Development*. IGI Global.

Glaves, H., Schaap, D., Arko, R., & Proctor, R. (2014). Ocean Data Interoperability Platform (ODIP): Supporting the development of a common global framework for marine data management through international collaboration. EGU General Assembly Conference Abstracts, 16, 14366.

Leadbetter, A., Cheatham, M., Shepherd, A., & Thomas, R. (2017). Linked Ocean Data 2.0. In *Oceanographic and Marine Cross-Domain Data Management for Sustainable Development*. Hershey, PA: IGI Global.

Oggioni, A., Tagliolato, P., Fugazza, C., Pepe, M., Menegon, S., Pavesi, F., & Carrara, P. (2017). Interoperability in marine sensor networks through SWE services. In *Oceanographic and Marine Cross-Domain Data Management for Sustainable Development*. Hershey, PA: IGI Global.

SeaDataNet. (2010). *Data quality control procedures*. Available at: http://www.seadatanet.org/content/download/18414/119624/file/SeaDataNet_QC_procedures_V2_(May_2010).pdf

UNESCO. (1993). *Manual of Quality Control Procedures for Validation of Oceanographic Data*. UNESCO, IOC-ICSU, Manuals and Guides No. 26. Retrieved from http://unesdoc.unesco.org/images/0013 /001388/138825eo.pdf

Vossen, R. (2011). *DTM exchange format specification*. Retrieved from http://www.emodnet-hydrography.eu/documents/emodnet_hydrography/html_page/euco-0901-002_dtm_exchange_format_specification_v1.5.pdf

Chapter 8
Repositioning Data Management Near Data Acquisition

Paolo Diviacco
Istituto Nazionale di Oceanografia e di Geofisica Sperimentale - OGS, Italy

Jordi Sorribas
CSIC, Spain

Karien De Cauwer
Royal Belgian Institute of Natural Sciences, Belgium

Jean Marc Sinquin
Ifremer, France

Raquel Casas
CSIC, Spain

Alessandro Busato
Istituto Nazionale di Oceanografia e di Geofisica Sperimentale - OGS, Italy

Yvan Stoyanov
Royal Belgian Institute of Natural Sciences, Belgium

Serge Scory
Royal Belgian Institute of Natural Sciences, Belgium

ABSTRACT

This chapter intends to propose a solution to the progressive paradigmatic drift that emerges when data is part of a workflow where processes and contexts are not accessible. In this case end users can have difficulties in identifying possible anomalies, or events that might be very important for the dataset and its interpretation. Since sustainable development is based on cross-disciplinary studies, this can revert in misunderstanding and difficulties to work collaboratively. To address this issue it is proposed to fill the gap between the data and its acquisition through a logging system named EARS that records underway data such as for example: meteo or swell, and events such as: anomalies or acquisition milestones. Once all these information are stored, they can be linked to the data through an OGC compliant metadata model and gathered as summaries as required by several data management initiatives.

DOI: 10.4018/978-1-5225-0700-0.ch008

INTRODUCTION

The study of sustainable development, especially in the marine sector, is a highly cross-disciplinary research domain. Biology, geology, chemistry and physics intermingle and interlace to create a complex network of skills and competencies. In this panorama the trend in Science that sees building knowledge as a collaborative endeavour becomes overwhelmingly important. This implies that data should be accessed and understood across domains, competencies and communities.

Unfortunately this is not easy to put in practice since data management tends to be biased by local practices, by technical limitations, by domain specific cognitive models and by socio-cultural trends; in other words by paradigms. This effect tends to increase during the steps that separate data acquisition from data usage, so that eventually this results in a speciation of environments that become incommensurable. In practice users living in one paradigm cannot understand what is produced within another one, yet their research interests compel them to do so.

BACKGROUND

Science serves to ground all activities that shape the study of sustainable development, and therefore science foundations are overwhelmingly important and condition all aspects of what is built on top of them. Any shortcoming in the foundations of Science can, therefore, potentially, undermine how sustainable development is addressed.

The main problem that can be envisioned is that there is a gap between what is needed by stakeholders, such as: directives, norms, decisions, protocols on one hand and, on the other, the underlying scientific theories. In the first case facts and objective positions are requested, while if we climb backward to the scientific research that is supposed to produce them we realize that things become complicated.

Objectivity is a complex concept to handle in itself, but in Science this complexity becomes even more difficult to deal with. A detailed review of the role of objectivity in Science and in particular in environmental sciences, can be found in Diviacco (2015), Diviacco (2015b) and Diviacco (2012). Here it is worth mentioning the main idea reported there, which is that scientific objectivity can be explained only as something built from the judgment of a group of sane, rational and sufficiently informed people. It is not based on an unbiased analysis of external facts; rather, it is something related to the convergence of a community. The subjective is private and the objective is public. This means that any position, idea, concept or representation has to be negotiated with other people who take part in the activity.

This perspective is evidently contrary to the popular viewpoint where Science grants facts that cannot be biased by human intervention.

We will not explore here the issue of how many other forms or definitions of objectivity are possible. We will instead highlight that the main hinge on which such distinction can be made is the role of the "experience", meaning the observation of events. Where experience is necessary to understand a process or a system, since each one of us can have a different experience, the public aspect of objectivity is raised.

Scientific research is of course based on observations and measurements, but while the standard view of the scientific method prescribes in the deductive/inductive loop a verification phase wherein an experiment could verify (or following Popper (2004) only falsify) a theory upon the result of the experiment, many recent thinkers have criticized this approach.

Following Hanson (1958, p19), for example, *Observation of x is shaped by prior knowledge of x.* Experiments themselves are devised by researchers upon a pre-existent set of theories and assumptions and that can condition them. This means that researchers cannot avoid projecting their backgrounds and way of thinking (in a word their paradigms) even on their very basic activities of observation.

This process applies at all levels and is called bias.

A wider view would see this phenomenon in a cognitive perspective, where researchers are able to understand concepts only if they match and fit their cognitive models. Studies on cognitive aspects of scientific research can be found in Diviacco (2015).

This raises the problem that since each scientific field, if not each school of thought or institute, tends to develop its own specific cognitive model, concepts derived within one field cannot be easily understood by researchers from another.

In environmental but also social and economic studies, which are at the base of sustainable development, often these effects are even stronger than in other fields. This is due to the fact that the verification/falsification phase, as known in the standard vision of science, takes place only with difficulty. In engineering it is often (albeit not always) possible to understand if a theory is true, or at least realistic, implementing it in a tool, gear or something similar, and checking if the expectations are met. In sustainable development this is not easily achievable since systems are complex and their parts are dependent on each other, so that the same effect can be triggered by different configurations of causes. This of course makes it simpler for different theories to co-exist, which conversely allows multiple paradigms, communities, and schools to live side by side, or more likely clash.

MAIN FOCUS OF THE CHAPTER

Progressive Paradigmatic Drift

The issue described above is evident in many aspects of the life of scientists. Instrumental measurements made during an experiment are progressively transformed into theories and theories are refined on the base of observations. During this process, researchers' backgrounds interfere heavily, producing intermediate artifacts that, at each step, add further bias and therefore introduce possible incommensurability. This is particularly evident in representation of observations. Representation can be anything that portrays something. It can be an artifact and abstraction such as images, maps, reports, or it can be a model, a mathematical entity or a computer simulation, even concrete objects can be representations.

Representation is interesting because, following Calender and Cohen (2006), representation can have features reality does and does not have. In their terminology, omission is the act of neglecting some possible relation to the phenomenon that is represented; commission is the act of deliberately changing the network of possible relations to the phenomenon. In fact scientists bias reality through representation by means of omissions and commissions.

Representation becomes sign and sign is such since someone interprets it as 'signifying' something (Sturrock, 1986, 22; Umberto Eco, 1976). Meaning is bound to its context while signification is both temporary and socially determined (Coward, & Ellis, 1977). Following Barthes (1957) signs are ideological and so is representation.

In this perspective, too many steps of representation imply high levels of bias. We can call this phenomenon "progressive paradigmatic drift".

In case of severe progressive paradigmatic drift collaborative research among people with different backgrounds can be very difficult.

SOLUTIONS AND RECOMMENDATIONS

The Approach

This chapter presents an approach, a practice and a software solution that address the issue of possible progressive paradigmatic drift in marine environmental research targeted for sustainable development.

The approach is based on the reduction of the distance in data management between data usage and acquisition that can be obtained though the availability of a larger set of information that can be browsed, and processed semantically across domains and of a large set of underway data recorded directly from sensors installed on the acquisition vessel.

It is important to understand that the solution we propose is not linked only with the marine case. On the contrary the proposed solution can be extended to any acquisition platform where progressive paradigmatic shift has to be compensated, or, as we are currently experimenting, even to the subsequent steps of data processing. Event logging during field research will be explored as the main use case.

Capturing the Context and History

When a boat is drifting at sea, to steer its route it is necessary to understand its positions. In the case of scientific research this means capture its context and history.

Metadata is data about data. Metadata of scientific research and of its representation can include information such as measured parameters, ranges or geographic position, but also much more abstract information such as aims of the research or quality of the data acquired.

Metadata, so far, have been compiled mostly manually, as a reasoned summary of ancillary information of data. Human reasoning of course is particularly prone to paradigmatic drift. In addition, the need to avoid lengthy and therefore costly manual metadata compilation often suggests reducing the amount of information available. The coupling of these two issues can severely increase the opacity of data.

The Initiative

The EU FP7 Projects Eurofleets and Eurofleets2 are an European wide alliance of marine research centers that aim to share their research vessels, meaning ships devoted to the acquisition of scientific data and observations to improve information sharing on planned, current and completed cruises, on details of ocean-going research vessels and specialized equipment, and to durably improve cost-effectiveness of cruises.

Within this context logging of information on how, when and where anything happens on board the vessel is crucial information for data users in a later stage. This forms a primordial step in the process of data quality control as it could assist in the understanding of anomalies and unexpected trends recorded in the acquired data sets. In this way completeness of the metadata is improved as it is recorded accurately at the origin of the measurement. The collection of this crucial information has been done in very different ways, using different procedures, formats and pieces of software in the context of the European Research Fleet. At the time that the Eurofleets project started, every institution and country had adopted different strategies and approaches, which complicated the task of logging general purpose information and events on-board whenever they accessed a different platform thereby loosing the opportunity to produce valuable metadata on-board the vessel during a research cruise survey.

Among the many goals identified by the Eurofleets project, a very important task is the development of an "event log software" called EARS (Eurofleets Automatic Reporting System) that enables scientists and operators to record what happens during a survey. EARS will allow users to fill, in a standardized way, the gap existing at the moment in metadata description that only very seldom links data with its history. Events generated automatically by acquisition instruments will also be handled, enhancing the granularity and precision of the event annotation. The adoption of a common procedure to log survey events and a common terminology to describe them is crucial to provide a friendly and successfully metadata on-board creation procedure for the whole the European Fleet. The possibility of automatically reporting metadata and general purpose data, will simplify the work of scientists and data managers with regards to data transmission.

An improved accuracy and completeness of metadata is expected when events are recorded at acquisition time. This will also facilitate multiple uses of the data as it allows verification of different requirements existing in different disciplines.

The System

The underlying idea of the system is that, during a survey, all underway data and occurring events should be recorded in a persistent database so that later, relevant information can be gathered semantically based upon domain specific criteria.

In the case of underway data various kinds of sensors with different wiring standards are bridged to a private IP Ethernet network where a software application can isolate datagrams and parse them. Most of the incoming datagrams are very likely to be in the National Marine Electronics Association (NMEA) format, but the system can handle any kind of format thanks to a datagram parser editor, that allows customization of the software for practically any kind of set-up.

Events occurring during a scientific cruise can be anything from instrumental anomalies, to observations not recorded by a dedicated sensor (Figure 1). Also the collection of any type of material, a 'sample', prior to analysis in a laboratory is considered to be an event. Events cover a wide range of domains. They generally refer to a human input and in this sense a specific Graphic User interface was developed. They can also be also generated by any means of automatic system sensing any interesting observation that will be then logged.

Logging of information on how, when and where an event happens on board of the vessel is crucial information for data users in a later stage to compute fields to get summary, metadata for the cruise (such as for example: number of lines, distance of trawl operations, location of events). An improved accuracy and completeness of metadata is expected when recorded at acquisition time. This will also enhance multiple use of the data as it allows verification of the different requirements existing in different disciplines.

It can be said that EARS is seated on top of the adoption of three basic components:

Figure 1. Concept of an event, and the special semantic design of it adopted in EARS

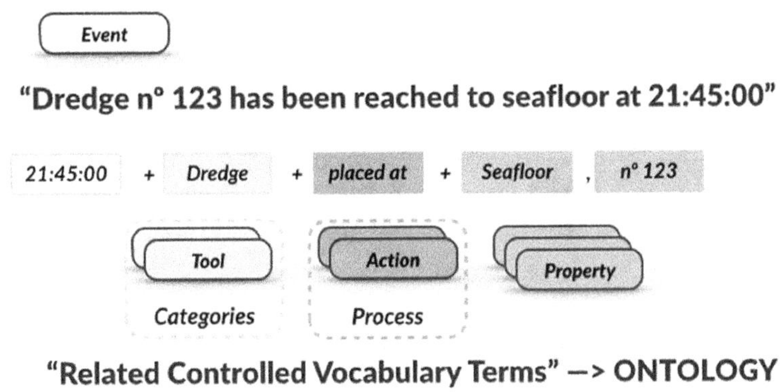

- The development of a common automatic event logging procedure,
- The development of a tool to handle manual events that would leverage the convergence of the designated communities on semantics.
- The development of a set of a common event access and retrieval procedures

AUTOMATIC EVENT LOGGING

Hardware and Software

EARS V1 is composed of a software suite designed and developed during the first edition of the Eurofleets project but also uses existing elements designed and used by Eurofleets partners and instrumentation commonly present on board. The main elements that compose EARS V1 are shown in Figure 2, where the arrows show the flow of information.

Automatic Event Acquisition

An initial, important choice that was taken by EARS developers was to ground the system on the transmission of information through an IP network. Traditionally sensors are connected to recording instruments through a dedicated link, such as, for example, a serial line. Connecting directly the logging system with such kind of links could have resulted in the need to rewire the vessels: something no vessel crew would have accepted easily. It was therefore preferred to have serial lines split close to the sensors in order to later tunnel the signal into the existent Ethernet network of the vessel through a serial to Ethernet device.

Serial to Ethernet Converters

Serial to Ethernet converters allow transformation of legacy devices into Ethernet devices, which can be monitored and controlled from any network location or, depending on the setting of the vessel network, even from the internet. Within EARS it was decided to use commercial serial to Ethernet converters, because the costs of these tools are very low and can be afforded by any partner of the project. In addi-

Figure 2. Schema of EARS system (hardware and software)

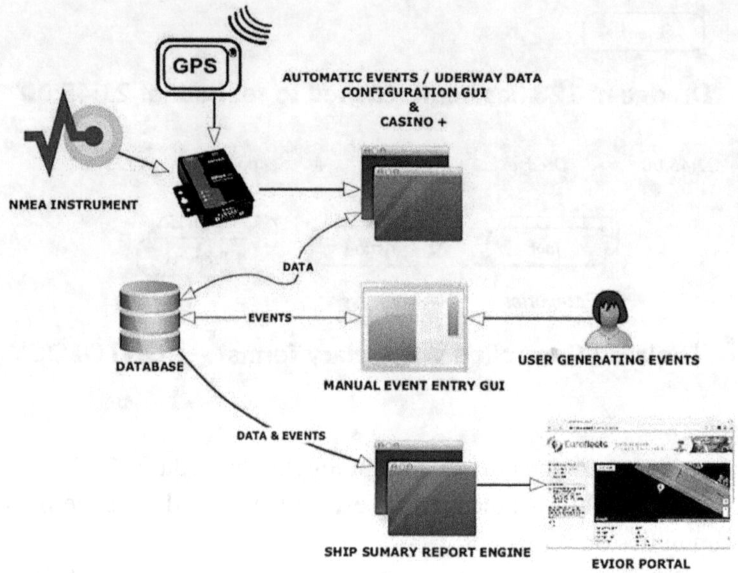

tion these converters are very easy to configure even from on-land headquarters, generally using web based configuration portals hosted by the device itself.

The Issue of Time Lags and the Introduction of CASINO+

The use of an Ethernet network was a big initial concern since time lags and delays in transmission were expected. It was therefore necessary to verify as soon as possible that the proposed paradigm could work and that the lags introduced will not become an unmanageable problem. This was not verifiable until the full EARS software was developed. It was therefore decided to take advantage of an already existing tool developed by Ifremer, named CASINO+ that offered some of the functionalities devised for EARS, allowing the problem of time lags and delays to be evaluated from the very beginning of the project. Besides, the introduction of CASINO+ was considered very important because it could have let the limits of the ideas behind EARS to emerge and therefore it could have helped in designing the extensions and evolutions needed by the original project.

To address the problem of time lags, CASINO+ is designed to handle data coming from the sensors and instruments as UDP packets. The UDP format uses a simple connectionless transmission model with a minimum of protocol mechanism. UDP allows computer applications to send messages (datagrams) to other hosts on an Internet Protocol (IP) network without the need of a specific set up. In addition, UDP avoids the overhead of error checking and it is, therefore, suitable for time-sensitive applications.

The initial tests revealed that the time lags introduced by the systems were not relevant or anyway less problematic than positioning problems. In fact since at the very core of any observation in the marine sciences is the geographic position of the recorded event, positioning indeterminacy dictate the accuracy of the whole system. If the time lags of event transmission and recording are smaller than positioning indeterminacy then we can say that time lags are irrelevant to the system.

Since these tests using CASINO+ were encouraging, it was decided to adopt it as a module of an initial version of EARS. This was meant to be further evolved later, but allowed at the beginning of the project to concentrate on the new functionalities that were expected from EARS. This chapter focuses on this first version of the software and mentions the new functionalities of the currently under development acquisition module in the future work section.

Datagram Parsing

After having received the datagram, CASINO+ parses it in order to extract the information contained.

Although, as mentioned above, CASINO+ allows any stream of data to be parsed, the most common datagrams it has to deal with follow NMEA protocols.

The NMEA has developed a specification that defines the interface between various pieces of marine electronic equipment. NMEA datagrams are sent as independent and self contained sentences. One of the most common data types that use NMEA encoding is GPS signal. An example of GPS sentence in NMEA is:

*$GPRMC,183729,A,3907.356,N,12102.482,W,000.0,360.0,080301,015.5,E*6F*

Each sentence begins with a '$' characters and end with a carriage return/line feed sequence and can be no longer than 80 characters. Data is contained within a single line with data items separated by commas. In the case above the payload is the geographic position of the vessel which is contained in the fields ranging from 4 to 8.

Using the conventions and formats mentioned above other types of data can be coded in NMEA.

Since multiple streams of data can be received from different sensors and instruments, it is necessary to instruct CASINO+ on how to parse each data type. This is done using two XML files:

- Dictionary.xml: this file holds the definitions of units, such as for example metres (see example below), kilograms etc.; sourcetypes, such as parameters and the relative unit, and possible further gathering of units and sourcetypes.

  Example of definition of the unit meter:

```
<unit major="1" minor="0" versiondate="2002-03-17T04:18:59Z"
versionauthor="Fabrice Lefebvre" versiondescription="initial version"
pluralform="false" name="kg" dimension="kilogram" definition="kilogram"
category="mass" description="is the symbol for the base unit kilogram"
reference="BIPM"/>
```

- Configuration.xml: this file describes the process of matching the parameters with the values contained in the datagrams.

  Example of definition of matching rules for vessel positioning (Box 1)

Box 1.


```
<device deviceid="NAVIGATION"
devicename="NAVIGATION" sourcetypelist="position"
firstusedate="1999" position="passerelle"
latestcalibrationdate="2000" calibrationparameters=""
installdate="1999" workingparameters="Diffu NMEA">
    <message
    name="GPSfix" major="1" minor="0">
        <frame
        name="position">
            <match>$PIFM,TSGEN,{acquisitiontime_msgen},NACOU,{acquisitiontime_
nacou},{EtatNav},{lat},{long},{surfcourse},{surfspeed},{gndcourse},{gndspeed},{driftcourse},{d
riftspeed},{heading},{FlagQualite},{OriginePoint},{SystemeGeodesie},MVATN,{acquisitiontime_
mvatn},{reserved},{headingMRU5},{roll},{Gite},{pitch},{AssietteDegres},{AssietteCm},{heave},{Acceleration},SONDE,{acquisitiontime_
sonde},{EtatSonde},{depth},{Identif
            icateur}</match>
        </frame>
    </message>
    <broadcastaddres port ="10200"/>
</device>
```


Once the two files are set it is possible to launch the acquisition module of the software that parses the datagrams, extract the useful information and send it to the designated database (Figure 3).

Figure 3. EARS listening to a datagram, parsing sentences, extracting values and sending them to the database

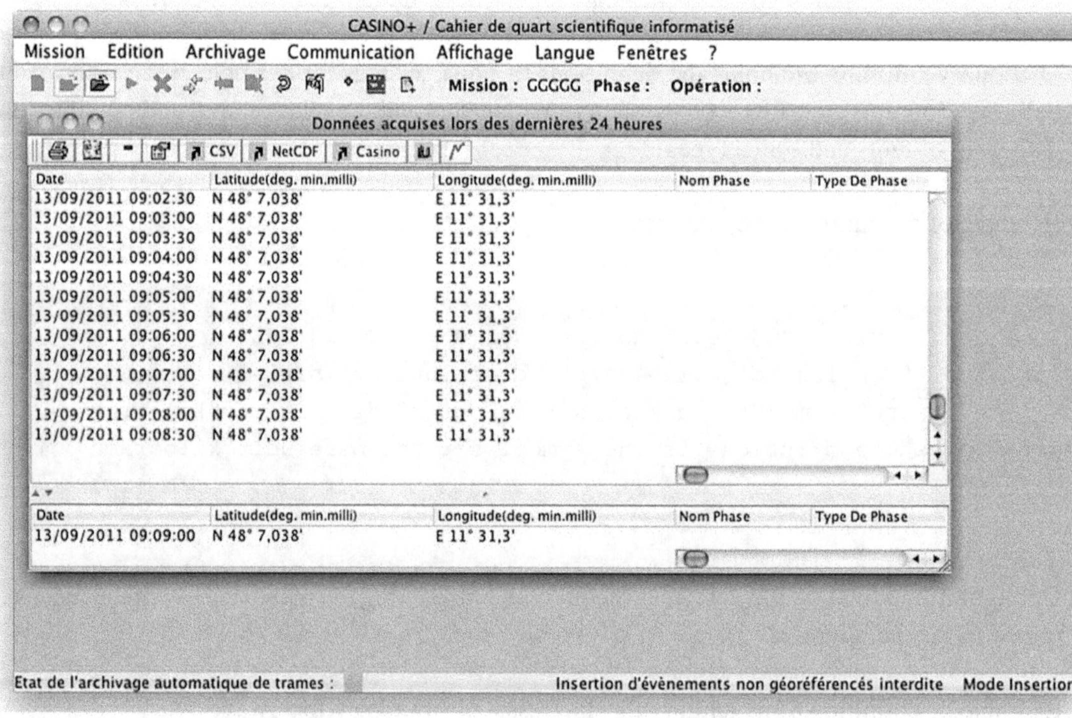

Acquisition Configuration Graphic User Interface

The two files had to be edited manually; which was not a trivial activity.

To solve the problems it was decided to develop a specific Graphic user interface that allow end users to setup EARS and in particular the CASINO+ acquisition module. The GUI hides the configuration details of EARS allowing for capture of the relevant information from the NMEA datagrams and production of the configuration files needed to set up the acquisition module that captures data and sends the information to the database. This is done graphically and interactively.

The GUI defines the ports where transmission is expected from the serial to Ethernet converters,listens to datagrams, isolates the fields graphically and links the values with parameters.

Since, often, several vessels use the same instrumentation, the configuration GUI offers a means to create a catalogue of them (Figure 4). This simplifies the work of technicians who need to recall from the catalog only that particular instrument setup and just set the transmission port number. Such catalogs can be freely shared among EARS users, while the partners that are in charge for EARS development can offer a "customization" service to easily match the instrument setup of the end users.

Figure 4. EARS configuration graphical user interface offers a catalogue of possible instrumentation to recall

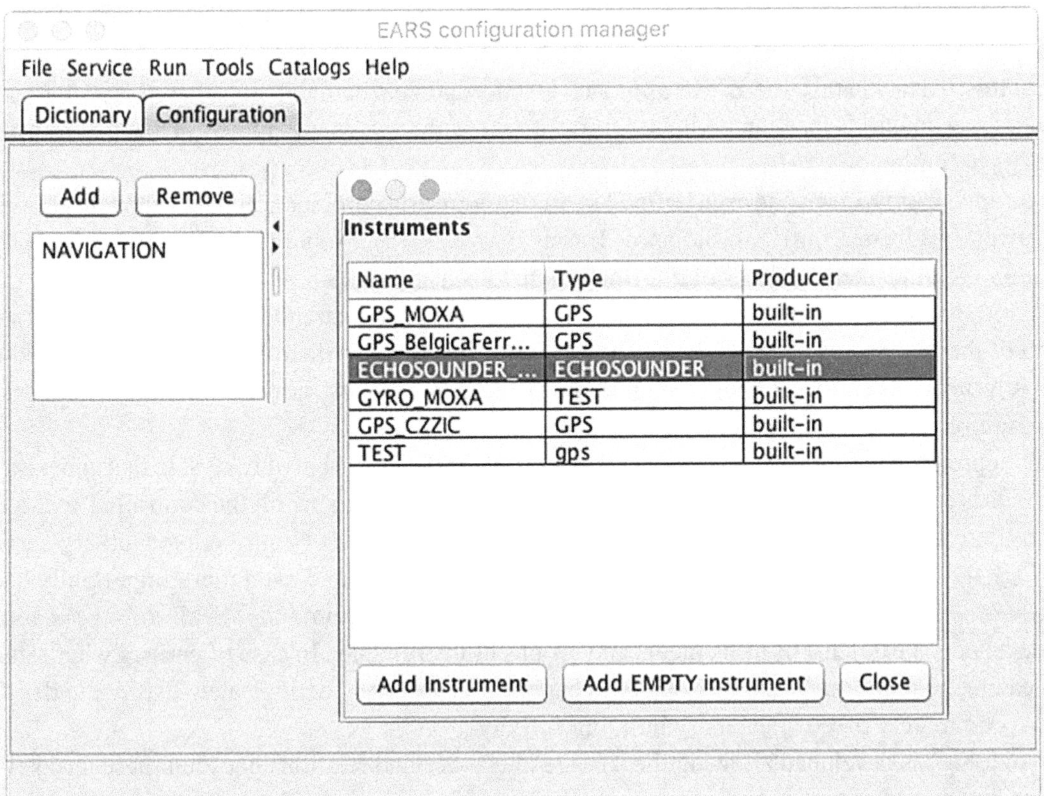

CASINO+

Once the setup is done, the EARS acquisition module based on CASINO+, captures, parses and eventually sends information from the datagrams to a database.

CASINO+ Database is based on a MySQL relational database that is often located on a secure space such as a Network Attached Storage (NAS) environment.

EARS v1 system embeds all the functionalities offered by CASINO+ among which we can recall:

- **Import/Export of an Existing EARS v1 Database:** This is commonly used before or after a cruise to recover information from different cruises or data sets.
- **Filtering Existing Data Set with Different Criteria:** time period, list of dedicated actions (list of coring actions, location of the trawling operations, number of CTD probes, …)
- **Processing of Several Abstracts:** The most important is obviously the CSR (Cruise Summary Report) which defines the most important elements of a cruise and is the baseline for several portals such as for example: Seadatanet or Evior. In addition, the software provides other type of summaries, such as: list of equipment used, or a daily summary report
- Export of the log book in CSV format in order to allow further filtering, sorting or editing with other software.

Manual Events and Semantics

It is easy to understand how important it is that the meaning of information is clear across, but even within domains. Issues such as progressive paradigmatic shift and tacit knowledge, blurring meaning, contrast this. To tackle this, the EARS approach is to formalize meaning using shared resources such as controlled vocabularies that can be linked to from within the information itself. This way information carries its definition.

The concept of an event has been defined in an ontology as a combination of components (Diviacco, De Cauwer, Leadbetter, Sorribas, Stojanov, Busato, Cova, 2015). From this perspective it was possible to identify the fundamental pieces of information that form an event, their relevance (if they are mandatory or not), the relationships between them (if existing) and the constraints or axioms that govern them. Several of these elements have been identified as being regulated in their content by a controlled list of terms, referred to as entities. Other elements, like 'Comment', 'Actor', are described by a free text entry and thus implemented as data properties.

This approach was followed throughout all aspects of the development of EARS. In fact underway data and events are always linked to external controlled vocabularies. Among all the controlled vocabularies we considered, a particular attention was devoted to the SeaDataNet controlled vocabulary served by NERC (BODC, 2015). The choice was dictated by its high quality and even more importantly because of the strong adoption of these vocabularies in the EU marine community. SeaDataNet vocabularies provide a good starting list of instruments and groups of instruments. In case of entities where standard terms cannot be found such as for example, in case of processes and actions, new lists are under design within a community based vocabulary building activity.

An ontology is developed allowing the storage of pre-defined relations between these terms (Figure 5). Querying this ontology, presents users a configuration for their scientific cruise. A common approach for storing events, using common terms, facilitates cross-domain data discovery.

Figure 5. EARS event model

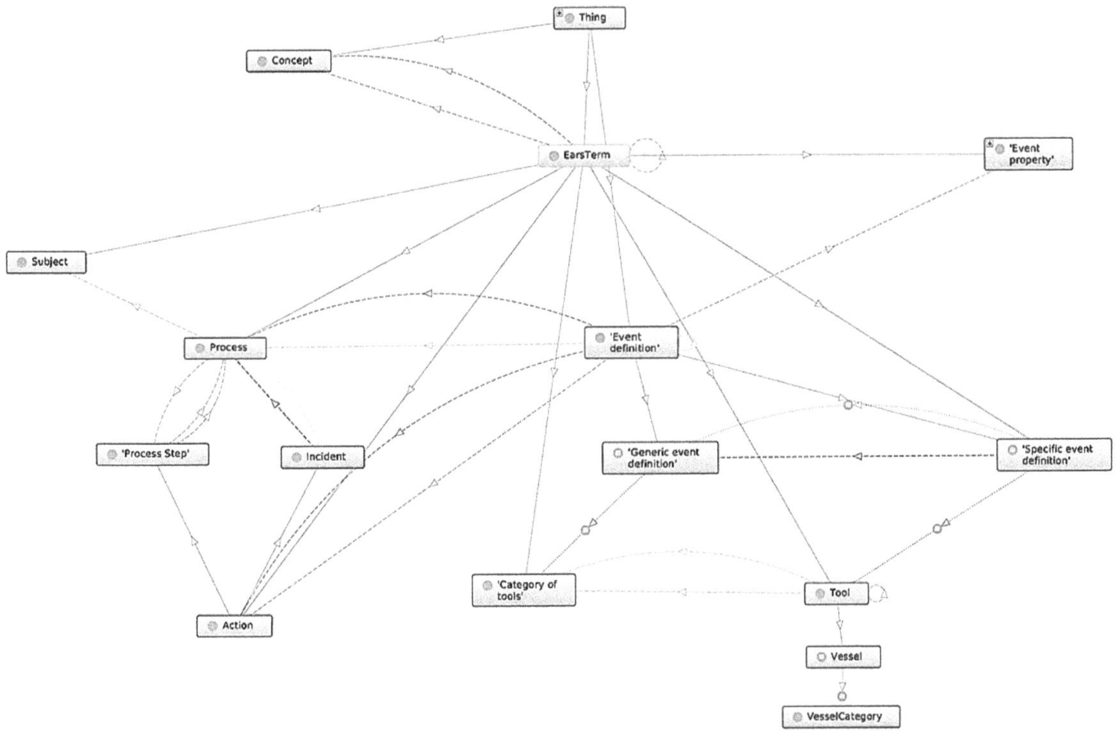

Manual Event Software

The EARS V1 manual entry software has been developed in order to offer a simple and straightforward tool to enter manual events according to the developed event model onboard research vessels. User friendliness and standardization have been identified as most important issues during the elaboration of the specifications. The resulting software is intended to be deployed at all EUROFLEETS partners' vessels.

Working conditions onboard tend to be difficult due to time constraints and weather conditions. Scientists and technicians often need to perform multiple tasks at different places on the vessel and in a tight cruise schedule. Weather conditions, like a bad sea state, can severely complicate the various operations. The user-friendliness of the software is thus of utmost importance. Event logging should be possible in different languages, using reference vocabularies in as few clicks as possible. The resulting software allows users to create a time-stamped manual event with a simple double-click or drag-and-drop starting from the configuration ('personal tree').

Personalization of the software is a very important aspect. This is granted by the configuration file that can be edited in order to tune the GUI differently from vessel to vessel or from discipline to discipline. It allows the user to have only a specific subset of terms. The configuration file can be reused and shared between institutes.

To ensure all operations can be captured by the tool, non-existing terms can be added whenever needed. The events and newly added terms created by EARS V1 will be used to validate and enrich the ontology.

At any time, additional properties (e.g. volume of water flowing through the centrifuge) or comment (e.g. the visual description of a sediment core) can be added as text description to the generated event.

By predefining these specific properties for an event, the software enables and assists scientists to log additional sampling or observation characteristics in a structured way providing a replacement of the paper logs traditionally used onboard. Moreover, a structured entry of these accompanying properties facilitates retrieval and further use of this information.

Another specific need that had to be considered is the offline use of the software. This functionality offers the possibility for the entry of observations anywhere on the deck which, for example, is required by scientists counting birds and/or mammals during tracks.

Manual Event Creation Using the GUI

The manual event GUI is written in java and maintained by BMDC (Royal Belgian Institute of Natural Sciences). It requires a recent Java virtual machine and has no particular hardware needs. The software has been designed by a team of developers and domain experts so that domain specific requirements and software coding are strongly intermingled having clearly in mind the target users, the aims of the software and the situations in which it is used. The tool can be used in offline modus. Once the device is connected to the network, all entries are centrally stored in a specific table added to the MySQL database of CASINO+.

This GUI allows the user to create time tagged events that describe any possible fact concerning the operation of a specific instrument (sensor or sampling device), the navigation of the vessel, planned or incidental observations or the course of the survey (Figure 6). Logged events can be modified and retrieved. The events are linked to the specific cruise during which they are created.

Figure 6. EARS manual event entry GUI

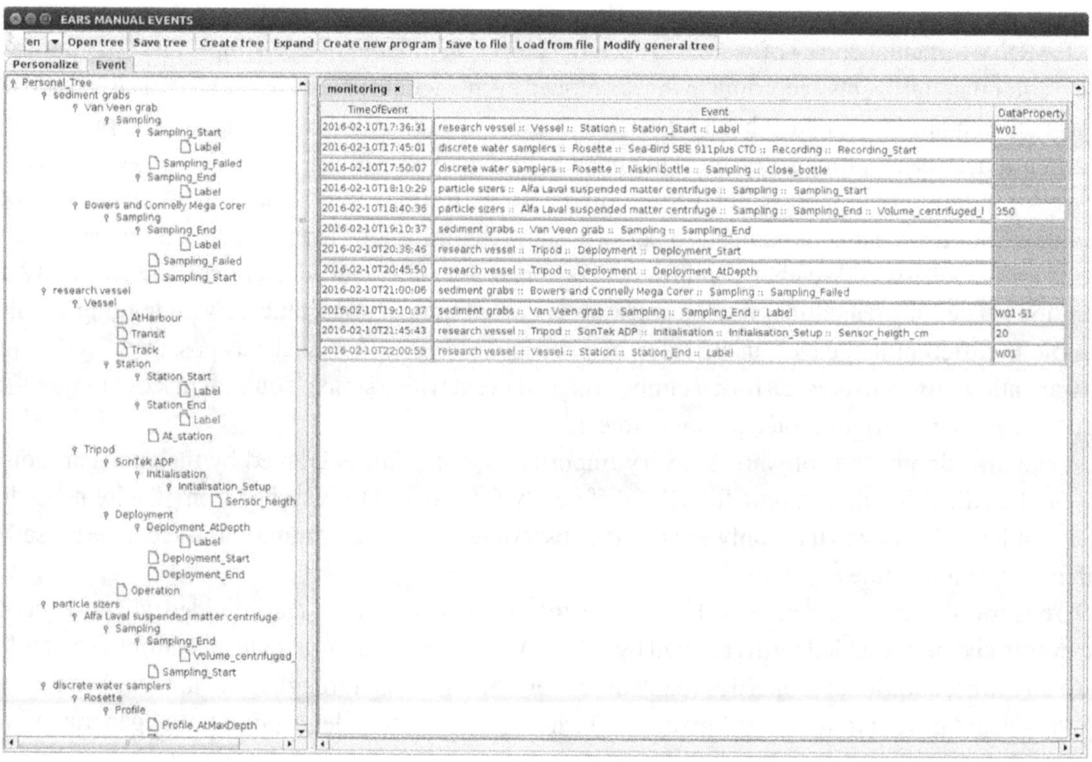

An advanced use of the Manual Event Entry GUI is the creation of the personal or vessel tree that consists of the set of tools and related terms with which the user is acquainted in his domain to describe relevant sampling and data acquisition operations. It consists of a subset of the different terms and predefined relations between them.

Firstly, a cruise setup needs to be done. During a given cruise, multiple teams active in different scientific

domains can perform measurements in the frame of different research projects or monitoring programs, resulting in the introduction of the notion of a 'program'. Events of one program are shown in one tab. This provides the user with a clear overview of those events that are relevant for them, together with update and delete possibilities. At the same time, the data collection activity is immediately situated in its context, providing useful information for data managers as well as users in a later stage. The different steps involved in the tool are shown in Figure 7.

The Graphical User Interface (GUI) offers lists of terms for every event component, using existing standard vocabularies where possible. For every individual term, the complete definition is shown in the GUI assisting the user in the choice of the correct term.

Next, a personal configuration can be created that consists of a sub selection of tools with its linked terms relevant for the user as some vocabulary lists contain a high number of terms. This configuration, displayed in the GUI as a tree, can be saved and reused. It will typically depend on the scientific domain and the vessel.

Properties, e.g. installation height of a sensor on a benthic lander, can be added in the tree, assisting users to enter relevant additional information for a specific event in a structured way.

Figure 7. The different steps required in manual event entry in EARS GUI

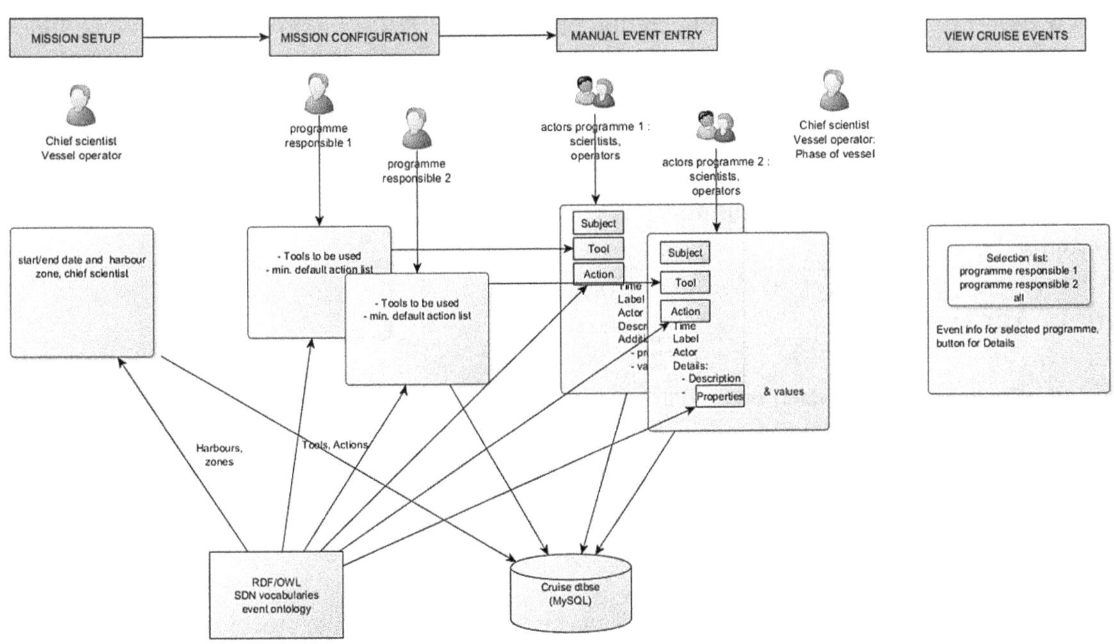

Before the first event can be created, a name for the program needs to be entered. For cases where the cruise is dedicated to more than one research program or project, a tab page will be presented for every program. An event is easily generated by double-clicking on or through drag-and-drop of the lowest level in the tree, e.g. an action or a property. A timestamp is automatically assigned, but it is possible to update it to allow events to be created afterwards. The event can be completed with free text input as comment or name of the actor. Besides that, the properties as defined in the configuration are presented in the GUI together with an empty field for free text input for that specific property. This information can be added at any time and gives the tool the full capacity for the replacement of a paper event log.

The configuration file was developed keeping in mind the flow of a common offshore survey. It contains terms for the different phases of the vessel such as: start of transit, start of a track or arrival at a given station where measurements will be carried out. Inclusion of a property 'label' allows the entry of the specific station name. Once the survey is started, operations are usually carried out using different instrumentation for which line start and end events can be noted, as well as all punctual samplings, each complemented with their specific characteristics. A catalogue of anomalies is proposed such as default instruments or weather difficulties.

In the configuration, the specific steps to be carried out and information to be noted is included. For a flow-through centrifuge, for example, start time and end time of sampling as well as information concerning the volume of water that passed through it are identified as key features. Different sensors (e.g. a CTD and ADV) can be attached to a seabed lander and the properties 'installation height' and 'serial number' are of interest and should be noted during setup, prior to deployment of the set of devices. To know the start of data acquisition, the time when the lander is put on the bottom needs to be recorded. Tools can be defined as 'nested tools' in the configuration tree, so that actions related to a set of tools need only be logged once.

Lists of controlled vocabulary entries are provided within EARS V1, but users should be able to add new individuals when necessary. As stated above, the manual events cover a wide range of disciplines and new tools are continuously put in operation or test phase. Users on board should not be blocked because a term is not available. Even if a new term did not yet pass the governance community, it is important to have the information on what happened and to have new terms and relationships created by the user to enrich the ontology. The terms existing in the ontology serve to generate a base tree of terms (the configuration file). EARS V1 enables event data storage according to the described event model. The use of it by many users in various domains will allow a thorough validation as well as extension of the proposed ontology.

Events Access

Once all information is collected in the database this can be retrieved, processed or summarized:

- Using Casino+ functionalities (already mentioned above)
- Exporting data in SensorML
- Using webservices, generally from outside the vessel.

It is possible to list data in its entirety or it is possible to process data to create summaries or subsets of them.

The production of event summaries could be done in many ways, from automatically compiled or on demand.

Event summaries can be provided both in a human readable format or in standardized and interoperable computer-to-computer formats.

SensorML Export

EARS can expose the event summary using a specific language from the Open Geospatial Consortium (OGC) called SensorML that allows EARS to produce fine description of any observation procedure, sensor or instrument, as well its history of associated events.

Events are exported as event lists as in the following example that describes a serial to Ethernet box: (Box 2)

Box 2.

```
<sml:history>
 <sml:EventList>
 <sml:member name=”event-2011-09-08” xlink:arcrole=”urn:ogc:def:property:OGC:event”>
 <sml:Event gml:id=”event20110908113000 “>
 <sml:date>2011-09-08</sml:date>
 <gml:description>Serial to Eth</gml:description>
                                    <sml:property name=“subject” xlink:role=”urn:x-ogc:def:property:OGC:Eurofleets_event:subject”>
                                     <swe:Text>
 <swe:value>UnderWaySystem</swe:value>
 </swe:Text>
                                    </sml:property>
                                    <sml:property name=“tool” xlink:role=”urn:x-ogc:def:property:OGC:Eurofleets_event:tool”>
 <swe:Text>
                              <swe:value>MOXA SYSTEM</swe:value>
                                    </swe:Text>
                               .....

 </sml:Event>
          </sml:member>
     </sml:EventList>
 </sml:history>
```

SensorML event lists can be very easily used within the Seadatanet metadata model. In fact this latter offers a slot to link via Observation and Measurement the Seadatanet Datset identifiers (CDI) with the SensorML eventlist (Diviacco, Lowry and Schaap, 2012). This way, every dataset can be immediately associated with the events happened during its acquisition.

EN-ROUTE INFORMATION AND EARS

A set of Web Service specific interfaces is prepared in order to get from the vessels different kind of information including the last 24h event registry. Since this service is mainly related to the ship we call it Ship Summary Report (SSR). SSR provides a snap-shot of the vessel situation. It gathers the most important information of the last 24 hours, and the vessel track.

The en-route data should contain up to date "operational information" from vessels to be linked and served by operational information portals like European Virtual Infrastructure in Ocean Research (EVIOR).

The purpose of producing en-route information is to produce a report with information such as: What ship and what cruise are we considering; what has been done today, where is the ship (track plot) and an inventory of data files of the day.

Minimal en-route information should report on: position of the vessel, owner and operator, and what the ship is doing at a given moment.

The SSR could be extended with a set of complimentary «latest records» like the last "happenings" or (events) related, sea-state and weather conditions and also with the list of the last data sets acquired using Common Data Index (CDI) files.

The SSR could be enhanced with the correspondent Cruise Summary Report (CSR). The purpose of such a tool is to report information on a cruise when it is over. Within EARS we can use it as a first description of the activities that are planed for the cruise such as, for example: scientific objectives, instrumentation used, working area or people involved in the survey.

SSR is conceptually different from CSR. The former reports dynamic metadata, as it contains information that is changing continuously while CSR reports static information.

The SSR can be linked easily with its correspondent CSR and associated CDI (metadata of the data sets acquired during a cruise) using naming conventions for Ship and Cruise identifiers into the data file names. However this relation can be better understood considering the previous versions of the CSR and CDI that can be produced on board as opposed to the final metadata files that finally will reside at the Data Centers after a process of data and metadata verification and cleaning. Those that can be produced on board as a preview are referred as EnRoute-CSR and EnRoute-CDI.

The production of both requires significant user intervention, and just a minimal portion can be filled out (at the end of the cruise) by automatic systems. Automatic input can be for example the bounding box or the start and end of surveys. In contrast, the production of SRR fits perfectly into the EARS architecture and the way EARS data are produced.

The SSR represents information about the real state of the ship (position and activity) independently from the fact that it is undertaking a cruise or not. Inside EARS a SSR data model is proposed for en-route information following two different schemas:

- Core expression that reports on the minimum necessary information,
- Extended expression that contains complementary information.

The core data model is essentially the navigational status (where the vessel is, and how it is moving). The mandatory information, in this case, is restricted to the minimum and contains:

- Ship identification,
- Status of the ship that is the time stamp of the last information actualization or revision date,
- Last geographic position.

The minimum navigation data should contain: geographical coordinates, courses and speeds, and water depth. Heading, speed over surface and water depths are not mandatory because they are not always provided by positioning systems.

An extra element (not mandatory) like the cruise identifier, can be included in order to reference the CSR when the vessel is on survey.

Since the SSR presents the information in a structured way, an XML schema was used to code the ship report. The easiest way to code SSR is to use a tailored schema and namespace to code the five main blocks of SSR information (Figure 8). Although some parts such as the daily event report or vessels' navigation tracks have been coded using specific standards and schemas that provide community accepted elements for their description.

Figure 8. Ship Summary Report "Core" elements and structure

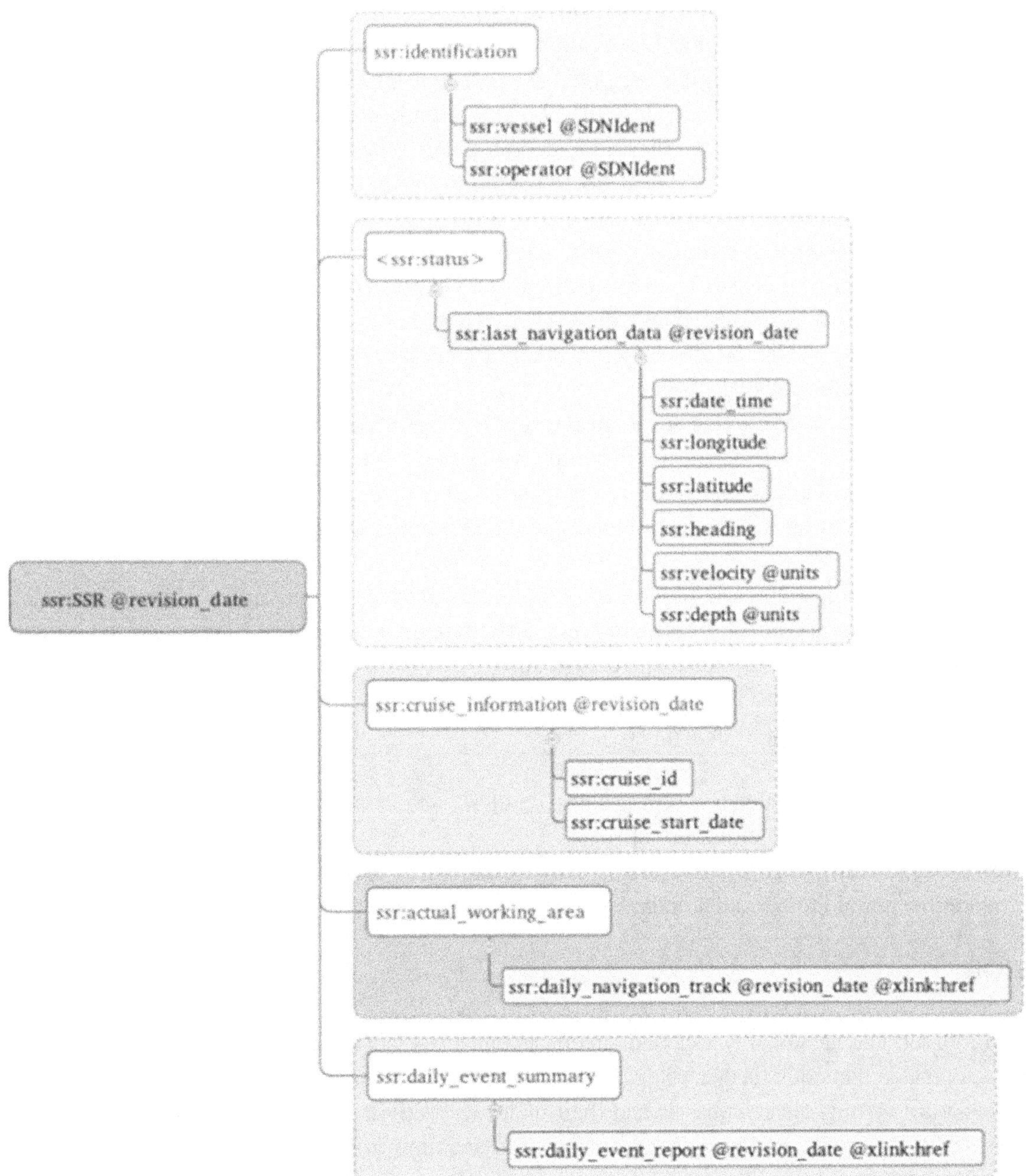

The EARS schema to code SSR has defined a specific namespace "*ssr*" (xmlns:ssr="http://www.eurofleets.eu/ssr) to code the identification, status and cruise_information by using "ad-hoc" tags, where the vessel and operator identification are coded using SeaDataNet vocabulary namespaces.

To code the 'last 24h vessel track' the GML 3.2 schema has been used. The key element is the use of gml:history tag, and the gml:MovingObjectStatus. Both elements provide a standard method to capture a record of the status of a moving object. A gml:MovingObjectStatus element allows the user to describe the present location, along with the speed, bearing, acceleration and elevation of an object in a particular time slice. Additional information about the current status of the object may be recorded in the gml:status or gml:statusReference property elements. The gml:history is complemented by a gml:boundingBox and a gml:temporalExtend to code the temporal and spatial extension of vessel track. The 24h vessel track can have a large number of positioning elements (1 entry by minute, 1440 by day) and should therefore be stored in a separated file and referenced in the SSR file using a "*xref*" attribute.

To code the 24h Event registry it is suggested to adopt the OGC SensorMLV2.0 standard. Each Event is coded using individual sml:Event elements of a sml:EventList container element. Every individual component of each event (subject, actor, tool, action, details, ...) is coded inside the sml:Event using sml:property tags where a specific role attribute is used to distinguish one from each others.

The xlink:role is used in conjunction with a URI pointing to the resource where the definition of the event element can be retrieved from the BODC vocabulary server.

As SSR file has five blocks of information in its core expression, each of them can be obtained independently and also at a different rates from EARS system. For example the ssr:last_navigation can be updated every minute while the 24h vessel track and event registry can be refreshed less often and the ssr:identification never updated.

The identification block, which never changes in time, is introduced at the SSR engine by the use of a template (it is necessary to define a SSR template file for each Eurofleets vessel) while an unlimited number of threads (ideally four) have the responsibility to update every block of information individually, and at a different rates. The main thread of the SSR engine saves the XML SSR file periodically to a file system.

There are different strategies to follow in order to transmit the en-route information from vessels to shore that should be valid not only for global vessels (with permanent satellite links) but also for regional ones or those lacking satellite broadband connections.

As CSR and SSR are linked but are generated independently, it is necessary to coordinate when both files are generated and sent to shore:

- SSR should be generated and sent to shore regularly and independently regardless of whether the vessel is conducting a cruise or not.
- CSR should only be generated and sent at the beginning of the cruise. Currently a link to the cruise identifier should to be coded in the SSR

EVIOR

EARS web services are not only accessed directly by end users, but also by the Eurofleets EVIOR portal. This is a service that adds further value providing a number of infobases among which is a prototype service for following research vessels and their underway activities based on the EARS web services access. EVIOR retrieves and harmonizes information acquired from multiple vessels at the same time and therefore allows for easy generation of a snapshot of the activities of all the vessels in the Eurofleet project.

FUTURE RESEARCH DIRECTIONS

EARS V1 was devised as a prototype for the Eurofleets1 project. It has been tested and proved to be robust enough, so that within the following Eurofleets2 project it was planned to install it on most of the partner's vessels. EARS was aimed to be operational but at the same time it was meant to highlight its current limitations and suggest possible extension or needs. By analyzing its use it was possible to identify several key points for the future developments.

In order to provide a flexible architecture to be deployed onboard different research vessels, we realized that the Data Logging System (where events, cruise and program information are stored jointly with other general purpose acquired data like position, depth, etc .) has to be decoupled from the Manual Event Log Graphical User Interface.

A specific software interface, called Eurofleets Web Service Layer (EWSL), is currently under design, that will act as an intermediate layer providing the necessary link between the onboard Data Logging System and the different pieces of software that act as data consumers and/or producers.

The EWSL will be formally defined and implemented as a W3C Web Service, providing a formal description of the interfaces, use cases and the set of rules necessary to use this intermediate layer. This is a way to standardize storage, modification and retrieval of Events, Cruise, Program and Navigation from and to the onboard Data Logging System through the different data consumer and producer applications that can be implemented onboard, and will allow for evolution and even to substitution of each individual part of the EARS system with a higher level of freedom while following the specifications of the interface.

Considering semantics and the manual event graphical user interface, it will be necessary to ease how end users can access the ontology and instruct them on the use of the ontology as a formal method to code events, which at the same time will help developers formalize the context of operations at sea. Similarly, we expect that from training activities the developer team will receive useful inputs for the improvement of EARS

CONCLUSION

This chapter aimed at providing a solution for the paradigmatic drift that can occur when data are not properly documented. This can cause serious issues in collaborative work, which is at the very kernel of any project in sustainable development.

We tackled the problem trying to enhance the quantity and quality of the metadata and ancillary data associated with the target data. This is a very problematic and tedious activity if done after the acquisition is ended. We adopted a different perspective where most of this information can be recorded directly during the acquisition. Information coming from devices such as GPSs, meteorological stations or other sensors can automatically be stored as underway data and subsequently be associated with any other logged event related to data acquisition through time and position. The same is also possible for events observed directly by the people involved in the acquisition such as for example in the case of anomalies or the collection of material during sampling. A specific suite of software named Eurofleets Automatic Reporting System (EARS) was developed that allows all these activities to be logged using standard terms, originating from reference vocabularies when possible. EARS stores all events in a database from which they can later be extracted and used. Events can be linked to the recorded data set through a standard O&M and SensorML, Open Geospatial Consortium (OGC) compliant metadata model, so that it is possible to associate any part of the target data with its history. Events can also be summarized in order to be submitted to international data management initiatives such as CSR (Cruise Summary Report). EARS offers a web service named Ship Summary Report (SSR) that can retrieve the most relevant events that have occurred in the last 24 hours. This web service can also be used by initiatives aiming to reach a wider community, such as EVIOR. This offers a web based geographic representation of the Eurofleets vessels' positions and the possibility to access each vessel SSR directly simply by clicking on the vessel route or icon.

REFERENCES

Barthes, R. (1957). *Mythologies*. London: Ebenezer Baylis & Son Ltd.

BODC. (n.d.). *NERC controlled vocabulary*. Retrieved from http://www.bodc.ac.uk/products/web_services/vocab/

Callender, C., & Cohen, J. (2006). There is No Special Problem About Scientific Representation. *Theoria, 21*(1), 67–85.

Coward, R., & Ellis, J. (1977). *Language and Materialism: Developments in Semiology and the Theory of the Subject*. London: Routledge & Kegan Paul.

Diviacco, P. (2012). Addressing Conflicting Cognitive Models in Collaborative E-Research: A Case Study in Exploration Geophysics. In *Collaborative and Distributed E-Research: Innovations in Technologies, Strategies and Applications*. Hershey, PA: IGI Global Press; doi:10.4018/978-1-4666-0125-3.ch012

Diviacco, P. (2015). Reconciling Knowledge and Collaborative e-Research. In P. Diviacco, P. Fox, C. Pshenichny, & A. Leadbetter (Eds.), *Collaborative Knowledge Scientific Research Networks*. Hershey, PA: IGI Global. doi:10.4018/978-1-4666-6567-5.ch001

Diviacco, P., & Busato, A. (2015b). Maps, graphs and annotations as Boundary Objects in Knowledge Networks, Distributed Cognition and Collaborative e-Research. In P. Diviacco, P. Fox, C. Pshenichny, & A. Leadbetter (Eds.), *Collaborative Knowledge in Scientific Research Networks*. Hershey, PA: IGI Global. doi:10.4018/978-1-4666-6567-5.ch019

Diviacco, P., De Cauwer,K., Leadbetter, A., Sorribas J., Stojanov,Y. Busato,A., Cova, A. (2015). Bridging semantically different paradigms in the field of marine acquisition event logging. *Earth Sci Inform, 8*, 135–146. DOI 10.1007/s12145-014-0192-0

Diviacco, P. Lowry, R., & Schaap, D. (2012). Marine seismic metadata for an integrated European scale data infrastructure: the FP7 Geo-Seas project. *Bollettino di Geofisica Teorica ed Applicata, 53*(2), 243-252. DOI 10.4430/bgta0051

Eco, U. (1976). A Theory of Semiotics. Indiana University Press. doi:10.1007/978-1-349-15849-2

Hanson, N. R. (1958). *Patterns of Discovery*. Cambridge, UK: Cambridge University Press.

Popper, K. (2004). Conjectures and refutations: the growth of scientific knowledge. London: Routledge.

Sturrock, J. (1986). *Structuralism*. London: Paladin.

KEY TERMS AND DEFINITIONS

Action (Event): Which is the actual task performed, for example start sampling, close bottle or start recording.

Action_property (Event): Any additional characteristic or parameter accompanying a given action for which the user needs to enter a value onboard f.e. volume of water centrifuged, installation height of a sensor or distance trawled.

Actor (Event): Is the person performing the action.

Category (Event): Describes the kind of event taking place, for example a malfunction, sampling, operation, or phase change.

Cruise: Describes the set of oceanographic research activities, which are developed within a research vessel.

Comment (Event): free text field that is left to the operator to enter more information like for example the visual description of a sediment sample.

Data Element: A unit of data that is considered, in context, to be indivisible.

Message: A basic unit of communication from one web software agent to another sent in a single logical transmission.

Program: Is a concept to organize in a practical way events to be linked to 'principal investigators' or 'data originator' and data originator's institute.

Protocol: A formal set of conventions governing the format and control of interaction among communicating functional units.

Registry: An enabling infrastructure that uses a formal registration process to store, catalogue, and manage metadata relevant to the service. A registry supports the search, identification, and understanding of resources, as well as query capabilities.

Subject (Event): Refers to the domain in which the event takes place, for example a seismic system or a sediment sampler.

Tool (Event): The device that produces the event, for example an anemometer or a Niskin bottle.

Uniform Resource Identifier (URI): A compact string of characters for identifying an abstract or physical resource.

Uniform Resource Locator (URL): A type of URI that identifies a resource via a representation of its primary access mechanism (e.g., its network "location"), rather than by some other attributes it may have.

Web Service: A self-describing, self-contained, modular unit of software application logic that provides defined business functionality. Web services are consumable software services that typically include some combination of business logic and data.

Chapter 9
Interoperability in Marine Sensor Networks through SWE Services:
The RITMARE Experience

Alessandro Oggioni
Institute for Electromagnetic Sensing of the Environment, National Research Council (IREA-CNR), Italy

Monica Pepe
Institute for Electromagnetic Sensing of the Environment, National Research Council (IREA-CNR), Italy

Paolo Tagliolato
Institute of Marine Sciences, National Research Council (ISMAR-CNR), Italy

Stefano Menegon
Institute of Marine Sciences, National Research Council (ISMAR-CNR), Italy

Cristiano Fugazza
Institute for Electromagnetic Sensing of the Environment, National Research Council (IREA-CNR), Italy

Fabio Pavesi
Institute for Electromagnetic Sensing of the Environment, National Research Council (IREA-CNR), Italy

Paola Carrara
Institute for Electromagnetic Sensing of the Environment, National Research Council (IREA-CNR), Italy

ABSTRACT

The rapid growth and development in different fields related to sensors has, together with the huge increase of devices due to the decrease of device costs, led to a shift from traditional monitoring, where the data collected is not subject to any management actions, to sensor/processing networks, where in the life cycle more stages are devoted to make the data accessible. Data integration is the first step in advanced environmental monitoring, but assuring that heterogeneous systems can interoperate is still a challenge. The Sensor Web Enablement (SWE) initiative defines a framework to address this issue, offering a set of standard models and interfaces to improve sensor interoperability and to face quality

DOI: 10.4018/978-1-5225-0700-0.ch009

issues in the reliability of sensors. The need for seamless access to observations from marine sensors has been the focus of several research projects. This chapter presents the actions taken in the development of the Spatial Data Infrastructure for project RITMARE to ease the adoption of SWE within the Italian marine community overcoming the main constraints in SWE adoption.

INTRODUCTION

The rapid growth and development in different fields related to sensors, such as in micro electro mechanical systems (MEMS), wireless communication, and digital electronics has, together with the huge increase of devices due to the decrease of their costs, led to a shift in environmental disciplines (Papp & Hakkesteegt, 2008): While in traditional monitoring data collections were not managed according to specific strategies, in sensor/processing networks more stages of data life cycle are devoted to trace data acquisition processes and distribution. Environmental Sensor Networks allow for increasing the number of observations and measurements that are made accessible: This improves the understanding of complex and fragile environments such as coastal zones. Indeed, in order to exploit the variety and volume of observations, it is necessary to adopt an integrated approach on distributed, mobile, fixed, and asynchronous sensors from different networks, as suggested by Hart and Martinez (2006). In the last few years, for the sciences of the environment, several initiatives were undertaken to allow for data discovery and accessibility, as ruled by standard protocols and application programming interfaces. In the past, the research work was characterized by a phase where the collected data were exchanged only after a painstaking formatting and harmonisation, according to a shared format. More recently, data integration is considered the prerequisite to advanced environmental monitoring and sustainable development, but the capabilities of different systems to work together has been realized only recently (Barnaghia, Ganza, & Abangara, 2011; Havlik et al., 2011) and, under many viewpoints, it is still a challenge. In particular, managing different and heterogeneous data sources, of several types, is an actual critical requirement for environmental monitoring, in particular in the field of marine sustainable development, for which the integration of information from different disciplines and domains (e.g. economic indicators, traditional GIS sources, output of oceanographic models, time-series of sensors monitoring the environment continuously) is a key point (Diviacco & Leadbetter, 2017, chapter 2 this book). In a vision for a Digital Earth in the modern era, Al Gore (1998) and, more recently, McKee (2015) argued for the creation of interoperable systems for environmental science, buisiness and policies and for free and open standards to implement them.

To achieve complete data integration, the actuation of concepts like interoperability are fundamental in order to realize a linkup among data and to enable across-the-board analyses for the advice of decision-makers and public stakeholders dealing with sustainable development policies. The dimensions of such linkup cover spatial (e.g. location, depth, footprint), temporal (e.g. time stamps, time zones), and thematic (e.g. parameter, measurement unit, domain, quality, methods) attributes. The involved domains and community of practices have their consolidated traditions in data management. The emerging sensor data deluge (Baraniuk, 2011) is requiring novel methods and tools for data management. Several marine observing systems (GOOS[1], EuroGOOS[2], IOOS[3], IMOS[4]), and marine research initiatives (ODIP[5], EuroFleets[6], SeaDataNet[7], FixO3[8], NeXOS[9], ScheMA[10], JericoNext[11], EnvEurope[12], RITMARE[13]) have focused on the requirements for interoperable access to observations from marine sensors. This topic has become more and more important with the emergence of climate change as a world-class issue, because the detection of sudden anomalies is now a priority for early warning systems, and the monitoring of environmental parameters is fundamental in order to understand ongoing phenomena and to create sustainable development scenarios.

The interoperable exploitation of observations coming from multiple, heterogeneous, distributed sensors (or sensor networks) still presents some open issues. A first one is the building of architectures based on distributed and independent nodes, which are in charge of data deploying. Service Oriented Architecture (SOA) allows services to be dynamically published, discovered, and finally invoked by clients (Jiang et al., 2013) - following the "find and bind" paradigm -, providing the framework for an integrated approach on distributed resources. In the sensor-based environmental research domain, the building of SOA is still a challenge to be won (Havlik et al., 2011); in fact, currently existing infrastructures seldom propose web services to exchange information or data. Though many authors (Chen et al., 2009; Woolf, 2008; Voigt et al., 2008; Di, 2007) have demonstrated how distributed services can be made interoperable, currently only a few examples have been implemented.

Another fundamental issue in performing environmental monitoring through networks of distributed and heterogeneous sensors is quality. This issue refers both to reporting the state and history of sensors and to assessing the quality of observations themselves. In fact, on the one hand, it is important to collect and make available details of each sensor such as its description, manufacturer, technical characteristics, classification, capability, position, owner or operator, input/output, components, history log to track changes and calibrations; these descriptions help assessing the reliability of each sensor and comparing measurements coming from different sensors, in particular with respect to users' purposes and needs. On the other hand, an interoperable and seamless search and access to the collected observations is the basis to automatically perform quality check operations such as detection of outliers.

Oggioni et al. (2014) proved how, in an interoperability perspective, the Sensor Web Enablement (SWE) technology (Reed et al., 2007) can efficiently deal with the integration of measurements of chemical/physical parameters of the sea deployed by autonomous, distributed and independent Web Services. In particular, this study demonstrated how SWE is capable of addressing the following challenges, as stated by Hart and Martinez (2006): i) heterogeneity of sites; ii) quality check and harmonization; iii) description and history of sensors.

In SWE, beside the main standard OGC Sensor Observation Service (SOS) (Na & Priest, 2007; Yongguo et al., 2013), a main role is played by one of the components of SWE, i.e. SensorML, a standard metadata format allowing for describing characteristics and capabilities of sensors, including history logs and quality features (Botts & Robin, 2004). Therefore, the richness of SensorML metadata offers a solution for discovering and comparing measurements coming from heterogeneous sensors, allowing a proper search for observations from different sensors.

However, such richness comes at the expense of the great effort required for an effective SWE implementation. In fact, the manual entry and update of several metadata items is generally considered as too labor and time-consuming. Moreover, after a correct and fully descriptive SensorML has been created for a sensor, its observations should be managed and distributed in a way allowing for their exploitation within the community of interest. This is also challenging, since the creation of OGC SOS and their population, but also their maintenance in time, are often too requiring for marine Researchers.

SWE can reveal a real solution for real Researchers, only if the building of its components is greatly facilitated by user-friendly tools, tailored on the daily practice of research data management (Buck & Lowry, 2017, chapter 1 this book).

In this Chapter, the Authors report on the activities that are carried out for the interoperable exploitation of observations from sensors in project RITMARE. They prove how a distributed, heterogeneous, asynchronous network of sensors connected to the Web can be searched on the basis of the meta-information contained in SensorML metadata, and how SWE components enable interoperability and sharing of sensed observations and measurements for the purposes of marine research. RITMARE has created the software tools to enable a proper metadata creation, data management and distribution according to OGC standards, and SOA. These tools are presented and discussed in this Chapter. The next Section briefly describes RITMARE, its spatial data infrastructure architecture; the third section introduces the challenges related to SWE implementation in the marine community. The fourth Section describes the main SWE exploitation scenarios and the set of tools developed in RITMARE to enable SWE. The fifth Section provides some use cases testifying the advantages of the proposal. The sixth Section discusses the approach and the final Section draws conclusions and outlines future work.

RITMARE: SENSOR DATA MANAGEMENT IN ITALIAN MARINE RESEARCH

RITMARE is a Flagship Project, promoted and funded by the Italian Ministry of University and Research, whose general goal is to address the integration of Italian marine research in an interdisciplinary and comprehensive fashion. The National Research Council of Italy (CNR) is in charge of project coordination.

Sub-project 7 (SP7) is aimed at creating the Spatial Data Infrastructure (SDI) for the whole project; such SDI is expected to integrate the different communities of Researchers involved. This includes enabling coordination and sharing of data, processes, and information produced by the other SPs. Specifically, the latter are covering the following disciplines: physical oceanography, chemical oceanography, geology, geophysics, coastal systems, ecosystems, fishing and aquaculture, molecular biology.

The RITMARE user base includes a broad and heterogeneous set of actors (more than 1500 Researchers, 12 research communities, 5 research institutions, and 2 university consortia) together with a variety of stakeholders (i.e. public administrations, private companies, and the general public).

These entities envisage a varied corpus of heterogeneous data, related annotations (metadata), processes, goals, visions, and requirements. Beside this, data providers featured different degrees of maturity with regard to the provisioning of data, different ICT capabilities, and different standards to comply with, depending on the projects, initiatives, and community of practices a given data provider contributes to.

In order to address this challenge, SP7 has designed a flexible infrastructure that adapts to the RITMARE specificities by: i) bringing the heterogeneous actors under the same umbrella while preserving their individual practices, formats, and workflows; ii) empowering users with standards-compliant Web services that afford for interoperability, in terms of discovery and access to the data and metadata that is collected by the project.

The approach adopted in RITMARE for designing and implementing the project flexible infrastructure is grounded on SOA and is based on the exploitation of INSPIRE-compliant, XML-based metadata and OGC-based services (European Commission, 2008; IOC Task Force for Network Services, 2011a; Initial Operating Capability Task Force, 2012; IOC Task Force for Network Services, 2011b).

Specifically, the RITMARE infrastructure can be categorized as a decentralized infrastructure, featuring:

- A set of peripheral nodes that expose OGC standards-compliant metadata and services;

- A centralized catalog, integrating the OGC Catalog Service for the Web standard (CSW) with more general-purpose access paradigms, that provides a single point of access to the resources made available by the project as a whole.

Because of the aforementioned heterogeneity in ICT expertise, capacity building by data providers has been the keystone in the creation of the RITMARE decentralized SDI. This attitude has led to the development of the Geoinformation Enabling Toolkit Starter Kit (GET-IT), a suite integrating existing and new software that is available under the Free and Open Source Software licence GPL v.3 and that was distributed as a ready-to-use virtual appliance to data providers with no pre-existing infrastructure. Figure 1 depicts GET-IT installations acting as peripheral nodes in the SDI.

Even when considering sensor data, which are the focus of this work, the heterogeneity and variety of the data provided in RITMARE is still apparent. In fact, even if the sensing stations and facilities belonging to Italian marine research organizations are meant to act as a network, they were not designed as a sensor network from the beginning. While Italian marine research offers excellence in the type of sensors that are used, its managing authorities are numerous and have different skills, resources, and IT expertise. Therefore, there is no consistency as regards to the collection, frequency, and distribution of the measured parameters (e.g., different temporal and spatial resolution, units, identifiers). This is the reason why the solutions to enable interoperability, as proposed by RITMARE SP7, can be seen as a paradigm of integration, not only among sensors at a national level, but also among sensor networks at large.

Figure 1. GET-IT nodes in the RITMARE spatial data infrastructure

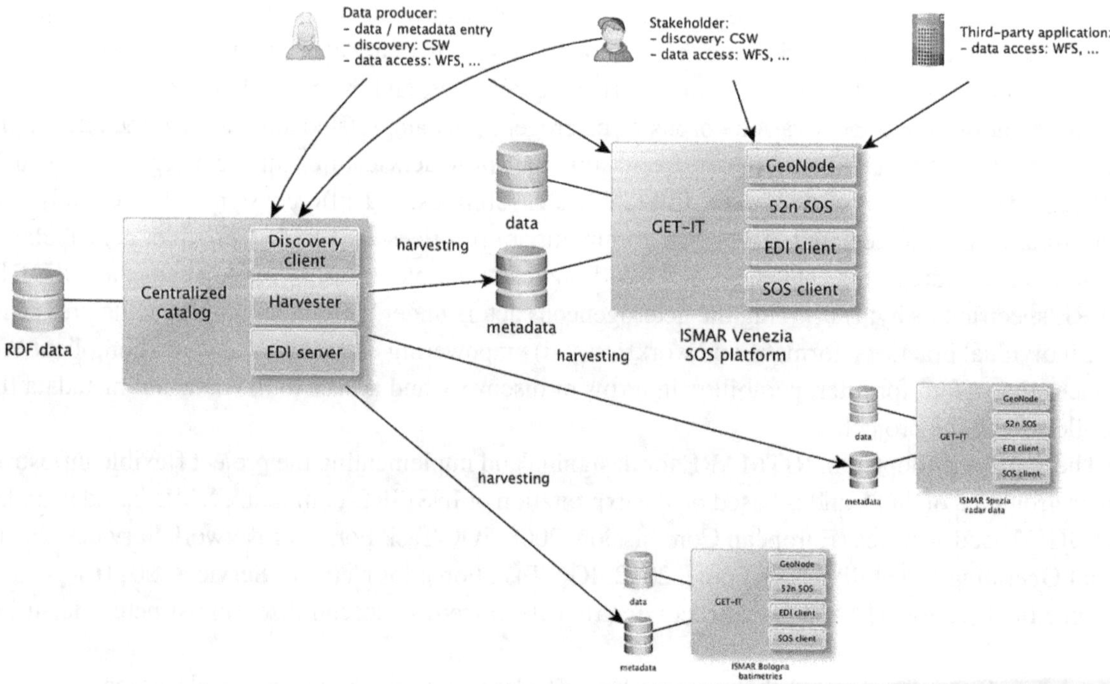

FOSTERING SWE IN THE MARINE COMMUNITY

The OGC Sensor Web Enablement (SWE) initiative constitutes a standard, service oriented solution for addressing interoperability of heterogeneous systems dealing with sensor observation data. The SWE initiative developed standard service interfaces (SOS) and data and metadata models, e.g., Observations and Measurements (O&M) and SensorML. On the basis of these specifications, free and open source applications have been developed, such as the implementation of a SOS server by 52°North that is bundled in the GET-IT suite.

Still widespread adoption of SWE by communities is hampered by the lack in user-oriented facilities for managing SWE implementations. Consider for instance the following actual scenarios.

1. A sensor is deployed in the field. The sensor manufacturer, as typically happens, did not provide the Researcher with technical specifications complying with any OGC standard. Still the availability of sensor information is crucial to making the collected data re-usable (e.g., for assessing reliability of a sensor, for comparing calibration of sensors of the same type or of the same sensor in different time periods). Also, in this perspective, a standardized way to access this information can make the difference in future integrations of the observing system. Providing this information, that is, producing XML documents complying with the SensorML standard, is challenging and error prone even for experienced users. In order to simplify creation of SensorML descriptions and to customize them to specific communities, *profiles* for SensorML have been discussed and developed (e.g. the one included in Jirka, Stasch, & Broering, 2014). A metadata profile can also restrict possible content values for a field to a specific controlled vocabulary or a semantic source under *authority control* (Haslhofer & Klas, 2010). Unfortunately, profiles address the structure and semantics of SensorML documents, not the user experience, so the lack in tools assisting users in SensorML composition persists.
2. Another possible scenario features a sensor, appropriately described with SensorML, that is discovered by a Researcher with little IT expertise. He retrieves an XML document but he can not evaluate if the characteristics of the sensor are of his interest. Human-readable representations of SensorML documents would ameliorate this situation.
3. The third scenario is particularly relevant to RITMARE and involves sensor observation data that are collected in tabular format and, when eventually digitized, stored as spreadsheets. In environmental research, this situation may arise in several cases: Historical datasets archived as files in personal computers; datasets of manually collected sensor observations (e.g. Secchi Disk Depth, specimen analyses, etc.); datasets saved in disparate file formats through proprietary software provided by sensor manufacturers for their specific sensor (often these software exist, making sensor data available to Researchers, but facilities to move data to an SOS endpoint and make the same data interoperable are missing).
4. The fourth scenario refers to the management of sensor systems for continuous real-time environmental monitoring. The system can consist of an individual sensors or can be a structured sensor network producing a significant amount of data at regular frequency. Data are usually stored and made available by proprietary or non-standard platforms. Researchers need to properly configure their sensor systems in order to send data to a previously configured SOS. At the same time, the researchers might need to upload to the SOS, all the archived data accumulated over time.

In the following we clarify the actions taken in RITMARE to address the issues emerging in these scenarios.

As introduced before, the GET-IT suite[14] is the basic building block of the SDI. By means of GET-IT, data providers can upload and manage their data (both maps and observations), and create the appropriate metadata for either datasets or sensors. The software suite provides discovery, visualization, and access interfaces according to OGC standards. Table 1 lists the web services that are exposed and the implementing software.

Three main strategies were followed: to simplify as much as possible the installation of the tools; to allow data providers to easily and directly manage, publish, and share their geospatial and observation datasets (without specific skills on SDIs); to foster reuse of the services provided by the GET-IT in projects and initiatives outside RITMARE.

The software that has been customized or created for GET-IT allows to keep data, metadata, and services tightly coupled. As an example, several metadata items are automatically extracted from datasets at upload-time and passed down to the metadata editor for insertion in the metadata record, thus partially relieving Researchers of the boring and error-prone activity of manual editing.

Beside managing geographic information, such as maps and layers[15], through a GeoNode instance, GET-IT is also capable of uploading, storing, and publishing sensor descriptions and sensor observations. For this purpose the implementation by 52°North Initiative (n.d.) of the SOS specification (Broering, Stasch, & Echterhoff, 2012) is bundled in the suite.

Among the components that have been developed for GET-IT, the following are those devoted to foster SWE:

- A web-based authoring tool for SensorML capable to insert and update sensor descriptions within the local SOS, based on EDI (the software described in the next section).

Table 1. Web services exposed by GET-IT

Service	Version	Implementing Software
OCG Web Map Service	1.1.1, 1.3.0	GeoNode (GeoServer)
OGC Web Feature Service	1.0.0, 1.1.0, 2.0	GeoNode (GeoServer)
OGC Web Coverage Service	1.0, 1.1	GeoNode (GeoServer)
OGC Styled Layer Descriptor	1.0	GeoNode (GeoServer)
OSGeo WMS Tiling Clients Recommendation		GeoNode (GeoWebCache)
OGC Web Map Tiling Service		GeoNode (GeoWebCache)
OSGeo Tile Map Service		GeoNode (GeoWebCache)
OGC Catalogue Services for the Web	2.0.2	GeoNode (pycsw)
INSPIRE Discovery Services	3.0	GeoNode (pycsw)
Search/Retrieve via URL (SRU) search protocol		GeoNode (pycsw)
OGC SOS	1.0.0, 2.0.0	52°North SOS
OGC Sensor Model Language (SensorML)	1.0.1	52°North SOS
OGC Observation and Measurement (O&M)	2.0.0	52°North SOS
OGC Geography Markup Language (GML)	3.1.1	52°North SOS

- A web-based transactional SOS client, by means of which the user can easily compose and upload observations collected in data-sheets, a typical use-case for historical sensor observations.
- Presentation layer for SensorML documents, that is graphical user interface realized by means of XSLT style-sheets (Clark, 1999), transforming the XML sensor information in a human readable web-page.
- An SOS GIS-Client allowing the user to easily interact with the SOS to retrieve, for each sensor, its position, the parameter list, its *offerings*[16], and the collected values. Notably this client provides users the ability of creating mashups of SWE information and traditional web-GIS layers (i.e., vector and raster maps). These data can reside in the local GET-IT node, but can also be collected through the invocation of remote OGC web-services (Figure 1 and Table 1).
- A shared repository of snippets, scripts and HOWTOs to facilitate and support the activation of data flow from sensor systems to SOS service.

The following subsections will describe in more details these components.

SensorML Editor

One of the key components in the RITMARE infrastructure is EDI, a customizable, template-driven metadata editor, providing users with web-forms to interact with. EDI was developed leveraging a variety of complementary techniques. The rationale for this is to employ the more appropriate technology for each of the distinct phases in the workflow for metadata creation. For the time being, the editing of INSPIRE, RNDT (the Italian profile to the former), and SensorML metadata are supported. However, the data model underlying template creation is expressive enough to support most of the metadata formats in the SDI domain. As an example, EDI is being currently employed for the management of NetCDF metadata in Unidata's NcML format (Nativi et al., 2005)[17]. The only input required by EDI to assist composition of metadata records in a given XML format is an appropriate template[18]. The template language introduces the primitives that are necessary for the definition of metadata elements, intended as high-level abstraction of the information that can be either provided by the user or pre-computed by the application itself (such as the geographic boundary, the so-called bounding box, that can be automatically computed at data upload-time). Beside abstracting from a specific XML format, the strength of EDI as a semantics-aware metadata editing tool resides in the generic RDF data structures that can be plugged in. This allows for annotating metadata with unique identifiers (e.g., the Uniform Resource Identifiers (URI) corresponding to terms from a controlled vocabulary). Such characteristic enables support for profiles following, to achieve interoperability at the semantic level, the strategy of authority control (Haslhofer and Klas, 2010). URIs will be used later on, when metadata is harvested by the central catalog collecting the contributions of the distinct nodes of the infrastructure. This information constitutes the fundamental step for providing advanced query functionalities by the overall infrastructure.

Listing 1 presents an excerpt from the EDI template governing creation of sensor metadata in RITMARE. It is based on the Lightweight SOS Profile to SensorML v2.0.0 (Jirka, Stasch, & Broering, 2014) and it adds specific semantic characterization of admissible values (e.g. for parameters, keywords, persons and institutes, sensor manufacturers). The template is organized in several sections: *settings* (Listing 1, line 03) defines configuration variables, e.g. the specification of an authoritative SPARQL endpoint; *datasources* (line 04) plugs in external RDF data structures by defining specific SPARQL queries; *group* sections (from line 65) will follow, in which, within *element* tags, both the definition of the user interface and the mapping of user inputs into the final XML document are defined. The essential information defining an *element* is the language-dependent textual representation to be displayed in the interface and the text describing its purpose more extensively (lines 67-70, 73-76, and 100-103), the compulsoriness of the metadata item and its multiplicity (defined on lines 77 and 78), and the XPath defining the position where multiple instances of the metadata field shall be rooted (line 79). Instead, the distinct *items* associated with the *element* define the specific XML nodes that shall be generated. In the HTML form interface generated by EDI, the user is presented with the form fields that correspond to *items* with a value of *isFixed* set to "false" (as on line 85). For these, attribute *hasDatatype* specifies the data type of the specific metadata content. In the listing (line 82 et seq.), the first *item* of the *element* "output_param" is of type "autoCompletion", meaning that the values that shall be recommended to the user shall be taken from the results of the SPARQL (Harris & Seaborne, 2013) query that is defined for the *datasource* "keywordsS". EDI processes the template and generates the actual HTML form that is displayed in the user's browser. Still, this interface is a dumb form that only distinguishes between input fields, text areas, dropdown lists, etc. but whose input fields and text areas have no predefined values, whose dropdown lists have no options (the values selectable by the user), etc. In fact, it is up to the JavaScript component of EDI to dynamically load these contents: For doing this, the JavaScript code accesses codelists and controlled vocabularies stored in the RDF (Brickley & Guha, 2004) triple store on the server-side, as described in (Fugazza et al., 2014a; Fugazza et al., 2014b). These categorizations of terms also provide all the accessory information that is required for creating the items that are either fixed or dependent on the user's choices (the items whose attribute *isFixed* is set to "true").

Listing 1. Portion of the XML EDI template for the Lightweight SOS Profile of SensorML v2.0.0

```
01 <?xml version="1.0" encoding="UTF-8"?>
02 <template>
03    <settings>…<sparqlEndpoint>http://sparql.get-it.it</sparqlEndpoint>…</settings>
04     <datasources>
05         <datasource>
06             <id>keywordsS</id>
07             <singleton>true</singleton>
08             <endpointType>virtuoso</endpointType>
09             <query>
10             <![CDATA[
11             PREFIX skos: <http://www.w3.org/2004/02/skos/core#>
12             SELECT ?c ?l ?a
13             FROM <http://ritmare.it/rdfdata/parameters>
14             WHERE {
15               ?c rdf:type skos:Concept.
16               OPTIONAL {
17                 ?c skos:prefLabel ?l.
18                 FILTER(LANG(?l) = "en")
19               }
```

continued on following page

Listing 1. Continued

```
          OPTIONAL {
            ?c skos:prefLabel ?a.
            FILTER(LANG(?a) = “it”)
          }
          FILTER(REGEX(STR(?c), “$search_param”, “i”) || REGEX(STR(?a), “$search_param”, “i”))
        }
        ORDER BY ASC(?l) ASC(?a)
        ]]>
        </query>
        <triggerItem>output_param_1_uri</triggerItem>
        <type>sparql</type>
    </datasource>
    <datasource>
        <id>uom</id>
        <singleton>true</singleton>
        <endpointType>virtuoso</endpointType>
        <query>
        <![CDATA[
        PREFIX skos: <http://www.w3.org/2004/02/skos/core#>
        SELECT ?c ?l ?urn
        FROM <http://ritmare.it/rdfdata/units>
        WHERE {
          ?c rdf:type skos:Concept.
          OPTIONAL {
            ?c skos:prefLabel ?l.
            FILTER(LANG(?l) = “en”)
          }
          OPTIONAL {
            ?c skos:prefLabel ?a.
            FILTER(LANG(?a) = “en”)
          }
          OPTIONAL {
            ?c skos:altLabel ?urn.
            FILTER(LANG(?urn) = “en”)
          }
          FILTER(REGEX(STR(?l), “$search_param”, “i”) || REGEX(STR(?a), “$search_param”, “i”))
        }
        ORDER BY ASC(?l) ASC(?a)
        ]]>
        </query>
        <type> sparql </type>
```

continued on following page

Listing 1. Continued

```
        </datasource>
        ...
    </datasources>
    ...
    <group>
        <id>output</id>
        <label xml:lang="it">Parametri misurati</label>
        <help xml:lang="it">Lista dei parametri misurati</help>
        <label xml:lang="en">Output signals</label>
        <help xml:lang="en">List of output signals</help>
        <element>
            <id>output_param</id>
            <label xml:lang="it">Parametro misurato</label>
            <help xml:lang="it">Parametro misurato dal sensore, sistema o
            piattaforma e informazioni relative. Si possono aggiungere
            pi&#249; parametri, cui associare le relative
            informazioni</help>
            <label xml:lang="en">Output signals</label>
            <help xml:lang="en">Parameter measured by the
            system/sensor/platform and related information. Several
            parameters, and related information, can be inserted.</help>
            <isMandatory>forAll</isMandatory>
            <isMultiple>true</isMultiple>
            <hasRoot>sml:PhysicalSystem/sml:outputs/sml:OutputList
            /sml:output</hasRoot>
            <produces>
                ...
                <item>
                    <hasIndex>1</hasIndex>
                    <hasPath>@name</hasPath>
                    <isFixed>false</isFixed>
                    <hasDatatype>autoCompletion</hasDatatype>
                    <datasource>keywords</datasource>
                    <label xml:lang="it">Nome parametro misurato</label>
                    <help xml:lang="it">Nome del parametro misurato.
                    Iniziare a digitare tre lettere per ottenere dei
                    suggerimenti da vocabolario controllato.</help>
                    <label xml:lang="en">Output signal</label>
                    <help xml:lang="en">Measured parameter. Start typing
                    the first three letters to obtain suggested
                    autocompletion from controlled vocabularies.</help>
                </item>
                ...
```

continued on following page

Listing 1. Continued

```
94                  <item>
95                      <hasIndex>4</hasIndex>
96                      <hasPath>swe:Quantity/swe:uom</hasPath>
97                      <isFixed>false</isFixed>
98                      <hasDatatype>autoCompletion</hasDatatype>
99                      <datasource>uom</datasource>
100                      <label xml:lang="it">Unit&#224; di misura</label>
101                      <help xml:lang="it">Unit&#224; di misura del parametro
                         misurato. Iniziare a digitare tre lettere per ottenere
                         dei suggerimenti da vocabolario controllato.</help>
102                      <label xml:lang="en">Unit of measurement</label>
103                      <help xml:lang="en">Unit of measurement of the ouput
                         parameter. Start typing the first three letters to
                         obtain suggested autocompletion from controlled
                         vocabularies.</help>
104                 </item>
105             </produces>
106         </element>
107     </group>
108     ...
109</template>
```

Figure 2 shows the section of the web form generated by EDI on the basis of the code in Listing 1 (lines 65-107), and in particular the autocompletion functionality: for the "Output signal" field the system provides the user with the range of observation parameters defined by the SeaVox vocabularies P01, P02, and P07.

Semantics-aware characterization of data items allows the interface to provide three distinct functionalities: the generation of dropdown lists on the basis of the mandated codelists, translated into the specific language that is chosen for the metadata; the aforementioned autocompletion of metadata fields whose values shall be taken from controlled vocabularies that are too large for rendering as a dropdown list; the automatic provision of values for metadata fields whose content depends on the choices that have been made for other metadata fields. The RDF data structures present in the infrastructure and any other SPARQL-compliant data source, can be used for assisting the editing of metadata. This is the primary means for relating metadata items to resources in the Semantic Web. Metadata fields can also be autocompleted on the basis of HTTP parameters in the call to the editing interface, can be generated on demand, can duplicate the content of another field, and even use generic XPath functions in order to mix-and-match values in other form fields. Once metadata is posted to the server-side component of the tool, the prescribed XML output is generated by creating XML elements and attributes following XPath expressions (such as that on line 84 or 96 of Listing 1). Besides, the server-side component also produces the set of triples (the atomic data elements in RDF) that constitute the counterpart to the specific XML metadata format. The latter representation elicits novel criteria for metadata management, such as those

Figure 2. EDI HTML user interface form for the Lightweight SOS Profile of SensorML v2.0.0 generated by the definitions in Listing 1

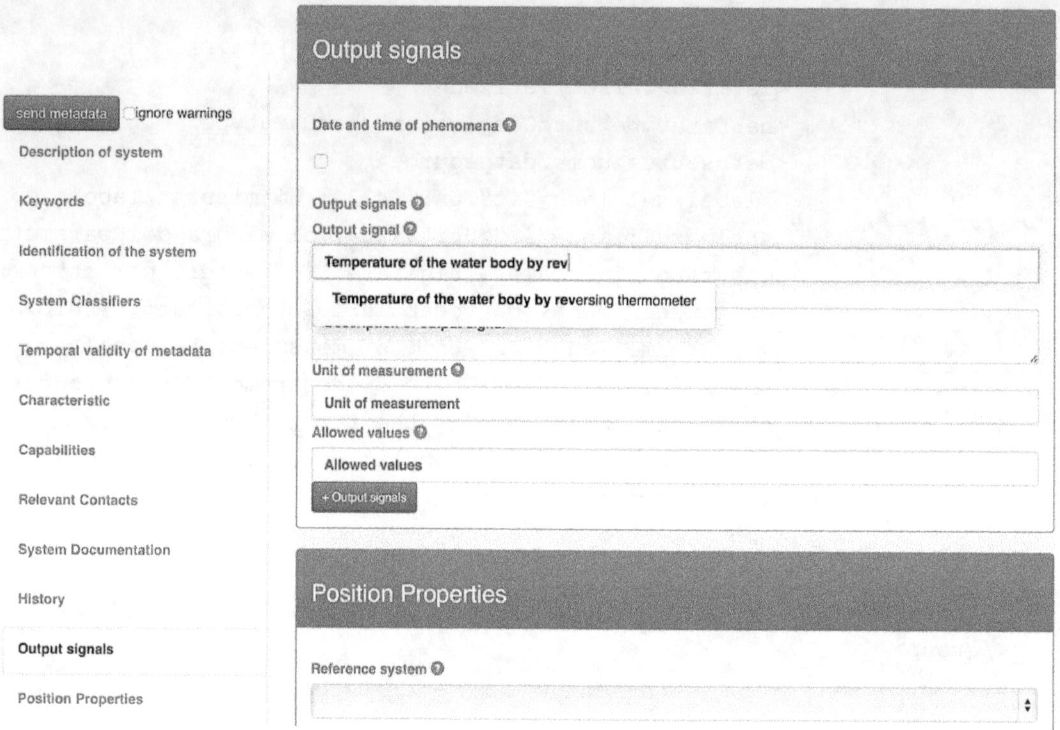

described in (Fugazza et al., 2015). On the one hand, plugging in external data sources allows system managers to easily tailor the metadata editing experience to the specificities of the working group or community of practice. On the other, this capability allows metadata items to refer to third-party data structures that are made available in the Semantic Web at large.

SWE Clients

The web-based transactional SOS client, developed by the Authors, allows the user to easily provide the information that is necessary in order to model the data according to the SWE standard Observation and Measurements (O&M) (Cox, 2013). The workflow for this use case is organized in three steps: i) The choice of a *procedure* among those available in the local GET-IT SOS endpoint; ii) the choice of the observation's *feature of interest* (FOI) (Cox, 2013); iii) The provision of the actual observation *results* via a simple table whose columns correspond to the sensor's *observable properties*. Particular attention has been devoted to the phase related to the choice of the FOI. In the Authors' experience, the notion of FOI is often debated and difficult to grasp by O&M users. The most recent versions of the 52°North SOS implementation supports the OGC *spatial sampling feature* schema, particularly the notion of *spatial sampling points*[19]. This type of feature of interest consists of a spatial point representing the point where the observation's sampling took place (this, in O&M terminology, is called the *sampling point*) and a sampled entity (*sampled feature*) representing the ultimate feature of interest of the observation. While it is clear how to specify a coordinate pair, more debated is what should be the content of the *sampled feature*. According to the definition (Cox, 2013) this should be the specific, concrete entity under observation (and not an abstract concept), e.g. the Venice lagoon as opposed to the concept of "lagoon". To ease the selection of these elements, a map-based GUI connected to the WFS and WMS services: After displaying the WMS representation of the WFS layer, the user can easily select one of the available features as the sampled feature. She can then click on a point (or input the numerical coordinates) related to the actual feature of interest in order to specify the sampling point. The final step, that is, inserting the actual numerical (or categorical) observation values (the O&M *result*) is eased by an editable HTML table: If a spreadsheet is the digital support for data, the user can copy the range of values she is interested in and paste it in the web form. At the end of the composition phase, when the user submits the data, an XML O&M observation is composed through an XSLT transformation invoked by a call to the browser's XSLT processor, and a transactional SOS *insertObservation* request is performed.

The software is modular and organized as separate javascript files and XSLT stylesheets in order to foster reuse, customization, and future improvements, such as providing further support to the O&M spatial sampling features (e.g. spatial sampling lines, spatial sampling surface).

Integrated with the GeoNode Map Composer, an SOS client executes requests to retrieve data from SOS endpoints (local and remote), so that users can display on a GIS-like interface this information. The SOS Client is a JavaScript library based on Open-Layers[20], ExtJS[21], and Flot[22]. Once the features of interest, only sampling points at the moment, of a SOS has been displayed on the map, the observing sensors can then be queried for time series of one or more observed properties. The user, by selecting a sampling point, interacts with the contents being able to: view the information of the procedure, the list of observed property including the extension of time; choose one or more observed property and and view trends over time; download observations displayed; compare different observed property with the same time extension. An important feature is also included to suit the specific needs of some users RITMARE: the real-time visualization of the observations. In fact, by selecting real-time mode and sampling frequency, users can get real-time updates from by the sensor selected.

Finally, the SWE Clients include a presentation layer for SensorML. This allows for a more intelligible representation of metadata with respect to the raw XML serialization. Sensor inserted through EDI client. Specifically, SensorML descriptions, both v1.0.1 and v2.0.0, are transformed into HTML pages through XSLT transformations (see Figure 3). The user can easily visualize information such as: name, description, information about the manufacturer, the owner and operator, the geographical position of the sensor, and especially the history of the sensor. Some of these fields are links to external resources describing the specific field value (e.g., the definition of a code or keyword from a controlled vocabulary).

Documentation and Shared Repository

In the context of the RITMARE project, SP7 has activated an experimental phase by supporting the implementation of 9 GET-IT instances. At least of 6 of them are hosted by research groups that deal with environmental sensors or manage sensor networks.

In this cases it has been necessary, not only the creation of sensors metadata, but also the configuration and activation of a data flow engine that moves data between sources (proprietary and non-standard data storages) and destinations (SOS server).

The data flow engine consists, mainly, in a pipeline of scripts that reads from the data-loggers or from local data storage and performs transactional SOS operations (e.g. *InsertObservation*, *InsertResult*) on the server.

Figure 3. HTML user friendly presentation layer for SensorML. The more important XML elements are made readable for the users (A), the semantic enrichment within SensorML it is made explicit by clicking the parameter name (B)

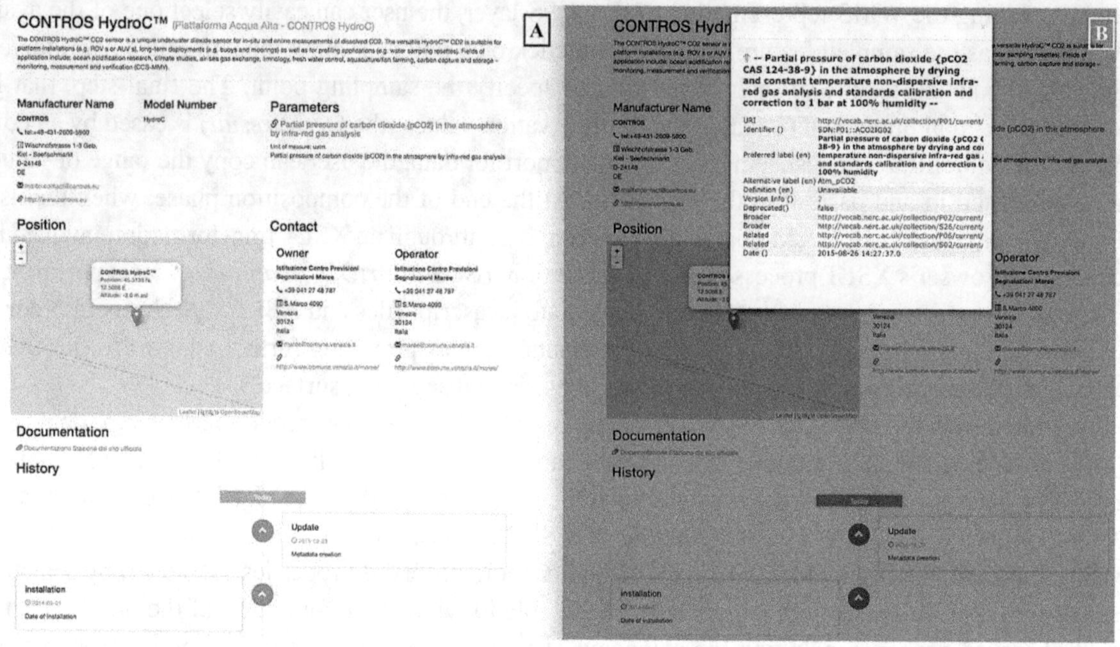

Generally, the data flow engines are case specific and difficult to generalize due to the great variability and heterogeneity of the sensors systems both in terms of commercial products (e.g. sensor hardware and data-loggers) but also in terms of local data storages (e.g. non-standard data models). However, sharing and re-using of scripts, snippets and experiences is strategically important to foster the implementation of other SWE infrastructure. In order to encourage this process the Authors have arranged a dedicated web solution to collect and publish experiences, examples and code. The solution is structured on three parts mutually interconnected: a) a source code repository where sharing scripts and snippets; b) a discussion area where sharing issues and solutions; c) an extension of the documentation/tutorials where publishing the consolidated examples. The solutions is implemented by the GitHub's and Read The Docs's facilities (Figure 4).

Use Cases

This subsection is devoted to describing three real-world use cases in order to check whether the aforementioned tools fully meet the needs of real users. We assume that the potential users of our application are operators involved in marine monitoring: the main actions they are going to perform for assessing sea-water quality through a network of heterogeneous, distributed stations of sensors can be summarized as follows:

- retrieve and display the description of the platform, of the available sensors, and of the measurement processes (e.g. calibration, gain, accuracy, offset, etc.) which may include quality control procedures for all sensors in a station;

Figure 4. Example of basic InsertObservation through a JSON binding

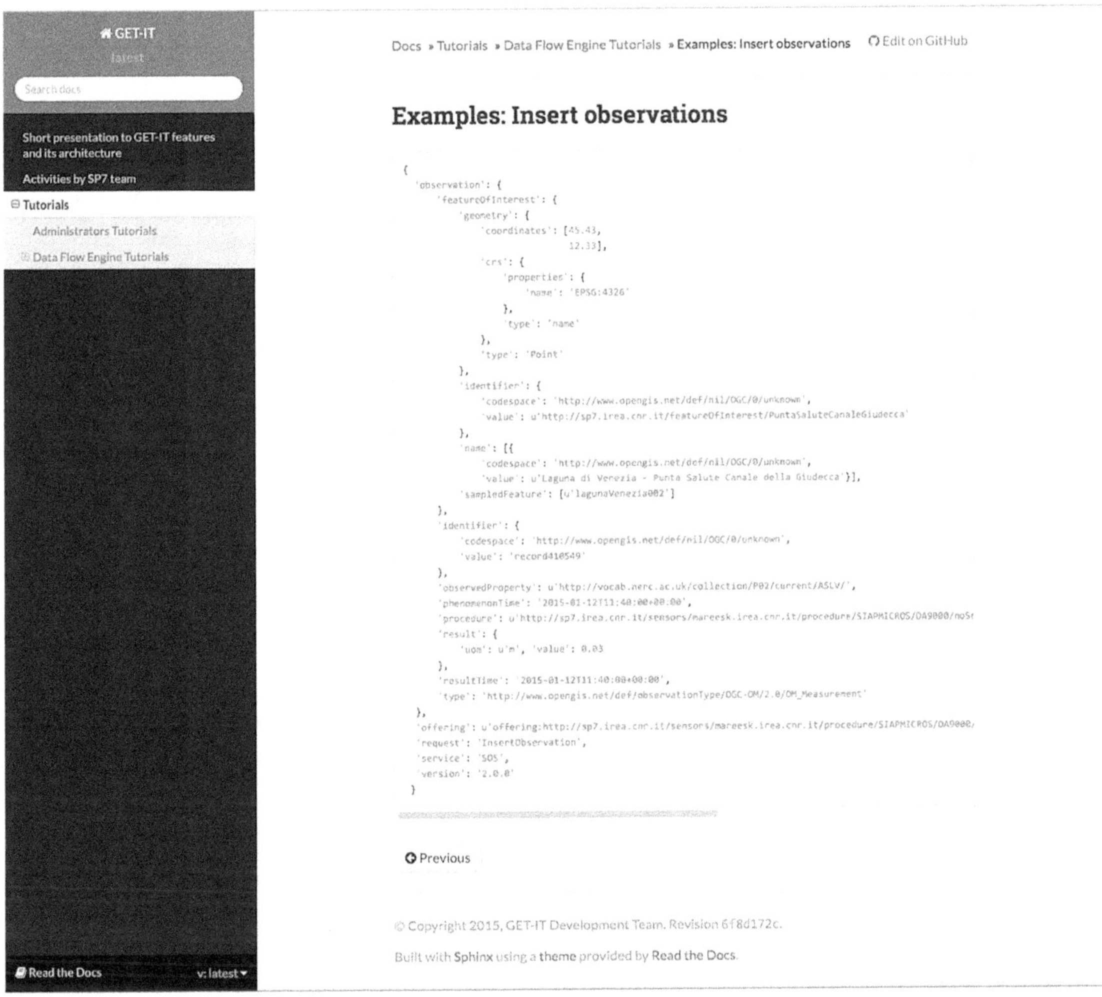

- retrieve and cross-reference observations for one selected parameter collected from multiple, distributed sensors and from other environmental measures;
- share observations and sensors information from all historical stations.

Below, we exemplify how these three use cases are made possible by using the GET-IT suite.

- **Use case a)** *The user wants to compare her conductivity dataset, collected by a CTD sensor, with existing ones in order to extend the geographical boundaries of her spatial analysis.* In the *sensors* tab of the GET-IT interface, the user can read the name and description of the instrument. By selecting the *sensor details* button, it is possible to obtain an HTML page containing all SensorML information: description, keywords, identification, classification, characterization of physical properties, electrical requirements, capability, contacts of manufacturer, owner or operator, output and components of the system, and moreover its history log to track any changes or calibration (Figure 5A).

- **Use case b)** *A novel remote sensing algorithm aims at providing an estimate of chlorophyll concentration in seawater. In order to evaluate the accuracy of the model, ground values are needed.* A raster image of the surface chlorophyll distribution, shared by remote or local OGC WMS endpoints, can be added in the GET-IT *views* section using the *add layers* button located in the tool bar of the map. In the same *view* the user can compare the values collected by a fieldwork overlap SOS sampling points choose between local and remote servers through *Add SOS* button. Finally, by *Identify* function, the user can obtain the values of pixels while SOS client provides the time series of the selected point (Figure 5B).
- **Use case c)** *The observations collected over the years constitute the starting point for evaluating the long-term evolution of the environment. Still it may be awkward to compare them with the data collected in real time because the documentation on the methodology that has been applied may be missing, as well as the people that may provide this information.* Instead, when creating a new sensor with the EDI SensorML template, this additional information can be formalized and stored for later use. When uploading observations through the functionalities provided by GET-IT, this information is related to the former and can be leveraged on at a later stage in order to assess the quality of historical observations, and eventually using them in series analysis, such as for consulting sensor information (e.g. for calibration, owner, operator, etc.) or for viewing the progress of the series (Figure 5C).
- **Use case d)** *The historical basement structure of Venice is fragile and worldwide concern is its preservation. Sustainable policies for maritime transport routes must take into account the impacts of transient waves generated by local ship traffic. Decision makers can exploit continuous monitoring from wavemeters in order to design the best spatial optimization of routes.* In GET-IT decision makers could share styled maps of designed routes with INSPIRE compliant metadata reporting their starting validity. In the same map users can visualize the effects of the changes of rules by inspecting the wavelength time-series at monitoring points. Through the *view* section mashups of linear routes and fixed point time-series can be shared and linked. More interestingly, the same information, now available through GET-IT standard services, could be discovered and exploited, for instance, by an oceanological model aimed at the aforementioned spatial optimization problem.

DISCUSSION

Throughout this Chapter, the Authors pinpointed the many issues that make adoption of SWE standards and practices awkward to marine Researchers. These issues encompass the whole life cycle of sensor information: from the provision of accurate sensor descriptions to browsing of the sensors integrated in the network; through the insertion of observations in the SOS implementation of choice; to displaying them in a GUI. This Chapter presented the solutions developed in project RITMARE for the purpose of easing sensor management by the marine research communities involved in the former. These are focused on the SWE approach, which allows for an interoperable management of sensor information whose complexity is managed by an SOS implementation behind the scenes.

In the presented solutions, the metadata editor EDI offers a friendly way to create sensor metadata whose consistency and openness is assured by a semantic lift; metadata created by EDI can easily be

Figure 5. SWE clients within GET-IT can reach out some needs of the marine community users. A list of sensors sensors stored in the nodes of SDI and a link with sensor details in order to improve knowledge

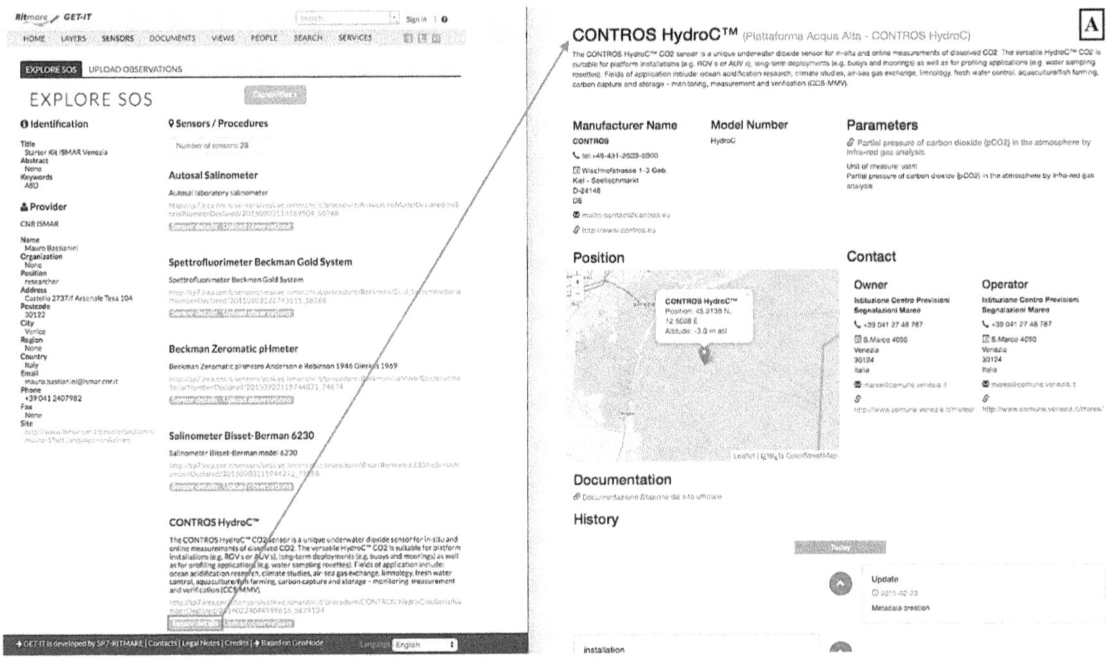

Figure 6. A view to overlap sensor observations, vector and raster images

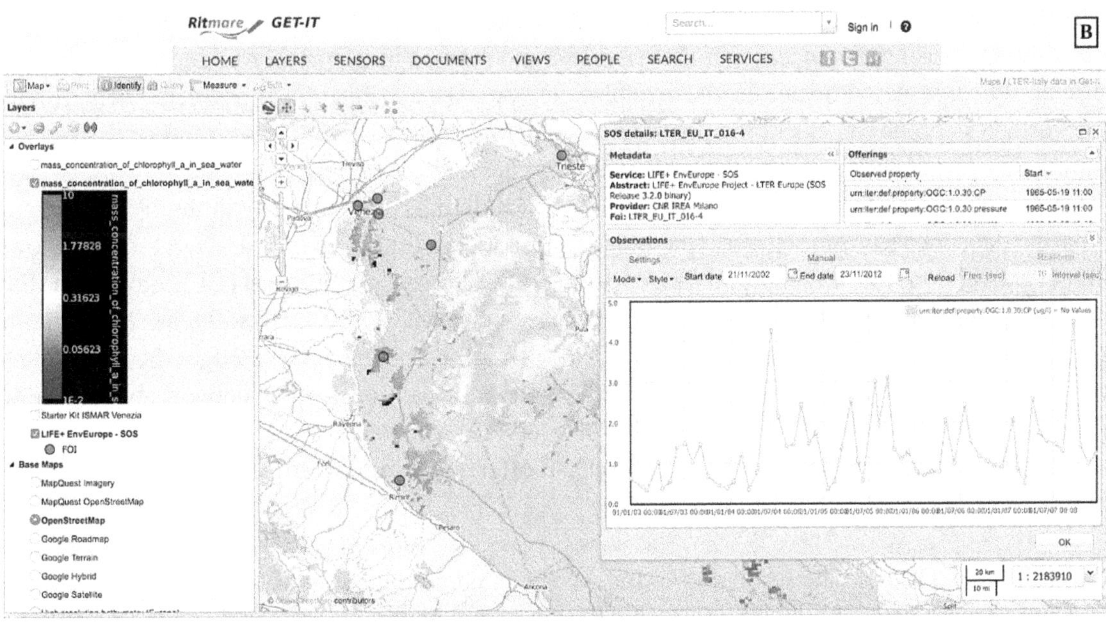

Figure 7. A chance to store, maintain and finally compare information and observations that might otherwise be lost and no longer be usable

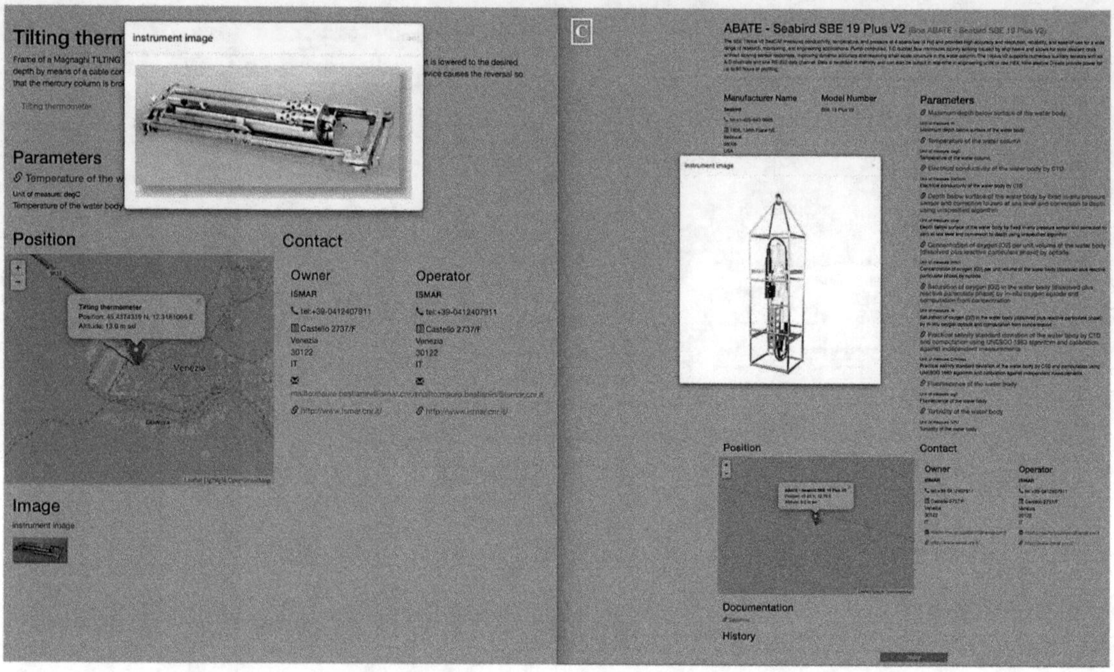

migrated to even more advanced paradigms for sensor representation, such as the Semantic Sensor Network Ontology (SSN)[23]. It should be noted that this kind of fine-grained categorization of sensors and observations is considered as a prerequisite to advanced functionalities, such as improving discovery capabilities and easing organization of these devices and outputs into workflows. These desiderata cannot be easily achieved with state-of-the-art solutions for sensor management but many data providers are nevertheless prone to neglect the possible novel applications that may arise in order to favor other features, such as network performance.

It is worth noticing that the adoption of the SWE paradigm is still a novelty in the field of practices for marine data derived from sensors, where the representation of data in the NetCDF format following different conventions for metadata and data encoding (e.g. Climate and Forecast (CF) convention (Eaton et al., 2011), SeaDataNet extensions of CF) and the use of the OCG THREDDS software service for their deployment are common. For this reason, the RITMARE infrastructure acknowledges the data providers adopting this last practice, by integrating THREDDS endpoints as nodes of the network, brokering[24] and harvesting their metadata and exposing their WMS services when available.

Still, in the Authors' opinion, SWE adoption can offer some improvements, which are synthesized here following:

- While NetCDF is a data format, O&M is a conceptual model and thus addresses a broader spectrum of data categories with respect to the former, which is more suitable for coverage data. Moreover, in theory, it would be possible to envelop in an O&M document exhaustively describing an observation the reference to a NetCDF source for the *result* of the observation. Interestingly, even in practice, tentative solutions in the direction of combining SWE and THREDDS/NetCDF technologies have been developed (e.g. ncSOS[25], IOOS i52n-SOS[26]).

- The concept of "sensor" is at the core of SWE and metadata of sensors, their characteristics, history, manufacturers, owners, etc. are a prerequisite for establishing the quality of collected observations and comparing different sources for them. This concept currently finds scarce support in other practices: for instance SeaDataNet introduced (Lowry et al., 2015; Schaap, 2017, chapter 7 this book) optional instrument attributes, extending CF, taking values in the SeaVox controlled vocabulary L22 where just textual descriptions of instruments can be found.
- Within SWE, metadata for sensed observations have a high degree of maturity, whereas in NetCDF/THREDDS the provision and subsequent retrieval of sensor observations metadata involves digging either in the NetCDF file itself (for the parameters that are passed down directly from the sensor) and the NcML metadata that describe, in the content disseminated across the THREDDS file system, the specificities of the individual data provider as well as of the specific dataset or series.
- SWE technologies harness fine-grained, non-redundant metadata that also allows for semantic annotation of datasets.

Within RITMARE, the two aforementioned approaches coexist: On the one hand, observations that are well represented by data coverage are deployed according to the NetCDF/THREDDS philosophy and the metadata are brokered in order to be discovered as CSW catalogs; on the other, the GET-IT suite allows for advanced management of sensor data and metadata via O&M and SensorML.

CONCLUSION AND FUTURE RESEARCH DIRECTIONS

In this Chapter, the Authors presented the set of tools and practices developed in RITMARE for the management of sensors and observations in the Italian marine research domain.

Some scenarios were sketched to highlight common needs of the marine community with respect to data management. Technological solutions were proposed to address some open issues, and their effectiveness was shown by use cases taking into account the practices of the community.

For SWE conceptual models and technologies to be used in environmental research daily practices, the development of easy-to-use tools was needed. Two novel SOS clients, respectively for insertion of data and for their visualization, and a metadata editor (EDI) have been created to be used in the data life cycle. The GET-IT open-source software suite bundles these tools and other components aimed at effective management of both sensor observations and traditional geographic information. Most of the tools originally developed for GET-IT can be easily used in isolation and repurposed for custom applications. As an example, referring to the aforementioned complementarity between SWE- and NetCDF-based sensor management paradigms, we are currently evaluating adoption of our metadata authoring tool, EDI, as a standalone tool for the management of NcML metadata for THREDDS servers. This practice is going to ease annotation of existing infrastructures of this kind but requires tighter integration with the underlying organization of the file system. For these reasons, we regard the SWE tools we provide in the GET-IT suite as the only all-round, user-friendly solution available in the sensor management landscape.

The solutions proposed by RITMARE not only reach the goal of providing interoperability under a technological point of view, but also create a semantics-aware framework including practices, interfaces and tools to minimize efforts, optimize metadata creation, and allow for data management enabling fine-grained discovery and exploitation of sensor data. Moreover this allows for the integration of sensor and manifold traditional data, a prerequisite for achieving sustainable development.

ACKNOWLEDGMENT

The activities described in this chapter have been funded by the Italian Flagship Project RITMARE and LifeWatch Italy.

REFERENCES

Baraniuk, R. G. (2011). More Is Less: Signal Processing and the Data Deluge. *Science*, *331*(6018), 717–719. doi:10.1126/science.1197448 PMID:21311012

Barnaghia, P., Ganza, F., & Abangara, H. (2011). *Sense2Web: A Linked Data Platform for Semantic Sensor Networks*. Retrieved from http://www.semantic-web-journal.net/sites/default/files/swj189.pdf

Benthall, B., & Gill, S. (2010). SDI Best Practices with GeoNode. In *Proceedings of the 2010 Free and Open Source Software for Geospatial Conference* (FOSS4G 2010).

Botts, M., & Robin, A. (2004). *OGC SensorML: Model and Xml Encoding Standard*. Open Geospatial Consortium.

Brickley, D., & Guha, R. V. (2004). RDF Vocabulary Description Language 1.0: RDF Schema. *W3C*. Retrieved from http://www.w3.org/TR/rdf-schema/

Broering, A., Stasch, C., & Echterhoff, J. (2012). *OGC Sensor Observation Service Interface Standard, Version 2.0*. OGC.

Buck, J., & Lowry, R. (2017). Oceanographic Data Management; Quills and Free Text to the digital age and 'Big Data. In *Oceanographic and Marine Cross-Domain Data Management for Sustainable Development*. Hershey, PA: IGI Global.

Chen, N., Liping, D., & Genong Yua, M. M. (2009). A flexible geospatial sensor observation service for diverse sensor data based on Web service. *ISPRS Journal of Photogrammetry and Remote Sensing*, *64*(2), 234–242. doi:10.1016/j.isprsjprs.2008.12.001

Clark, J. (1999). *XSL Transformations (XSLT) Version 1.0*. W3C Recommendation. Retrieved from http://www.w3.org/TR/xslt

Cox, S. (2013). *Geographic information – Observations and measurements*. OGC and ISO 19156.

Di, L. (2007). (GMU): A General Framework and System Prototypes for the Self-Adaptive Earth Predictive Systems (SEPS). In *Dynamically Coupling Sensor Web with Earth System Models* (AIST-05-0064). ESTO-AIST Sensor Web PI Meeting.

Diviacco, P., & Leadbetter, A. (2017). Balancing Formalization and Representation in Cross-Domain Data Management for Sustainable Development. In *Oceanographic and Marine Cross-Domain Data Management for Sustainable Development*. Hershey, PA: IGI Global.

Eaton, B., Gregory, J., Drach, B., Taylor, K., Hankin, S., Caron, Signell, R., Bentley, P., Rappa, G., Hoeck,. H, Pamment, A., Juckes, M., Raspaud, M. (2011). *NetCDF Climate and Forecast Metadata Conventions, v 1.6*. Retrieved from http://cf-pcmdi.llnl.gov/documents/cf-conventions/latest-cf-conventions-document-1

European Commission (2008). *Commission regulation (EC) no 1205/2008 of 3 December 2008 implementing Directive 2007/2/EC of the European Parliament and of the Council as regards metadata*. Author.

Fugazza, C., Menegon, S., Oggioni, A., Pepe, M., & Carrara, P. (2014a). The RITMARE Starter Kit: Bottom-up capacity building for geospatial data providers. In *Proceedings of ICSOFT 2014 (9th International Conference on Software Paradigm Trends)*.

Fugazza, C., Pepe, M., Oggioni, A., Pavesi, F., & Carrara, P. (2014b). A holistic, semantics-aware approach to Spatial Data Infrastructures.*3rd International Conference on Data Management Technologies and Applications (DATA)*, Vienna, Austria. doi:10.5220/0004997603490356

Fugazza, C., Pepe, M., Oggioni, A., Tagliolato, P., & Carrara, P. (2015). *Streamlining geospatial metadata in the Semantic Web*. 9th Symposium of the International Society for Digital Earth (ISDE), Halifax, Canada.

Harris, S., & Seaborne, A. (2013). *SPARQL 1.1 Query Language W3C Recommendation 21 March 2013*. Retrieved from http://www.w3.org/TR/sparql11-query/

Hart, J. K., & Martinez, K. (2006). Environmental Sensor Networks: A revolution in the earth system science? *Earth-Science Reviews*, *78*(3-4), 177–191. doi:10.1016/j.earscirev.2006.05.001

Haslhofer, B., & Klas, W. (2010). A survey of techniques for achieving metadata interoperability. *ACM Computing Surveys*, *42*(2), 1–37. doi:10.1145/1667062.1667064

Havlik, D., Schade, S., Sabeur, Z. A., Mazzetti, P., Watson, K., Berre, A. J., & Mon, J. L. (2011). From Sensor to Observation Web with Environmental Enablers in the Future Internet. *Sensors (Basel, Switzerland)*, *11*(4), 3874–3907. doi:10.3390/s110403874 PMID:22163827

Initial Operating Capability Task Force. (2012). *Technical Guidance for the Implementation of INSPIRE download services*. Author.

IOC Task Force for Network Services. (2011a). *Technical Guidance for the Implementation of INSPIRE discovery services*. Author.

IOC Task Force for Network Services. (2011b). *Technical Guidance for the Implementation of INSPIRE view services*. Author.

Jirka, S., Stasch, C., & Broering, A. (2014). *OGC Best Practice for Sensor Web Enablement Lightweight SOS Profile for Stationary In-Situ Sensors - OGC 11-169r1*. Open Geospatial Consortium.

Lowry, R., Fichaut, M., & Schlitzer, R. (2015). *SeaDataNet data file formats - deliverable D8.5*. Retrieved from http://www.seadatanet.org/content/download/16251/106283/file/SDN2_D85_WP8_Datafile_formats.pdf

McKee, L. (2015). *OGC Information Technology Standards for Sustainable Development*. OGC® White Paper 14-095. Open Geospatial Consortium.

Na, A. & Priest, M. (2007). *Sensor Observation Service*. English, (OGC 06-009r6).

Nativi, S., Caron, J., Davis, E., & Domenico, B. (2005). Design and implementation of netCDF markup language (NcML) and its GML-based extension (NcML-GML). *Computers & Geosciences*, *31*(9), 1104–1118. doi:10.1016/j.cageo.2004.12.006

52. *North*. (n.d.). Retrieved from: http://52north.org/

Oggioni, A., Bastianini, M., Carrara, P., Minuzzo, T., & Pavesi, F. (2014). Sensing real-time observatories in marine sites: a proof-of-concept. *Proceedings of the 3rd International Conference on Sensor Networks 2014* (SENSORNETS '14).

Papp, Z., & Hakkesteegt, H. (2008). Sensor Web, Sensor Networks: New possibilities and new challenges. In M. Grothe & J. Kooijman (Eds.), *Sensor Web Enablement. Delft: NCG, Nederlandse Commissie voor Geodesie* (pp. 21–39). Delft, The Netherlands: Netherlands Geodetic Commission.

Reed, C., Botts, M., Davidson, J., & Percivall, G. (2007). OGC® sensor web enablement: overview and high level architecture. In 2007 IEEE Autotestcon (pp. 372–380). IEEE. doi:10.1109/AUTEST.2007.4374243

Schaap, D. (2017). SeaDataNet – Towards a Pan-European Infrastructure for Marine and Ocean Data Management. In *Oceanographic and Marine Cross-Domain Data Management for Sustainable Development*. Hershey, PA: IGI Global.

Voigt, T., Tsiftes, N., & He, Z. (2008). Remote Water Monitoring With Sensor Networking Technology. *Ercim News*, 39-40.

Woolf, A. (2008). Building the Sensor Web - Standard by Standard. *Ercim News*, 24-25.

Yongguo, J., Zhongwen, G., Keyong, H., Feifei, S., & Feng, H. (2013). Using Sensor Web to Sharing Data of Ocean Observing Systems. In Advances in Wireless Sensor Networks. Berlin: Springer-Verlag.

ENDNOTES

1 The Global Ocean Observing System (http://www.ioc-goos.org/).

2 The European Global Ocean Observing System (http://eurogoos.eu/).

3 The U.S. Integrated Ocean Observing System (http://www.ioos.noaa.gov/).

4 The Australian Integrated Marine Observing System (http://imos.org.au/).

5 FP7, HORIZON 2020 Project ODIP - Ocean Data Interoperability Platform (http://www.odip.eu/).

6 FP7 Project EUROFLEETS - Towards an Alliance of European Research Fleets (http://www.eurofleets.eu/).

7 FP7 Project SeaDataNet - Pan European Infrastructure for Ocean and Marine Data Management (http://www.seadatanet.org/).

8 FP7 Project FixO3 - The Fixed point Open Ocean Observatory network (http://www.fixo3.eu/).

9 FP7 Project NeXOS - Next generation Low-Cost Multifunctional Web Enabled Ocean Sensor Systems Empowering Marine, Maritime and Fisheries Management (http://www.nexosproject.eu).

10 FP7 Project SCHeMA - In Situ Chemical Mapping Probes (http://www.schema-ocean.eu).

11 FP7 Project JericoNext - Towards a joint European research infrastructure network for coastal observatories (http://www.jerico-fp7.eu)

12 LIFE+ Project EnvEurope (http://www.enveurope.eu)

13 Flagship Project for Italian Marine Research RITMARE (http://www.ritmare.it)

14 GET-IT is partly based on the results of the CIGNo project (http://cigno.corila.it), a collaborative, federated system to share scientific and geographic data.

15 The management of geographic information is primarily delegated to GeoNode (Benthall & Gill, 2010), a widely acknowledged open-source software that is capable of dealing with maps, geographic layers, and also with geotagged documents.

16 Within SOS, an Observation Offering groups collections of observations produced by one procedure, e.g., a sensor system, and lists the basic metadata for the associated observations including the observed properties of the observations (Broering, Stasch, Echterhof, 2012).

17 The NetCDF Markup Language (NcML): http://www.unidata.ucar.edu/software/thredds/current/netcdf-java/ncml/

18 An EDI template is an XML document following the EDI template language, which is defined by XMLSchema.

19 http://www.opengis.net/def/samplingFeatureType/OGC-OM/2.0/SF_SamplingPoint_

20 OpenLayers: http://openlayers.org/

21 ExtJS: http://www.sencha.com/products/extjs/

22 Flot: https://code.google.com/p/flot/

23 W3C – Semantic Sensor Network Ontology: http://www.w3.org/2005/Incubator/ssn/ssnx/ssn

24 by means of GI-Cat: http://lab.usgin.org/applications/gi-cat

25 https://github.com/asascience-open/ncSOS_

26 http://ioos.github.io/i52n-sos/_

Chapter 10

Building a 4D Voxel-Based Decision Support System for a Sustainable Management of Marine Geological Resources

Vera Van Lancker
Royal Belgian Institute of Natural Sciences, Belgium

Frederic Francken
Royal Belgian Institute of Natural Sciences, Belgium

Lars Kint
Royal Belgian Institute of Natural Sciences, Belgium

Nathan Terseleer
Royal Belgian Institute of Natural Sciences, Belgium

Dries Van den Eynde
Royal Belgian Institute of Natural Sciences, Belgium

Lies De Mol
FPS Economy, Belgium

Guy De Tré
Ghent University, Belgium

Robin De Mol
Ghent University, Belgium

Tine Missiaen
Ghent University, Belgium

Vasileios Hademenos
Ghent University, Belgium

Marcel Bakker
Geological Survey of the Netherlands, The Netherlands

Denise Maljers
Geological Survey of the Netherlands, The Netherlands

Jan Stafleu
Geological Survey of the Netherlands, The Netherlands

Sytze van Heteren
Geological Survey of the Netherlands, The Netherlands

DOI: 10.4018/978-1-5225-0700-0.ch010

ABSTRACT

For sustainable management of marine geological resources, a geological knowledge base is being built for the Belgian and southern Netherlands part of the North Sea. Voxel models of the subsurface are used for predictions on sand and gravel quantities and qualities, to ensure long-term resource use. The voxels are filled with geological data from boreholes and seismic lines, but other information can be added also. The geology provides boundary conditions needed to run environmental impact models that calculate resource depletion and regeneration under various scenarios of aggregate extraction. Such analyses are important in monitoring progress towards good environmental status, as outlined in the Marine Strategy Framework Directive. By including uncertainty, data products can be generated with confidence limits, which is critical for assessing the significance of changes in the habitat or in any other resource-relevant parameter. All of the information is integrated into a cross-domain, multi-criteria decision support system optimised for user-friendliness and online visualisation.

INTRODUCTION

Mineral and geological resources, including fossil fuels and aggregates, can be considered to be non-renewable on time scales relevant for decision makers (e.g. Reid et al., 2010; Mudd, 2010). In the vast majority of places, they are more rapidly exhausted by humans than they can be replenished by nature, meaning that truly sustainable resource management is not possible (Prior et al., 2012). Comprehensive knowledge on the distribution, composition and dynamics of geological resources, therefore, is critical for developing long-term strategies for optimised resource use.

With growing land-use constraints and depletion of terrestrial aggregate resources, marine sand and gravel have gained importance (ICES, 2016). On a global scale, the stock for these widely demanded bulk resources may be assumed infinite (USGS, 2016). Regionally, though, exploitation is limited by a diversity of interacting social, economical, technological, geological and political factors (Habert et al. 2010). In a further constraint, legislation for the protection of the environment requires the assessment of a large number of conditions that need to be fulfilled for sand and gravel extraction to be permitted (Radzevičius et al., 2010; ICES, 2016). However, thorough assessment frameworks are most commonly lacking, and environmental impact assessments are rarely based on best available knowledge (Velegrakis et al., 2010).

In this chapter, an innovative way of dealing with a multitude of cross-disciplinary information in a resource-management context is presented. It discusses the workflow developed within the research network TILES (Transnational and Integrated Long-term Marine Exploitation Strategies, http://odnature.naturalsciences.be/tiles/). The aim of TILES is the development of a 4D (3D and time), voxel (3D pixel or cuboid) -based resource decision support system (DSS) containing tools that link 3D geological models, knowledge and concepts to numerical environmental impact models. The (DSS) will primarily serve as a tool to generate resource suitability maps. These maps are instrumental in answering questions regarding application-specific extraction demands and preferences, showing if these can be met and, if so, highlighting well-suited areas.

The study area comprises the Belgian and southern Netherlands part of the North Sea (further abbreviated as BPNS and sNPNS), in total ± 7400 km^2. Today, most of the Belgian resources are extracted from sandbanks (Degrendele et al., 2010; Degrendele et al., 2014). In the Netherlands, sand is also extracted in other, usually deeper, morphological settings (Stolk & Dijkshoorn, 2009). From south to north, the area is marked by increasing resource availability. The BPNS is relatively depleted of extractable sand and gravel, with Quaternary sediments being only 10 m thick on average (Le Bot et al., 2005), and less than 2.5 m thick or even absent in many of the swales separating the tidal sandbanks. The sNPNS has only few areas with such limited resource availability, although gravel is rare. Just north of the study area, in the extraction pit for the latest Rotterdam Harbour extension, sand was extracted to about 20 m below the seabed (de Jong et al., 2014).

In the following sections of the chapter, the background of and need for marine geological resource management are presented first from a European perspective. Next, issues and problems are discussed that need to be tackled to go from unrelated single-discipline databases to an integrated knowledge base

Figure 1. The Belgian and southern Netherlands part of the North Sea, a typical sandbank environment in the Southern Bight of the North Sea, northwestern Europe

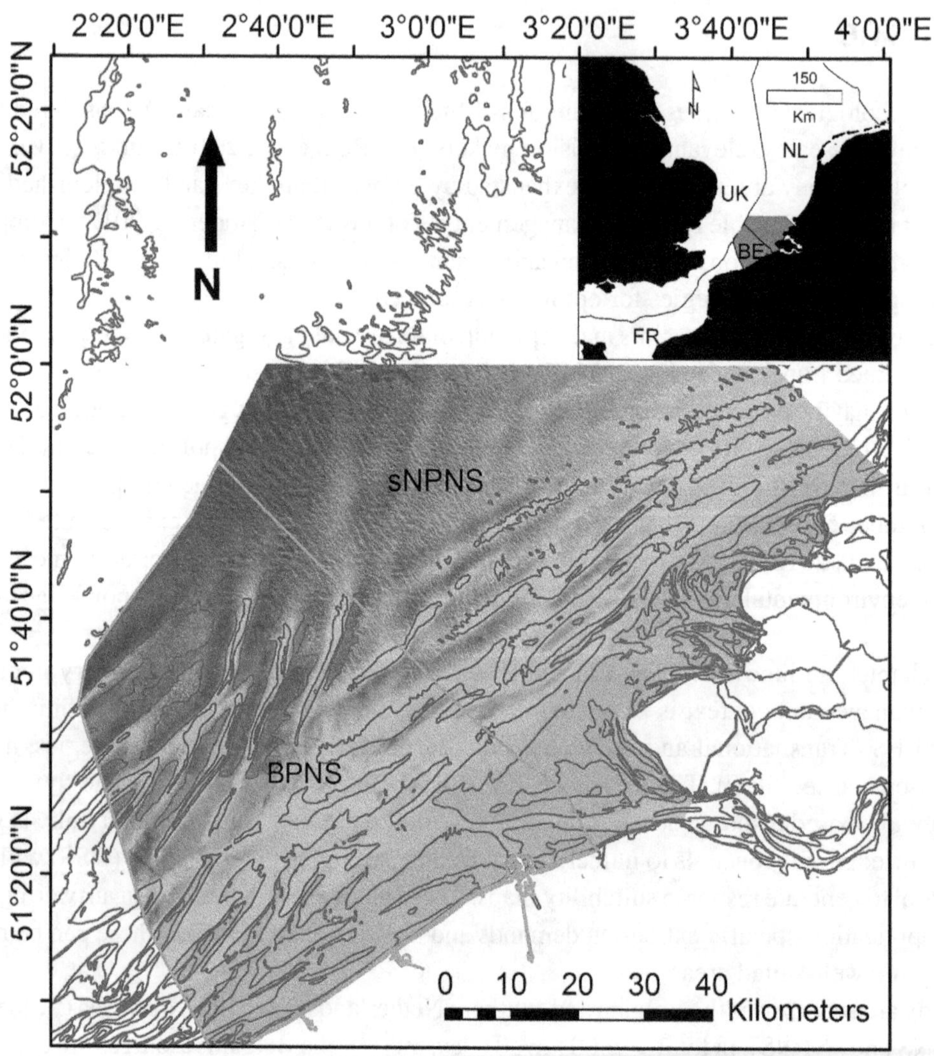

as part of sustainable resource management. The core of the chapter addresses the TILES approach, bringing together geoscientists and cyber infrastructure experts (cf. Mookerjee et al., 2015), presenting solutions for holistic resource management, and formulating recommendations related to data and metadata harmonisation, data integration, definition of sustainability criteria, and cross-domain model and tool development. Finally, future research directions are given in light of recent developments that will allow further optimisation.

BACKGROUND

Sustainable development, as originally defined in the Brundtland Report (WCED, 1987), involves three equally important primary dimensions: safeguarding long-term ecological sustainability, satisfying basic needs, and promoting intragenerational and intergenerational equity (Holden et al., 2014).

To secure the basic needs for the expansion of Europe's Maritime Policy and Blue Growth initiatives, the European Commission provided recommendations in 2008 (with later updates in 2011 and 2012) on resource use through its Raw Materials Initiative. Improvement of the EU knowledge base was recognised to be a key condition to maximise sustainable supply from within the EU (EC Enterprise and Industry, 2010). Comprehensive information on all geological resources was regarded particularly important, not only for stock assessment, but also to underpin the preparation of marine spatial plans (European Commission, 2010; 2011) (e.g. Solar et al., 2012). Recognising that improved resource assessments by themselves are not enough for extraction sustainability, the flagship initiative Europe 2020 targets an efficient use of these resources, emphasising the need for structural and technological improvements to reach the 2020 milestones (including proper valuation and protection of natural capital and ecosystem services, and resource efficiency as a shared pan-European objective). In both initiatives, scientific support is regarded an essential element in making informed decisions on the exploitation of most-suitable resource supplies that meet the demands of society and member state economies.

In parallel with the stimulation of economic growth, Europe also recognises the importance of long-term ecological sustainability and the necessity for marine regions to safeguard a good environmental status (GES). Maintaining or achieving GES is the main 2020 goal of EU's Marine Strategy Framework Directive (MSFD; European Union, 2008). To facilitate monitoring the evolution of any particular region towards GES, a series of descriptors was defined. Related to physico-chemical seabed attributes, seafloor integrity and hydrographic conditions are the most relevant descriptors in the context of the sustainable management of marine sand and gravel (ICES, 2015). MSFD Article 9 stipulates that GES for seafloor integrity refers to the structure and functions of the ecosystems that need safeguarding, without adversely affecting benthic ecosystems, whilst GES for hydrographic conditions means that permanent alteration of hydrographical conditions, for example due to exploitation, does not adversely affect marine ecosystems. 'Not adversely affected' should be interpreted as meaning that impacts may be occurring, but all impacts are sustainable such that natural levels of diversity, productivity, and ecosystem processes are not degraded (Rice et al., 2012).

SUSTAINABLE MANAGEMENT OF MARINE GEOLOGICAL RESOURCES

Issues and Problems

Lack of Structured Marine Resource Information

Sustainable management of marine geological resources requires sound knowledge on their distribution, composition, and dynamics. By their very nature, geological resources have three dimensions. One vertical and two horizontal dimensions introduce complexity into the processing and modelling of the associated point (mostly boreholes; 1D) and line (mostly seismic profiles; 2D) data. As a further complication, there is also a fourth dimension, which is related to the evolution of the resource in time. On the one hand, the natural dynamics of the geological resource (in terms of provenance and deposition) should be considered, especially in case of sand and gravel located at or near the ever-changing seabed. On the other hand, a thorough understanding of changes in resources due to human exploitation is indispensable, as it links directly to the prediction of extraction-related environmental impacts.

Except for oil and gas reserves (e.g. Feblowitz, 2012), minerals (e.g. Cassard et al., 2015) and rare earth elements (e.g. Cassard et al., 2014), which are assessed qualitatively and quantitatively for direct economic reasons, only few examples exist of systematically structured marine resource information (e.g. Judge, 2015). Existing tools for easy and flexible data querying in function of societal and economic resource needs are available from on-land applications (van der Meulen et al., 2005; Maljers et al., 2015) and have been applied and modified on a small scale in marine environments, though only where harmonised data were available in the required formats (van Heteren et al., 2012). However, despite great efforts in data collection, most European countries have neither a quantitative estimate of the shallow marine resources available, nor reliable figures on their quality and their depletion rates. Fully digital geological knowledge bases allowing for the incorporation of cross-domain information are quasi non-existent, even though they are the core of any future tool for assessing the environmental impact of resource exploitation.

Despite the lack of adequate information, annual sand and gravel extraction in the North-East Atlantic region (OSPAR) has exceeded 100 million m^3 in some of the past five years (ICES, 2015 and earlier annual reports of the ICES Working Group on the Effects of Extraction of Marine Sediments on the Marine Ecosystem). With the advent of Blue Economic Growth, as laid down in various marine spatial plans, many new initiatives are foreseen that require much larger volumes of sand and gravel than ever before. First, increasing volumes of nourishment sand are needed to replenish our beaches and nearshore zones as accelerating sea-level rise will leave our coastlines ever more vulnerable (Stive et al., 1991). Secondly, vast quantities of sand and gravel will have to be extracted to realise the large infrastructural works that are the key components of many visions on coastal-zone and offshore development (e.g. energy islands, harbour extensions, offshore airports, mega-terps (artificial dwelling mounds)). To date, there is only little awareness on how the feasibility of these plans depends on the availability and sustainable supply of the right quality of sand and gravel.

With increasing pressure on finite marine sand and gravel stocks, especially in resource-depleted areas, there is a need to define sustainability criteria that can be used in resource assessments. These help in evaluating whether or not supplies are secure for future generations, and in impact assessments, and need to ensure that exploitation is not adversely impacting on the environment. Detailed studies, quantifying environmental impacts of resource exploitation (e.g. overview papers in Van Lancker et al.,

2010b), are mostly restricted in space and time, and are usually addressing only one ecosystem component. Broadening existing impact concepts, and extrapolating results from regional studies in space and time, is a difficult task.

Main Challenges

Ideally, sustainable resource management is based on a single platform that integrates all available data, information and knowledge from multiple sources and research disciplines in a systematic way. Such integration is highly challenging, as marine data are very diverse in structure and content. Numerous formats are used, and national standards are the rule rather than the exception. Fragmentation across multiple heterogeneous databases, not all publicly available, provides an additional challenge. On a worldwide basis, many initiatives are taken to address data-standardisation and -harmonisation issues and to tackle other concerns. In the field of marine geological and oceanographic data, collaborative research networks such as Geo-Seas (EU FP7, www.geoseas.eu), SeaDataNet (EU FP7, www.seadatanet.org), and ODIP (EU FP7, www.odip.org) have developed useful protocols and provided common vocabularies. Furthermore, to cross-link data and data products such as maps, grids and models, Europe has been investing in a European Marine Data and Observation network (EMODnet, www.emodnet.eu) that compiles information from disciplines that are most critical for managing the marine environment: hydrography, geology, seabed habitats, chemistry, biology, physics and human activities. To date, however, the exchange of data, information and knowledge between the disciplines has been limited, and integration of socio-economic aspects has not been addressed in a resource context. At best, data and information are valorised in other domains by only few open-minded experts. Most challenging is their upgrade to cross-disciplinary knowledge, with quantified links that can be fed back into the network and its analytical tools.

When it comes to the practical usage of a central information system in the day-to-day resource management of member states, more challenges need to be tackled. Before user-friendly and reliable data products can be delivered, it is key to optimally cope with: (1) varying data density; (2) differing horizontal and vertical resolutions of resource models; (3) uneven data quality (including incomplete metadata); and (4) typical hardware limitations hampering the visualisation, querying and analysis of complex datasets by marine resource managers and decision makers.

Unlike for the onshore, data density in the offshore drastically reduces with distance from the coast. Concerning geological data, this is especially true for boreholes. They are mostly restricted to areas of economic interest, with site surveys commonly producing clustered data. Within the BPNS and sNPNS, there are extensive areas for which no borehole descriptions are available. Some borehole-density differences are compensated by line data from interpreted seismic profiles. In seismic data acquisition, reflected acoustic waves form a useful though imperfect image of the subsurface, which is limited by the resolution of the acoustic source-receiver system, and inherently subject to interpretation rather than description (Bond et al., 2012). Additional supporting data are needed to add meaning to seismics, as well as appropriate interpolation techniques that allow the integration of point and line data in combination with expert geological knowledge. Any resulting model of the geological subsurface will have a large degree of subjectivity (Bond et al., 2007; Polson and Curtis, 2010). Adding expert knowledge is especially important where there is variable data density. Limiting the uncertainty effects of the latter in resource models remains a challenge.

Regarding discretisation or model resolution, it is most often perceived that the higher the resolution is, the better the product will be. However, a model should not show more resolution than is warranted by the underlying data, unless a measure of uncertainty is provided. The optimal resolution will depend on data availability, geological complexity, the space and time scale over which the resource needs to be analysed, and the stakeholder's specific needs. Once data and information are collated in an information system, flexibility and speed of the querying process are critical. Hence, the optimal resolution is also a function of the required performance of the information system. Furthermore, in a resource context, one needs to be aware that the choice of the resolution has a direct bearing on the outcome of volume calculations, especially in non-flat areas (e.g. irregular slopes of sandbanks). Another important factor to consider relates to loss of information when too coarse a resolution is chosen. For example, thin sand or gravel layers or laterally limited sedimentary units (< 200 m) may easily be lost in a 200 x 200 x 1 m resolution model. On the other hand, only lower-resolution data will allow long-term predictions on resource availabilities to be generated rapidly. The associated inherent uncertainty, which is high, is a necessary consequence. Hence, in an ideal world, a resource-assessment model would consist of multi-resolution spaces tuned automatically to minimise loss of important details where small-scale information is available, and, simultaneously, to avoid generating large sets of small yet uninformative information cells in areas where data are scarce. Although it is still difficult to work routinely with multi-resolution models today, existing limitations will likely be overcome in the near future. In light of the anticipated progress, it is recommended to develop products that can be updated easily, and be improved when technologies are ready for it.

Data and data products are often assumed to represent the truth, especially by non-specialists. However, time and experience have shown that neither measurements nor descriptions and interpretations are entirely objective (e.g. van Heteren & Van Lancker, 2015). They include various sources of bias and error, and are merely an imperfect representation of the actual situation or parameter to be portrayed for a specific point in time. The discrepancy between the observed 'truth' and the real truth is due in part to data quality, and in part to the hypothesised indicative meaning of the underlying observations and measurements. The past has shown that many hypotheses on representativeness for a particular purpose or parameter do not stand the test of time. Where possible, data quality and known uncertainty concerning indicative meaning should be quantified and incorporated in the resource models. The end user needs to be made aware of the degree in which model output may differ from reality. Traditionally, this uncertainty has typically been addressed only in generic disclaimers or by incorporating standard safety factors (e.g. related to risk in engineering works, Gattinoni et al., 2014). The past decade, however, has shown an increasing appreciation for data-quality assessment, to the point where non-inclusion of confidence indicators in data products is perceived as an important shortcoming. It is now considered essential when it comes to data sharing and when data are used in other contexts than the originally intended purpose (Diesing & Stephens, 2015; Lark et al., 2015b). Aside from interpolation-related uncertainty, which is now mostly incorporated into geological models, an indication of data-related uncertainty needs to be provided. Incorporation of adequate metadata is therefore essential, and should include: vintage, sampling technique, exact moment of sampling, and sample positioning system. Where possible, interpretation-related uncertainty should be considered also. Analysis and interpretation of geological data are generally complex, and final assessments should ideally be based on group decisions (Bond et al., 2007). A consensus opinion or view can be reached by tracking the opinions and judgments of individual experts in a group environment in a structural way (Polson & Curtis, 2015, and references therein). Such a technique, together with increased computing power to propagate and weight the various uncertainties mentioned above, sheds a different light on the quality and reliability of data products. Quantifying uncertainty is still at its infancy in the marine realm, though in a seabed-mapping context, mapped boundaries between adjacent geological units are increasingly represented with confidence limits (e.g. Lark et al., 2015a).

Regarding the development, analyses and visualisation of resource models, sophisticated software packages (e.g. Isatis (www.geovariances.com), GoCad (www.mirageoscience.com), Petrel (https://www.software.slb.com)) are still predominantly used. These software packages are expensive, difficult to operate, and neither available to the wider user community nor usable on their computers. Once built, the models require dedicated visualisation tools for their further exploration. Even when free, these tools have to be downloaded, along with any content that needs to be visualised. Standard GIS software, more common in the user community, still lacks functionality to process and query complex resource models at satisfactory speed.

Beyond the challenges related to data collection, management, exploitation and visualisation, the dynamic nature of marine geological resources and their human exploitation call for numerical simulations of their evolution through time. Acknowledging that sediment dynamics is controlled by the interplay of diverse sediment fractions, behaving differently (e.g. coarser fractions may limit the availability of finer sediments for suspension and transport by forming an impenetrable layer at the seabed surface, armouring underlying fines; Wiberg et al., 1994), and that there is a need for assessing the effects of human interventions on the seabed composition (e.g. Cooper, 2012), advanced numerical models that can account for different sediment classes and multi-layered seabeds have emerged (e.g. Reed et al., 1999; van Ledden et al., 2006; Warner et al., 2008; Le Hir et al., 2011). Typically, these models represent processes such as lateral sediment transport, vertical sediment exchange between water column and seabed (through erosion and deposition), and resulting morphological and sediment-compositional changes. Whilst these processes can be simulated with varying detail using realistic sediment parameterisation, real-life applications are too often hampered by a poor model initiation (t_0) of seabed depth, structure and composition. Fully digital sediment-parameter databases are best suited as input for model initiation, but are only slowly becoming available (e.g. Van Lancker & van Heteren, 2013). Visual descriptions of geological layers, provided by borehole-description databases, are less easily implemented in environmental impact models. Still, this qualitative information is critical for the prediction of habitat changes under varying scenarios of extraction depths. Above all, it is needed to incorporate geological boundary conditions that affect the amounts of transportable and extractable sand, and control the spatial distribution of non-erodible layers. To date, such information is only rarely taken into account in models (cf. van Ledden et al., 2006; Blaas et al., 2007; Warner et al., 2010). Most often, superfluous sediment supply is assumed when regeneration after extraction is predicted (Roos & Hulscher, 2003; Van den Eynde et al., 2010). Hence, a high-accuracy quantitative and qualitative estimation of the initial seabed state and of various boundary conditions is required to optimise numerical model output generated to assess the temporal evolution of marine geological resources. The coupling of 3D resource models to 4D numerical models represents one of the major challenges in sustainable resource management.

For decision making, all data, models (3D and 4D) and information need to be combined in a knowledge base that is both functional and easily queryable. With the example resolution of 200 x 200 x 1 m, the region of the BPNS and sNPNS is represented by little over 5 million cuboids, each filled with enough information to make the resource model a clear example of Big Data. Querying such a large model in real time is not trivial. The science of data mining dictates that big knowledge bases often contain unapparent patterns of useful information (Fayyad et al., 1996). Advanced mathematics combines techniques from fields such as artificial intelligence, machine learning and statistics to extract non-explicit data and present them to the data analyst in a form that can be understood more easily. A DSS is a specific type of an *ad hoc* data-mining tool designed to facilitate decision making. As an example, a dedicated DSS applied to a geological knowledge base of the subsurface can be used to identify best-suited locations for specific applications. However, current DSSs show two shortcomings (Fick & Sprague, 2013): (1) they are not user friendly and content-specific enough for a typical decision maker, as they are generally aimed at experts in the field of the knowledge base; and (2) they do not take data quality into consideration as a measure of uncertainty. As offshore data are seldom plentiful and never purely objective, these shortcomings have severe implications for existing DSSs using interpolation-based subsurface models as their key input.

New Protocols and Workflow for Marine Sustainable Resource Management

Research Network TILES

With its interdisciplinary and transnational approach, TILES develops a new tool for the sustainable management of marine sand and gravel, addressing all issues and problems discussed in the previous section. As part of a flexible approach, several solutions are tested and compared, and adjustments to protocols and tool elements are made as experience grows. By paying specific attention to temporal change, related to natural dynamics and extraction activities, to environmental impact, and to various sources of uncertainty, TILES represents a leap forward in resource assessment and in the associated decision making.

In a key step, state-of-the-art 3D geological models are developed for the BPNS and sNPNS by transforming a geological layer model, which defines stratigraphic unit boundaries, into a so-called voxel model (consisting of '*tiles*' or volume elements) using geostatistical software. As part of this step, qualitative or quantitative lithological or other characteristics are assigned to each voxel (e.g. Stafleu et al., 2011; Maljers et al., 2015). All primary data that are added to the voxels are subjected to uncertainty analyses, a necessary step ensuring that data quality and other uncertainty factors are incorporated into voxel-based resource calculations. The voxel model is highly suitable for quantitative post-processing, needed to integrate 4D data and to generate realistic and user-friendly data products. To quantify the environmental impact of extraction over time, the geological boundary conditions from the 3D geological models are coupled to 4D output from numerical impact models. Importantly, the final voxel model will be queryable through a multi-criteria DSS that assesses geological, environmental and socio-economical parameters. On the basis of application-dependent queries, the best-suited regions can be identified and viewed as 2D or 3D suitability maps, which will greatly assist in the optimisation of resource management and are invaluable for outreach and educational purposes.

More detailed information on all critical steps needed to move from data to a knowledge base in the resource-assessment context is provided below. Since sustainability is the common thread throughout the TILES workflow, it is discussed first.

Definition of Sustainability Criteria

The importance of defining practical sustainability criteria is highlighted by MSFD, which puts forward that recovery from human perturbations in the marine environment should be rapid and secure. In other words, levels of diversity, productivity, and ecosystem processes should not show persistent decrease to values outside the natural range (Rice et al., 2012). In constraining the envelope of natural variability, knowledge on the dynamics of the geological resource (e.g. sediment dynamics) is critical. Since sediment type is closest related to predictions of biodiversity (e.g. Degraer et al., 2008), some member states (e.g. Belgium) have put forward that the distribution and extent of EUNIS level 3 habitats (e.g. sandy mud to mud, muddy sand to sand, coarse sediments, gravel beds) should remain constant within the accuracy limits of maps presented in the Initial Assessments. Applied in a resource exploitation context, this means that extraction is only allowed in geological units where the least variability in lithological characteristics is expected.

Building upon the definition of mineral resource potential (Taylor & Steven, 1983), the TILES geological knowledge base is built around the concept of 'resource potential'. In the context of marine resources, it is based on three main components (Figure 2). The first encompasses a detailed inventory of resource quality and quantity in which the link between geological knowledge and both seabed habitats and biodiversity is concretised (e.g. Degraer et al., 2008; Van Lancker & van Heteren, 2013). The second relates to defining sustainable exploitation practices. Here, knowledge on natural sediment dynamics is indispensable to advise on where to extract and what areas to avoid (e.g. Van Lancker et al., 2010a) if environmental impacts are to be minimised. The third component holds information on competing use of marine space that constrains resource availability, both now and in the future. Regular updates of this information are needed, as marine spatial plans are revised on a regular basis. Even though it is difficult to look too far ahead, with short-term gain ruling politics, this final component provides new insight into resource availabilities for the next generations, a cornerstone of sustainability.

Figure 2. Definition of the resource potential. Three information pillars provide input for a DSS on sustainable sand and gravel extraction: resource availability, sustainable exploitation, and resource constraints. Integration of cross-domain information yields sustainable extraction thresholds to be used in scenario analyses that will help to ensure that extracted volumes stay within sustainability limits, a prerequisite within Europe's MSFD

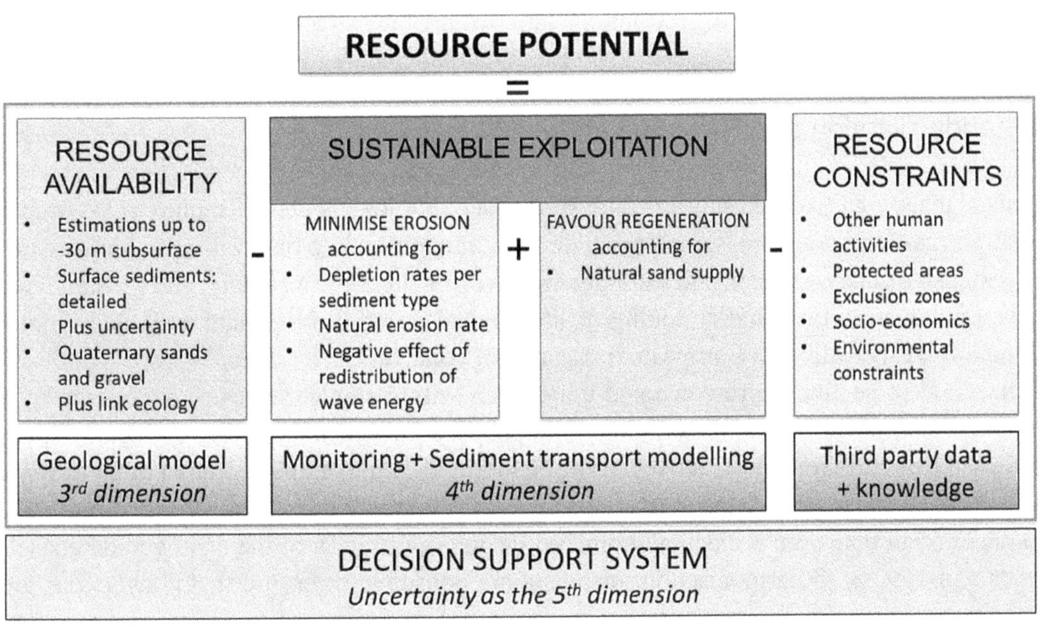

In practice, the geological knowledge base will deliver the following data products that directly support decision making in relation to the sustainable exploitation of marine sand and gravel:

1. Seabed-sediment maps, which are a proxy of habitat biodiversity; and hence help in avoiding sensitive areas;
2. The natural envelope of sediment composition (ratio of mud, sand and gravel fractions, derived from full distribution grain-size data), as to assist in the evaluation whether or not habitat changes result from extraction activities;
3. Base maps of the different lithostratigraphic subsurface units, as they help assessing where the likelihood of a lithological (and hence habitat) change, is high;
4. Probability maps of long-term natural erosion and deposition magnitudes and rates, to favour extraction in depositional areas, and to avoid naturally erosive areas, hence minimising environmental impacts;
5. Statistical models of depletion rates per substrate type, to gain knowledge on rates of resource depletion under the current extraction practices, which is critical for long-term predictions of resource availabilities.

The spatial and temporal resolutions of these data products differ. The first two products are derived from spatially very-high-resolution mapping of seabed sediments, based on sediment sampling and acoustic measurements. Temporally, resolution is typically low, with data collected over many decades contributing to a single product. The lithostratigraphic base maps originate from 3D geological modelling of the subsurface, using all available data. Temporal resolution is low, and spatial resolution decreases as a rule with depth and with distance to the coast. The final two data products result from a 16-year hindcast modelling study of hydrodynamics and sediment transport, validated with available bathymetric monitoring series from FPS Economy, Belgium (Francken et al., 2014). These model outcomes are further statistically analysed following the approach of Dalyander et al. (2013) and cover different time spans (seasonal, annual, multi-annual variation) and spatial scales. The spatial resolution of the models is relatively low, but their temporal resolution is high. A key challenge of TILES is the translation of all of these data products with differing resolutions into sustainable extraction thresholds (e.g. volumes per area that can be extracted without causing persistent habitat changes), to be used in scenario analyses.

Data Standardisation and Harmonisation

Harmonised mapping of seabed substrate and subsurface lithological classes requires (1) compatibility of all data formats and projections, preferably through integration into one centralised information system; (2) extensive quality control, and subsequent improvement of the metadata; and (3) standardisation in the data parameterisation and/or coding of lithological descriptions. These actions are not always straightforward, especially in a European, transnational context.

Formats and projections have come and gone, even within single countries, depending in part on location (near-coastal *versus* farther offshore) and purpose for which data were collected. The marine borehole data from the Geological Survey of the Netherlands, for example, come in three formats: fully digital and coded (post-2000 data), fully digital but only partially coded (mostly text; pre-2000 data) and on paper (data that were accidentally omitted during digitisation of the analogue database during the 1980s and 1990s). In harmonisation, metadata are consulted to ensure that conversions are done correctly. Incorporation of metadata on sediment and borehole data, such as vintage, time of sampling, positioning, sampling technique and analytical procedures, is critical for data-quality assessment. In an important optimisation step, time of sampling, neglected during the original digitisation, is now being recorded where available in original shipboard documents. This information is used for conversion of measured water depths to a common reference datum such as mean sea level, improving the vertical positioning of sample descriptions by eliminating uncertainty related to tidal phase.

Available borehole data have been standardised following European guidelines (e.g. EU-FP7 Geo-Seas for geological and geophysical data). Main lithology (e.g. gravel, sand, clay), Wentworth classes (e.g. fine, medium, coarse sand) and (adjusted) Folk classes (e.g. sandy mud, muddy sand) are mapped consistently for the BPNS and sNPNS. All terms used in the conversion are linked in the metadata to INSPIRE-compliant vocabulary organised in different hierarchical levels. Analogue and digital seismic data, collected during the past 45 years, were catalogued and assigned with a quality indicator following the same Geo-Seas guideline (values range from 0 (no data) to 5 (good data)). This indicator was used during the collation of seismic interpretations to provide an insight into the uncertainty associated with seismics-based layer models covering specific areas.

2D Geological Layers to 3D Voxels

Voxel modelling was first applied in the earth sciences in the 1990s (Marschallinger, 1996), but its full-scale application to large subsurface databases did not start until about five years ago. Applications were mostly for land data and based on closely spaced borehole data (e.g. Stafleu et al., 2011). In TILES, this is now done, for the first time, on a large marine dataset. However, compared to land, offshore borehole data are scarce, and therefore seismic data are needed to produce the layer models that form the basis of the voxel models. These seismic data are well-suited to the mapping of unit boundaries, which commonly produce strong reflections of the sound waves used in this method. The combination of data from single locations and data collected along transects, along with the large variation in associated data quality and density (especially for seismic data), calls for significant adjustments to the developed workflow.

Figure 3. Seismic (left) and borehole (right) data available for the BPNS and sNPNS

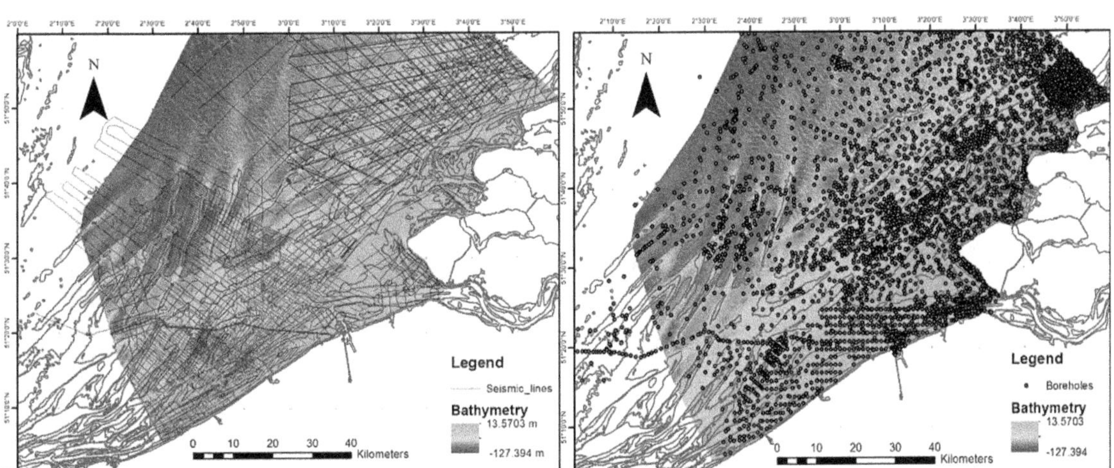

An important step in the TILES workflow is using the borehole-description database in conjunction with the seismic database to create the velocity model needed to convert time (seismic methods measure the time needed for a sound wave to travel from the transceiver to a unit boundary in the subsurface and back: the so-called two-way travel time) to depth. Thus, 2D geological surfaces are modelled that represent the boundaries between superimposed sedimentary units as accurately as possible. This procedure is mostly manual because it requires expert knowledge from geologists. A good insight in and control over the quality of the input data minimises the errors that are introduced through time-depth conversion. Detailed metadata and quality flags are taken into account during this process.

The surfaces created from the combination of borehole and seismic data outline the geological units (e.g. Holocene, Formation X or Y) that are the first primary attribute of the voxel model. Each voxel belonging to such a unit, having a midpoint falling between its top and base, is assigned with the same value. Other primary attributes are lithology and probability of occurrence. Lithology concerns only the most likely lithological class for a specific voxel (Soares, 1992). Probabilities of occurrence, provided for each lithological class, are a measure of interpolation-related uncertainty. They are calculated from 100 interpolations on each dataset (Stafleu et al., 2011) (see section '*Dealing with uncertainty*' below).

In 3D voxel modelling, both discretisation problems and issues related to a low and uneven model certainty (possibly caused by varying data density) can be addressed by using irregular grids, composed of voxels of differing resolution. Creating smaller voxels where model certainty is high and larger voxels where model certainty is low provides detail, where possible, and omits detail elsewhere. The dimensions of the voxels can be adjusted in such a way that the detailed geometry of the layer-based model is preserved (Figure 4) (Jørgensen et al., 2013). However, many interpolation algorithms, as well as their implementations in software packages such as Isatis and SGeMS, do not support irregular grids. For the time being, they are generated by post-processing Isatis output and performing visualisation and analysis in other subsurface viewers (as mentioned below).

Figure 4. Standard voxel model (left) and irregular-sized model (right) (northern coastal area in the Netherlands). The algorithm that constructed the right-hand model starts with large voxel sizes and checks if voxel size can be reduced given certain homogeneity and uncertainty criteria. By going through several iterations until the smallest voxel size is reached, a clearer, less ambiguous and less voluminous model is created

Free subsurface viewers are available allowing users to download and visualise both the geological layer-based models and the voxel models on their desktop computers. TILES provides access to the SubsurfaceViewer (INSIGHT GmbH and the British Geological Survey; Kessler et al., 2009; Sobisch, 2011), which offers a map view and an interactive 3D view in separate windows. In addition, the software enables the creation of synthetic boreholes, as well as vertical cross-sections for random points and transects. The viewer supports both regular and irregular voxel models (Stafleu et al., 2013).

Dealing with Uncertainty

Interpolation-related uncertainty is included in the voxel models created by Isatis. It is translated into probabilities, evaluated on the basis of 100 equally likely 3D spatial interpolations to fill the empty voxels. These interpolations use available sediment data (from boreholes and seabed surface samples) and simulated neighbouring voxels in 100 random areas around the target voxel. Using all simulation results, every voxel is attributed to the most likely lithological class, on the basis of the number of simulations that assign a voxel to each different class. There is a probability, and thus an uncertainty assessment, for all sediment classes considered, which can be visualised separately (Figure 5).

In order to capture the variability of the simulations for all sediment classes, the information entropy approach of Wellmann & Regenauer-Lieb (2012) is used to obtain a measure of overall uncertainty in 3D models. The information entropy of a voxel is a single value ranging from 0 to 1 that can easily be

Figure 5. Preliminary voxel model of Quaternary sediments in the BPNS (3600 km²; 65 km of coastline), showing the probability of medium sand. Along with coarse sand, it is the most important aggregate resource available. Note that the occurrence of medium sand is very patchy, implying also that extraction will be very localised. The central white area is devoid of Quaternary sands. Horizontal dimensions of the block are 79.4 x 89 km (x, y), height of the block is 75 m. Vertical scale is exaggerated 100 times

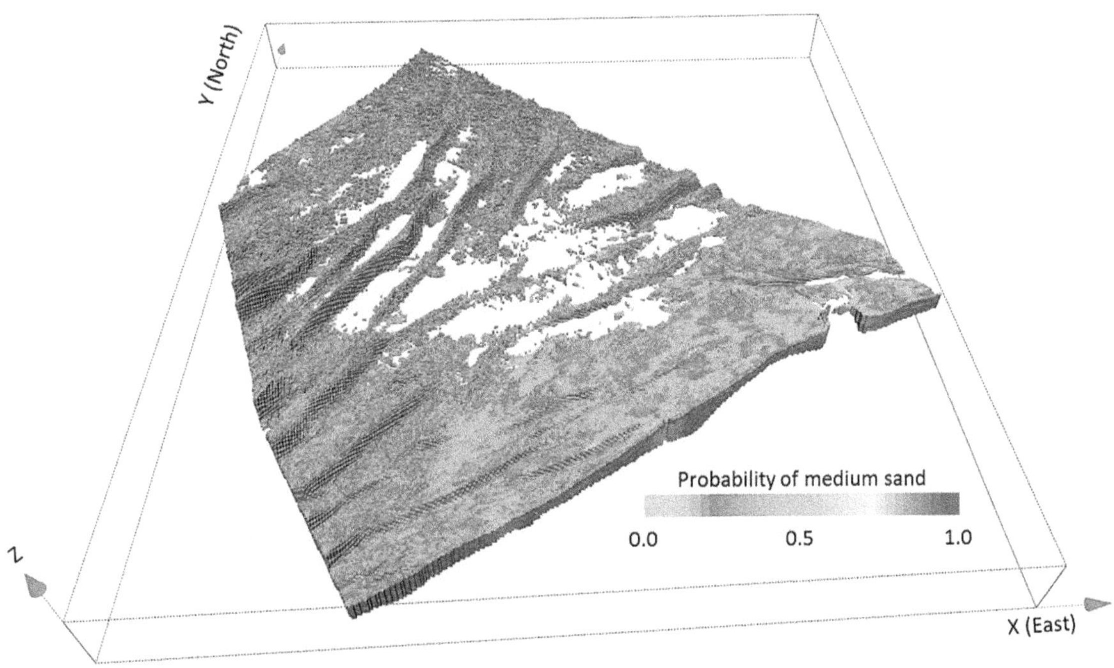

calculated from each of the probabilities of lithological classes. An entropy value of 0 means that there is no uncertainty (every simulation yields the same lithological class for the voxel), whilst a value of 1 occurs when all lithological classes have the same probability. Values in between 0 and 1 account for both the number of lithological classes with a probability higher than 0 (the more classes, the higher the entropy) and the differences amongst the probabilities (the greater the differences, the lower the entropy). The method was applied successfully in voxel models for onshore parts of the Netherlands to reflect uncertainty on the lithological class attribute for each voxel (Stafleu et al., 2014).

Aside from this interpolation-related uncertainty, the quality of the underlying borehole and seismic data and metadata (Table 1) needs to be incorporated into the assessment tool during post-processing of Isatis output. At present, this is not yet possible. First, appropriate attributes must be defined. They should be readily available in the metadata, and easily classifiable into relevant quality-related categories.

Building Multi-Parameter Voxel Models

In essence, the voxel model is a 3D mesh of the subsurface and as such, each voxel represents a finite cuboid of space with internal properties that are assumed to be homogeneous. The backbone of the model is the geometry of the geological units steering the interpolation of lithological classes within these units. Traditionally, only other relevant geological attributes are added. However, voxels can be used to store any *ad hoc* parameter, providing a highly suitable but still unexplored platform for quantitative cross-disciplinary analysis. Within the context of sustainable resource management, the research domains addressing seabed mobility, ecology, marine chemistry and human activities are particularly relevant. They provide information on the permanence or non-permanence of voxels, which may disap-

Table 1. Uncertainties associated with sediment data and data products (van Heteren & Van Lancker, 2015)

Step	Causes of uncertainty	Actions needed for harmonisation
Remote sensing	Survey design, method of acquisition (*incl.* calibration) and data processing, vintage, timing	Filtering data on the basis of methods and vintage
Sampling	Survey design, method of acquisition (*incl.* calibration) and data processing, vintage, timing	Filtering data on the basis of methods and vintage
Subsampling	Representativeness	Adherence to minimum sample sizes for sediments of different grain size
Description	Subjectivity, gross simplification	Description of standard parameters
Analysis	Pre-treatment, systematic error due to particle properties and imperfect conversion models, simplification	Intercalibration of granulometry methods and protocols, inclusion of methods and protocols in metadata
Classification	Simplification, inconsistency of standards, definition of class breaks, poor metadata	Development of mapping protocols for translating between standards, inclusion of standards in metadata
Interpolation	Spatial heterogeneity, temporal variability, poor system knowledge, validity of statistics, imperfect relation with external-drift variable	Use of standard methods, calculation of uncertainty
Interpretation	Subjectivity, poorly known surrogacy, poor system knowledge	Multidisciplinary and transnational collaboration and knowledge exchange

pear because of erosion and appear because of deposition, on the presence or absence of key benthic species or communities, on pollutants, and on availability for future extraction.

Once generated, the voxel model can easily be used to perform spatial analyses constraining any of the defined attributes. For example, volume calculations, crucial for the management of raw materials in the marine environment, can be performed by selecting and counting the voxels that meet certain criteria (for example all voxels within 5 m of the seabed with a greater than 60% probability of coarse sand). Voxels are also highly suited to quantitative analyses of relationships among different attributes. As an added advantage, the creation of secondary products is straightforward. Real-time construction of cross sections on any chosen location, previously impossible because of the time needed to construct such sections by hand (Stafleu et al., 2011), is a valuable visualisation tool. More importantly, the creation of customized 2D raster maps using voxel-stack analysis (Dambrink et al., 2015) simplifies 3D data products that are commonly difficult to interpret.

Coupling of the Geological Models with Environmental Impact Models

In the previous paragraphs, it has been demonstrated that the 3D voxel models by themselves provide an invaluable tool to assess data and information on the current state of the resource. However, for prospective resource management, evolution of resource volumes and qualities through time needs to be addressed as well. In a new step, the 3D voxel models are being coupled to 4D environmental impact models, which has not been attempted before. Although voxel models are now being linked to land subsidence in a 4D pilot study in the Netherlands, the incorporation of temporal change in voxel models is far from solved. Marine resources are subject to both natural processes (bedform migration, erosion and deposition; Dyer & Huntley, 1999) and human influences (e.g. dredging activities, extraction; Degrendele et al., 2010; Van Lancker et al., 2010a) that cause their quantity and quality to change. To understand past and predict future behaviour of the resource in this 4th dimension (time), it is necessary to couple the 3D voxel model to a suite of environmental impact models.

The model suite used in TILES contains a hydrodynamic model of the continental shelf (COHERENS; Luyten et al., 1999), a sediment-transport model (simulating erosion, transport and deposition) and a seabed-morphology model (simulating changes in seabed composition and bathymetry). The model suite is coupled to a third-generation wave model (SWAN; Booij et al., 1999) in order to realistically account for the wave constraint on hydrodynamics and sediment transport and, in return, to evaluate the prospective effect of profound seabed changes on wave dissipation. This suite explains the static top of the 3D voxel model in terms of driving forces, and is crucial for a quantitative understanding of past as well as future resource changes resulting from natural seabed dynamics and human impact. A coupling framework including some or all of these components is not unique (e.g. Liang et al., 2007; Tang et al., 2009; Van den Eynde et al., 2010; Warner et al., 2010), and was already used to investigate the effect of sand extraction on sediment transport in the BPNS (Van den Eynde et al., 2010). Here, additional coupling with the voxel model offers an opportunity to initiate and parameterise the model suite in order to investigate the aggregate resource at an unprecedented level of detail and accuracy.

There are several ways in how to transfer the information contained in the 3D voxel model into the environmental model suite. The sediment-transport and seabed-morphology modules include the possibility to simulate different sediment fractions and seabed layers. Relevant information contained in the voxel model can be transferred to the numerical model suite (Figure 6): various voxel attributes will be used to define initial conditions (as a measure of expected sediment concentrations and erodibility) and to distil the most suitable parameters (e.g. particle size, settling velocity, cohesion, or any other relevant parameter) that characterise the different sediment fractions and stratigraphic units in the seabed sediment and shallow subsurface.

A fundamental advantage of this coupling is that sand and gravel in the seabed subsurface, now realistically parameterised through the voxel modelling (Figure 6), can be analysed for the first time as finite, exhaustible volumes. For the time being, two surfaces are used to constrain the aggregate resource (Figure 6): an upper surface represented as the present-day bathymetry and a lower surface below where no erosion or sand extraction is expected (i.e. erosion and sediment concentrations are set to zero in the environmental model). The definitive volume in between is subdivided into layers characterised by predefined distributions of the different sediment fractions. Each layer has a unique set of quantitative and qualitative attributes.

The geological voxel models will be adapted to optimise the 3D-4D workflow. In the standard 3D approach, the assignment of attribute values to initially empty voxels (i.e. having no original data in the 3D voxel model) is based on equally probable realisations of a stochastic interpolation procedure. Thus, probability of occurrence is generated for each lithological class, and the class with the highest likeli-

Figure 6. Coupling between the 3D voxel model (left) and the suite of environmental impact models (right). The voxel model provides percentages (%) of the different lithological classes (Litho), which are translated into concentrations (C) of the different sediment fractions to initialise the seabed of the numerical model. Properties of the sediment fractions (e.g. particle size) are provided as well to parameterise the numerical model. Owing to computational constraints, the different lithological classes of the voxel model must be grouped into broader sediment fractions in the numerical model (i.e. $f \leq c$). Also, highly similar superimposed voxels should be merged before feeding into the numerical model (i.e. $k \leq i$)

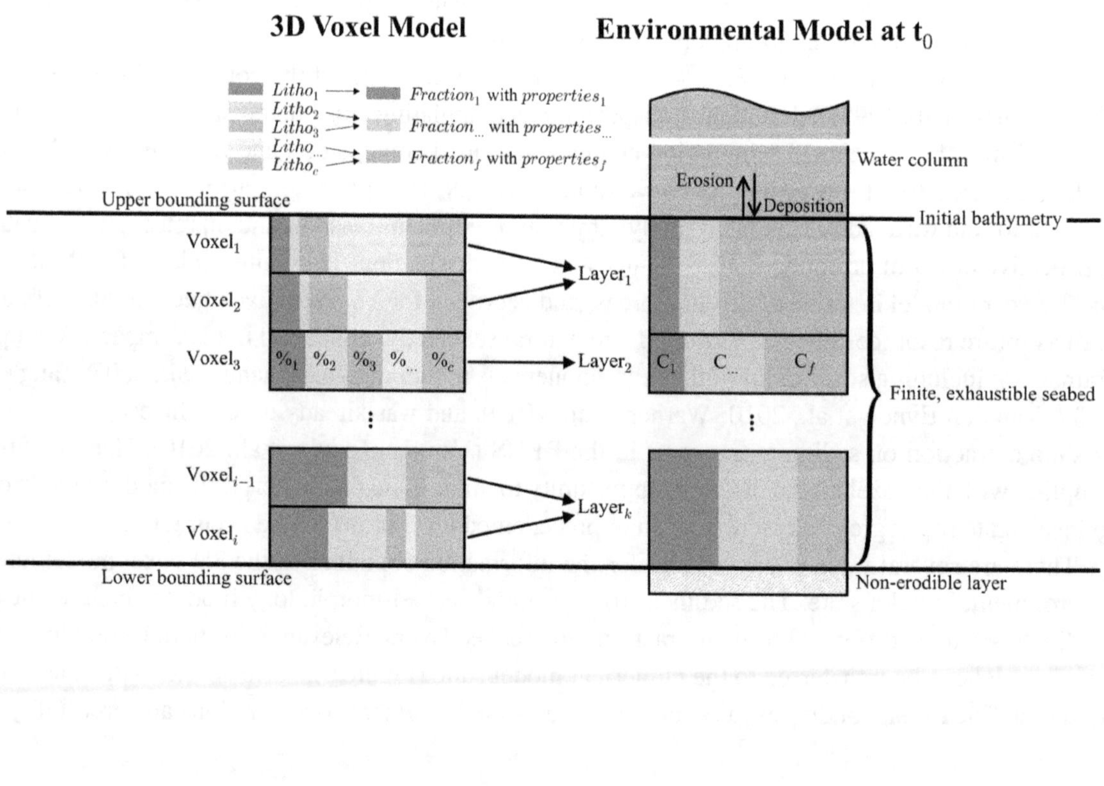

hood is assigned to each voxel (see section *'Dealing with uncertainty'* above). While valuable for the analysis of interpolation-related uncertainty, it cannot be used directly to initiate the t_0 seabed for runs of the suite of environmental impact models; these all require proportions of each sediment fraction (to establish initial concentrations), rather than probabilities, as input. A voxel model filled on the basis of a deterministic statistical approach (such as a linear kriging; e.g. van der Meulen et al., 2005) is therefore required here. Furthermore, the resolutions of the voxel model and the numerical model suite need adaptation. Presently, computational constraints impose limitations on the resolution attainable by the numerical environmental impact models. Thus, a voxel model at lower resolution is needed. Owing to the spatial autocorrelation structure involved in the interpolation procedure used to fill the empty voxels (i.e. kriging with fitted variograms), simple aggregation of an existing high-resolution voxel model into larger volume blocks is undesirable. Coupled voxel and environmental impact models need to be of equal resolution from the very beginning, calling for a separate, lower-resolution version of the initial voxel model. The adaptations made in this particular version of the voxel model (providing proportions and working at lower resolution) need to be done only from the perspective of compatibility with the environmental impact model suite. 4D output can be fed back into the initial voxel model, and does not impose any limitation to the development and further exploitation of the original 3D data product.

Cross-Domain, Multi-Criteria Decision Support System

Following the previous steps, all relevant information, including the 4D model output, is stored in a voxel model that represents the current state of knowledge on the subsurface, including inherent uncertainties. For flexible and easy data analysis, a multi-criteria DSS is required. This system, which harvests input by post-processing the voxel model, makes it possible to enter criteria on multiple attributes, specify their individual importance, and define a bonus to the concurrent satisfaction of any combination of attributes. A key product of the DSS is the suitability map, a visual representation of the degree to which each voxel satisfies the entered criteria. User-specific suitability maps make it a simple task to identify suitable locations for specific activities or measures.

Unlike other DSSs, TILES incorporates uncertainty associated with data quality in its calculations as a primary dimension. A measure of data quality is added as a secondary attribute to each voxel. It is a direct indicator of the reliability of data that form the basis of satisfaction calculations. As the combined result of degree of satisfaction and degree of reliability, the final suitability map can be used not only to find regions where the requirements are met, but also to identify regions where data are too scarce or poor to draw accurate conclusions.

The strength of the DSS is largely dependent on the expressiveness when entering criteria, which is inversely correlated to its ease-of-use. Finding the optimal balance between simplicity and expressivity profits from the application of linguistic, computational intelligence techniques (e.g. Dujmović, 2007). Through the possibility to enter criteria using words rather than numbers or codes, data analysis becomes intuitive yet extremely powerful. An example could be: "Find all voxels that are *close to* the seabed, *absolutely* do not contain any fine sand and *preferably* contain *a lot of* coarse sand". Invisible to the decision maker, this seemingly simple information request is automatically translated into a complex mathematical model that evaluates all voxels.

Performance and flexibility are other factors determining the user-friendliness of the DSS. Both the voxel model and the DSS are specifically designed to be easily extendable with new attributes or updated attribute values when new data, information or knowledge become available. In light of the fact that a successful DSS has to be available online, specific attention is paid to minimising the generation time of each suitability map. To that end, the voxel model is stored in a binary, machine-readable format, and the algorithms used to evaluate the voxels exploit multithreading paradigms to optimally use all available hardware resources. This solution will scale seamlessly with time as more powerful hardware becomes available.

On the Visualisation of 3D Suitability-Reliability

Suitability maps, as synthesised by a DSS, can be seen as tabular representations of evaluated data instances (independent groups of attributes, in this case stored in voxels) and their calculated suitability scores for a specific application. If the data instances can be represented in either 2D or 3D space, the mapping can be visualised using a colour gradient. First, each data instance is rendered in the 2D or 3D space according to its geometrical properties (e.g. a dot or a cuboid). Secondly, each data instance is coloured according to its corresponding suitability. Typically, a colour scale of green (ideal data instance, suitability equals 100%) to red (unfit data instance, suitability equals 0%) is used. Data products using this colour scale can intuitively be interpreted by decision makers: green clusters represent well-suited areas for the application under consideration, and red clusters represent unsuitable areas. To further simplify the visual representation, especially in 3D, completely unsuitable clusters may be omitted altogether to avoid visual cluttering and to reduce the demand on computational resources.

The addition in TILES of a secondary reliability attribute, representing the quality of data used in the computation of the suitability of each voxel, requires additional visualisation techniques. By combining a transparency scale and a colour scale, it would become possible to display both outputs simultaneously. Initial experiments have shown that the visually most appealing maps and grids are those where transparency is inversely related to suitability. In such data products, clusters of well-suited voxels are accentuated whilst unsuitable clusters are masked. This experimental result corroborates the previously mentioned intuition that it is best to leave out unsuitable regions.

FUTURE RESEARCH DIRECTIONS

With the transition from traditional mapping to digital geomodelling during the past decade, further optimisation of this process will be minimal. Some progress may be expected from the anticipated collection of high-resolution 3D seismics, providing great detail (metre-scale lateral resolution) for small areas. Integrating such volume data with single-location and transect data will further increase the contrast between data-rich and data-poor areas, providing additional challenges when building voxel models.

Partially in light of this data-density contrast, dedicated research is needed on dynamic voxel modelling. It is important to develop automated and autonomous procedures for partitioning the subsurface in an optimal, spatially varying resolution, based not only on data density and subsurface heterogeneity (measures of interpolation-related uncertainty), but also on data quality. Ideally, a combination of multiple mathematical models can be used to produce a minimal discretisation of the subsurface, so that no unnecessary or uninformative voxels are generated. Decision makers will see a reproducible and realistic visualisation of the resolution offered by the available data. Also, with fewer voxels to calculate, DSS performance will increase.

Future developments in visualisation head for 3D display and analysis via the internet rather than through the 'thick-client' solution offered by some currently used subsurface viewers (van der Meulen et al., 2013). A web solution has the great advantage that users no longer need to download and install software, and load and manage the data to be visualised. It will only be successful, however, when its performance is good, with rapid generation of suitability maps and grids. Luckily, there is a tendency amongst leading GIS vendors to move to full 3D functionality; 3D raster analysis and visualisation may be available to decision makers in the near future.

Voxels models are ideally suited to generating knowledge with knowledge. Their visualisation power, especially the possibility to generate multiple maps and grids for a single attribute or a combination of attributes, render them accessible to researchers from various disciplines. Each researcher, and even end user, may recognise patterns that were not recognised by the model developers. Recently, it has been shown, for example, that 3D voxel models provide new insights into the geological development of the corresponding region, with commonly subtle gradients offering much-needed information on depositional processes (Busschers et al., 2011; Stafleu & Busschers, 2014).

In TILES, coupling resource-related 3D voxel models with a suite of numerical environmental impact models, and including data quality as an uncertainty indicator, are major steps forward. However, the focus is still very much on geological attributes. Biology, of equal importance, is not yet included, and should be part of future valorisation efforts. It is clear that marine aggregate extraction affects the whole ecosystem and in particular its living organisms (e.g. benthic or fish populations; Cooper et al., 2008; Stelzenmüller et al., 2010; De Backer et al., 2014). The resilience of marine ecosystems to human stressors (such as aggregate extraction) remains poorly quantified (Elliott et al., 2007), and 4D modelling frameworks can contribute to filling this gap in knowledge. Whilst model studies of the effect of extraction on living organisms (e.g. Barry et al., 2010) are increasing in number, a full suite of numerical models including physical drivers (e.g. hydrodynamic conditions, human disturbance through dredging) and biological components (e.g. benthic assemblages) remains to be established. The quantitative biotic-abiotic links established through such an integrated approach will allow more targeted monitoring of the environmental consequences of sand and gravel exploitation, and will help the decision maker or other end user to select the most informative attributes and associated value ranges for visualisation. A holistic approach linking sand and gravel extraction to natural systems' behaviour and specified future human needs (e.g. Elliott et al., 2007; Atkins et al., 2011), and thus improving knowledge-based decision making, is possible only when data and models are shared freely (Sutherland et al., 2014), a key condition recognised by the European Commission.

From a resource-management perspective, future research will have to focus on the development of long-term adaptive strategies for the exploitation of marine sand and gravel. Adaptation of extraction plans is only possible when a mechanism is in place for structured incorporation into resource-assessment models of additional data and information from multiple sources and research disciplines, including actualised resource supplies and needs. An ultimate valorisation of the geological knowledge base is the proposition of most optimal aggregate concession zones, based on up-to-date knowledge on the nature and dynamics of the geological resources, the environmental impact of their removal, and constraints imposed by other users of the marine space. A successful DSS will ensure that the relevant EU Directives and recommendations can be adhered to.

CONCLUSION

This chapter provides an overview of the main challenges and envisaged solutions associated with mapping and quantifying, from a sustainability perspective, actual and future marine sand and gravel supplies. It contributes to the development of long-term adaptive management strategies for the exploitation of geological resources, and has direct applicability in Marine Spatial Planning, EU's Marine Strategy Framework Directive, and Integrated Coastal Zone Management.

The development of a transnational, harmonised geological knowledge base is a critical precondition for the exchange of data, information and knowledge on a finite, exhaustible resource between expert and decision maker. Inclusion of various measures of uncertainty, and integration with 4D environmental impact models, both highly innovative, will herald a new age in local as well as transnational resource management. Relevant data and model output will be made freely available, and the decision-support tool will be accessible online. Ease of access and user-friendliness will stimulate further integration of cross-domain information and expertise, and thus improve the resource models and their generic applicability in diverse fields.

ACKNOWLEDGMENT

This chapter is a contribution to the Brain-be project TILES (Transnational and Integrated Long-term marine Exploitation Strategies), funded by Belgian Science Policy (Belspo) under contract BR/121/A2/TILES. The research is fully supported by the ZAGRI project, a federal Belgian programme for continuous monitoring of sand and gravel extraction, paid from private revenues. Contributing EU projects have been EMODnet Geology (MARE/2008/03; MARE/2012/10), Geo-Seas (FP7, Grant 238952) and ODIP (FP7, Grant 312492). Underlying data have been acquired during numerous campaigns on RV Belgica, RV Simon Stevin, and many Dutch research and monitoring vessels. For Belgian waters, shiptime was granted by Belspo / RBINS ODNature and Flanders Marine Institute. Data and metadata are stored in and made available through national data repositories: the Belgian Marine Data Centre and the Dutch Subsurface Data and Information system DINO. Peter-Paul van Maanen provided ideas on generating variable voxel sizes, and prepared Figure 4.

REFERENCES

Atkins, J. P., Burdon, D., Elliott, M., & Gregory, A. J. (2011). Management of the marine environment: Integrating ecosystem services and societal benefits with the DPSIR framework in a systems approach. *Marine Pollution Bulletin*, *62*(2), 215–226. doi:10.1016/j.marpolbul.2010.12.012 PMID:21215974

Barry, J., Boyd, S., & Fryer, R. (2010). Modelling the effects of marine aggregate extraction on benthic assemblages. *Journal of the Marine Biological Association of the United Kingdom*, *90*(1), 105–114. doi:10.1017/S0025315409990737

Blaas, M., Dong, C., Marchesiello, P., McWilliams, J. C., & Stolzenbach, K. D. (2007). Sediment-transport modeling on Southern Californian shelves: A ROMS case study. *Continental Shelf Research*, *27*(6), 832–853. doi:10.1016/j.csr.2006.12.003

Bond, C. E., Gibbs, A., Shipton, Z. K., & Jones, S. (2007). What do you think this is? 'Conceptual uncertainty' in geoscience interpretation. *GSA Today*, *17*(11), 4–10. doi:10.1130/GSAT01711A.1

Bond, C. E., Lunn, R. J., Shipton, Z. K., & Lunn, A. D. (2012). What makes an expert effective at interpreting seismic images? *Geology*, *40*(1), 75–78. doi:10.1130/G32375.1

Bond, C. E., Shipton, Z. K., Gibbs, A. D., & Jones, S. (2008). Structural models: Optimizing risk analysis by understanding conceptual uncertainty. *First Break*, *26*(1288), 65–71. doi:10.3997/1365-2397.2008006

Booij, N., Ris, R. C., & Holthuijsen, L. H. (1999). A third-generation wave model for coastal regions: 1. Model description and validation. *Journal of Geophysical Research: Oceans (1978–2012)*, *104*(C4), 7649-7666.

Busschers, F. S., Stafleu, J., Maljers, D., Gunnink, J. L., & Sobisch, H.-G. (2011, October). *New geological insights from stochastic 3D voxel models.* Paper presented at the Annual Meeting of the Geological Society of America. Minneapolis, MN.

Cassard, D., Bertrand, G., Billa, M., Serrano, J.-J., Tourlière, B., Angel, J.-M., & Gaál, G. (2015). ProMine Mineral Databases: New Tools to Assess Primary and Secondary Mineral Resources in Europe. In P. Weihed (Ed.), 3D, 4D and Predictive Modelling of Major Mineral Belts in Europe (pp. 9-58). Mineral Resource Reviews Book Series. Cham, Switzerland: Springer International Publishing.

Cassard, D., Tertre, F., Bertrand, G., Schjøth, F., Tulstrup, J., Heijboer, T., & Vuollo, J. (2014). EURARE IKMS: An Integrated Knowledge Management System for Rare Earth Element Resources in Europe. In *ERES 2014 - 1st European Rare Earth Resources Conference* (pp. 292-300). Retrieved from http://www.eurare.eu/docs/eres2014/fifthSession/GuillaumeBertrand.pdf

Cooper, K. M. (2012). Setting limits for acceptable change in sediment particle size composition following marine aggregate dredging. *Marine Pollution Bulletin*, *64*(8), 1667–1677. doi:10.1016/j.marpolbul.2012.05.031 PMID:22721693

Cooper, K. M., Froján, C. R. B., Defew, E., Curtis, M., Fleddum, A., Brooks, L., & Paterson, D. M. (2008). Assessment of ecosystem function following marine aggregate dredging. *Journal of Experimental Marine Biology and Ecology*, *366*(1), 82–91. doi:10.1016/j.jembe.2008.07.011

Dalyander, P. S., Butman, B., Sherwood, C. R., Signell, R. P., & Wilkin, J. L. (2013). Characterizing wave-and current-induced bottom shear stress: US middle Atlantic continental shelf. *Continental Shelf Research*, *52*, 73–86. doi:10.1016/j.csr.2012.10.012

Dambrink, R. M., Maljers, D., Schokker, J., & Stafleu, J. (2015, November). *Tailor-made maps generated from 3D geological voxel models.* Paper presented at the Annual Meeting of the Geological Society of America. Baltimore, MD.

De Backer, A., Van Hoey, G., Coates, D., Vanaverbeke, J., & Hostens, K. (2014). Similar diversity-disturbance responses to different physical impacts: Three cases of small-scale biodiversity increase in the Belgian part of the North Sea. *Marine Pollution Bulletin*, *84*(1), 251–262. doi:10.1016/j.marpolbul.2014.05.006 PMID:24889315 de Jong, M. F., Baptist, M. J., van Hal, R., de Boois, I. J., Lindeboom, H. J., & Hoekstra, P. (2014). Impact on demersal fish of a large-scale and deep sand extraction site with ecosystem-based landscaped sandbars. *Estuarine, Coastal and Shelf Science*, *146*, 83–94. doi:10.1016/j.ecss.2014.05.029

Degraer, S., Verfaillie, E., Willems, W., Adriaens, E., Vincx, M., & Van Lancker, V. (2008). Habitat suitability modelling as a mapping tool for macrobenthic communities: An example from the Belgian part of the North Sea. *Continental Shelf Research*, *28*(3), 369–379. doi:10.1016/j.csr.2007.09.001

Degrendele, K., Roche, M., De Mol, L., Schotte, P., & Vandenreyken, H. (2014). Synthesis of the monitoring of the aggregate extraction on the Belgian Continental Shelf from 2011 till 2014. In: L. De Mol & H. Vandenreyken (Eds.), *Which future for the sand extraction in the Belgian part of the North Sea?Proceedings of the Study Day Sand Extraction* (pp. 3-29). Blankenberge, Belgium: FOD Economie.

Degrendele, K., Roche, M., Schotte, P., Van Lancker, V. R., Bellec, V. K., & Bonne, W. M. (2010). Morphological evolution of the Kwinte Bank central depression before and after the cessation of aggregate extraction. *Journal of Coastal Research SI*, *51*, 77–86.

Diesing, M., & Stephens, D. (2015). A multi-model ensemble approach to seabed mapping. *Journal of Sea Research*, *100*, 62–69. doi:10.1016/j.seares.2014.10.013

Dujmović, J. J. (2007). Continuous preference logic for system evaluation. *IEEE Transactions on Fuzzy Systems*, *15*(6), 1082–1099. doi:10.1109/TFUZZ.2007.902041

Dyer, K. R., & Huntley, D. A. (1999). The origin, classification and modelling of sand banks and ridges. *Continental Shelf Research*, *19*(10), 1285–1330. doi:10.1016/S0278-4343(99)00028-X

EC Enterprise and Industry. (2010). *Improving framework conditions for extracting minerals for the EU. Exchanging best practice on land-use planning, permitting and geological knowledge sharing*. Retrieved October 31, 2015 from http://www.google.nl/url?sa=t&rct=j&q=&esrc=s&source=web&cd=1&ved=0CCEQFjAAahUKEwjo4PqEh-3IAhWGXRoKHUh4AZM&url=http%3A%2F%2Fec.europa.eu%2FDocsRoom%2Fdocuments%2F9986%2Fattachments%2F1%2Ftranslations%2Fen%2Frenditions%2Fnative&usg=AFQjCNGOCjuWZYABqUCMw8HqzXXOB43bBw&sig2=sNL54AEbIEG32vmHIfEJ3w

Elliott, M., Burdon, D., Hemingway, K. L., & Apitz, S. E. (2007). Estuarine, coastal and marine ecosystem restoration: Confusing management and science – a revision of concepts. *Estuarine, Coastal and Shelf Science*, *74*(3), 349–366. doi:10.1016/j.ecss.2007.05.034

European Commission. (2010). *Maritime spatial planning in the EU - Achievements and future developments* [EC COM2010_771]. Publications Office of the European Union, Luxembourg. Retrieved October 31, 2015 from http://ec.europa.eu/maritimeaffairs/documentation/publications/documents/com_2010_771_brochure_en.pdf

European Commission. (2011). *Communication from the EC to the European Parliament, the Council, the European Economic and Social Committee and the Committee of the Regions. Roadmap to a Resource Efficient Europe* [EC COM2011_571]. Publications Office of the European Union, Luxembourg. Retrieved October 31, 2015 from http://ec.europa.eu/environment/resource_efficiency/about/roadmap/index_en.htm

European Union. (2008). Directive 2008/56/EC of the European Parliament and of the Council of 17 June 2008: Establishing a framework for community action in the field of maritime environmental policy (Marine Strategy Framework Directive). *Official Journal of the European Union*, L 164/19-40.

Fayyad, U., Piatetsky-Shapiro, G., & Smyth, P. (1996). From data mining to knowledge discovery in databases. *AI Magazine*, *17*(3), 37–54.

Feblowitz, J. (2012). The big deal about big data in upstream oil and gas. *IDC Energy Insights*, 1-11. Retrieved October 31, 2015 from https://www.hds.com/assets/pdf/the-big-deal-about-big-data-in-upstream-oil-and-gas.pdf

G. Fick, & R. H. Sprague, (Eds.). (2013). Decision Support Systems: Issues and Challenges.*Proceedings of an International Task Force Meeting*. Oxford, UK: Pergamon Press.

Francken, F., Van den Eynde, D., & Van Lancker, V. (2014). On-demand assessment of spatial and temporal variability of sediment transport parameters, Belgian and southern Dutch part of the North Sea. In: L. De Mol & H. Vandenreyken (Eds.), *Which future for the sand extraction in the Belgian part of the North Sea?Proceedings of the Study Day Sand Extraction* (pp. 73-80). Blankenberge, Belgium: FOD Economie.

Gattinoni, P., Pizzarotti, E. M., & Scesi, L. (2014). Geological Risk Management. In *Engineering Geology for Underground Works* (pp. 143–159). Dordrecht, The Netherlands: Springer. doi:10.1007/978-94-007-7850-4_5

Habert, G., Bouzidi, Y., Chen, C., & Jullien, A. (2010). Development of a depletion indicator for natural resources used in concrete. *Resources, Conservation and Recycling*, *54*(6), 364–376. doi:10.1016/j.resconrec.2009.09.002

Holden, E., Linnerud, K., & Banister, D. (2014). Sustainable development: Our common future revisited. *Global Environmental Change*, *26*, 130–139.

ICES. (2015). *Interim report of the Working Group on the Effects of Extraction of Marine Sediments on the Marine Ecosystem (WGEXT). ICES CM 2015/SSGEPI:06.* Copenhagen, Denmark: ICES.

ICES. (2016). *Effects of extraction of marine sediments on the marine environment 2005– 2011.* ICES Cooperative Research Report No. 330. ICES.

Jørgensen, F., Møller, R. R., Nebel, L., Jensen, N.-P., Christiansen, A. V., & Sandersen, P. B. E. (2013). A method for cognitive 3D geological voxel modelling of AEM data. *Bulletin of Engineering Geology and the Environment*, *72*(3-4), 421–432. doi:10.1007/s10064-013-0487-2

Judge, M. (2015). *European Marine Observation and Data Network (EMODnet): Making Fragmented Marine Data Relevant and Accessible.* Retrieved October 31, 2015 from http://earthzine.org/2015/10/08/european-marine-observation-and-data-network-emodnet-making-fragmented-marine-data-relevant-and-accessible/

Kessler, H., Mathers, S., & Sobisch, H. G. (2009). The capture and dissemination of integrated 3D geospatial knowledge at the British Geological Survey using GSI3D software and methodology. *Computers & Geosciences*, *35*(6), 1311–1321. doi:10.1016/j.cageo.2008.04.005

Lark, R. M., Lawley, R. S., Barron, A. J. M., Aldiss, D. T., Ambrose, K., Cooper, A. H., & Waters, C. N. et al. (2015a). Uncertainty in mapped geological boundaries held by a national geological survey: Eliciting the geologists' tacit error model. *Solid Earth*, *6*(2), 727–745. doi:10.5194/se-6-727-2015

Lark, R. M., Marchant, B. P., Dove, D., Green, S. L., Stewart, H., & Diesing, M. (2015b). Combining observations with acoustic swath bathymetry and backscatter to map seabed sediment texture classes: The empirical best linear unbiased predictor. *Sedimentary Geology*, *328*, 17–32. doi:10.1016/j.sedgeo.2015.07.012

Le Bot, S., Van Lancker, V., Deleu, S., De Batist, M., Henriet, J. P., & Haegeman, W. (2005). Geological characteristics and geotechnical properties of Eocene and Quaternary deposits on the Belgian continental shelf: Synthesis in the context of offshore wind farming. *Netherlands Journal of Geosciences*, *84*(2), 147–160. doi:10.1017/S0016774600023027

Le Hir, P., Cayocca, F., & Waeles, B. (2011). Dynamics of sand and mud mixtures: A multiprocess-based modelling strategy. *Continental Shelf Research*, *31*(10), S135–S149. doi:10.1016/j.csr.2010.12.009

Liang, B., Li, H., & Lee, D. (2007). Numerical study of three-dimensional suspended sediment transport in waves and currents. *Ocean Engineering*, *34*(11), 1569–1583. doi:10.1016/j.oceaneng.2006.12.002

Luyten, P. J., Jones, J. E., Proctor, R., Tabor, A., Tett, P., & Wild-Allen, K. (1999). *COHERENS – A coupled hydrodynamical-ecological model for regional and shelf seas: user documentation.* Unpublished Report of the Management Unit of the Mathematical Models of the North Sea. Brussels, Belgium: MUMM.

Maljers, D., Stafleu, J., van der Meulen, M. J., & Dambrink, R. M. (2015). Advances in constructing regional geological voxel models, illustrated by their application in aggregate resources. *Netherlands Journal of Geoscience*, *94*(3), 257–270. doi:10.1017/njg.2014.46

Marschallinger, R. (1996). A voxel visualization and analysis system based on AutoCAD. *Computers & Geosciences*, *22*(4), 379–386. doi:10.1016/0098-3004(95)00093-3

Mookerjee, M., Vieira, D., Chan, M. A., Gil, Y., Goodwin, C., Shipley, T. F., & Tikoff, B. (2015). We need to talk: Facilitating communication between field-based geoscience and cyberinfrastructure communities. *GSA Today*, *25*, 34–35. doi:10.1130/GSATG248GW.1

Mudd, G. M. (2010). The environmental sustainability of mining in Australia: Key mega-trends and looming constraints. *Resources Policy*, *35*(2), 98–115. doi:10.1016/j.resourpol.2009.12.001

Polson, D., & Curtis, A. (2010). Dynamics of uncertainty in geological interpretation. *Journal of the Geological Society*, *167*(1), 5–10. doi:10.1144/0016-76492009-055

Polson, D., & Curtis, A. (2015). Assessing individual influence on group decisions in geological carbon capture and storage problems. In P. Diviacco, P. Fox, C. Pshenichny & A. Leadbetter (Eds.), Collaborative knowledge in scientific research networks (pp. 55-75). IGI Global. doi:10.4018/978-1-4666-6567-5.ch004

Prior, T., Giurco, D., Mudd, G., Mason, L., & Behrisch, J. (2012). Resource depletion, peak minerals and the implications for sustainable resource management. *Global Environmental Change*, *22*(3), 577–587. doi:10.1016/j.gloenvcha.2011.08.009

Radzevičius, R., Velegrakis, A. F., Bonne, W. M., Kortekaas, S., Garel, E., Blažauskas, N., & Asariotis, R. (2010). Marine aggregate extraction regulation in EU member states. *Journal of Coastal Research SI*, *51*, 15–37.

Reed, C. W., Niedoroda, A. W., & Swift, D. J. (1999). Modeling sediment entrainment and transport processes limited by bed armoring. *Marine Geology*, *154*(1), 143–154. doi:10.1016/S0025-3227(98)00109-1

Reid, W. V., Chen, D., Goldfarb, L., Hackmann, H., Lee, Y. T., Mokhele, K., & Whyte, A. et al. (2010). Earth System Science for Global Sustainability: Grand Challenges. *Science*, *330*(6006), 916–917. doi:10.1126/science.1196263 PMID:21071651

Rice, J., Arvanitidis, C., Borja, A., Frid, C., Hiddink, J. G., Krause, J., & Norkko, A. et al. (2012). Indicators for sea-floor integrity under the European Marine Strategy Framework Directive. *Ecological Indicators*, *12*(1), 174–184. doi:10.1016/j.ecolind.2011.03.021

Roos, P. C., & Hulscher, S. J. (2003). Large-scale seabed dynamics in offshore morphology: Modeling human intervention. Reviews of Geophysics, 41(2), 1010. doi:10.1029/2002RG000120

Soares, A. (1992). Geostatistical estimation of multi-phase structures. *Mathematical Geology*, *24*(2), 149–160. doi:10.1007/BF00897028

Sobisch, H.-G. (2011, October). *One viewer to publish 3D subsurface models of any modelling origin.* Paper presented at the Annual Meeting of the Geological Society of America, Minneapolis, MN.

Solar, S. V., Demicheli, L., & Wall, P. (2012). Raw Materials Initiative: A contribution to the European Minerals Policy Framework. In R. Sinding-Larsen & F.-W. Wellmer (Eds.), *Non-Renewable Resource Issues: Geoscientific and Societal Challenges* (pp. 21–33). Dordrecht, The Netherlands: Springer. doi:10.1007/978-90-481-8679-2_2

Stafleu, J., & Busschers, F. S. (2014). Lithological, grain-size and architectural trends in the Holocene Rhine-Meuse delta - insights from 3D voxel models. In *Proceedings of the 76th European Association of Geoscientists and Engineers Conference and Exhibition: Experience the Energy* (pp. 832-836). Houten, The Netherlands: EAGE. doi:10.3997/2214-4609.20141505

Stafleu, J., Maljers, D., Gunnink, J. L., Menkovic, A., & Busschers, F. S. (2011). 3D modelling of the shallow subsurface of Zeeland, the Netherlands. *Netherlands Journal of Geosciences*, *90*(4), 293–310.

Stafleu, J., Maljers, D., Hummelman, J., & Gunnink, J. L. (2014, April). *Visualization of uncertainty in 3D stochastic voxel models of the Netherlands.* Paper presented at the EGU General Assembly, Vienna, Austria.

Stafleu, J., Sobisch, H.-G., Maljers, D., Hummelman, J., Dambrink, R. M., & Gunnink, J. L. (2013, October). *Visualization and dissemination of 3D geological property models of the Netherlands.* Paper presented at the Annual Meeting of the Geological Society of America, Denver, CO.

Stelzenmüller, V., Ellis, J. R., & Rogers, S. I. (2010). Towards a spatially explicit risk assessment for marine management: Assessing the vulnerability of fish to aggregate extraction. *Biological Conservation*, *143*(1), 230–238. doi:10.1016/j.biocon.2009.10.007

Stive, M. J. F., Nicholls, R. J., & de Vriend, H. J. (1991). Sea-level rise and shore nourishment: A discussion. *Coastal Engineering*, *16*(1), 147–163. doi:10.1016/0378-3839(91)90057-N

Stolk, A., & Dijkshoorn, C. (2009). Sand extraction Maasvlakte 2 Project: License, Environmental Impact Assessment and Monitoring. In *Proceedings of the EMSAGG 2009 Conference: A wave of opportunities for the marine aggregates industry*. London, UK: CIRIA.

Sutherland, J., Townend, I. H., Harpham, Q. K., & Pearce, G. R. (2014). From integration to fusion: the challenges ahead. In A. T. Riddick, H. Kessler, & J. R. A. Giles (Eds.), Geological Society, London, Special Publications: Vol. 408. *Integrated environmental modelling to solve real-world problems*. doi:10.1144/SP408.6

Tang, H. S., Keen, T. R., & Khanbilvardi, R. (2009). A model-coupling framework for nearshore waves, currents, sediment transport, and seabed morphology. *Communications in Nonlinear Science and Numerical Simulation*, *14*(7), 2935–2947. doi:10.1016/j.cnsns.2008.10.012

Taylor, R. B., & Steven, T. A. (1983). Definition of mineral resource potential. *Economic Geology and the Bulletin of the Society of Economic Geologists*, *78*(6), 1268–1270. doi:10.2113/gsecongeo.78.6.1268

U.S. Geological Survey. (2016). *Mineral commodity summaries 2016*. U.S. Geological Survey; doi:10.3133/70140094

Van den Eynde, D., Giardino, A., Portilla, J., Fettweis, M., Francken, F., & Monbaliu, J. (2010). Modelling the effects of sand extraction, on sediment transport due to tides, on the Kwinte Bank. *Journal of Coastal Research SI*, *51*, 101–116.

van der Meulen, M. J., Doornenbal, J. C., Gunnink, J. L., Stafleu, J., Schokker, J., Vernes, R. W., & van Daalen, T. M. et al. (2013). 3D Geology in a 2D country: Perspectives for geological surveying in the Netherlands. *Netherlands Journal of Geosciences*, *92*(4), 217–241. doi:10.1017/S0016774600000184

van der Meulen, M. J., van Gessel, S. F., & Veldkamp, J. G. (2005). Aggregate resources in the Netherlands. *Netherlands Journal of Geosciences*, *84*(4), 379–387. doi:10.1017/S0016774600021193

van Heteren, S., & Van Lancker, V. (2015). Collaborative seabed-habitat mapping: uncertainty in sediment data as an obstacle in harmonization. In P. Diviacco, P. Fox, C. Pshenichny & A. Leadbetter (Eds.), Collaborative knowledge in scientific research networks (pp. 154-176). IGI Global. doi:10.4018/978-1-4666-6567-5.ch008

van Heteren, S., Vonhögen-Peeters, L., & Maljers, D. (2012). *Multiple querying on geological databases to visualize offshore sand resources in the Netherlands part of the North Sea*. Unpublished TNO report TNO 2012 R10097. Utrecht, The Netherlands: TNO.

Van Lancker, V., Bonne, W., Uriarte, A., & Collins, M. B. (Eds.). (2010b). EUMARSAND: European Marine Sand and Gravel Resources. Special Issue Journal of Coastal Research Royal Palm Beach, FL: CERF.

Van Lancker, V., & van Heteren, S. (2013). Case Study 4: Revisiting the spatial distribution of EUNIS Level 3 North Sea habitats in view of Europe's Marine Strategy Framework Directive. In V. Van Lancker & S. van Heteren (Eds.), *Standardisation and harmonisation in seabed habitat mapping: role and added value of geological data and information* (pp. 86-93). Retrieved October 31, 2015 from http://www.geoseas.eu/documents/geoseas/html_page/d10-5A_gs_wp10_seabed-habitat-mapping_part-Sediment_final.pdf

Van Lancker, V. R. M., Bonne, W., Garel, E., Degrendele, K., Roche, M., & Van den Eynde, D., Bellec, V., Brière, C., Collins, M. BVelegrakis, A. F. (2010a). Recommendations for the sustainable exploitation of tidal sandbanks. *Journal of Coastal Research SI, 51*, 151–164.

van Ledden, M., Wang, Z. B., Winterwerp, H., & De Vriend, H. (2006). Modelling sand–mud morphodynamics in the Friesche Zeegat. *Ocean Dynamics, 56*(3-4), 248–265. doi:10.1007/s10236-005-0055-9

Velegrakis, A. F., Ballay, A., Poulos, S. E., Radzevičius, R., Bellec, V. K., & Manso, F. (2010). European marine aggregates resources: Origins, usage, prospecting and dredging techniques. *Journal of Coastal Research SI, 51*, 1–14.

Warner, J. C., Armstrong, B., He, R., & Zambon, J. B. (2010). Development of a coupled ocean–atmosphere–wave–sediment transport (COAWST) modeling system. *Ocean Modelling, 35*(3), 230–244. doi:10.1016/j.ocemod.2010.07.010

Warner, J. C., Sherwood, C. R., Signell, R. P., Harris, C. K., & Arango, H. G. (2008). Development of a three-dimensional, regional, coupled wave, current, and sediment-transport model. *Computers & Geosciences, 34*(10), 1284–1306. doi:10.1016/j.cageo.2008.02.012

WCED. (1987). *Our Common Future. World Commission on Environmental and Development.* Oxford, UK: Oxford University Press.

Wellmann, J. F., & Regenauer-Lieb, K. (2012). Uncertainties have a meaning: Information entropy as a quality measure for 3-D geological models. *Tectonophysics, 526-529*, 207–216. doi:10.1016/j.tecto.2011.05.001

Wiberg, P. L., Drake, D. E., & Cacchione, D. A. (1994). Sediment resuspension and bed armoring during high bottom stress events on the northern California inner continental shelf: Measurements and predictions. *Continental Shelf Research, 14*(10), 1191–1219. doi:10.1016/0278-4343(94)90034-5

KEY TERMS AND DEFINITIONS

Big Data: Datasets of such volume that traditional data-processing techniques cannot be used to process them. Typically, Big Data require context-specific exploitation of the underlying information model to tailor custom algorithms that can cope with the sheer volume of data.

Data Quality: Description of a degree in which observed or measured data are considered to be an accurate representation of reality. Descriptions or measurements can be different from reality for numerous reasons. Data are said to be of high quality if they can be used to derive reliable solutions in decision-making applications.

Decision Support System: Flexible digital platform for knowledge synthesis, enabling decision making on the basis of large or complicated datasets.

Entropy: Measure of uncertainty with respect to a probability distribution. If multiple outcomes of a specific process are observed, the resulting entropy of these results reflects how hard it is to predict the next outcome.

Environmental Impact Models: Predictors of the environmental consequences (positive or negative) of a plan, policy, program, or project prior to the decision to move forward with the proposed action.

Marine Strategy Framework Directive: European Directive aimed at achieving or maintaining Good Environmental Status (GES) in European seas. GES is described by eleven descriptors, including seafloor integrity and hydrographic conditions, which are highly relevant in geological resource management.

Resource Potential: Measure of the likelihood of occurrence of valuable resources, now and in the future. It can only be judged by integrating many diverse factors, some of which can be measured and some of which will depend on the specific experience and knowledge of the individual making the judgement.

Seafloor Integrity: The ability of the seafloor to function healthily, continue to provide natural goods and services, and maintain biodiversity.

Suitability Maps: Data products derived from a mapping between data instances and the degree to which they satisfy specific criteria. It is possible to evaluate a dataset multiple times with different criteria to generate different suitability maps.

Sustainable Resource Management: Management based on sustainable development principles, hence accounting for three equally important primary dimensions: safeguarding long-term ecological sustainability, satisfying basic needs, and promoting intragenerational and intergenerational equity.

Voxel: A unit of graphic information that defines a cuboid volume in three-dimensional space. Whilst a pixel (picture element) is defined in two-dimensional space with its x and y coordinates, a voxel has a third coordinate (z), representing depth and thus creating a volume element.

Chapter 11
Providing Oceanographic Data and Information to Pacific Island Communities

James Potemra
University of Hawai'i, USA

John Maurer
University of Hawai'i, USA

Echelle S. Burns
University of Hawai'i, USA

ABSTRACT

The Pacific Islands Ocean Observing System (PacIOOS) is a partnership of Federal, State and local agencies, academia, industry and various other groups that provides data and information about the ocean environment to a wide array of users in the insular Pacific. By collecting measurements and running numerical model forecasts, conducting scientific analyses and then providing these to the public through common interfaces, PacIOOS enables science-based decision making with data services and products. This chapter describes how data collection from a wide-array of sensors can support decision-making in the region if the data are properly (easily, effectively, consistently) served.

INTRODUCTION

The scientific process has always been tightly coupled to data collection and analysis. Traditionally, scientific analyses have been a necessary, and usually final, step in the flow from data to information. With the recent improvements in technology, this process is changing. Data collection and delivery has become more commonplace, and there are almost limitless applications that rely on and make use of data. This "new age" has been described as a new paradigm termed "data-driven science" [Gray, 2009].

Coincident with this have been improvements in data access. Personal computers, mobile devices, and any other "internet enabled" devices can now directly access data. Because of this, different efforts

DOI: 10.4018/978-1-5225-0700-0.ch011

have been designed to improve the efficiency by which users get data, a process referred to as "data transport." Many of these efforts stress the importance of standard services that promote interoperability.

The Pacific Islands Ocean Observing System (PacIOOS) is one such data-serving group that promotes and relies on data interoperability. The PacIOOS "enterprise" is part of a large, US National effort aimed at providing information about the ocean environment to a wide range of users. These users range from casual beach-goers, who are interested in the latest weather forecast or wave conditions, to federal agencies that are responsible for public safety. The governance structure of PacIOOS includes management and outreach components, regional representatives in outlying islands, and an oversight committee (Governing Council). Working together, these groups identify possible new data sets and desired products.

In an effort to bridge the gap between the scientific community, who are responsible for making measurements and running forecast models, and the wide-ranging end-users, the data management group in PacIOOS has developed the infrastructure to host and distribute ocean-related data. The efficiency of this system has also allowed the group to build web-based tools to further help users. In this chapter we describe these efforts in more detail.

BACKGROUND

The Integrated Ocean Observing System (IOOS) was established as the US contribution to the Global Ocean Observing System (GOOS; Levy et al., 2011). The IOOS mission developed after a series of meetings and workshops in the early 2000's. A more formal establishment occurred following the Integrated Coastal and Ocean Observation System (ICOOS) Act of 2009. The ICOOS Act provided the structure and foundation for the development of a US IOOS built upon a national-regional partnership. This partnership is working to provide new tools and forecasts to improve safety, enhance the economy, and protect the environment.

The IOOS premise of a national-regional partnership is manifest through a collection of regional associations (RA) that link in a virtual sense into a national backbone [Basset et al, 2010; de La Beaujardière et al., 2010a; de La Beaujardière et al., 2010b]. IOOS is composed of 11 different RA's (see Figure 1), each locally managed and responsive to a certain geographic area.

PacIOOS is the RA designed to monitor and serve the US Affiliate islands, including the State of Hawai'i, Guam, American Samoa, the Federated States of Micronesia (FSM), the Republic of the Marshall Islands (RMI), the Commonwealth of the Northern Mariana Islands (CNMI), and several minor outlying islands.

The initial planning document for IOOS sought to develop an end-to-end, "user driven" system with four subsystems: observations; analysis and models; data communications and management; and user groups and applications [Ocean.US, 2002]. These four foci provide the basis for PacIOOS as well. While PacIOOS is independent from the other ten Regional Associations, all of the RAs work toward a common goal of interoperability. The IOOS Program Office provides guidance and recommendations as to data services that the regions should employ. Additionally, the Program Office provides nation-wide integrated catalogs [O'Brien, 2008] and services end-points, for example the GeoIDE "clean catalog" of THREDDS servers (currently at http://ferret.pmel.noaa.gov/geoide/geoIDECleanCatalog.html).

PacIOOS started with a grant from the National Oceanographic and Atmospheric Administration (NOAA) in 2005, but significant advancements in the observing system assets did not begin until several

Figure 1. Distribution of IOOS Regional Associations. The IOOS regional associations are marked with yellow dots. The RA for the Pacific Islands, PacIOOS, includes the main US islands and territories in the Pacific Ocean, including the State of Hawai'i.

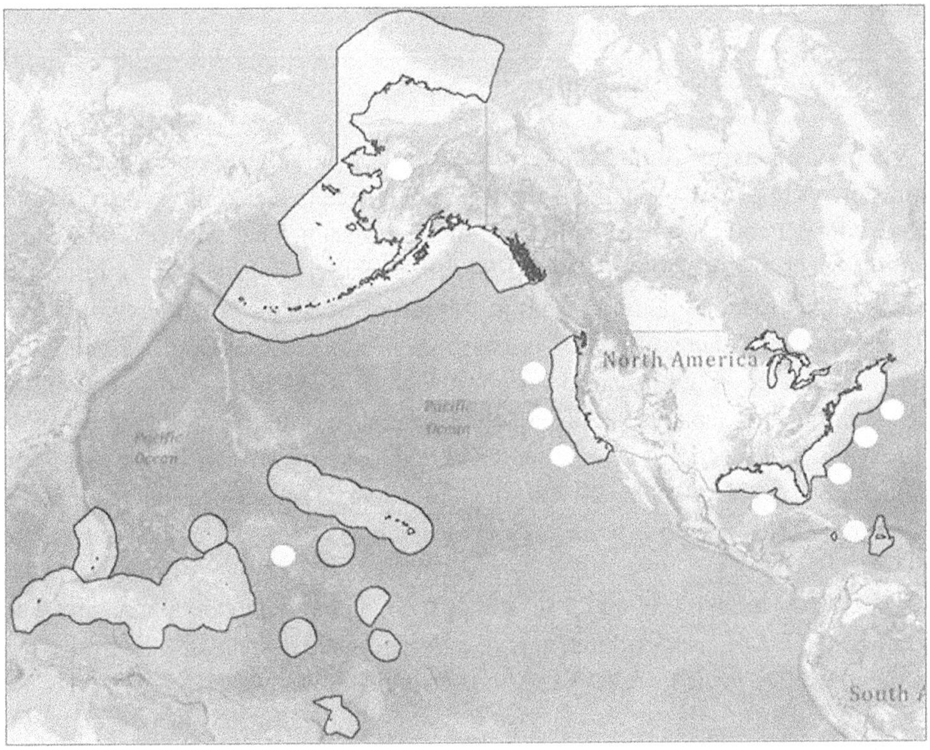

years later. The University of Hawai'i (UH) is the lead group for PacIOOS, and through the School of Ocean and Earth Science and Technology (SOEST), PacIOOS leverages scientific research activities with the school including equipment, computing facilities, personnel and infrastructure. Similar to the broader IOOS, PacIOOS can be described as the integration of several different subsystems that connect data to users. These include an observing system (instruments), numerical models, data management and, finally, users and applications. These are described in more detail as follows.

Observations Subsystem

The PacIOOS observing system consists of a variety of *in situ* and remote devices. The main *in situ* measurements are made using so-called "near-shore sensors." These are typically fixed to a pier or other structure and report temperature, salinity, turbidity and florescence (an estimation of primary productivity in the water column) at a single point through time. PacIOOS also maintains bottom-anchored buoys that have instruments measuring these same variables, but the buoys also include measurements of carbon dioxide, oxygen, and pH. Autonomous gliders are also employed to measure water properties, again mainly temperature and salinity, along pre-described tracks. Finally, one of the most popular data streams comes from a system of surface wave buoys. These buoys are moored in relatively shallow water and measure incoming wave height, period and direction.

Remote observations are conducted using high-frequency radios (HFR). These are land-based but scan the ocean surface to detect surface ocean movement (currents). HFR measure currents along "beams" to/from the device, and so a pair of instruments with overlapping coverage are used to get north/south, east/west currents.

Modeling and Analysis Subsystem

Observing systems are generally limited in the sense that they provide measurements in discrete places, as either points or swaths, or at a specific time (*e.g.*, a one-time ocean bottom survey). To enhance the PacIOOS observing network, a suite of numerical models are run. These not only provide estimates of ocean conditions over large, island-scale areas, but they can also produce forecast information. An additional benefit of the models is putting the point measurements into a wider geographic context.

PacIOOS currently runs three different types of models. The Weather Research and Forecast (WRF) model produces estimates of atmospheric variables such as wind speed and direction, precipitation, surface temperature, *etc*. A separate set of model integrations is run to produce forecasts of the ocean wave field (wave height, direction, period). PacIOOS employs WaveWatch-III for basin-scale forecasts and the Simulating Waves Near-shore (SWAN) model for coastal forecasts. Finally, the Regional Ocean Modeling System (ROMS) is used to produce estimates of the ocean state, including temperature, salinity and currents.

These three different models are run for several different geographic domains. In general, most have a relatively coarse grid over a wider domain, *e.g.*, a Pacific-wide domain at 100 km grid spacing, and then finer-scale resolution grids for island, or single coastline domains.

Data Management Subsystem

Both IOOS and PacIOOS aim to provide users with free, open access to data and information about the ocean environment to users. To realize this goal, PacIOOS has configured a variety of data servers that provide access to the observations and model output previously described. These data servers are all run on local, Linux-based machines on the University of Hawai'i campus and the data are stored on local RAID systems. Each serves a different purpose, such as accessing and providing geospatial data, or aggregating netCDF files, *etc*. Different web-based tools are also maintained in order to provide users with different options. Since all are accessing the same data, the maintenance of the servers becomes less of a challenge.

In addition to the observations and model data sets, PacIOOS also has a wide array of geospatial layers, including maps, features, aerial imagery, and so on. The data services at present are mainly divided along two lines that separate the point and gridded data sets from the geospatial layers.

The PacIOOS data management structure can be viewed as a series of layers (Figure 2). At one end are the data, either taken directly from instruments or model output, or saved as digital maps, images, *etc*. (Input Layer). Data from instruments are handled using an Open Source tool called DataTurbine [Fountain *et al.*, 2009] that allows users direct access, but also sends the data to a database.

For the most part, PacIOOS stores geospatial data in a PostgreSQL database, while fluid earth data are stored on disk in a flat file system (File System Layer).

Figure 2. PacIOOS data management structure. The PacIOOS data management system layers, from raw data (Input Layer) to end-users (Viewers & Clients) are shown as a schematic, with lines representing the flow of data from sensor to client application.

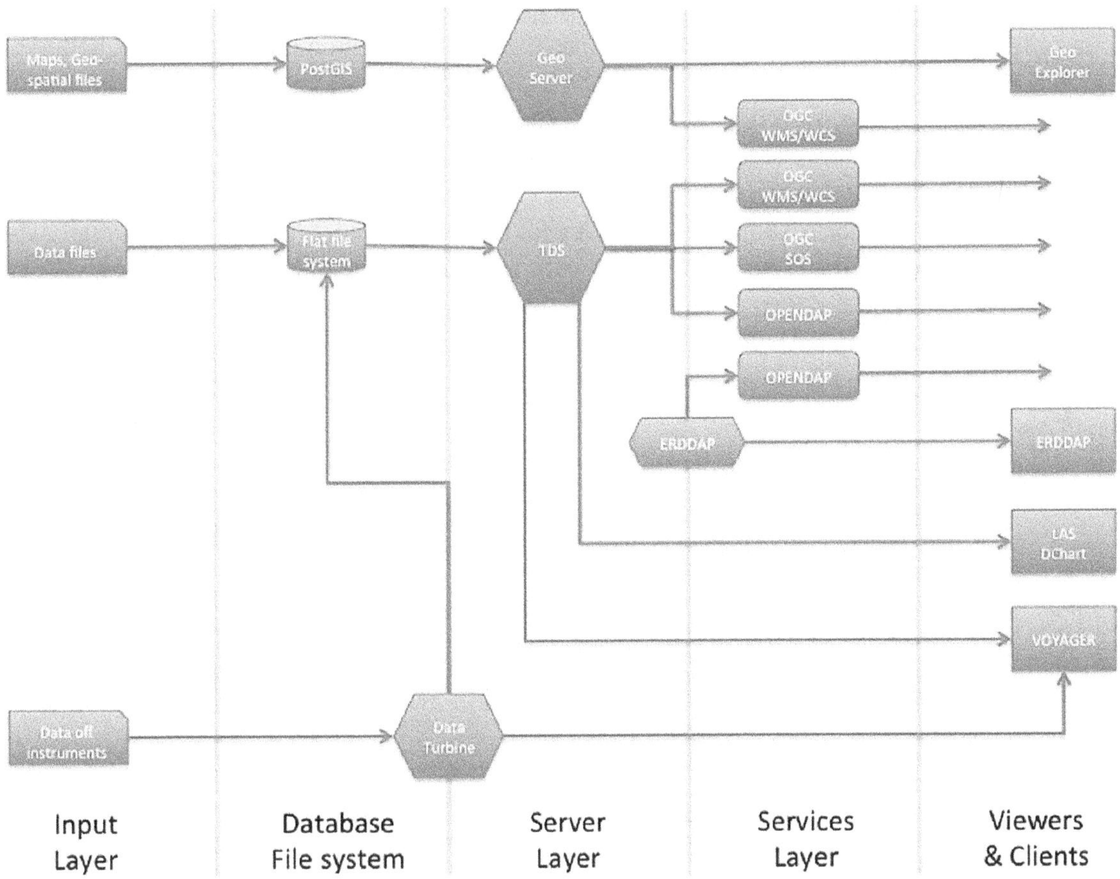

Central to the PacIOOS data services for *in situ* measurements and model output is a java-based utility called OPeNDAP (Open-source Project for a Network Data Access Protocol; Cornillon *et al.*, 2009). OPeNDAP allows users with appropriately enabled tools to directly access data from PacIOOS without having to transfer the data set from remote (PacIOOS) to local (user) machines. The specific type of OPeNDAP server that PacIOOS runs is the Thematic Real-time Environmental Distributed Data System (THREDDS[1]). The THREDDS server is referred to as TDS. TDS works best with netCDF[2] files as input, so most of the PacIOOS *in situ* data and model output are stored in this format. It should be noted that while TDS is the main OPeNDAP server at PacIOOS, other utilities provide OPeNDAP services as well, and these will be discussed later.

Geospatial data sets are managed and served using another java-based utility called GeoServer. GeoServer is an Open Source tool that allows for the management of geospatial layers in a database, and GeoServer publishes these data in Open Geospatial Consortium (OGC) standards such as Web Map Service (WMS), Web Coverage Service (WCS), and Web Feature Service (WFS).

While TDS and GeoServer provide the backend server layer that allows client software to access data, PacIOOS also maintains some web-based data browsing tools (Viewers Layer in Figure 2). These tools allow users to access and browse (including subset) data, and then make plots on-the-fly. These tools also provide downloading capabilities (save to local disk).

The Live Access Server (LAS) and DChart are two such example tools. These were both developed by different groups at the NOAA Pacific Marine Environmental Laboratory (PMEL). These two web-based tools read data from OPeNDAP servers and allow users to make lat/lon/depth/time "slices" or subsets through the data set and then save the results as either an image or file. Figure 3 shows an example LAS page.

The main entry page for LAS has an option (upper left) to first select the data set, and then select the variable of interest. A map is displayed to show the geographic extent of the data set, and pull-down menus provide a method for otherwise sub-setting the data (*e.g.*, by time or depth). These menus are automatically generated by LAS depending on the chosen data set. For example, the satellite data set shown in Figure 3 does not have a depth-dependency (surface values only), so no depth selector is shown.

Figure 3. PacIOOS Live Access Server (LAS). The LAS provides a web-based browse capability for PacIOOS data. In this example, sea surface temperature anomalies are shown. Controls for the plot are shown on the left and include variables, plot type, time/depth/lat/lon ranges, etc.

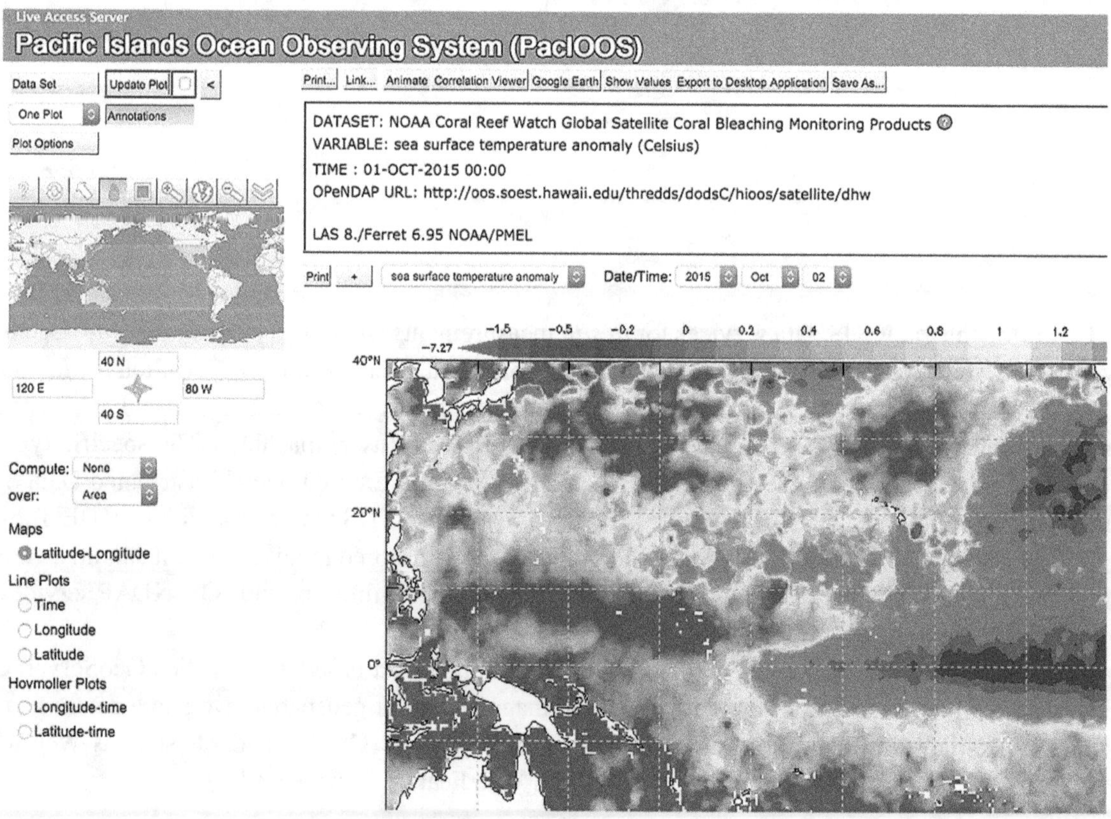

In addition to selecting the variable and ranges, LAS (and DChart, not shown) let the user select a plot type; the example shown is a lat/lon map, but the user could select a line or property plot instead.

Along the top of an LAS window are a series of buttons that let the user print, save, or otherwise export the data. DChart works in a very similar way, and was originally designed for point measurements (while LAS was developed for gridded data), but both LAS and DChart work on a variety of point and gridded data.

In addition to LAS and DChart, PacIOOS runs an instance of the Environmental Research Division's Data Access Program (ERDDAP). ERDDAP combines both the data transport of OPeNDAP and the web browsing functions of LAS. It has many additional features, such as the ability to save data, and user-defined subsets, in a large number of formats (including netCDF). Like TDS and GeoServer, PacIOOS relies on ERDDAP to provide data to some applications. Further benefits of ERDDAP are that it includes RESTful data access API and GeoJSON data formats that are extremely useful for many web applications.

PacIOOS also runs a geospatial browser called GeoExplorer. GeoExplorer accesses data from GeoServer (similar to how LAS and DChart access data from OPeNDAP). As a geospatial viewer, GeoExplorer's purpose is to display static GIS layers. Thus, GeoExplorer does not have all the functionality of LAS/DChart. An example application is provided in the next section.

At present, PacIOOS has several terabytes of data in its on-line storage system. The largest area of growth is the output from numerical models since new forecasts are produced each day. The data are cataloged with standard FGDC and/or ISO metadata and are fully discoverable via a Python Catalog Service for the Web (pycsw). Ultimately, all PacIOOS observations and the model output will be preserved at one of the NOAA National Centers for Environmental Information (NCEI). The process of long-term archive of all IOOS data is still ongoing, and at present, individual data providers are encouraged to submit their data. Eventually the RAs will be able to provide this service.

User and Applications Subsystem

PacIOOS has an administrative structure in place that allows for continuous feedback between end users and the PacIOOS staff. This helps to identify new data sets of interest and new ways to integrate these data into products. In turn, this ensures the long-term success of the program while helping PacIOOS develop an ever-growing list of "stake holders."

Through an iterative process of meeting with data providers, users, management agencies, *etc.*, the PacIOOS outreach staff helps provide input to the data management group. In the early days of PacIOOS, this engagement was more directed to promoting PacIOOS activities, or educating various groups about the capabilities of PacIOOS. As PacIOOS, specifically the PacIOOS data system, has matured over the past few years, PacIOOS has become better equipped to respond to a wider variety of requests. These requests include hosting and serving data for groups that either lack the capacity to do so or that are looking to reach a wider audience, as well as developing customized web-based tools to help with data interpretation and presentation.

The following sections describe the PacIOOS effort in more detail, with a focus on the unique aspect of the region, the data sets and data services used, and several example case studies that demonstrate how these capabilities can integrate and provide utility to end-users.

PACIOOS STAKEHOLDERS

The Pacific Islands are uniquely connected to the ocean environment. Typically small in terms of land area, this connection is obviously geographic but extends to cultural and economic aspects. Concerns about the near-shore environment range from recreational (*e.g.*, when will the surf be up), to economic (*e.g.*, where is the best place to fish), to safety (*e.g.*, search and rescue), and more. The information needed to address these concerns typically spans different domains, from near-real-time and short-term forecasts to long-term climate change impacts. Ironically, the Pacific Islands are generally considered to be under extreme threat from climate change mainly due to the combined effects of sea level rise and projected increases in tropical storm intensities, yet at the same time there are relatively few direct observations in the region. Additionally, many islands are either "sub-grid scale" in most coupled climate models, and thus do not appear as land masses, or they are inaccurately resolved vertically, and thus things like orographic effects are not well modeled. Therefore, it is critical to make best use of the data and information that are available to help Pacific Island communities deal with issues surrounding the coastal environment.

Many of the Pacific Islands can be categorized as low-lying atolls (*e.g.*, the Republic of the Marshall Islands) or "high-islands" with steep topography (*e.g.*, Hawai'i). In both cases, there is a strong connection between land-based processes and the coastal zone. For example, excessive rainfall over the land will lead to runoff and degradation of water quality in the near-shore ocean environment. In addition to this "geospatial" connection, there are also tensions between different sectors. With tourism driving much of the economy in places like Hawai'i, there needs to be a balance between marine ecosystems, sustainable fisheries, ocean safety, *etc.*, while providing access to ocean users at the same time.

PacIOOS is working with a number of groups in these regions. In order to facilitate this, PacIOOS has regional liaisons in each of the main islands. These liaisons live and work in the region and understand the local issues, cultures, and infrastructure. They work within PacIOOS to ensure that all the data services and tools are appropriate and helpful. This relationship also helps identify existing data sets in the regions. PacIOOS can then acquire these data sets and put them on the servers.

PACIOOS DATA FLOW AND SYSTEM ARCHITECTURE

The data stored and served by PacIOOS can be broadly classified into two categories: geo-referenced static maps (GIS data) and multi-variate measurements (fluid earth data). PacIOOS responds to data needs throughout the insular Pacific. For the most part, the data needs and requirements outside of the State of Hawai'i are non-real-time GIS layers, while those within the State are real-time (and forecast) fluid-earth data.

For example, in many Pacific Islands, managers need to know the locations of reefs and infrastructure elements. Maps of such assets address these needs. Conversely, many users in Hawai'i want real-time or forecast information about waves (*e.g.*, for surfing) or currents (*e.g.*, for recreation or search and rescue activities). Thus, the data types are different and the required data servers are different.

PacIOOS handles all vector geospatial (GIS) data via a PostgreSQL database (rasters and images are stored in a flat file system). The Open Source GeoServer is then used to query the database and provide data via an OGC standard. PacIOOS has configured on top of this a GeoExplorer instance to provide a web-based browser for these layers.

Data from observing platforms and models are stored using the netCDF data model, a platform-independent, self-describing format. Data from *in situ* sensors (*e.g.*, near-shore sensors and buoys) are retrieved from the sensor using DataTurbine (DT). Local scripts then retrieve the ASCII data from DT and create netCDF files on a regular (*e.g.*, daily) basis.

The data are saved in netCDF onto a local, flat file system. A catalog is then used to "expose" these data as time-aggregated datasets via the PacIOOS TDS. Users and/or client tools[3] can then access these data via an OPeNDAP call (access data via a URL). The TDS is also used to supply data to the web-browsing tools, LAS and DChart. Simultaneously the data are read and published by an ERDDAP server.

Users can then access PacIOOS data either from DataTurbine, typically using "wget" or screen-scrape, from an OPeNDAP call to TDS or ERDDAP, or from an OGC call (WMS/WCS/WFS). Alternately, users can browse and subset data with LAS, DChart, and/or ERDDAP and then save a custom data set to their local machine.

It should be noted that while PacIOOS typically separates geospatial data and model/observations with GeoServer and TDS, the TDS also provides data via OGC standards. Users can, therefore, access data from a numerical model served by TDS as a WMS. To help illustrate these servers, the following examples are provided.

Many software tools in the geosciences are "OPeNDAP enabled" and can thus access data directly from OPeNDAP servers. This is typically done by first accessing the OPeNDAP page for a specific data set, and then using the appropriate URL. In most cases, users enter the URL into the data selection tool or "load data" option. Most PacIOOS users that access data via OPeNDAP are research scientists, but PacIOOS takes advantage of OPeNDAP by using it in conjunction with other tools, such as LAS.

Similarly, while GIS data are served by GeoServer, they are more often accessed using the on-line tool GeoExplorer. An example case would be a coastal manager in Majuro in the Marshall Islands looking to produce a map of assets by making different overlays.

The Republic of the Marshall Islands (RMI) is a freely associated state of the United States. RMI is comprised of many coral atolls stretching from the dateline westward for about 15,000 kilometers, between 5° and 15°N. The main population center and capital city is Majuro with a population of about 25,000. An important economic sector is tourism, but the natural environment is important for many other socio-cultural reasons. In order to help preserve the environment, planners have designated certain regions as "conservation targets," or areas with current conditions, including ecosystem health, biodiversity, and population levels that are to be maintained. A coastal manager can quickly see where these conservation target areas are with respect to, for example, dive sites to match tourist areas with monitoring areas. Figure 4 shows such an example from the PacIOOS GeoExplorer, which renders images from GeoServer.

PACIOOS EXAMPLE PRODUCTS

The PacIOOS user-base is extremely varied. Accordingly, the data needs and requests are wide-ranging. Here we attempt to broadly categorize these requests based on a particular user group and then present example case studies of how the PacIOOS data servers and tools can be used.

Since IOOS is generally focused on providing information to users free of charge and as close to real-time as possible, there are potentially unlimited use cases. This also means that RA's like PacIOOS cannot easily track how and by whom data are being used. Nevertheless, PacIOOS can identify a variety

Figure 4. Example GeoExplorer display. The example shown is an aerial image with conservation target areas outlined in yellow and dive sites marked with flag icons.

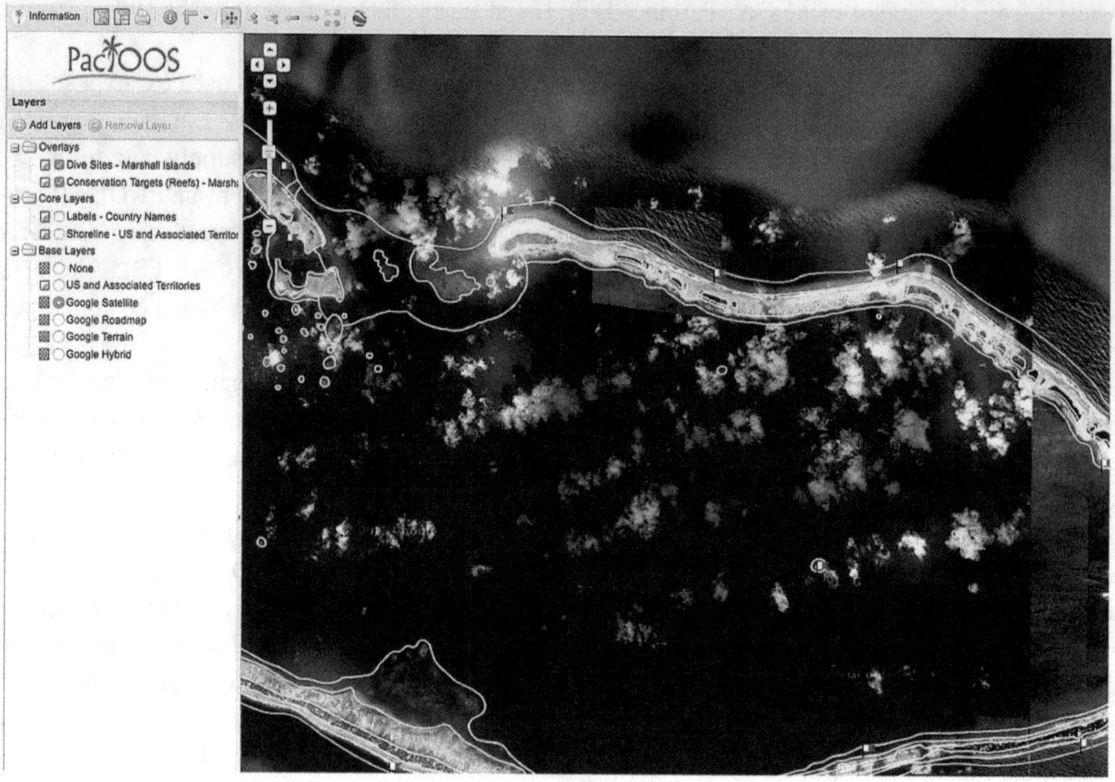

of users. These include research scientists who typically want direct access to the data; Federal, State and City and County agencies that need processed or interpreted data and products; individual ocean users (surfers, boaters, *etc.*) who view web-based maps or graphs; and students or educators who are looking for synopses or written descriptions. Some example cases follow.

Use Case 1: Research Scientist

Since PacIOOS is mostly managed within the University of Hawai'i, and since the modeling and observing system assets are configured, deployed, and maintained by University scientists, these individuals represent a large part of the PacIOOS user base.

Scientists typically want direct access to data in order to perform analyses or to run calculations. PacIOOS provides all data either through an OGC standard service (for maps and other GIS applications) or via OPeNDAP. The main OPeNDAP page is a TDS catalog of all the PacIOOS data sets presented as a list (see Figure 5). To get the particular URL for a data set, the user follows a series of links, first selecting the category (Wave Model Output in this case), then the file (Wave Watch III global). The final page gives a series of options for accessing this file, including OPeNDAP links. The page also supplies the metadata for the particular file.

Table 1. User access

User	Application	Data Access	Solution
research scientist	data analysis and plots	direct binary	OPeNDAP server to OPeNDAP enabled client tool
operational agency	forecast spill trajectories	direct binary	ERDDAP, data download, client tool
planning	shoreline change tool	N/A	web-based custom tool
local agency	brown water advisory	web-based tools	project page(s)
risk management	inundation forecast	N/A	automatic alerts via graphic display
general public	boating	web display	PacIOOS Voyager

Figure 5. THREDDS catalog layout. The THREDDS catalog contains a series of links to the PacIOOS data. The specific data page gives metadata and data service end-points.

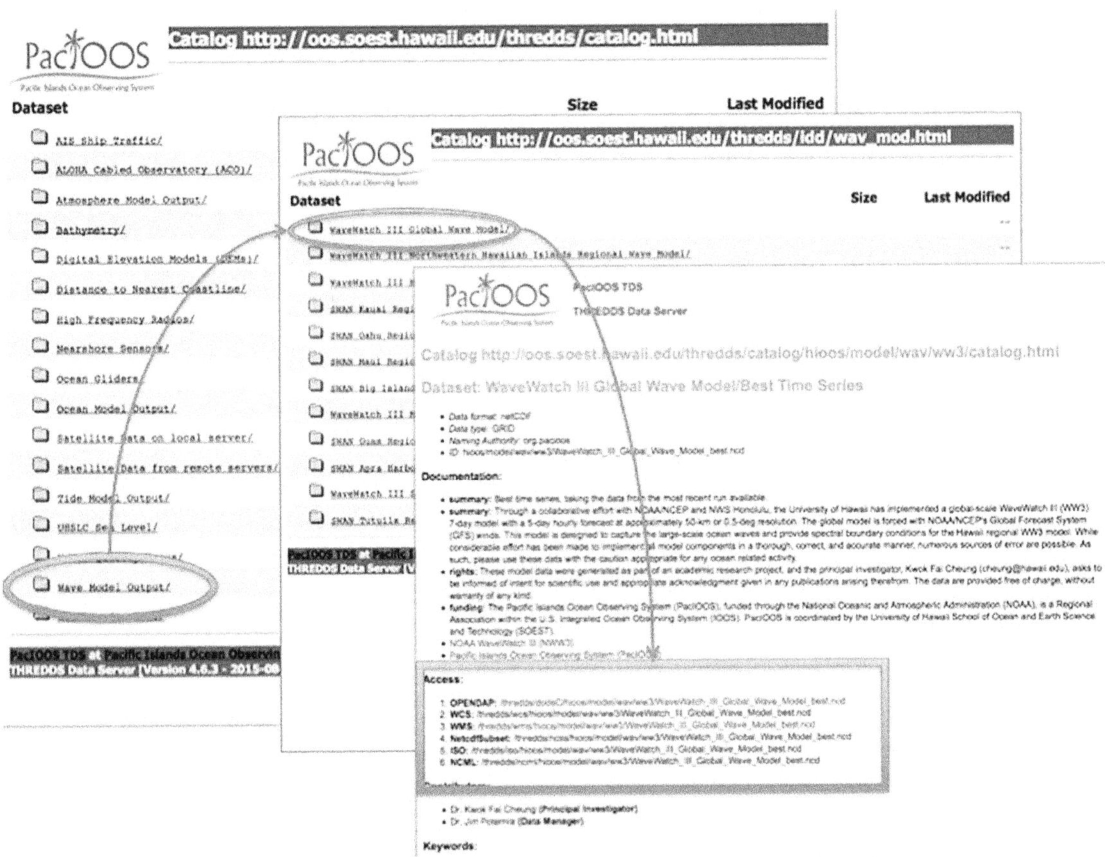

Figure 6. Example figure from OPeNDAP enabled client tool. The figure above was generated using the Matlab ® script given in the text which accessed the data via an OPeNDAP call.

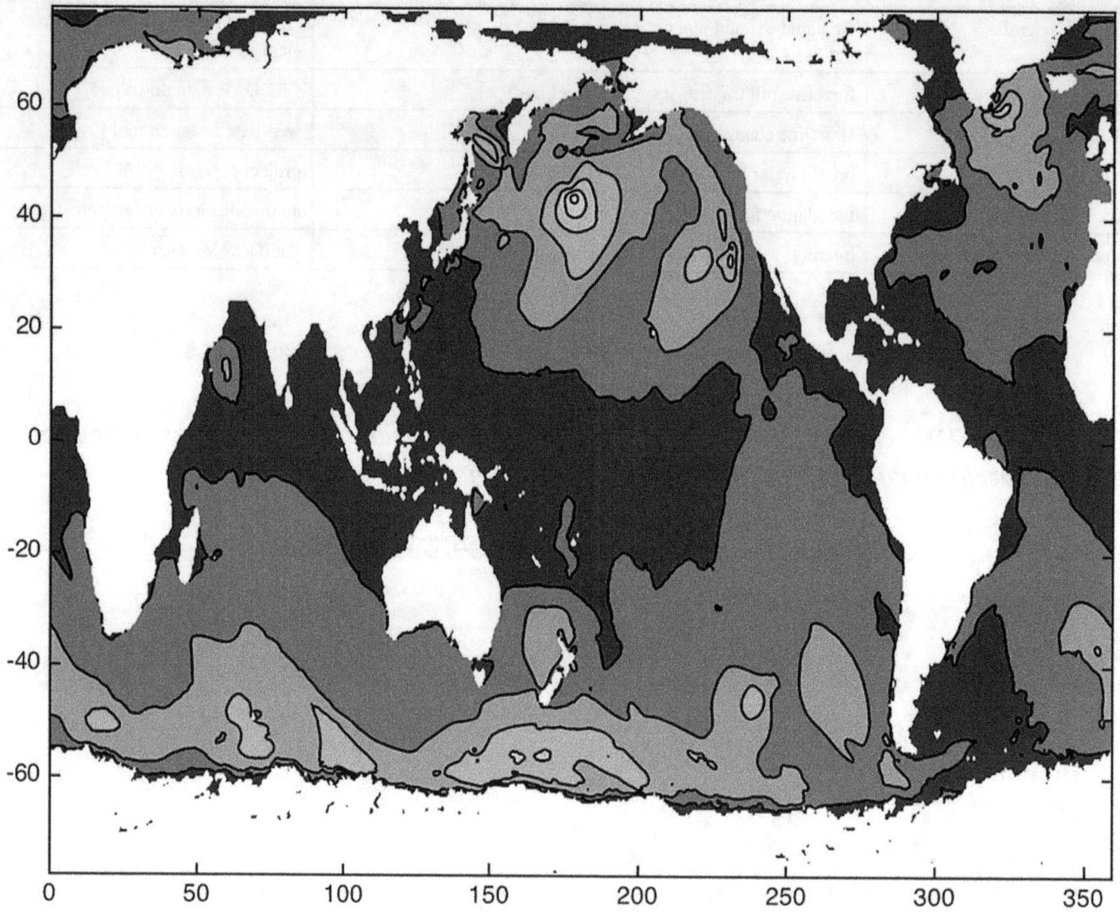

One example access via OPeNDAP is using Matlab®, and it serves as an example of an OPeNDAP-enabled client tool that is widely used in the geosciences. Instead of the traditional method of loading a data set from local disk, the user can address the URL to the file.

As an example, the following Matlab® code snippet will access surface wave height and display a color-shaded plot (Figure 6):

```
% Main PacIOOS TDS catalog:
top_url = 'http://oos.soest.hawaii.edu/thredds/dodsC/hioos/'

% URL for wave model output (obtained from PacIOOS TDS catalog):
data_url = 'model/wav/ww3/WaveWatch_III_Global_Wave_Model_best.ncd'

% load data:
lat = ncread([top_url,data_url],'lat');
lon = ncread([top_url,data_url],'lon');
```

```
H = ncread([top_url,data_url],'Thgt',[1 1 1 43328], [Inf Inf 1 1], [1 1 1 1]);

% make plot
contourf(lon,lat,H')
```

The first line specifies the upper-level TDS catalog URL, and the second is the specific path to the results from the WaveWatch-III ocean wave model. This is one of the models run every day at PacIOOS and provides a several-day forecast of wave heights (and other variables) for the Pacific. The next set of lines loads data from this data set via an OPeNDAP connection. First latitude data are loaded, then longitude, and finally wave height (Thgt). Note the final definition makes use of a subsetting feature whereby all values of latitude and longitude (1 to "infinity") are chosen at one depth level (1 to 1) and one time value (43328).

This is a simple example, but it highlights the utility of OPeNDAP access; a user can access a huge amount of data without having to transfer files from one machine to another.

In a similar way, geospatial data can be accessed from the PacIOOS servers. This is done, for example, with a WMS call. The same plot as above can be done with a GIS program such as QGIS (Figure 7). Here, the URL supplied by TDS is to a Web Map Service, http://oos.soest.hawaii.edu/thredds/wms/hioos/model/wav/ww3/WaveWatch_III_Global_Wave_Model_best.ncd?service=WMS&version=1.3.0&request=GetCapabilities, and this can be entered into QGIS via "add WMS layer."

Use Case 2: Operational Agency

One of the main users of PacIOOS real-time and forecast information are operational agencies within the Federal Government, including the National Weather Service (that provides coastal warnings), NOAA's Office of Response and Restoration (ORR, that works with spills in the ocean and coastal regions), the

Figure 7. Example figure from OGC enabled client tool. The figure above was generated using QGIS and access data from PacIOOS via a WMS call.

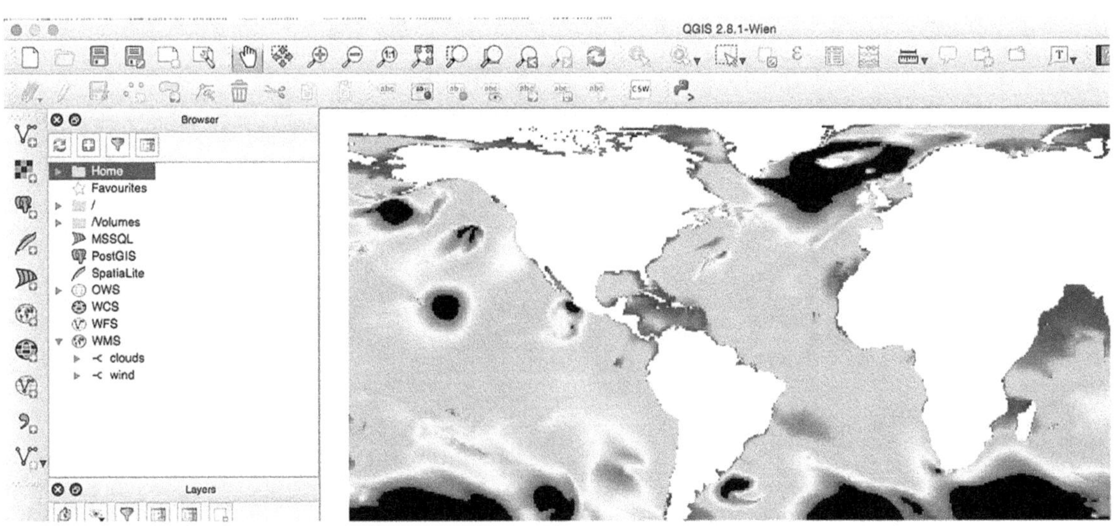

US Coast Guard (that conducts search and rescue missions) and so on. These agencies typically access data directly from the PacIOOS servers via OPeNDAP in an automatic way.

In the case of ORR, however, the data are downloaded using ERDDAP and then used in their own custom tool. In the case of an oil spill, ORR would be called upon to first provide an estimate of the spill dispersal pathway and areas of impact. To do this, ORR officials employ a desktop tool called the General NOAA Operational Modeling Environment (GNOME). GNOME allows the forecaster to select a region of interest, supply input parameters, such as spill amount, type and location and then run a simulation to show where the spill will disperse.

Two key parameters in the GNOME simulation are ocean currents and winds. Both act to disperse pollutants such as oil or marine debris. ORR can, therefore, use the PacIOOS oceanic and atmospheric model forecasts as input to GNOME. To do this, the ERDDAP server can be used.

Figure 8 shows the ERDDAP page following a selection of ocean currents for CNMI. ERDDAP allows for sub-setting by selecting ranges (either with sliders or text entry). In this example, a region surrounding Guam was chosen, so a vector plot of surface currents is displayed (the model has higher resolution than vectors displayed). In addition to geographic range, the user can select time and depth ranges. Once the desired limits are set, the user can get a quick-look plot ("Redraw the graph") and download the data. GNOME requires netCDF files as input, so the appropriate selection would be ".nc." This will then save the output to a file on the client's machine.

Figure 8. PacIOOS ERDDAP example. ERRDAP provides a way for users to display, subset, and save data via a web interface. The example above shows ocean currents from a forecast model around the island of Guam.

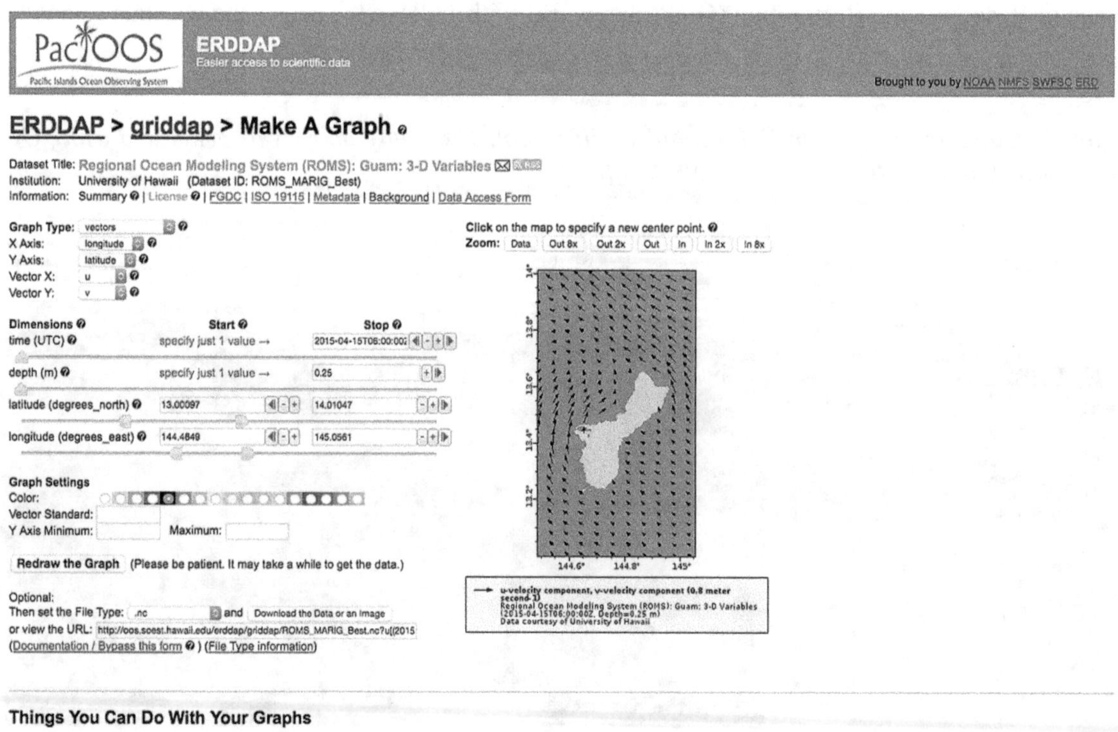

Once the ocean current data set is downloaded, a similar process can be done to get the wind field. Together, the ocean currents and winds are used as input to the GNOME simulation. Since PacIOOS supplies all model output via ERDDAP, it is relatively straightforward to access, subset, and download for uses such as this.

Use Case 3: Planning/Permitting Agency

PacIOOS helps various local, State, and county agencies with ocean-related inquiries and data applications. In some cases this goes beyond data services and involves the development of a custom tool. For example, a tool has been designed to help local, Hawai'i land managers and agencies properly plan for existing and projected shoreline changes.

PacIOOS worked with NOAA and the Hawai'i Department of Land and Natural Resources (DLNR) to develop a web-based tool to help with permitting properties and land setbacks. Much of the real-estate property in the State is close to the coastline, and different agencies in the State are responsible for assessing regions for new development and drafting proposals for developing existing properties.

The DLNR makes these assessments based on information from property maps (Tax Map Key, TMK), flood risk areas prepared by the Federal Emergency Management Agency (FEMA), and coastal inundation maps from NOAA. Each of these data sets is in a different format and in a different repository. In some cases these data sets are coarse and/or sparse for particular regions.

The customer, in this case the DLNR, wanted to combine the property, flood risk, and coastal inundation maps into an easy-to-use portal in which anyone could search for land parcels, either through key-entry or map view, and overlay different features. One feature of interest was the expected shoreline change projected over the next several years. This is an active area of research in the Coastal Geology Group (CGG) of the Geology and Geophysics Department at UH. In addition to studying shoreline erosion forecasts, the CGG has made *in situ* transects of the shore to get more precise measurements of the existing slopes perpendicular to the coast.

The PacIOOS task was then to integrate the various data sets provided by the State and Federal governments, along with the research products being developed by the CGG. The tool needed to be web-based and include a variety of features, including interactivity, documentation, and reporting (users needed to be able to create a map, for example, and then create a report automatically based on the selected features and domain). From the development side, the tool needed to be fast, reliable and somewhat intuitive.

The solution was to build a lightweight map utility based on the Leaflet JavaScript library. The tool (http://pacioos.org/shorelinechangetool) reads data and layers from the PacIOOS GeoServer and renders maps on the page. The page uses a map as the main feature, which has buttons for typical controls, such as pan and zoom. Other controls include searching by address, TMK or transect number along with a variety of overlays.

Figure 9 shows an example session. The main part of the page shows an area of the island of O'ahu in Hawai'i. The red lines and numbers are the parcel boundaries and TMK's. The yellow lines and numbers are transects along which shorelines were measured. A user can select a specific parcel (indicated by the large red dot) and transect (indicated by the large yellow dot). The tool then displays the corresponding flood assessment (so-called "DFIRM," or digital flood insurance rate map; in this case the value of "VE" indicates "subject to inundation by the 1-percent-annual-chance flood event with additional hazards due to storm-induced velocity wave action"). Additional information includes the measured shoreline change (-0.1 feet per year in this case) and erosion hazard.

Figure 9. The PacIOOS Shoreline Change Tool. This tool, developed by PacIOOS, allows users to generate customized maps and make estimates of setbacks based on sea level changes.

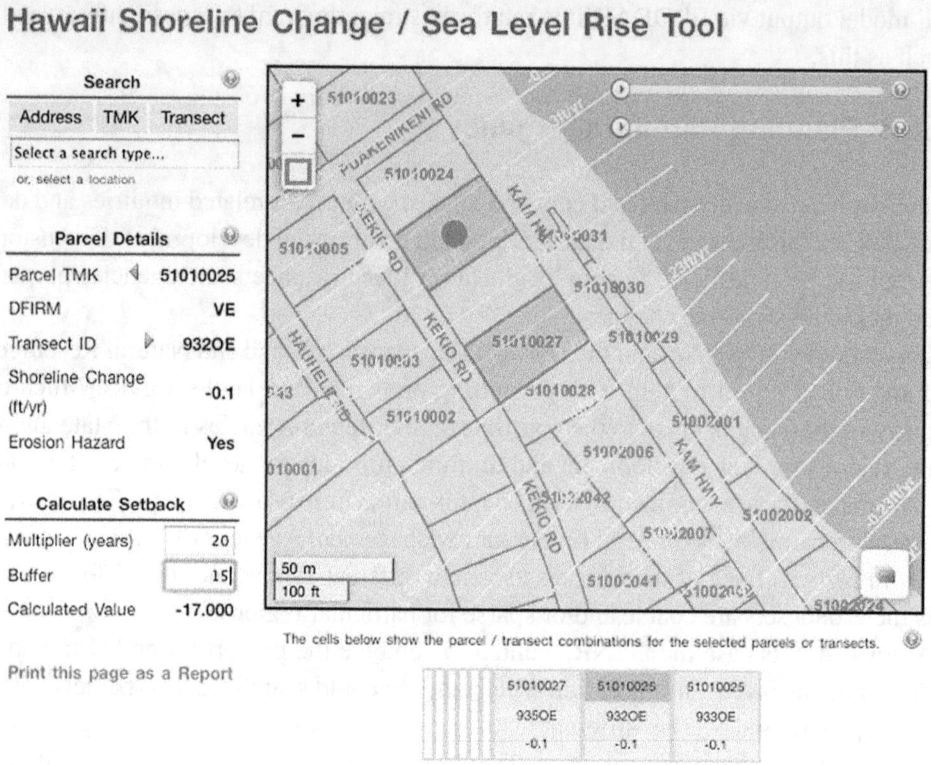

Finally, a user can "forecast" or estimate the appropriate property setback based on an input number of years and buffer zone. In this example, a local official may be interested in planning over the next 20 years and wants to include a mandated 15-foot buffer zone. The input is, therefore, 20 for "multiplier (years) and 15 for "buffer." The result, computed and displayed on-the-fly, is a 17 foot setback.

All of these results can then be saved in a document (Portable Document Framework, or PDF) by selecting "Print this page as a report" in the lower left.

The tool also provides a display based on forecast changes to sea level. The two sliders at the top right of the map allow users to select different years. The upper slider provides the confidence map and the lower slider provides the sea level rise inundation line based on projected sea level of one to six feet. Figure 10 shows the resulting inundation estimated from a four-foot sea level rise. The green areas are dry land, while the orange and blue areas are inundated regions, with blue above 80% confidence.

The tool itself contains many different features and overlays and can become quite complex. To aid users, metadata and information are provided in various places. The small, blue icon with a question mark near many of the sections on the page (*e.g.*, next to the "Parcel Details") will provide an explanation of that particular feature via a popup window when it is clicked. In addition, the small icon in the lower right corner (green folder icon) provides information about all of the layers, legends, and metadata (see Figure 11).

Figure 10. Example output from Shoreline Change Tool. The map shows estimates of inundation based on a four-foot rise in sea level. The blue colors have greater than 80% chance of becoming inundated, while orange colors show less confidence.

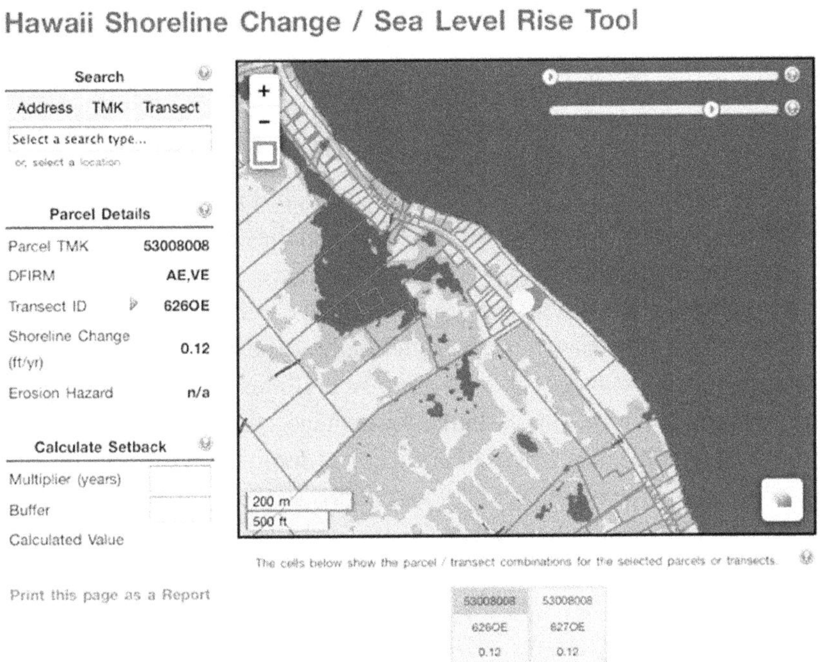

Figure 11. Example metadata from Shoreline Change Tool. This page, accessed via the green folder icon on the shoreline change tool, provides an explanation of colors and layers in the plot.

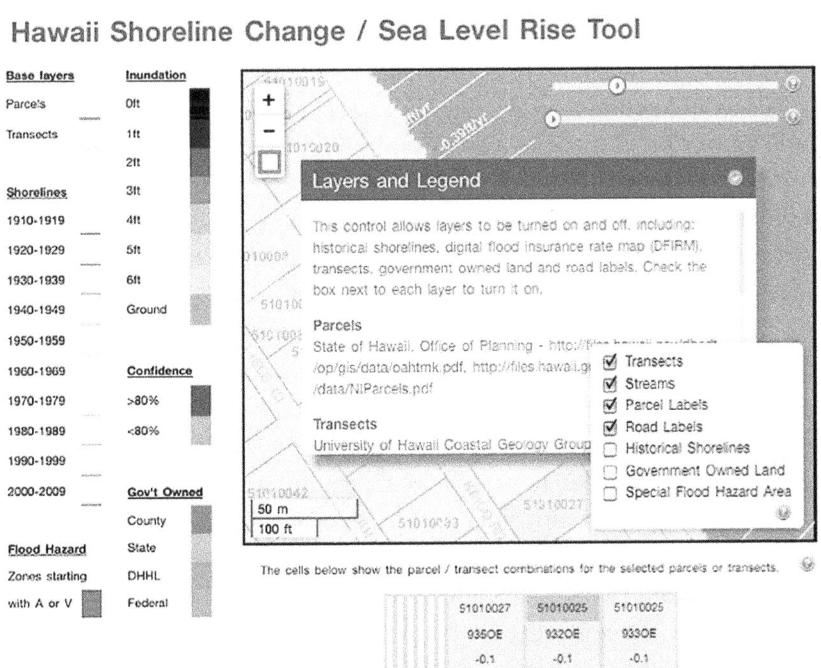

Use Case 4: Local Planning/Advisory

Turbidity is a measure of water clarity and is defined by the number of tiny solids suspended in the water column. Near-shore regions are particularly susceptible to changes in turbidity, as significant storm events create terrestrial runoff that can drastically increase turbidity levels. Runoff that contains pollutants and contaminants is referred to as "brown water" and can adversely affect citizen health. Brown water contaminants include, but are not limited to: sewage, harmful microorganisms, and chemicals from residential, commercial, and recreational sources.

Many near-shore regions that are susceptible to instances of brown water runoff may also be areas that are used extensively by the public. The Ala Wai canal on the southern shore of O'ahu, Hawai'i, for example, is often used for rowing crews to practice and for recreational kayakers to explore. While the canal itself is extensively used, the water flow from the canal enters the waters surrounding the popular tourist destination, Waikīkī. After heavy rainfall events, unaware beach-goers and swimmers may become sick after inadvertently coming into contact with harmful pollutants and contaminants originating from the Ala Wai canal. With this in mind, it is essential for managers and communication officers to be able to react to predictions of possible heavy rainfall events, in order to prepare for instances in which brown water may occur. Within the process of preparation, the officers must also be able to inform the public of the potential safety hazards to which they may be exposed. The PacIOOS Water Quality Focus Area provides a means to accomplish this goal, as illustrated by the following example.

The National Weather Station (NWS) has reported flash floods and heavy rainfall in Honolulu and a Public Information Officer at the Clean Water Branch (CWB) must be prepared to issue a "Brown Water Advisory" if the flow of runoff into Honolulu beaches becomes unsafe for recreationalists.

The PacIOOS Ala Wai Turbidity Plume model is updated daily and provides hourly estimates of turbidity levels in southern O'ahu, spanning from the northern tip of Sand Island to the southern end of Waikīkī. The model displays plume position and turbidity values that are determined by using the results from the Regional Ocean Modeling System (ROMS) ocean circulation model, which has been configured for O'ahu's south shore (Figure 12). The model simulates the plume advection and dispersion by using near real-time river runoff and turbidity measurements for the Ala Wai canal, Simulating Waves Nearshore (SWAN) wave model forcings, and Weather Research and Forecasting (WRF) model forcings. By checking the hourly animation of the Ala Wai Turbidity Plume Model during the heavy rainfall event, the CWB officer can visualize the turbidity predictions and determine whether the turbidity levels in the canal exceed the threshold that designate brown water events.

Once turbidity levels breach the safety threshold within the canal, the officer's concern shifts to the near-shore regions by the canal that are often used by the public. PacIOOS has a water quality platform composed of near-shore sensors (NSS) and water quality buoys (WQB) that provide real-time data on oceanographic conditions including salinity, temperature, chlorophyll, turbidity, and depth [Tomlinson et al., 2011]. One sensor (NS02) is located directly at the mouth of the Ala Wai canal. Since the data are real-time, the officer will be able to constantly monitor the sensor and determine whether unsafe turbidity levels are beginning to spread to near-shore recreational regions. If turbidity levels continue to rise, it may be a cause for public concern and an advisory may be required.

If high turbidity levels are recorded at a particular near-shore sensor, the officer may wait to release a warning until he is sure that the conditions will continue to spread. In this case, the officer will revert his attention to the PacIOOS Water Quality Buoy (WQBAW) located further offshore from NS02. This buoy also provides real-time data similar to the near-shore sensors (salinity, temperature, chlorophyll,

Figure 12. PacIOOS dispersion model. PacIOOS runs daily simulations initialized by in situ measurements. The model provides a forecast of the distribution of turbid waters from a coastal canal (Ala Wai) into the ocean.

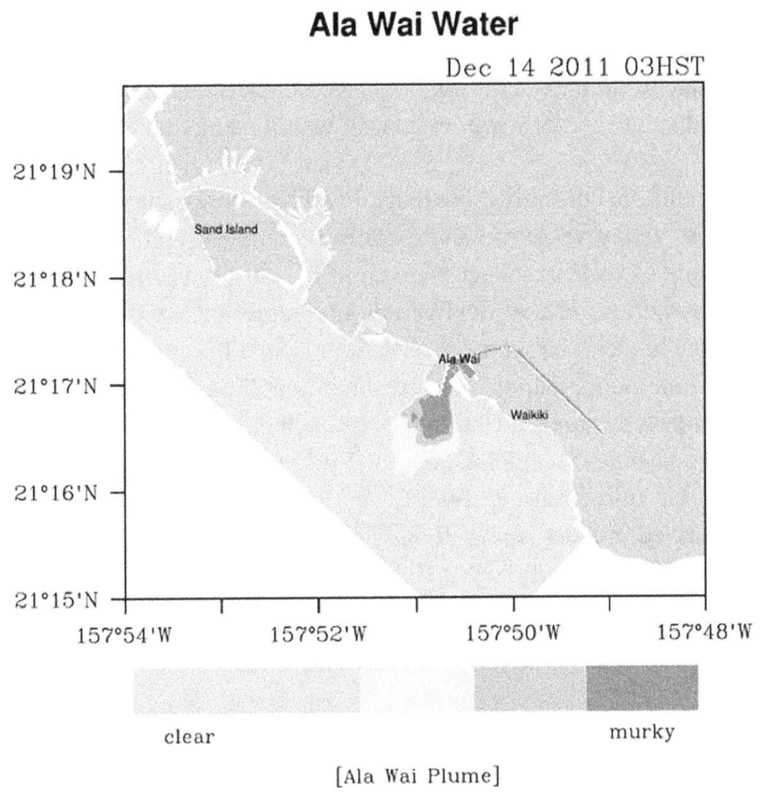

turbidity), in addition to dissolved oxygen levels. With these data, the officer can continue tracking the diffusion of high turbidity levels and determine the degree of concern for near-shore water quality conditions following the heavy rainfall and runoff event. If the turbidity levels remain high and continue to expand, the officer may choose to issue a CWB warning. Warnings and advisories are easily accessed by members of the public on the PacIOOS website, through the Water Quality Focus Area.

In this instance, the CWB officer was able to take PacIOOS water quality data from a variety of models, sensors, and platforms, and use these data to determine the appropriate protocol required to keep beach-goers and ocean recreationalists as safe as possible. Accurate, real-time data are essential for effective management; these data allow officers to post advisories when they are relevant and to abstain from posting advisories when oceanographic conditions pose a negligible threat to ocean users. The PacIOOS Water Quality Focus Area provides these essential data through an expansive array of sensors and models that are easy to locate, view, and interpret. Therefore, while this Focus Area may be particularly useful to officers, it is also accessible to any who are interested in water quality conditions.

Use Case 5: Risk Management

While PacIOOS is not an advisory agency, it does provide data products that can be used to help in forecasting events. As an example, extreme sea level events in the Marshall Islands can happen throughout the year. These may have catastrophic impacts, not only to property, but to human safety as well. The RMI consists of low-lying atolls, so any inundation event can have severe consequences. With proper warnings, however, people can evacuate the very near-shore regions and take other precautions.

Researchers at the University of Hawai'i were able to determine that these extreme sea level events were due to a combination of factors. First, particular large-scale ocean circulation patterns would elevate the background state. Extreme tides (king tides) would also contribute to extreme sea level events. Finally, storm-generated surge, and/or waves and swell would add to the elevated background state and create inundation.

All of these components are currently modeled at PacIOOS. Since the information comes from both observations (tide gauges and wave buoys) and models, both a near-real-time and forecast product can be generated. An example of such a product is given in Figure 13. The plot shows height above mean sea level over a nine-day period. The vertical yellow line represents "today." The curves to the left of this line show a three-day hindcast based on observations, while the curves to the right of the line show a six-day forecast based on model output. The red shading at the top indicates the extreme inundation level; when the curves reach this area, a flooding event is likely to occur. The different curves are representative of different coastline orientations, *e.g.*, the magenta line is for a coastline that runs at 110° from north (a coastline that runs almost east-west).

This type of data-derived product is posted daily on the PacIOOS web site. Similar plots are made for locations in Hawai'i (Waikīkī and the North Shore) and the RMI (Majuro and Kwajalein). Since the data are available in a data service, once the algorithm is derived for a specific site, the plots can be completely automated. This provides a visual tool that is easy to understand, in hopes that it will help reduce the risk for these events.

Figure 13. PacIOOS inundation forecast. These figures are generated daily ("today" indicated with the yellow line), and show a three-day hindcast based on observations and a six-day forecast based on models (see text for explanation).

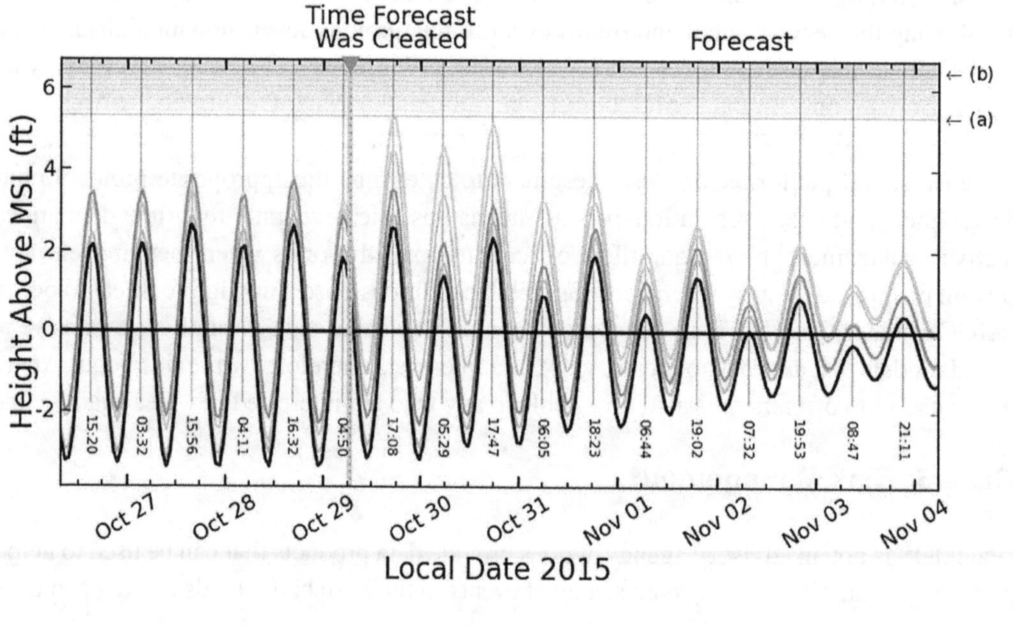

The example plot shows that the sea level along a coastline oriented at 10° from north (the cyan line) will approach a previous extreme event (red shading indicated with (a)) on the 29th of October.

PacIOOS will continue to develop these products for individual locations in the region. Additional data and finer resolution models will allow for more site-specific analyses, as well as information within harbors and other semi-enclosed basins.

Use Case 6: General Public

Perhaps the most visible and highly-used PacIOOS tool is the web-based "Voyager" data viewer. Voyager has an interactive map interface that can be used for visualizing and downloading a plethora of different datasets. It is publicly accessible in any Internet browser and can even be used on mobile devices. While many sites provide links for researchers to download raw data sets, PacIOOS Voyager allows users of all discourses to view and interpret different data sets that span across the Pacific and across different oceanographic and biological facets.

Since the expansive array of data that are available on PacIOOS Voyager incorporates many different marine, terrestrial, and atmospheric topics, the data sets have been grouped into eight categories: hazards, observations, remote sensing, forecasts, biology, navigation, seafloor, and land. Instead of providing available data sets in lists according to contributors and sources, PacIOOS Voyager gathers data in a way that can be easily manipulated by the general public, by researchers, and by managers alike. These main data headings appear as a collapsed list on the left side of the Voyager main page (see Figure 14).

Figure 14. PacIOOS Voyager. Voyager, the main data viewer for PacIOOS, allows users to view and interact with data (list in the left panel).

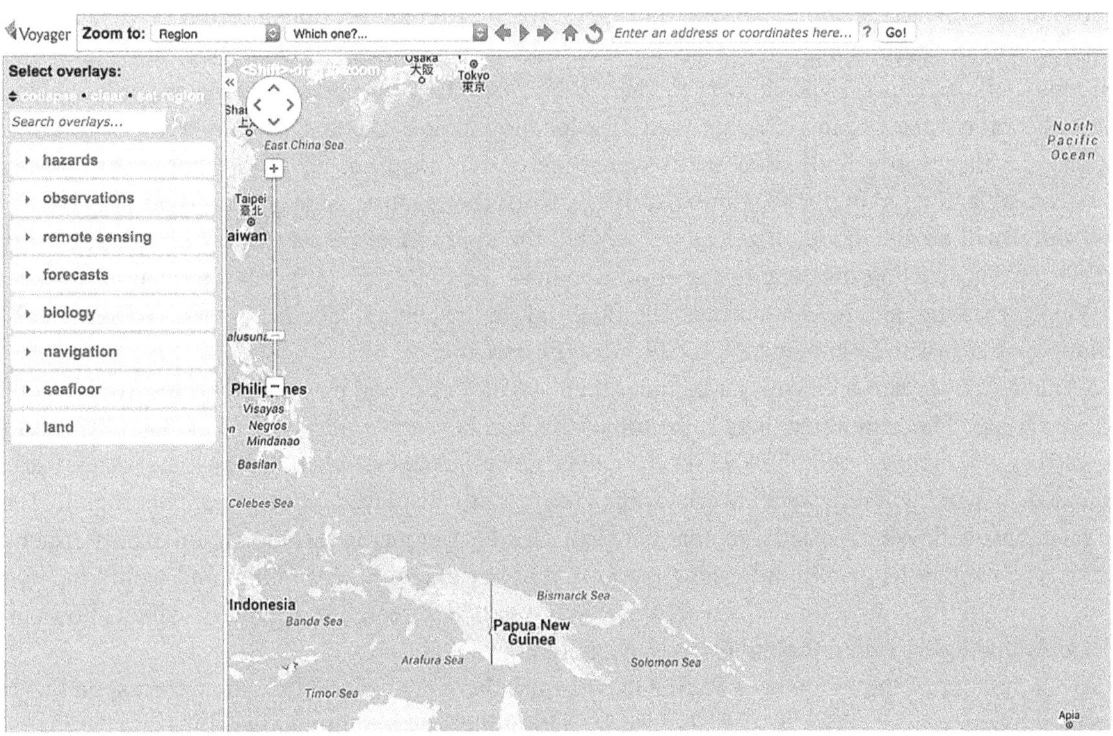

There are three subcategories within the "hazards" macro-category, including observations (*e.g.*, air quality, earthquakes, *etc.*); forecasts (*e.g.*, harbor surge, hurricanes and tropical cyclones, and wave run-up); and basemaps (*e.g.*, emergency shelters, flood hazard zones, *etc.*).

Within the "observation" macro-category, there are several subcategories that feature datasets regarding waves, tide and water levels, water quality, water resources, ocean acidification, ocean gliders, and ocean observatories.

The third macro-category available to PacIOOS Voyager users, "remote sensing," includes data regarding Doppler radars, surface currents, and satellite data. The "forecasts" macro-category contains a wide array of data relating to weather, surf, tide, ocean, and sea level forecasts, in addition to sea level rise data and forecast discussions.

The "biology" macro-category contains two subcategories: "fish and wildlife" and "protected areas."

The sixth macro-category available to PacIOOS Voyager users is "navigation" and provides NOAA nautical charts, maritime boundaries, and geographic boundaries. The final macro-category available to users is "land," which includes elevation hillshades and contours that have been derived from USGS land surface Digital Elevation Models. Data are available on a global scale or on more regional scales for islands throughout the Pacific.

While the amount and diversity of data available on PacIOOS Voyager is substantial, much of the usability of this web-based tool comes from its unique features. The initial startup page presents a popup that describes PacIOOS Voyager, and provides information on the tool itself, as well as tips to using the tool, PacIOOS regions, credit allocation, information on the mobile version of Voyager, Voyager News, and contact information.

Beginning at the top left of the page (see Figure 14), one can select a specific region, city, or island in which to zoom. Users can also jump to different locations within the zoom list, cycle through each location in the zoom list, return to a designated home region's original zoom, or return to the last zoomed-to region. In addition to the ability to move quickly between different locations, PacIOOS Voyager allows users to save their current map view. Clicking the "link" icon on the top right provides users with a short and long URL and an HTML element to share their map. This link preserves the zoom location, the selected overlays, the basemap (background), the location of open popups, and any other unique styling options. In other words, it allows users to share their current map *exactly* as they see it. Alternatively, users can generate a map image with most, if not all of the features displayed on their map. Clicking this button will create and automatically download the appropriate portable network graphics (PNG) image file. Those using this feature can choose to have the image contain only the map, or to include the Voyager sidebar and header as well. The final link in the header, labeled "about," reintroduces the initial popup that is present at the PacIOOS Voyager start-up.

The left side of PacIOOS Voyager includes the list of overlays and macro-categories from which users can choose. There are three links at the top of this bar, labeled "collapse," "clear," and "set region," respectively. "Collapse" will close all macro-categories in instances when one or many categories are expanded. "Clear" will remove all overlays that are currently displayed on the map. "Set region" filters the available overlays by a specific region. For example, if a user were to set the region to only American Samoa, the overlay list would only show layers that are available in that region and would hide other layers. Beneath these three links is a search tool with predictive search suggestions, which allows users to quickly and easily find layers in which they are interested.

At the bottom of the left side of PacIOOS Voyager, there are several links that correspond to map visualization parameters. The "lat/lon" link shows latitudinal and longitudinal gridlines on the map. If a user clicks the "range" link, then clicks a specific point on the map, he or she will see concentric range rings that correspond to distance from the initial point. Alternatively, a user can determine the distance between two different points by clicking the "distance" link, then by clicking two regions on the map. This will create a ruler that shows the distance between the two points. The points can be clicked for coordinate information or dragged to new locations. If dragged, the distance ruler will adjust accordingly. "Area" can be used similarly, in which users can click many points on the map and get a value for the selected area. Double clicking on one point while selecting the area will stop the selection tool from creating new points. These points can also be moved and the area value will adjust automatically. Clicking the respective links a second time will remove these features. To remove more than one of these features at once, users can click the "clear all" link directly below the feature links. Users can also change the units for which these features present their data, by clicking the "change units" link. Finally, by clicking the "show map dimensions" users can view the coordinates for the center of the map, as well as its bounds, zoom level, spanning distance, image size, and map projection.

Since PacIOOS Voyager is served on a Google Maps platform, there are many ways for users to create unique, stylized basemaps on which to display their data. There are three boxes on the top right of the map itself: "More…," "Map," and "Satellite." By clicking "More…," users can add various terrestrial components to their map, including Day/Night differences, traffic levels, transit routes, and cycling routes. Below these options, there are many different map styles from which users can choose. These include bright colored maps, grayscale maps, black and white maps, monochrome maps, and topographical maps, to name a few. Users can also choose basemaps with or without labels, depending on their preference. The "Map" box switches the basemap to a default map style in which the water is a light blue and land masses are a light gray and green color. Users can also select "Terrain" within this box's drop down menu to see a more three-dimensional version of the map that includes major mountain ranges and rivers. Clicking "Satellite" will automatically switch the basemap to one derived from satellite imagery. Once again, users can choose to display or hide labels. In addition to the features that allow users to zoom to distinct locations, the map also has panning and zooming options on the top right. To quickly zoom to a rectangular region, users can simultaneously hold down the 'Shift' key while clicking and dragging their mouse to enclose the desired area.

Most of the layers that are available on Voyager are also clickable. That is, when viewed on the map, a user simply needs to click on a desired location and a popup will appear. This popup will contain information specific to the location and the active layer. Each popup will display latitudinal and longitudinal coordinates for the selected location, as well as the date and time of the data displayed on the map. The coordinates and dates can be changed to different formats, like decimal degrees, or time zones, like Coordinated Universal Time. Variables that are unique to the displayed layer can also be reported in a variety of different units. In addition, popups provide users with a way to download the raw data that they are viewing. For temporal data sets, users can select to download data from the last 24 hours, from the last three days, from the last week, from the last month, or they may choose to view the archived data.

When navigating through the layers on PacIOOS Voyager, users may find themselves with questions pertaining to a specific data set or to an entire data category. Each layer name and subcategory name is clickable within PacIOOS Voyager. Clicking a subcategory name will provide a general synopsis of the data that are in this particular group. Clicking a layer name will provide a more cohesive summary of the data, while also providing links to the data sources and external links to sites where users can learn more.

As mentioned earlier, PacIOOS Voyager is also available on mobile devices. This allows users to access the real-time and near-real-time data that are available on PacIOOS Voyager anywhere with a cellular connection. This is particularly useful for users who are constantly on the go and make plans according to specific oceanographic or atmospheric parameters.

Although PacIOOS Voyager is a user-friendly data visualization tool with many data sets and useful features, its functionality is rather complex. As mentioned previously, PacIOOS Voyager's foundation is that of the Google Maps API, but it also includes many third-party software packages and data servers. Most layers are pulled directly from the PacIOOS THREDDS, ERDDAP, and GeoServer data servers.

Some other datasets have unique third-party software packages that allow their visualization in PacIOOS Voyager. This includes ship routes and locations, which are possible through the use of AIS message parsing of data acquired by the University of California San Diego (UCSD) and the Southern California Coastal Ocean Observing System (SCCOOS).

Data allocation is not the only component of PacIOOS Voyager that uses an array of different software to run smoothly; the overall appearance of PacIOOS Voyager is also the product of many different technologies and scripts running simultaneously. The aesthetically pleasing animations and popups that PacIOOS Voyager uses are possible through the use of jQuery and jQuery UI. The charts within these popups, like those for PacIOOS wave buoys, are the result of incorporating a JavaScript plotting library for jQuery called Flot. CasperJS makes it possible for users to save the image of their current screen, while PhoneGap gives PacIOOS Voyager the capability to be viewed on any mobile smart phone.

In summary, PacIOOS Voyager is far from a simple, easily reproducible tool. It takes a wide variety of data, technology, and software packages to create a truly unique platform for which members of the general public and the scientific community can view and interpret oceanographic and atmospheric data. The complexity in the underlying mechanism of creating this tool is what makes it so user-friendly, customizable, and valuable. To put this into a user context, the following is one example "session" using Voyager.

One of the many popular ocean activities off Waikīkī Beach is off-shore sailboat racing. These events range from officially sponsored and monitored events to the more casual "Friday Night Regattas." Typically, the sailors want information about the course, forecast winds, currents, and waves. Voyager allows users to aggregate these data into a single map. Figure 15 shows such an example.

The map shows the south shore of O'ahu ("Zoom to: Island; O'ahu"). Navigational buoys, marking channels, are displayed as green and red icons (selected from the list on the left). An example course is drawn, showing distances, by using the "distance" tool on the lower left. The upper panel shows the ocean currents with vectors, color-shaded by speed, while the lower panel shows the winds in similar fashion. Both currents and winds can be displayed at different times; each overlay selection shown on the left has a calendar and time selector.

With very few keystrokes, a user can thereby get a quick indication of current conditions (or forecast). In doing so, they are actually accessing the PacIOOS data archive via a data service. Voyager brings together many of the aspects of the PacIOOS data system, and it provides an easy-to-use, web-based interface to the large PacIOOS data holdings.

CONCLUSION AND FUTURE DIRECTIONS

The Pacific Islands Ocean Observing System (PacIOOS) has developed a robust data management system that allows users to access data and data-derived products in a straightforward way. The core system is built around two main services: OPeNDAP for point, swath, and gridded data and model output; and

Figure 15. Example overlay maps from PacIOOS Voyager. The upper panel shows surface currents from HFR; the lower panel shows forecast winds (both as vector overlays). Additionally, channel markers and distance lines are shown.

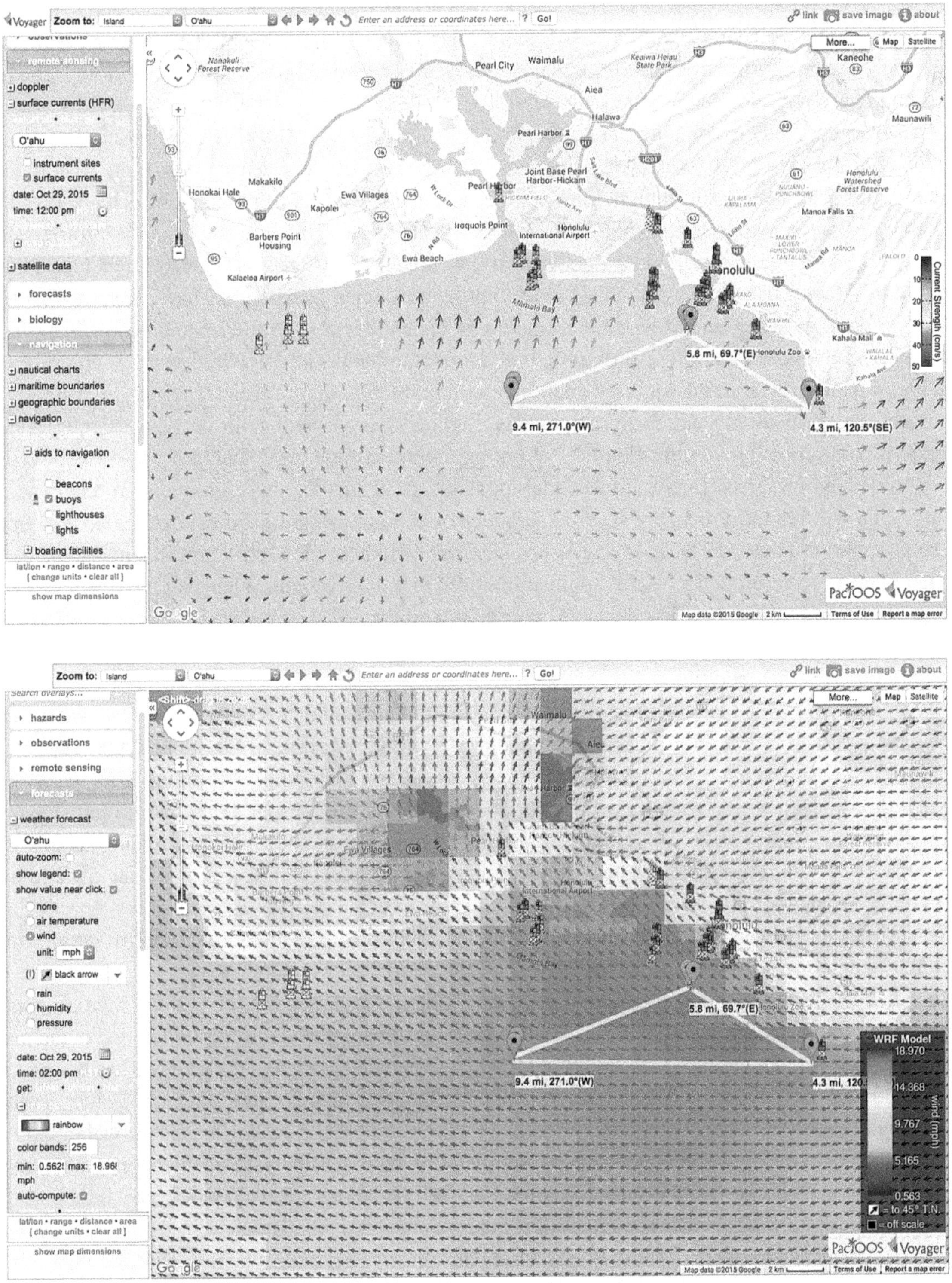

GeoServer for geospatial (time-invariate) data. On top of these two servers are a variety of web-based browsing tools (LAS, ERRDAP, GeoExplorer) that allow users to interact with data and produce custom plots and/or download subsets of the data.

The PacIOOS data system relies on standardization in both the background data (*e.g.*, netCDF, shapefile, *etc.*), and the data services (OPeNDAP and OGC). This has many benefits, including making the PacIOOS data archive interoperable with the national IOOS network, but also allows for straightforward development of data-derived products and tools.

PacIOOS began by developing the servers and collecting data. As the program continues to evolve, the focus will likely shift to product development. With mature data services in place, more custom tools and applications can be built. Integration with the National network will remain a priority, and as new technologies and services are promoted, PacIOOS will integrate them into the existing infrastructure. However, custom products will likely derive from stakeholder input. These products can include a range of activities, from customized web-based tools to project-oriented web pages. These, along with all the PacIOOS data, will continue to be made available on the PacIOOS site (http://pacioos.org).

These products are identified through close interactions with the PacIOOS user community that includes various government, industry and university partners. The governance structure of PacIOOS includes management and outreach components, regional representatives in outlying islands, and an oversight committee (Governing Council). Working together, these groups identify possible new data sets and desired products. Important considerations when identifying these products are the expected utility (*e.g.*, will it appeal to a very specific audience or be more general) and the desired access. In the case of the latter, PacIOOS faces severe issues with bandwidth and infrastructure in some areas of the insular Pacific, so developed tools and applications need to be lightweight. Whenever possible, therefore, server-side processing is used. For example, any derived products that involve calculations or geographic sub-setting are done by the PacIOOS servers and then supply the result, rather than have the client's machine perform the operation.

The success to which PacIOOS has developed a fully integrated data system that allows users to access data in a seamless fashion can be attributed to a few different things. First, PacIOOS was able to build the system from the beginning, knowing the goal of interoperability with other RAs. In contrast, some regions began with a wide array of *in situ* measurements from disparate providers, and some had data systems already in place. So, rather than try and maintain an existing system with an existing user-base, PacIOOS could design the data system from scratch. Related to this, a choice was made to have OPeNDAP services at the core. This has proven to be an efficient way to not only serve data, but also to transport data between different internal servers, as well as to provide a guideline for data formats (netCDF) and conventions.

Challenges still exist, however. Aside from getting legacy data sets into the system, there remains a lag between traditional methods of data access and the use of data services described here. For example, many existing workflows do not incorporate OPeNDAP-enabled clients and PacIOOS has been trying to educate and train users to adapt to this change. Bandwidth issues are also a key challenge, as many island regions lack the cyber-infrastructure to handle large data sets, such as forecast model output.

Consequently, as PacIOOS and its data system mature, more groups have been encouraged to make use of the servers and distribution mechanisms. For example, many local agencies have data that they would like to share with a wider audience (in some cases they are mandated to do so), but they lack the infrastructure to do so. This is another area of expansion for PacIOOS, wherein external data can be added to the process and distributed using the existing system.

REFERENCES

Bassett, R., Beard, R., Burnett, W., Crout, R., Griffith, B., Jensen, R., & Signell, R. (2010). Implementing the National Integrated Ocean Observing System (IOOS®)—From the Federal Agency Perspective. *J. Mar. Tech. Soc*, *44*(6), 32–41. doi:10.4031/MTSJ.44.6.8

Cornillon, P., Adams, J., Blumenthal, M. B., Chassignet, E., Davis, E., Hankin, S., & Sirot, J. et al. (2009). NVODS and the development of OPeNDAP - an integrative tool for oceanographic data systems. *Oceanography (Washington, D.C.)*, *22*(2), 116–127. doi:10.5670/oceanog.2009.43

de La Beaujardière, J. (2009). *IOOS Data Management Activities*. IEEE. Retrieved from http://ieeexplore.ieee.org/xpls/abs_all.jsp?arnumber=5422388

de La Beaujardière, J., Beegle-Krause, C. J., Bermudez, L., Hankin, S., Hazard, L., Howlett, E., ... (2010a). Ocean and Coastal Data Management. In *Proceedings of OceanObs'09: Sustained Ocean Observations and Information for Society*. European Space Agency. doi:10.5270/OceanObs09.cwp.22

de La Beaujardière, J., Mendelssohn, R., Ortiz, C., & Signell, R. (2010b). Building the IOOS® Data Management Subsystem. *J. Mar. Tech. Soc*, *44*(6), 73–83. doi:10.4031/MTSJ.44.6.10

Fountain, T., Tilak, S., Hubbard, P., Shin, P., & Freudinger, L. (2009). The Open Source DataTurbine Initiative: Streaming data middleware for environmental observing systems. *International Symposium on Remote Sensing of Environment.*

Gray, J. (2009). Jim Gray on eScience: A Transformed Scientific Method. In The Fourth Paradigm: Data-Intensive Scientific Discovery. Microsoft Research.

Haines, S., M. Subramanian, E. Mayorga, D. P. Snowden, R. Ragsdale, C. A. Rueda, & M. Howard. (2012). *IOOS Vocabulary and Ontology Strategy for Observed Properties*. IEEE. doi:.10.1109/OCEANS.2012.6405083

Hankin, S., Bermudez, L., Blower, J. D., Blumenthal, B., Casey, K. S., & Fornwall, M. et al.. (2010). Data Management for the Ocean Sciences - Perspectives for the Next Decade, 570–79. *European Space Agency, 2010*. doi:10.5270/OceanObs09.pp.21

Hankin, S. C. (2005). *United States National Office for Integrated, Sustained Ocean Observations, United States National Office for Integrated, Sustained Ocean Observations Data Management, and Communications Steering Committee. Data Management and Communications Plan for Research and Operational Integrated Ocean Observing Systems: Interoperable Data Discovery, Access and Archive.* National Office for Integrated and Sustained Ocean Observations. Retrieved from http://www.ioos.gov/library/dmac _plan_2005.pdf

Howlett, E., R. P. Signell, D. Wilson, D. P. Snowden, & K. R. Knee. (2014). *Data Management Update for the Integrated Ocean Observing System (IOOS®)*. IEEE. doi:10.1109/OCEANS.2014.7003284

Levy, J. M., Snowden, D., Clark, C., Crane, K., Diamond, H. J., Goodrich, D., & Stanitski, D. et al. (2011). The Global Ocean Observing Component of IOOS: Implementation of the Initial Global Ocean Observing System for Climate and the Path Forward. *J. Mar. Tech. Soc.*, *45*(1), 9–18. doi:10.4031/MTSJ.45.1.9

O'Brien, K. (2008). *Moving Toward Climate Data Integration: The Observing System Monitoring Center*. Paper presented at the American Geophysical Union, Fall Meeting, San Francisco, CA. http://adsabs.harvard.edu/cgi-bin/nph-bib_query?bibcode=2008AGUFMIN52A.05O

Ocean.US. (2002). An Integrated and Sustained Ocean Observing System (IOOS) for the United States: Design and Implementation. Ocean.US.

Signell, R., & Snowden, D. (2010). Advances in a Distributed Approach for Ocean Model Data Interoperability. *J. Mar. Sci. and Eng*, *2*(1), 194–208. doi:10.3390/jmse2010194

Snowden, D., Belbeoch, M., Burnett, B., Carval, T., Graybeal, J., Habermann, T., . . . Woodruff, S. (2009). *Metadata Management in Global Distributed Ocean Observation Networks*. Community White Paper for OceanObs09. Retrieved from https://abstracts.congrex.com/scripts/jmevent/abstracts/FCXNL-09A02a-1745513-1-OceanObs09_MetadataManagement_20May2009.doc

Tomlinson, M. S., De Carlo, E. H., McManus, M. A., Pawlak, G., Steward, G. F., Sansone, F. J., & Ostrander, C. E. et al. (2011). Characterizing the effects of two storms on the coastal waters of O'ahu, Hawai'i, using data from the Pacific Islands Ocean Observing System. *Oceanography (Washington, D.C.)*, *24*(2), 182–199. doi:10.5670/oceanog.2011.38

US IOOS Program Office. (2010). *Data Integration Framework (DIF) Final Assessment Report, November 22, 2010*. Retrieved from http://www.ioos.gov/library/ioos_dif_assmnt_report_final.pdf

US IOOS Program Office. (2010). *Guidance for Implementation of the Integrated Ocean Observing System (IOOS®) Data Management and Communications (DMAC) Subsystem, March 12, 2010*. Retrieved from http://www.ioos.gov/library/dmac_implementation_2010.pdf

U.S. IOOS Program Office. (2010). *U.S. Integrated Ocean Observing System: A Blueprint for Full Capability, November 2010*. Retrieved from http://www.ioos.gov/library/us_ioos_blueprint_ver1.pdf

KEY TERMS AND DEFINITIONS

Integrated Ocean Observing System (IOOS): The US contribution to the Global Ocean Observing System. It is composed of eleven Regional Associations (RA) that provide information about the ocean environment to a wide range of users.

Open-Source Project for a Network Data Access Protocol (OPeNDAP): A data transport architecture and protocol based on HTTP. OPeNDAP includes standards for encapsulating structured data, annotating the data with attributes, and adding semantics that describe the data.

Pacific Islands Ocean Observing System (PacIOOS): The IOOS Regional Association for the insular Pacific.

ENDNOTES

1 THREDDS is distributed by Unidata at the University Corporation for Atmospheric Research (UCAR).

2 netCDF, or network Common Data Format, is a self-describing platform-independent data model developed by Unidata.

3 Henceforth the term "user" will be used for any client application, including direct download by a customer as well as an access call by some software program.

Chapter 12
Marine Environment Data Management Related to the Human Activity in the South-Eastern Baltic Sea (The Lithuanian Sector)

Algimantas Grigelis
Institute of Geology and Geography, Lithuania

Nerijus Blažauskas
Klaipėda University, Lithuania

Leonora Živilė Gelumbauskaitė
Institute of Geology and Geography, Lithuania

Saulius Gulbinskas
Klaipėda University, Lithuania

Sergej Suzdalev
Klaipėda University, Lithuania

Christian Ferrarin
Institute of Marine Sciences-National Research Council (ISMAR-CNR), Italy

ABSTRACT

The chapter provides an overview of the activities fostering the blue growth and measures supporting the sustainable development and maintaining the good status of the marine environment in Lithuania. The recent findings and developed practical management tools facilitating the effective management of the marine resources support the implementation of EU Strategy for the Baltic Sea Region. Preparation of the first Lithuanian integrated maritime spatial plan, development of innovative solutions for handling dredged soil in ports, the new approach for selection of new dumping sites and prediction of possible adverse effects are among the activities that provided the most valuable results for the further development of marine activities also in the wider region.

DOI: 10.4018/978-1-5225-0700-0.ch012

INTRODUCTION

Growing maritime related business – the "blue" economy has fostered the demand for maritime research and data. Along with growing economy, we recognize the need of a proper management of the marine space and resources. The concept of the integrated maritime spatial planning and recently (July 2014) adopted EU Directive (2014/89/EU) encourages the Member States to implement the planning of the Seas. A number of international cooperation projects has been initiated in order to develop, adopt and implement the principles of maritime spatial planning. The main principles of the planning of maritime activities are embedded in the international strategic documents. Among those are EU Integrated Maritime Policy; HELCOM Baltic Sea Action Plan (Recommendation 28E/9 on the "Development of Broad-scale Maritime Spatial Planning Principles in the Baltic Sea Area"); VASAB Long Term Perspective for the Territorial Development of the Baltic Sea Region and recently endorsed by European Parliament the new Framework Directive for Maritime Spatial Planning (Directive 2014/89/EU, adopted in July, 2014). The Directive sets the deadline for the Member States to establish the Maritime Spatial Plans (MSP) by 31st of March 2021. This should foster implementation of the principles of sustainable development and sustainable utilisation of the maritime space and resources. Furthermore – ensure the proper management of the economic developments and potential as well as existing conflicts between the different sea users in the marine areas. To achieve this a fundamental knowledge, and new, applicable for the planning, data sets and innovative solutions for management of the marine resources are required.

The south-eastern sub-region of the Baltic Sea stretches from Poland–Russia border in the south to Latvia–Estonia border at the Irben Straight, in the north. The Lithuanian sea sector occupies a mid part of the sub-region and borders with waters of Russia, Sweden and Latvia. EEZ area of the Lithuanian sector is just about 4560 km^2, the territorial waters – about 1810 km^2, and the length of the open sea coastline is about 91 km.

The demand for maritime space in Lithuania increased dramatically during the past several years. Emerging new maritime uses such as an offshore wind energy or/and marine aquaculture along with development of a new port facilities; a concepts of the underwater electricity and underwater oil pipelines; development of a liquefied natural gas (LNG) market requires comprehensive maritime space planning (MSP) and proper management of the marine uses. The main nature and civil objects on the seacoast are the Curonian Spit [UNESCO Heritage list], Klaipėda city, Palanga and Šventoji resorts. The main marine industry is related to the Klaipėda State Seaport, the Marine Cargo handling companies, the Western Shipyard, the Klaipėda LNG terminal, the Klaipėda and the Būtingė oil import-export terminals. The Lithuanian marine transport is growing, the new sea uses (such us an offshore energy) are emerging and requiring not only the proper regulation, space and natural resources, but also having strong pressure on the sea floor.

Following the mentioned above, the Lithuanian Academy of Sciences has identified several marine environment data management tasks that are relevant for the south-eastern Baltic Sea (the Lithuanian sector) (Grigelis, 2012):

- Eutrophication: the marine pollution by the nitrogen and phosphorus rich compounds transported from the mainland area and resulting in the sea blooming caused by the blue algae.
- Negative impact on marine fauna and flora caused by the marine transport, economic activities in the harbours and port terminals, sea pollution by the oil products, chemical and synthetic materials.

- Risks to the fish gene pool related to the submerged chemical weapons, toxic (yperite, arsenic) substances including radionuclides.
- Insufficient sea water exchange with the Atlantic Ocean.
- Impacts related to the invasive species.
- Increasing frequency of the storms, hurricanes, excess rainfalls, deluges, acceleration of the sea level rise, coastal erosion, and degradation of the natural sandy beaches.

Meeting the identified ecological threats and in order to support the blue growth on the national scale as well as the Baltic Sea scale quite a number of international cooperation projects has been initiated. Among the EC-funded projects that have been identified and reported by EGS Marine Geology Expert Group (Grigelis, Gelumbauskaitė, 2013), some were dealing specifically with the geological and oceanographical investigations in Lithuania:

- EMODNET-GEOLOGY and EMODNET-2 (European Marine Observation and Data Network);
- GEO-SEAS (Pan-European Infrastructure for Management of Marine and Ocean Geological and Geophysical Data);
- COPAF (Coastline changes of the Southern Baltic Sea);
- SINCOS (Baltic Sea sinking coasts);
- EUROSEISMIC (European seas seismic metadata network);
- MASS (Marine environmental assessment in Gdansk basin).
- KAMO (Curonian Lagoon geo-system model).
- SURVAS (European vulnerability to impacts of sea-level rise);
- BASYS (Baltic Sea system studies: Palaeoenvironment).
- GEOBALT (Baltic Sea bathymetry and bottom sediments digital maps).
- GOBEX (Gotland Basin experiment).
- CHEMSEA (Chemical munitions search and assessment);
- GENOTOX-CG (Environmental genotoxicity studies);
- BALTCICA (Climate changes impact to the coastal area);
- SUBCOAST (Subsidence hazards in coastal lowland around Europe as a part of GMES);
- DEVOTES (Human activities impacts on marine biodiversity);
- NATO SPS for Oil Spill in SE Baltics (Effective oil spill management);
- PLANCOAST, BALTCOAST, BALTSEAPLAN, PARTISEAPATE, BALTSPACE (Maritime spatial planning throughout the Baltic Sea);
- SUBMARINER (Sustainable uses of Baltic marine resources);
- ARTWEI (Action for the Reinforcement of the Transitional Waters' Environmental Integrity).

Several key institutions dealing with and providing the scientific knowledge on the marine environment represent the Lithuania. Those are Nature Research Centre (NRC, Vilnius) and two units of Klaipėda University – Open Access Centre for Marine Research and Marine Research and Technology Centre. The results of the projects have been achieved in cooperation with the Marine Research Department and Lithuanian Geological Survey (both working under the Ministry of Environment).

HANDLING OF THE MARINE DATA AND INFORMATION

Studies of the marine environment proves how complex marine system is. Understanding the dynamics and multi-dimensional nature of the changes poses specific challenges for the proper data collection and representation. This is especially relevant for the maritime spatial planning (MSP) process, when obtained data should represent the temporal dimension and be incorporated into the planning and decision support tools. BaltsSeaplan was amongst the first projects analysing the existing data and information sources around the Baltic Sea and dealing with applicability of those data for specific – MSP purposes. The project has focused on two main sets of questions – data availability and data gaps. Monopolization of the information by different stakeholders and data holders, scale and resolution, interdisciplinary nature of the required information were among the main challenges to overcome by the data users (Schultz-Zehden & Gee, 2013). It was also clear that spatial attribution of ecological information is insufficient for the planning purposes – scattered and incomplete. This requires appropriate modelling and generalization of the data to take place.

Today we may add one more concern – how efficient available data can be transformed into the proper knowledge in order to make the sustainably sound decisions while planning the economic developments at the sea. MSP integrates the knowledge of very different aspects, asks for the data on very complex and varying environments and systems. Geological data on the seabed; biological information on benthic communities, fish, seabirds, marine mammals; hydrological data covering the dynamics of entire water column; meteorological conditions and also spatial distribution of anthropogenic developments at the sea – those are only basic informational data sets that are required in order to organize the sea space wisely. Despite the high variety of the thematic data sets, planners also deal with even greater variety of scales, format and timing of the collected data. For example, ecological data is quite abundant, but usually is presented as a general descriptions of habitats, bird and fish distribution. Descriptive character of such data has to be transformed into graphical, parametric and spatially georeferenced information layers in order to be useful for the planning. Another important concern - legal status of the data they deal with. Planners usually are forced to ground final decisions on the official data layers published or mapped by the relevant authorities or is stored in official data portals.

In order to overcome the above mentioned aspects BaltSeaPlan project has developed a concept of specific Baltic scale MSP data storage and quality controlled infrastructure - a transnational network for MSP data exchange. The MSP data infrastructure should consist of MSP data providers, offering their data to the regional/national MSP Data Contact Points; national MSP data contact points. The latest are responsible for making national MSP relevant data available to the MSP infrastructure. Pan-Baltic MSP Data Coordinating Group should be responsible for managing the Baltic MSP Infrastructure and creating harmonised pan-Baltic data sets.

Current study is a good example of integration of two main parties of the common MSP data Infrastructure to be developed on pan-Baltic scale – integration of data provided by research (data owners) and MSP relevant datasets developed by the planners at national level.

THE USE OF ENVIRONMENTAL DATA IN PLANNING THE HUMAN ACTIVITIES

The planning of the Lithuanian maritime space was the first attempt to integrate the environmental, economic and social needs into one comprehensive plan. Prepared spatial solutions creates the preconditions for the future developments at the sea (see also Blažauskas *et al.,* 2015). At the same time, new developments require new quality of the scientific research while investigating the marine resources and evaluating the economic effect as well as the environmental consequences. Maritime spatial planning (MSP) is the horizontal measure allowing controlling the potential conflicts at the very early stage. Therefore, the potential conflicts can be easier managed and allows avoiding the negative impact for socio-economic as well as natural environment. Planning creates the conditions for the development of existing activities and fostering development of the new ones. However, prior to that, proper environmental and policy regulations need to be followed. Considering further, the most important prevention measure remains the environmental impact assessment – EIA (including the collection of the missing data sets). The considerations of the EU environmental policies, regulations and directives as well as law of Lithuania are the main measures in order to avoid the negative impact to the environment. The particular attention should be paid on the impact related to the shipping, port activities, marine pollution from ships and other economic activities.

The Law of Territory Planning issued in 1995 and post-statutory legal acts regulate the planning procedure of the Republic of Lithuania. According to the Law of Territory Planning, the general plans of whole territory of the Republic of Lithuania, territories of counties, municipalities and their parts are the obligatory planning documents in o to ensure the long-term sustainable development and reasoned use of the area, finances and natural resources. The concept applied in this study introduces a simplified approach of combining different types of information in order to create the background for the planning purposes and conceptual allocation of the maritime activities. Planning of the sea space was implemented via integrated, cyclic and iterative process, which included:

- Identification of the demand for planning. This involved the analysis of the ongoing initiatives and targets defined in the maritime related strategic documents as well as investigation of their coherence, *i.e.* conflict analysis among the different stakeholders and potential sea users;
- Investigations of the natural and socio-economic assets. The action that required various data sets to be developed, analysed, harmonized and interpreted in order to consider the environmental conditions, pre-define environmental limitations; assess social implications important for the maritime sector development; evaluate the economic development trends, opportunities and threats. And, the introduction of the possibilities for filling data gaps by using the existing modelling results when data from direct investigations is not yet available or too expensive to carry out;
- Analysis of the institutional set up, capacities and legal framework, the responsibilities among the involved/responsible authorities;
- Acquiring the public acceptance and ensuring the stakeholder involvement. The planning involved visions of the future developments and maritime space usage, which should be discussed among the stakeholders and supported by the public. Therefore, the proper scientific grounding and research based evidences were of the very high relevance and importance.

Current study took an attempt to systemize the available environmental data developed during number of scientific projects carried out at former Coastal Research and Planning Institute of the Klaipėda University, Institute of Ecology and Institute of Geology and Geography of Nature Research Centre, Lithuanian Energy Institute. The next step was to transform it into the format that is the most suitable for the planning. The main datasets used for the development of maritime spatial plan for Lithuania were: morphology of the seabed; geological conditions, valuable bottom habitats, nursery and spawning grounds, areas important for wintering birds, hydrodynamic conditions, allocation of mineral resources (sand and oil).

Coastal zone, including the territorial waters of the Republic of Lithuania, is distinguished as one of the areas of specific functions. Recreation, nature conservation and fishery as well as transport are among the prioritised activities here. Furthermore, we have identified the open sea as area having a specific functional role. Defined priorities here are shipping, fishing and development of the marine infrastructure for oil survey and ocean energy projects.

During the planning process, two alternative concepts have been developed. The conceptual solutions supplementing the General Plan with marine part are based on the MSP principles developed by the Cross-border Oceanographic Commission of UNESCO, adopted by VASAB – HELCOM and followed by the actions defined in the EU Baltic Sea Region Strategy Action Plan. Both were based on the specifics of natural framework and conditions required to facilitate the developments of future uses as well as optimize the existing ones. The priorities set for the spatial use in both alternatives considers the possibility for exploration and exploitation of the marine mineral resources in the marine area of Lithuania, except the near shore and marine protected areas. The exploration of the marine mineral resources in Lithuania is not started yet, therefore the specific, most suitable or reserved areas could not be defined at this stage. The exploitation of the mineral resources is possible after the suitable places are grounded by geological investigations and the environmental impact is properly assessed. Both concepts consider the nearshore as prioritized for recreation, fishery and development of the port infrastructure (according to the regulations set in the Law of Coastal Stripe). Finally, the common spatial concept of the marine areas development have been prepared which combines the spatial solutions of the both alternatives and was accepted by the stakeholders and the governmental institutions as well as the public participating in the planning process (see also Milerienė *et al.,* 2014).

As a result, four main functional zones have been distinguished. (1) Nearshore zone up to 20 m water depth – an important zone for the land-sea interaction, ports development, area where the most of the biological assets are concentrated, the zone is important for recreation and coastal stability. (2, 3) Sea floor elevations: (2) Klaipėda-Ventspils Plateau and Klaipėda Bank in the north and (3) Curonian-Sambian Plateau in the south are characterized by the relatively shallower waters in the open sea; here we have favourable conditions for the marine infrastructure development and extraction of the mineral resources as well as potentially suitable for the development of valuable bottom habitats. (4) The deepest parts of the marine area – Gdansk Basin, palaeovalley of Nemunas River and slopes of the Gotland Basin are reserved for the navigation, fishing and future needs.

The final spatial solutions have been justified, approved and mapped (Figure 1). Seven functional regions have been distinguished where certain prioritization for the marine activities developments were identified. Those are:

- Decentralized development;
- Use of renewable energy sources;
- Shipping;
- Military training and ecosystem conservation;
- Mixed purposes;
- Port development;
- Protection of coastal ecosystem.

Figure 1. Spatial solutions for the development of the maritime activities

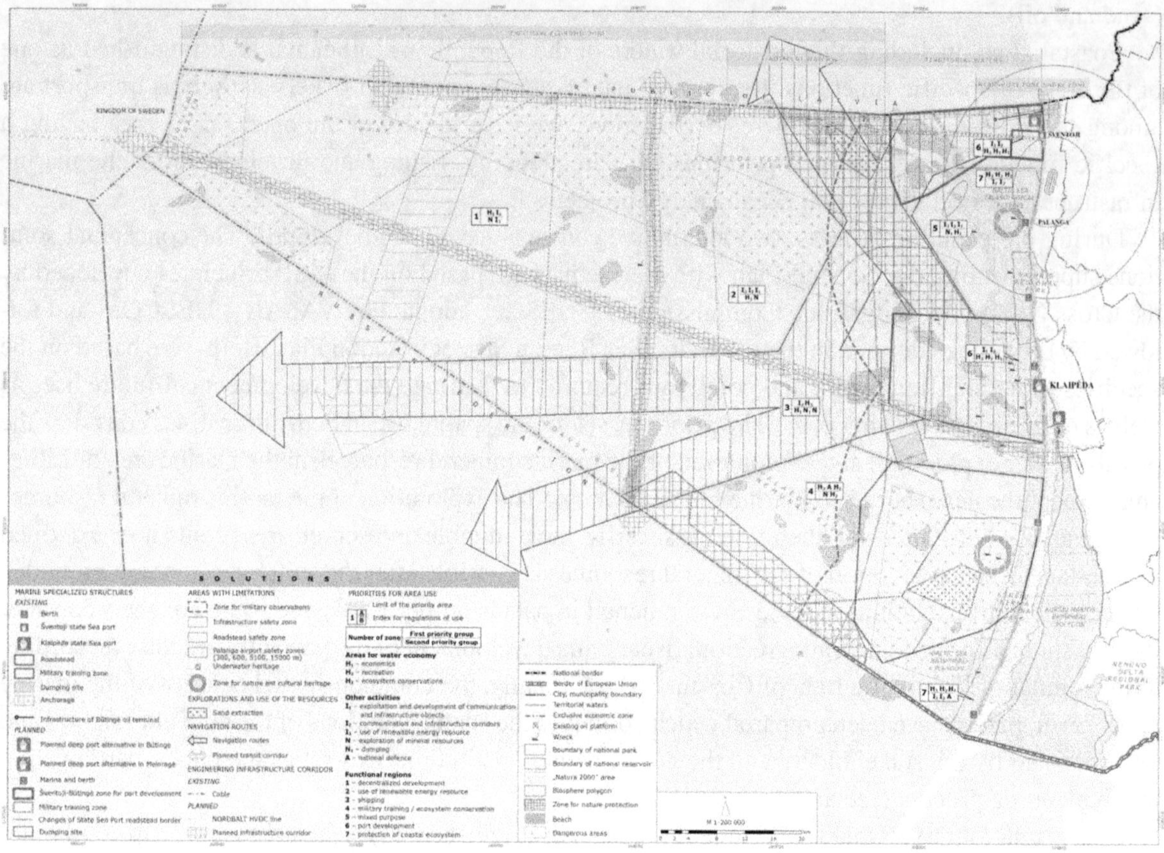

Priorities in each of the region differ. Those are indexed in order to reflect the level of prioritization. As a result the primary and secondary groups of uses have been distinguished in each of the delineated region (Figure 1).

Relevant for planning data is research based, limited in time and funding. This often results in a discontinuity and fragmentation of the knowledge. Dispersal of the data is a problem at both national as well as the pan-Baltic scale. The spatial and temporal resolution of the available data is also highly influencing the resolution and quality of the plan (Zaucha, 2014). On the other hand, there is a lack of knowledge when talking about the cumulative impacts on the marine ecosystem and interactions of the human activities. Data helping the establishment of so-called ecological corridors is very much complicated, often without proper indications of a susceptibility of the species of concern to the natural and anthropogenic pressures.

In order to facilitate the planning, complex environmental information needs to be translated into the parameters, which can be used in the assessment. However, it is not always clear which parameters are actually needed or even suitable for the planning purpose. Usually, planners cannot use the basic ecological information unless there are clear procedure for translating ecological and hydrographic data into the relevant planning information – generalized and integrated maps and schemes.

Although the marine spatial planning integrates the biological, hydrological, geological and even socio-economic data sets, the most important for the examples provided in the current chapter are the physical features of the area – the sediment distribution, stability of the coastline, lithological pattern of the superficial layer as well as deeper structure of the sea bottom.

BASIC GEOLOGICAL SETTING

The geological setting presented in this paper is based on the analysis of geological-geophysical data from different sources:

- Seismo-acoustic profiling made in 1997–1999 (R/V "Doctor Lubecki"),
- Geological mapping at a scale 1:50'000 (1997, 2000, 2004);
- Geological coring in 1997 (R/V "Kimberlite");
- Geological mapping at a scale 1:200'000 (1975–1978; 1989);
- Corings in 1982 ("Word Ocean" programme by ABIO RAS);
- Geological coring in 1989–1990 ("PETROBALTIC");
- Geological coring in 1995 (R/V "YUNIKON").

The geological metadata have been stored in the NRC database developed during the implementation of EMODNET-GEOLOGY and GEO-SEAS projects. The collected data are post-processed using the ARC/GIS tools and are available at a scale of 1:200 000. Different map layers have been fully integrated and stored into the One-Geology Europe (1G-E) portal (Stevenson and Glaves, 2012). Deeper knowledge covering the sea bedrock geology including the map of pre-Quaternary geology (age, lithology, and tectonics) is published (Grigelis, 2011).

The basic geological conditions of the Lithuanian part of Baltic Sea bottom are presented (Figures 2-5). The set of illustrations is composed of the detailed bottom topography map, the geo-seismic sections and the simplified map of seabed surface sediments. The palaeosurface of the sedimentary bedrock contains some denudation levels inserted in the Devonian / Triassic / Jurassic / Cretaceous rock surface (at 95–70, 65–50, 45–35 m b.s.l.). The thickness of the Quaternary deposits offshore, at the depression flat, varies from 40 m to 5–15 m; yet at the coast it increases up to 50–60 m. The Holocene sediments offshore are up to 3 m thick. However, the Baltic Ice Lake sedimentary complex (early development stage of the Baltic Sea) at the depression flat exceeds 17 m. Very uneven Late–Glacial–Holocene cover along the traced geo-seismic profiles evidents that eustatic / isostatic / tectonic dynamic in the postglacial basins are connected with the glacial till deformations. The deposition rate at offshore and near shore differs considerably (Gelumbauskaitė, 2009).

The Detailed Bottom Topography

The basic marine and coastal zone morphological units are presented on the detailed bathymetry map (Figure 2). The area of investigations accommodates the Curonian and Klaipėda plateaus, the gentle slope of the Gdansk depression; the Curonian Lagoon and the barrier island Curonian Spit; the nearshore zone and the palaeovalley of the Nemunas River.

Figure 2. Bottom topography of the south-eastern sector of the Baltic Sea (Lithuanian waters). Geological sections: A-A – Klaipėda traverse (see Figure 4), B-B – Nida traverse (see Figure 3). After Gelumbauskaitė, 2009.

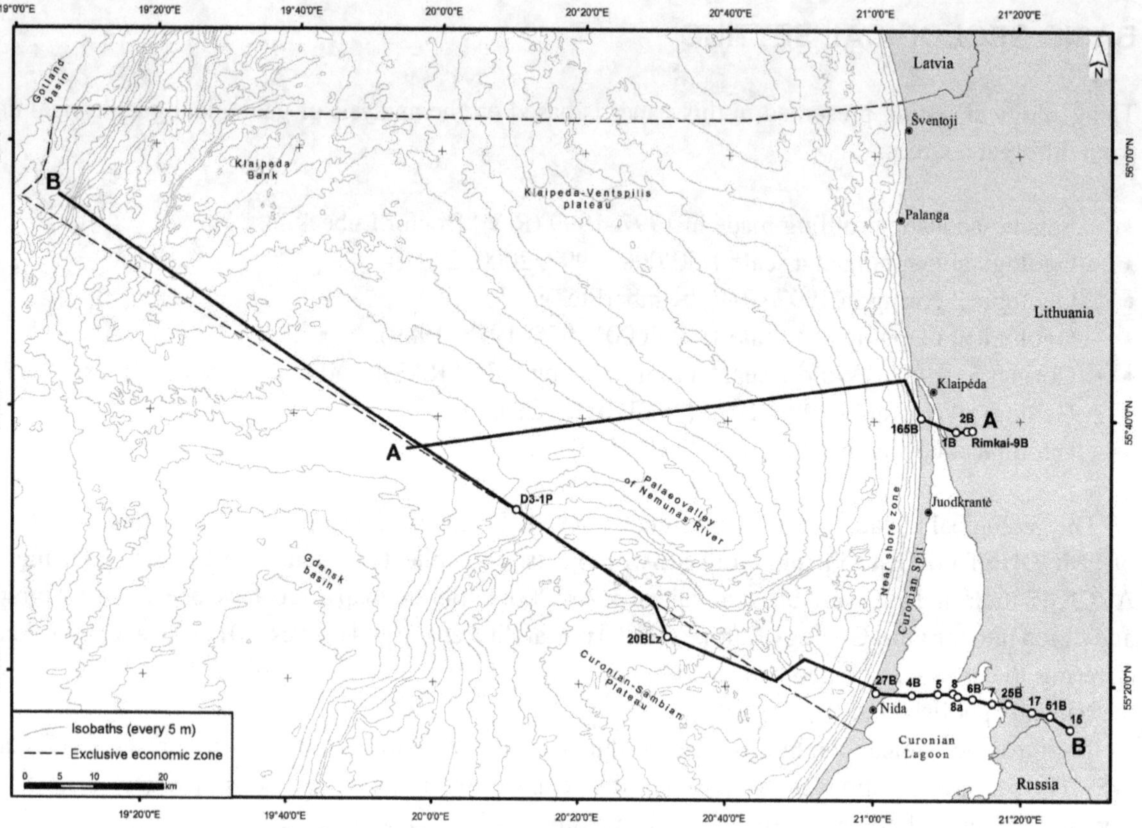

The Palaeomorphology and Internal Structure of the Basin

The geological sections (Figures 3 and 4) reveal the palaeomorphology and internal structure of the Lithuanian part of the Baltic basin. The sections (Nida and Klaipėda traverse) reveal the evidences of the former transgressions and regressions. The interpretation is based on the curves of the relative sea level fluctuations related to Post-glacial time. The geomorphometry and morphogenetic features of the modern relief are examined using the detailed bathymetry (at 2 m intervals) and the seismo-acoustic records. Lithological composition and sorting of the sediments, the identified ancient coastal bars and scarps as well as the fragments of the terraces allows tracing different hypsometric levels of the subaqueous ancient shore formations. Seismic records of strong reflectors penetrating the Late–Glacial and Holocene sequence clearly indicate the differences in lithology, allows distinguishing the seismic units and correlate the litho–biostratigraphy of the profiles with the cores and/or boreholes.

The sections explain palaeogeography and depositional–erosional history of the fluctuations of the Late–Glacial–Holocene basins in the south-eastern part of the Baltic Sea (Gelumbauskaitė, 2009). Large complex of the geomorphologic, geo–seismic and radiometric methods have been used and surveys of the subsiding SE Baltic coast carried out. This allow to recognise and investigate the fragments of the

Figure 3. Geological section on the Nida traverse (Gelumbauskaitė, 2009). Legend: 1–aeolian sand; 2–peat; 3–deposits of Postlitorina Stage; 4–deposits of Litorina$_{1\text{-}2\text{-}3}$ Sea Stage; 5–deposits of Postlitorina and Litorina Stages without internal stratification; 6–deposits of Ancylus$_{1\text{-}2}$ Lake stage; 7–deposits of Yoldia Sea Stage; 8–deposits of Ice Lakes and Baltic Ice Lake Stage without internal stratification; 9–limnoglacial deposits of Baltija (Pommeranian) Stage; 10–till of Baltija–Grūda Stage (Upper Nemunas–Late Weichselian); 11–till of Medininkai (Warthe) Stage; 12–till of Žemaitija (Warthe) Stage; 13–fliuvioglacial deposits of Grūda Stage; 14–transgression–regression peaks (marked by arrows) of different Baltic Sea stages

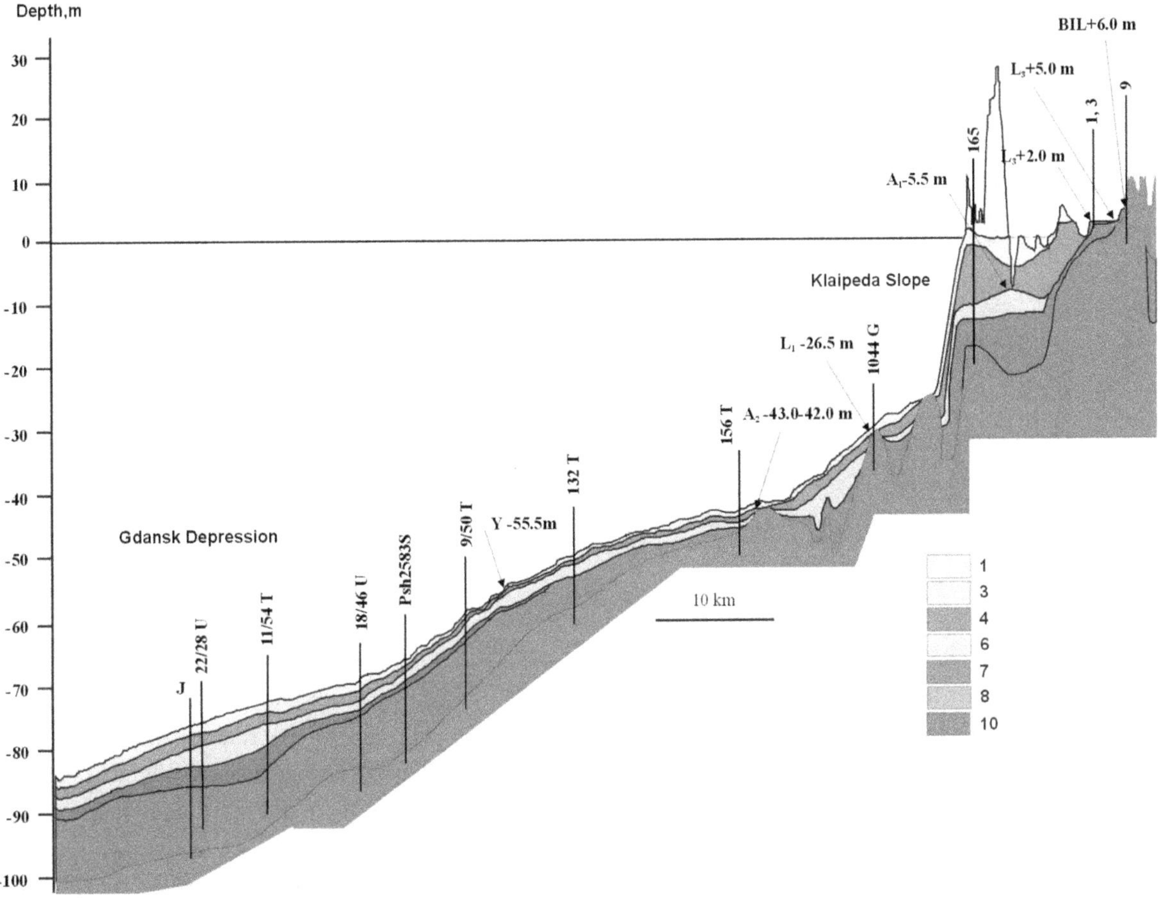

ancient shore levels and, even more - to reconstruct an internal structure and morphology of the ancient shore formations, to identify the boundaries and oscillation peaks of the Late–Glacial–Holocene basins. According these data the model of the relative sea level changes on the Lithuanian coast following from south to north has been constructed.

The Lithology of the Seabed Surface Sediments

The lithological composition of surface sediments deposited in the coastal and also open sea waters is presented (Figure 5). The map is compiled following the EUSeaMap requirements developed by

Figure 4. Geological section on the Klaipėda traverse.
(Gelumbauskaitė, 2009). For a legend explanation see Figure 3

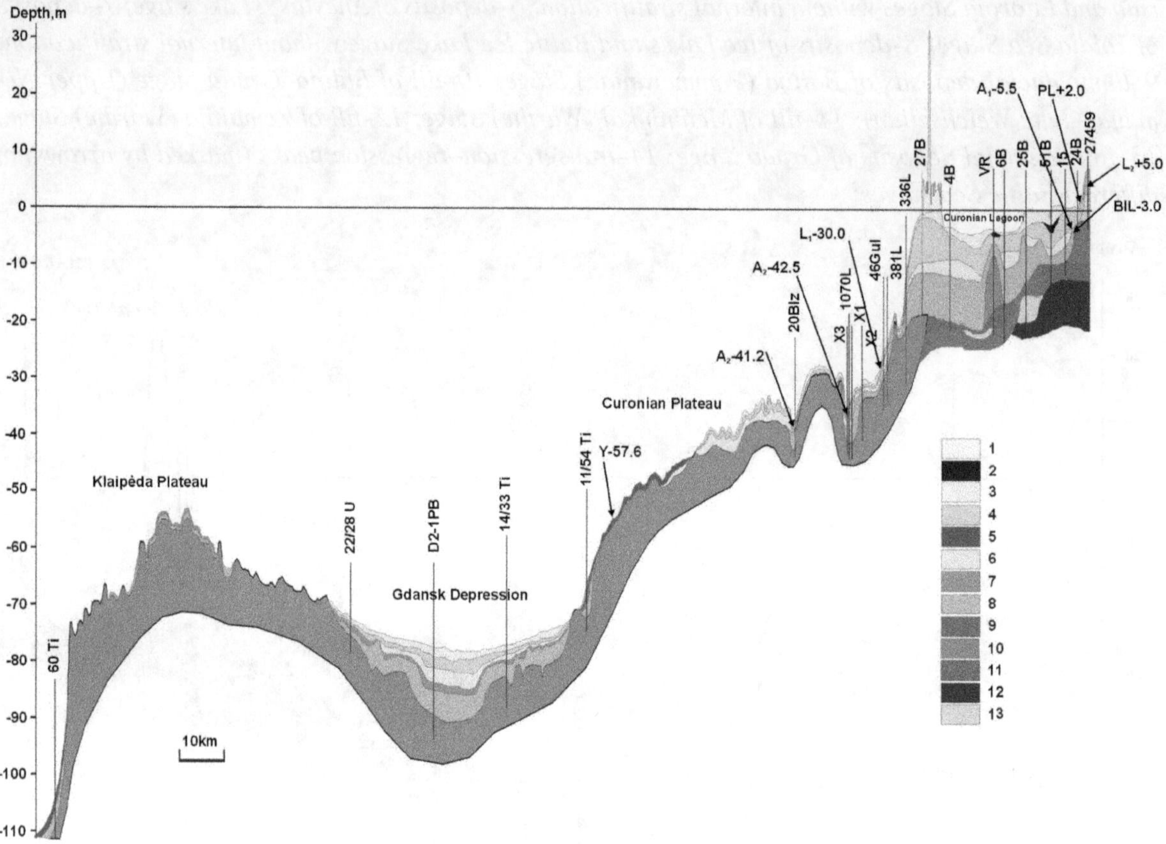

the European Marine Observation and Data Network. Data harmonisation and mapping of the seabed substrate includes an evaluation of the different classification schemes of the seabed sediments to be integrated with a broad-scale habitat mapping (Kaskela *et al.*, 2010). The approach was adapted and seabed surface material reclassified accordingly. Six substrate classes were distinguished based on the modified Folk triangle classification system (mud to sandy mud; sand to muddy sand; coarse-grained sediment; mixed sediment) and taking into account two additional classes (bedrock and till). The substrate reclassification scheme provides an estimate of the substrate from the uppermost 30 cm of the sediment column. One seabed substrate map was developed for the study area. Estimated (from sediment core data) sediment accumulation rate (SAR) for the open sea waters is 0.01–0.50 cm/year and 0.5–1.5 cm/year for the Curonian Lagoon.

Lithological composition of the nearshore zone has a direct influence to the resistance capabilities of the coastline to resist the erosion caused by storm events and rising sea level. According to the coastal typology the Lithuanian Baltic coastal area can be described as the combination of wave dominated sediment soft rock coast and micro-tidal dunes. Two different parts of the coast can be distinguished – sandy coast of the Curonian Spit and mainland coast (north from the Klaipėda state seaport entrance channel) represented by the accumulative-erosional sandy beaches mainly (except the coastal scarps and bluffs of glacial drift deposits present at the coast between Giruliai and Nemirseta settlements). The main

Figure 5. Seabed surface sediments map, modified after the Folk classification system. Legend: 1 – mud to sandy mud; 2 – sand to muddy sand; 3 – coarse-grained sediment; 4 – mixed sediment; 5 – bedrock; 6 – till.
After Gelumbauskaitė, 2010

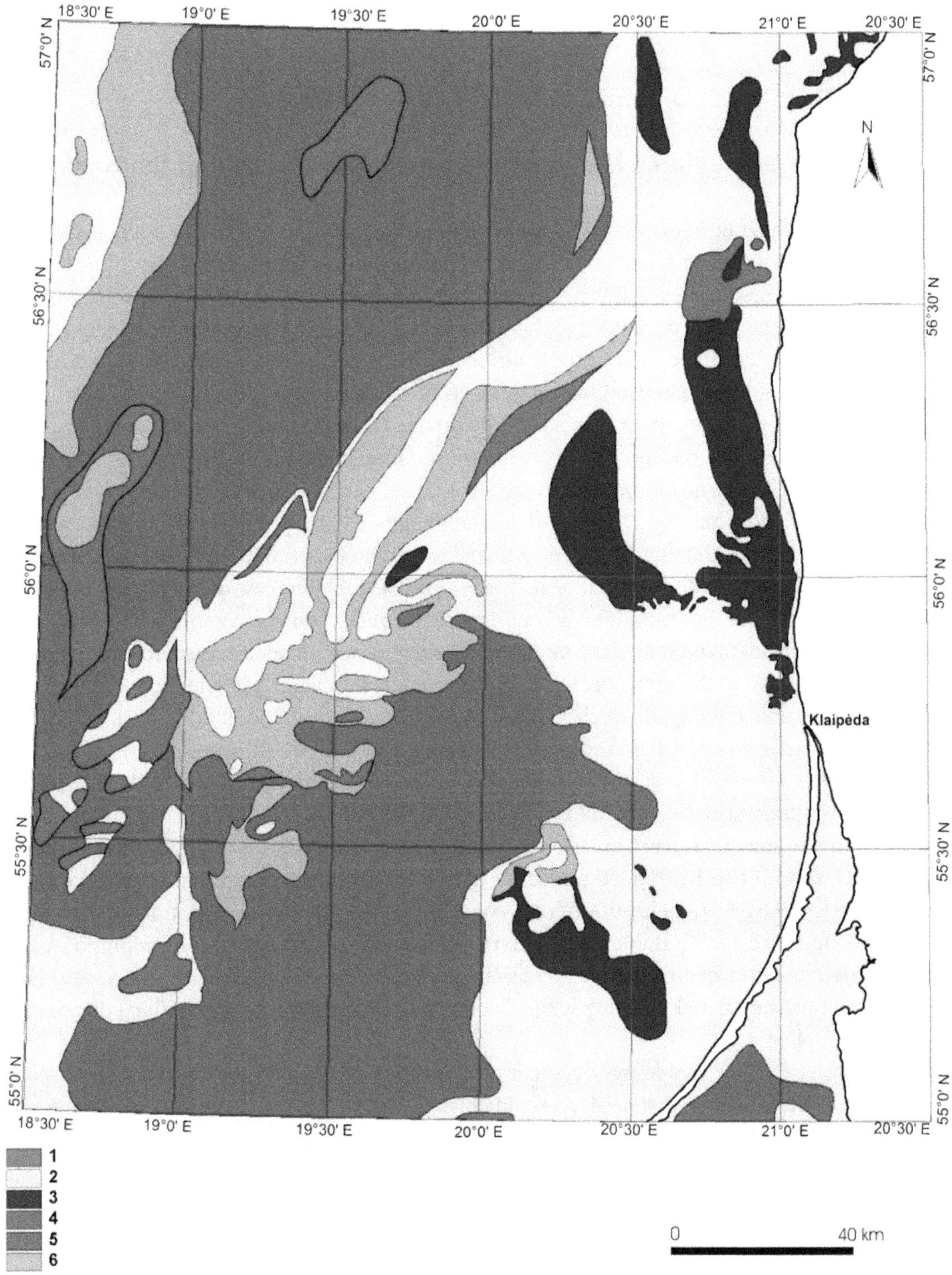

threat to the coast is imposed by the natural and artificial reasons. On the one hand, sea level rise and becoming more often stormy weather events are continuously affecting the coastal behaviour. On the other hand, the engineering developments in the port and particularly prolongation of the breakwaters at the entrance channel of Klaipėda port are blocking the sand migration along the shore and there why the sediment deficit is increasing in the northern part of the coastal area.

SUSTAINABLE DEVELOPMENT OF PORT AND FIREWAYS

The Options of Sustainable Management and Beneficial Use of the Sediments

Growth of the maritime business comes along with increased demand for proper navigation and port facilities. Maintenance of the navigable depths in ports and entrance waterways is one of the key activities of successful operation. This results in huge amount of material that has to be removed and handled. Taking into account the problems related to the sediment balance at the nearshore new innovative solutions have been introduced and efficiently contributes to the sustainable management and beneficial use of the valuable material. Dredged material has to be recognized as a valuable resource and not a waste. As the unconfined open water disposal of the dredged sediments deemed not appropriate for the contaminated material, the concept of the beneficial use of the contaminated sediments have been developed and tested. Clean sandy sediments are suitable for the beach nourishment and coastal protection if dumped in eco-friendly manner. Disposal of non-contaminated dredged material at the sea was and is considered as unwelcome and non-beneficial activity still. If no beneficial use of the dredged material is possible, using of the maritime spatial principles while selecting the new disposal sites might create favourable conditions for the economically and environmentally sound operations. On one hand the dumping of non-contaminated sediments might be considered as a secondary resource for the recovery of the sediment balance in the nearshore zone, on the other hand the proper selection of the site might have positive economic effects, which also contributes to the less environmental impact.

The application of ecosystem based concept proves to be a proper mechanism to ensure the sustainable development, maintenance of good environmental status and therefore contributes to the intelligent use of the sea as a common resource. This allows minimizing the costs of the disposal operations as well as serves as additional source of sandy matter contributing to the maintenance of good status of the sandy beaches. The results of two EU funded projects – SMOCS (Sustainable management of contaminated sediments) and ECODUMP (Application of ecosystem based principles for the location and management of offshore dumping sites in SE Baltic Sea Region) promotes the integration of the findings of scientific research, modelling of the hydrodynamic conditions and behaviour of the disposed mater. Additionally the principles of marine spatial planning while finding the suitable places for beneficial disposal at the sea have been applied.

The concept of the sustainable management of contaminated sediments (SMOCS) the disposal or beneficial use of the contaminated dredged sediments. A number of actions must be performed to ensure that there are no adverse effects on the environment, human health or other living species and the infrastructure. These measures include a risk assessment and a planning of sustainable management, considering ecologic, economic and social issues. The process of sustainable management requires the application of Best Environmental Practise (BEP) as described in the HELCOM Guidelines for the Disposal of Dredged Material at Sea as well as in accordance to the EU and national legislation throughout all stages of the management process.

The aim of the SMOCS partnership was to present a guideline on sustainable management of the contaminated sediments in the dredging projects in the Baltic Sea. It covers the whole process from identifying the need of dredging to the serviceability state after the handling of the dredged sediments and give support to the decisions to be made. The guideline provides an improved comprehensive and easy to use guidance starting from the problem definition and during the all stages of the project development. It also gives the descriptions on "how to do" in the different phases and situations. The focused toolbox comprising tools for the assessment of sustainability and decision support tools have been prepared and some stages tested in the different pilot cases. The special attention is dedicated to the beneficial use of the dredged sediments in *e.g.* constructions for the new port areas and a sediment stabilization/solidification method as one of the new emerging technologies.

The Pattern of Sediment Distribution in the Klaipėda Port

The one of the pilot areas to test the developed/promoted approach was the multipurpose, universal, deep-water Klaipėda State Sea Port. The port is developed in the Klaipėda Strait on the eastern coast of the Baltic Sea. The strait is an inlet connecting the freshwater Curonian Lagoon with the Baltic Sea. This is a very specific and dynamic natural system with permanent water circulation (Stakėnienė *et al.*, 2011) and intensive sedimentation (Trimonis *et al.*, 2010). The port accommodates 15 big stevedoring companies, ship repair and shipbuilding yards which provide the cargo-handling and recreational services. As a consequence of the port development, this water body is regularly deepened and due to the physical alterations was substantially changed and recently designated as a "heavily modified" water body. During the port development the depth of the navigational channel has changed from 12 m in 1976 (Jokšas *et al.*, 2003) to 14.8 m in 2009 (Galkus *et al.*, 2012). The composition of surface sediments and their distribution in the study area are determined by the natural hydro- and lithodynamical processes and technogenic reconstruction, taking place in the port (Trimonis, & Gulbinskas, 2000).

Constantly changing sedimentary environment, intensifying activities of the port enterprises and a variety of potential pollution sources was the main reason to carry out the extensive sampling of the port area, identify and clarify the character of some hazardous substances distribution in the surface sediments and distinguish the geochemical anomalies. The sampling of the surface sediments and throughout bathymetry survey was carried out during the implementation of international project SMOCS (www.smocs.eu). Altogether one hundred seventy-one samples have been collected in the natural environment and sites of the very intensive port activities (Figure 6A). The grain size composition of the sediment samples was analysed by the laser diffraction method using the laser particle analyser Analysette 22 MicroTec Plus, Fritsch. Lithological types of the sediments were distinguished in accordance with the Lithuanian classification standards (Gaigalas, 1995; LAND 46A-2002, 2002; Petelin, 1967). Five sediment types were determined: 1) medium grained sand (0.5 – 0.25 mm fractions >50%); 2) fine grained sand (0.25 – 0.1 mm fractions >50%); 3) silty sand (<0.063 mm fractions – 10-50%); 4) sandy mud (2–0.063 mm fractions 30–50%, <0.063 mm fractions – 50–70%); 5) silty mud (<0.063 mm fractions >70%). Glacial till sediments, including the boulders and pebbles detected in the study area, were described visually.

The biggest part of the study area is covered by the fine grained sediments with clear dominance of the silty and sandy mud sediments. Fine silty mud zone occupies almost all eastern part of the study area, including the small bays of low hydrodynamic intensity and favourable conditions for the accumulation

Figure 6. Lithological types of the sediment (A) and bottom morphology (B) in the Klaipėda Strait
Compiled by Gulbinskas and Suzdalev, 2015

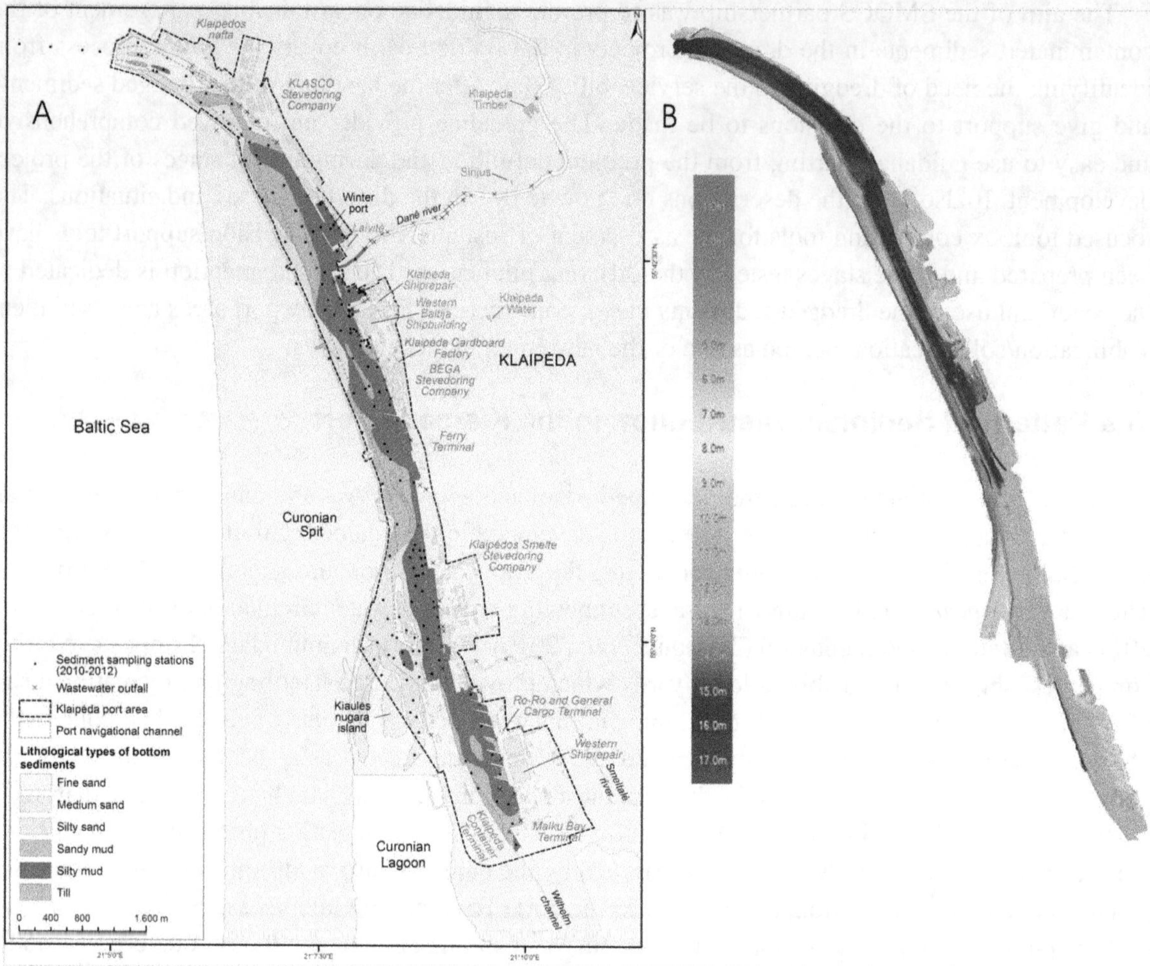

of fines. Silty mud was also identified locally in the central part of deepened navigational channel at the depths of 14.5 m.

According to the results of the current study, the areas of silty mud occupy part of the port to the north from Danė River mouth and large part of the navigational channel in the central area. Muddy sediments with the various amounts of sand fractions (defined as sandy mud) are deposited in the relatively small zones of the area, bordering the accumulation zones of the fine silty mud and silty sand. Silty sand is typical for the western shore of the study area as well as for the southern part, where the strait transits into the Curonian Lagoon and at the northern entrance channel of the Klaipėda Port.

Various-grained sand and coarser debris prevail in the sediment transit zones, e.g. the strait-lagoon junction in the south and at the entrance to the port in the north. The outcrops of the glacial till deposits identified only locally in the recently deepened areas. Most of the sediments accumulating in the study area contain high percentage (70-93%) of the fine particles (< 0.063 mm). This determines a high sorption property of the sediments and this is important to consider when assessing the potential contamination as well as the further management (dredging and disposal) of the sediments in the port area.

The Application of the MSP Principles for the Selection of the New Dumping Sites

In practice, the selection of a disposal site is an integral part of the complete application for the dredging and further utilization of the dredge material. If no beneficial use is applicable, dredge material has to be evaluated to ensure a suitability for the sea disposal. And this was the main focus of the ECODUMP project. Following has to be considered prior to the looking for the place for sea disposal:

- Amount and composition of the material;
- State, physical properties and lithological type of the material (especially grain size distribution and degree of consolidation);
- Chemical and biochemical properties;
- Radioactivity and toxicity level.

Existing legislation in the south-eastern part of the Baltic Sea does not address the establishment of the disposal sites at the sea. Although there are no officially approved planning procedures, countries have clearly identified general principles of selection the disposal sites and also have an obligation to full fill the national procedures of the environmental impact assessment before the permission is granted. The main principles address both – environmental and economic concerns. The main economic driver is the distance between site of dredging and disposal. Maximum allowed contamination level, the maintenance of sea floor integrity, avoidance of sensitive and valuable natural assets were amongst the main environmental preconditions to ensure the ecological balance and prove the economic viability. Currently, selection of new sites for offshore disposal of dredged material has to pass the appropriate planning procedure in order to avoid the potential conflicting with other sea users. When successfully established, MSP is the first document to consider while looking for the most sustainable alternative for a new disposal site. This would save the time for planning and permitting efforts.

If MSP is not in place, the identification of the new site for disposal has to follow the main principles and steps of MSP:

- Identification of the relevant maritime stakeholders interested in using the same marine area and/or the resources;
- Analysis of the existing ecological and socio-economic conditions – state of the art report on the existing natural, cultural and social assets;
- Mapping of the spatial distribution of the current uses;
- Screening of the existing regulations and legal framework of different sea uses in order to identify the existing and potential conflicts;
- Evaluation of the economic viability. The distance of the proposed dumping site from the dredging site is a major factor to be considered. A considerable distance between the dumping and dredging sites is not beneficial from the point of view of environmental protection and safety, since it increases the marine traffic, which, in turn, causes an increase in exhaust and noise emissions and difficulties for the shipping;

- The size and location of the disposal area(s);
- Development the alternative spatial solutions and ensure that consultations with involved stakeholders are in place.

The Šventoji Port was selected as a study object in order to test the proposed approach. Applying MSP principles, four potential places have been identified for the new offshore dumping sites (Figure 7B). All the alternatives are located in the territorial waters of the Lithuanian part of the Baltic Sea on the south-eastern slopes of Klaipėda-Ventspils Plateau, which is characterized by slightly hilly terrain. Alternative places for the disposal of the sediments were selected according to the current and planned uses of the sea, environmental conditions, geological-geomorphological structure of the sea bottom, lithodynamic conditions, distribution of ecosystem components, and the composition and pollution level of the sediments in Šventoji Port in order to minimize the impact on marine environment.

Figure 7. (A) Computational finite element grid of the Lithuania coastal waters and the Klaipėda Strait and (B) alternatives of the offshore dumping places and details of numerical mesh in the dumping area

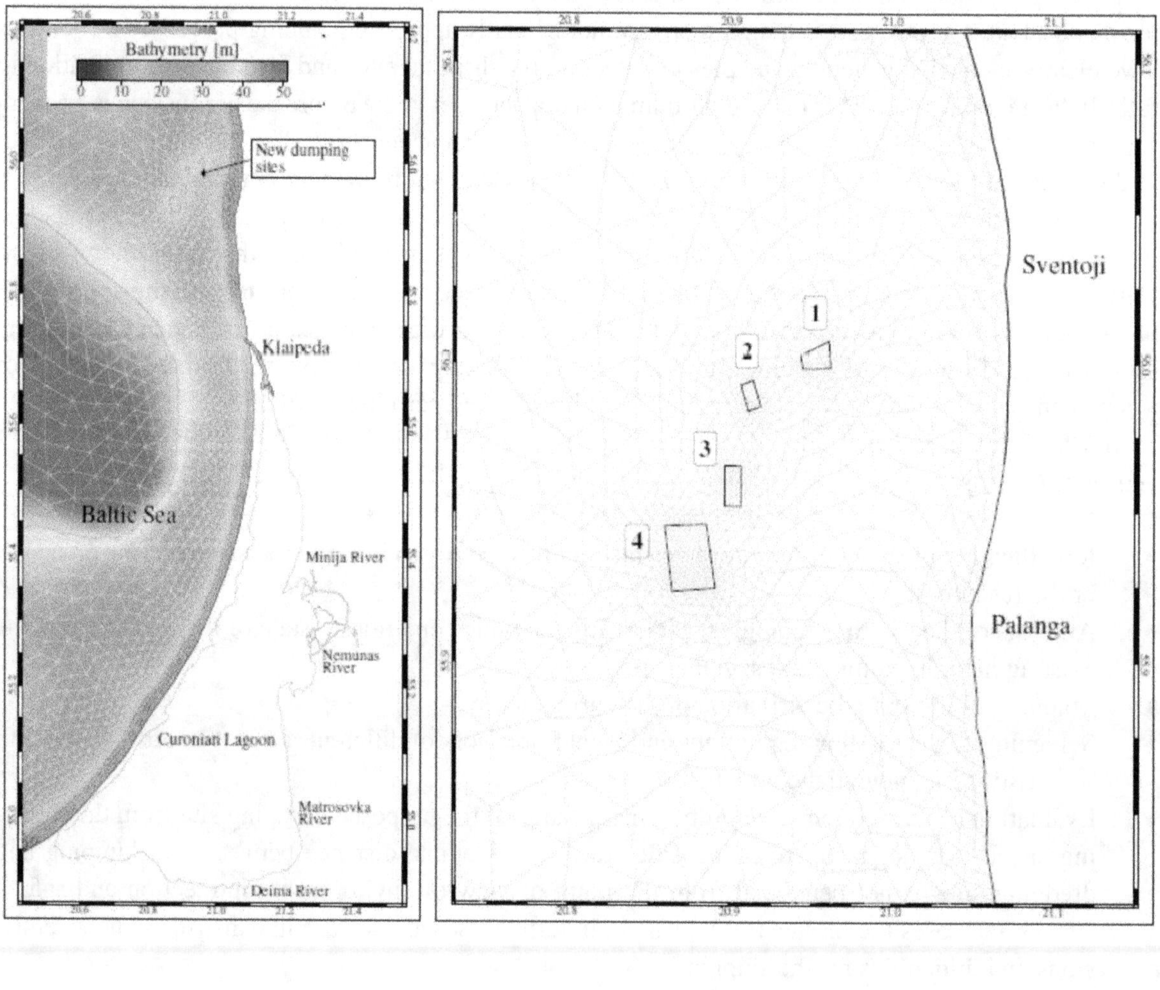

- 1st dumping site – in a distance of 7 km to the west from the Šventoji harbour entrance with the depth of 20.4–22.5 m. The site is selected in the coastal zone in the area of Klaipėda-Ventspils Plateau morainic relief formed by recent marine erosion-accumulation processes;
- 2nd dumping site – in a distance of 10 km to the west from the Šventoji harbour entrance with the depth of 17.5–22.5 m. The site is located in the deep formed in the surrounding of relict derivatives;
- 3rd dumping site – in a distance of 12 km to northwest from the Šventoji harbour entrance with the depth of 21–26 m. The site is in the neighbourhood of west edge of morainic relief with 20 m isobath;
- 4th dumping site – in a distance of 15 km from the Šventoji harbour entrance with the depth 22.5–27 m. In this sea bottom area southern end of the Klaipėda-Ventspils Plateau borders with northern slope of the Gdansk Deep.

The Modelling of the Sediment Transport

The selection of the best alternative was supported with an assessment of the potential spread of dumping material (bedload, suspended) dredged in the nearshore zone of the Šventoji harbor using the 3D sediment transport model SHYFEM (Ferrarin *et al.*, 2008, 2010). The modelling involve the spread of the dredged sediments during the disposal and after the sediments being settled with a given scenario of a multiple dumping events for one year. All necessary data for the simulations (dumping site locations, dumping material amounts and properties, areas of interest for sediments spread, bathymetry maps, maps of bottom sediment types, suspended sediments concentrations in the area of interest, atmospheric forcing data, river discharges) were developed by the Coastal Research and Planning Institute (CORPI), Klaipėda University. The numerical computation was carried out by the scientists of CNR-ISMAR, Venice and Klaipėda University.

Selected spatial domain represents the Klaipėda Strait and the Lithuania coastal area of the Baltic Sea until the 70 m depth contour through a finite element grid. As is shown (Figure 7A), the finite element method gives the possibility to follow faithfully the morphology and the bathymetry of the system and better represent the zones where hydrodynamic activity is important to consider. High spatial resolution was used to describe the areas of the four potential dumping sites. In these areas the model resolution (considered as the distance between two adjacent nodes) is about 150 m (Figure 7B). The water column is discretized into the maximum 16 vertical zeta levels with progressively increasing thickness varying from 1 m for the first 12 m to 18 m for the deepest layer of the outer continental shelf. The open boundaries of the considered system are the edges of the Baltic Sea area and the Klaipėda Strait. Open sea boundary water temperature, salinity, water levels and water velocity were obtained by the spatial interpolation of 1 nautical mile spatial resolution forecasts by the operational hydrodynamic HIROMB (Funkquist, 2003) provided by the Swedish Meteorological and Hydrological Institute. The temperature and salinity initial fields were also spatially interpolated from the data of model HIROMB while spatially uniform water level was used for the initial condition.

The Klaipėda Strait water fluxes, water temperature and salinity were obtained by a numerical simulation of the SHYFEM model over a domain which comprise both the Curonian Lagoon and the Baltic Sea (Zemlys *et al.*, 2013). Meteorological forcing fields were obtained by forecasts of the operational meteorological model HIRLAM (www.hirlam.org) provided by the Lithuania hydro-meteorological service. The simulations were carried out with a variable time step with a maximum value of 20 s for the time period between 1 January and 31 December of the year 2010. The model results show that the new dumping site area is located in the centre of a surficial circulation cell and therefore the average surface currents have intensity lower than 2 cm/s. In the numerical simulation, the dumping material was considered as a mixture of the sediment types, ranging from clay to coarse sand.

The applied methodology allows to describe the principal processes involved in the sediment spreading and deposition and to reproduce the fate of sediments during disposal and after the sediments being settled (Figure 8). Generally, in all cases sediments tend to spread at the surface towards north and south-est, while in the bottom they concentrate inside or nearby the dumping area.

The pilot studies have proved that adopted methodology is a powerful tool for selecting the place and investigating the fate of dumping material. Introduction of so called "eco-dumping" concept, among mentioned above, goes along with modelling exercises and therefore ensures that disposed sediments are in balance with the geological, hydrodynamic and ecological conditions on and adjacent to the disposal site. This makes dumping activity to be, in some extent, recognized to be beneficial too. However, the final decision of selection of the best alternative will depend on the results of comprehensive EIA study.

Figure 8. Maximum suspended sediment concentration computed by the model at the surface (left panel) and at the bottom (right panel) in the case of sediment release at the 4th dumping site

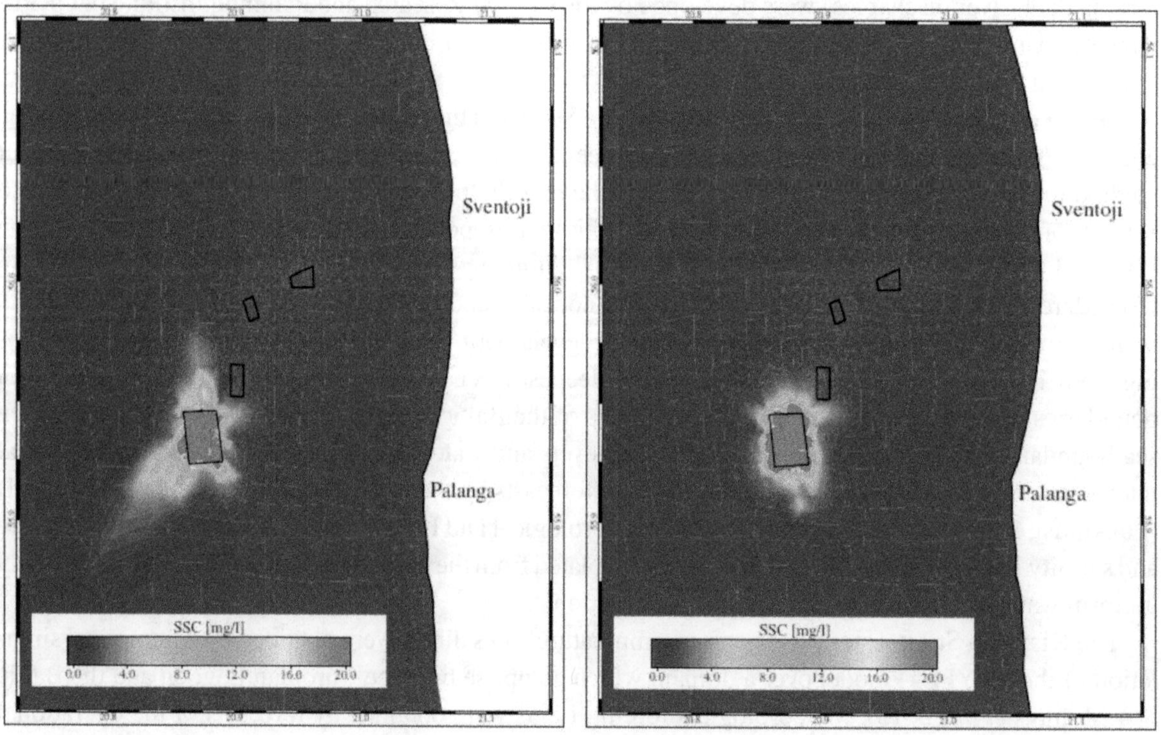

CONCLUSION

The study illustrates how the developed spatial solutions might influence the future developments at the sea and at the same time asks for new quality of the scientific data and management approaches for the effective and sustainable use of the marine resources. The sampling of the sediment distribution in the Port area was carried out using modern strategy of the sampling, which was developed during the implementation of international projects partly funded by the EU. The results of the MSP principles application for the selection of new sites for offshore disposal of dredged material from the port were provided. Applied approach promotes the integration of the findings of scientific research, modelling of hydrodynamic conditions and behaviour of the disposed mater. Obtained results, the prepared and tested methodology proves that MSP is a helpful for the planning and facilitating the economic developments at the sea. On the other hand application of sustainable management of dredge material at the ports and fireways at the sandy coasts might be attractive practical guide for similar areas of the South Baltic Sea and elsewhere.

ACKNOWLEDGMENT

Authors are grateful to Dr. Paolo Diviacco (OGS, Trieste, Italy) for a scientific guidance of this work. Positive evaluation of this chapter and useful remarks made by anonymous referees are appreciated very much.

REFERENCES

Blažauskas, N., Grigelis, A., Gelumbauskaitė, L. Ž., Gulbinskas, S., Suzdalev, S., & Ferrarin, Ch. (2015). Towards sustainable use of marine resources in the south-eastern Baltic Sea (Lithuania): A review. *Baltica, 28*(2), 179–188. doi:10.5200/baltica.2015.28.15

EC. (2000). Directive 2000/60/EC of the European Parliament and of the Council of 23 October 2000 establishing a framework for community action in the field of water policy (Water Framework Directive). *Official Journal of the European Communities,* L327/1.

Ferrarin, C., Cucco, A., Umgiesser, G., Bellafiore, D., & Amos, C. L. (2010). Modelling fluxes of water and sediment between the Venice Lagoon and the sea. *Continental Shelf Research, 30*(8), 904–914. doi:10.1016/j.csr.2009.08.014

Ferrarin, C., Umgiesser, G., Cucco, A., Hsu, T.-W., Roland, A., & Amos, C. L. (2008). Development and validation of a finite element morphological model for shallow water basins. *Coastal Engineering, 55*(9), 716–731. doi:10.1016/j.coastaleng.2008.02.016

Funkquist, L. (2003). A unified model system for the Baltic Sea. In H. Dahlin, N. C. Flemming, K. Nittis, & S. E. Petersson (Eds.), *Elsevier Oceanography Series* (Vol. 69, pp. 516–518). Elsevier. doi:10.1016/S0422-9894(03)80082-X

Gaigalas, A. (1995). *Granulometric classification of clastic deposits and rocks*. Vilnius: Vilnius University Press.

Galkus, A., Jokšas, K., Stakėnienė, R., & Lagunavičienė, L. (2012). Heavy metal contamination of harbor bottom sediments. *Polish Journal of Environmental Studies*, *21*(6), 1583–1594.

Gelumbauskaitė, L. Ž. (2009). Character of sea level changes in the subsiding south-eastern Baltic Sea during Late Quaternary. *Baltica*, *22*(1), 23–36.

Gelumbauskaitė, L. Ž. (2010). Interim report of the EMODNET-Geology project. The Nature Research Centre.

Grigelis, A. (2011). Research of the bedrock geology of the Central Baltic Sea. *Baltica*, *24*(1), 1–12.

Grigelis, A. 2012. Report to European Academies of Sciences Advisory Council EASAC. The Lithuanian Academy of Sciences.

Jokšas, K., Galkus, A., & Stakėnienė, R. (2003). *The Only Lithuanian Seaport and its Environment*. Vilnius: Institute of Geology and Geography.

Kaskela, A., Kotilainin, A., Alanen, U., Stevenson, A., Cooper, R., Green, S., … Verbruggen, K. (2010). Emodnet geology: combining and harmonising sea-bed sediment information. *Marine geological and biological habitat mapping: GeoHab conference: Programme and Abstracts*. Retrieved from http://geohab.org/wellington.html

Milerienė, R., Blažauskas, N., & Gulbinskas, S. (2014). Integration of marine research results into a maritime spatial plan for Lithuania. *Baltica*, *27*(Special Issue), 65–72.

Petelin, V. P. (1967). *Granulometric analysis of the marine bottom sediments*. Moscow: Nauka. (In Russian)

Schultz-Zehden, A., & Gee, K. (2013). BaltSeaPlan Findings – Experiences and Lessons. Academic Press.

Stakėnienė, R., Galkus, A., & Jokšas, K. (2011). Pollution of Klaipėda port waters. *Polish Journal of Environmental Studies*, *20*(2), 445–459.

Stevenson, A., & Glaves, H. (2012). *EMODnet-Geology.5th EMODNET Progress Meeting*, Brussels, Belgium.

Trimonis, E., & Gulbinskas, S. (2000). Bottom sediments of the Klaipėda Strait. *Geologija*, *30*, 20–27.

Trimonis, E., Vaikutienė, G., & Gulbinskas, S. (2010). Seasonal and spatial variations of sedimentary matter and diatom transport in the Klaipėda Strait (Eastern Baltic). *Baltica*, *23*(2), 127–134.

Zaucha, J. (2014). The key to governing the fragile Baltic Sea. Maritime spatial planning in the Baltic Sea region and way forward. VASAB Secretariat, Riga.

Zemlys, P., Ferrarin, C., Umgiesser, G., Gulbinskas, S., & Bellafiore, D. (2013). Investigation of saline water intrusions into the Curonian Lagoon (Lithuania) and two-layer flow in the Klaipėda Strait using finite element hydrodynamic model. *Ocean Science*, *9*(3), 573–584. doi:10.5194/os-9-573-2013

Chapter 13
Data and Operational Oceanography:
A Review in Support of Responsible Fisheries and Aquaculture

Enrique Wulff
Andalusian Institute for Marine Sciences (CSIC), Spain

ABSTRACT

Data obtained within the operational oceanography infrastructures denote a possibility of significant analysis of fisheries in the progress of the implementation of the 1995 FAO Code for Responsible Fisheries. Data products for fisheries and aquaculture cater for research needs, thereby ensuring a capacity to meet temporal resolution, time frame, reporting frequency and interoperability formats requirements. The existence of this legislation and the costs of its enforcement make that the present data infrastructures are employed under relatively precarious circumstances. This chapter provides with an overview, not exhaustive but demonstrative, of what has been achieved for data within operational oceanography addressing the needs of fisheries and aquaculture scientists.

INTRODUCTION

Intensive use of the ocean has led to ongoing pollution of marine spaces. Pollution is affecting 60% of the world's major marine ecosystems, which are increasingly threatened by acidification caused by unsustainable aquaculture practices, pollution, waste, invasive species and climate change (Bijma et al., 2013). Data collection, data analysis and research incorporate the contribution of fisheries and aquaculture to society (and should not be used to expand its metrics to a broader range of industrial activities). Additionally, substantial net benefits can be derived from management of data on responsible fisheries and aquaculture with a focus on economics (World Bank & FAO, 2009, p.81).

The precautionary approach concerning the harvesting, handling, processing and distribution of fish and fishery products has received considerable attention since 1995, when FAO proposed an international Code of Conduct for Responsible Fisheries (Food and Agriculture Organization of the United Nations, 2000). Over the intervening twenty years of progress, fisheries management policies (including those

DOI: 10.4018/978-1-5225-0700-0.ch013

for aquaculture) have taken into account the fact that uncertainty (Pontecorvo, 2003) and absence of data for estimation of fishery parameters (e.g. length at first capture, fishing season) and aquaculture parameters (alkalinity, carbon dioxide, dissolved oxygen, hardness, pH and salinity) should not justify a failure to apply rules, especially in the presence of indicators pointing to overexploitation of the fish stocks. These indicators show that the fisheries and aquaculture sector relies on assumed efficient traceability schemes to determine the origin of wild-caught fish and farmed fish, both because aquaculture production might lead to environmental challenges and because persistent overexploitation of fish stocks, and illegal fishing, pose a major threat to adequate future availability of wild fish, an important natural resource (Saitoh et al., 2011).

Fulfilling the obligations suggested by the Code requires the stage to be set for satellite or In situ-based operational oceanography applied to fisheries data. Inputting in-situ or altimetry data for the open oceans and coastal systems require data resources including sea surface temperature, vorticity, and ocean color, and data products by merging satellite/environment data with catch data and biological data. These are stored in a variety of different formats and are dispersed across a network of data centres. As fishery managers integrate more data in the spatial dimension, and data delivery becomes more operational, product requests are likely to change rapidly (Berx et al., 2011). So the visibility of operational oceanographic data resources for fisheries has a direct influence on the ability of data managers to fulfill the obligations of the FAO Code that is concerned with responsible fisheries.

Countries are not complying with UN Code of Conduct for Responsible Fisheries (Pitcher et al., 2009). As an open access resource, any fishery faces the reluctance of States to transfer sovereign powers to the international level. To improve science for the purposes of fisheries management and to help managers minimize the adverse effect of aquaculture, data collection has been made a high priority, and economic data on commercial fish processing and recreational fisheries are now required, a framework that is as objective as possible for evaluating the quality of by-catch data to support assessments and conservation of these resources. The 1995 FAO Code legal regime gave coastal States rights on the fishery resources in the Exclusive Economic Zone, an area that can expand 200 miles from the national coastlines. These national jurisdictions embrace more than 90 percent of the world marine fisheries. To protect and preserve the aquatic environment taking into account the best scientific evidence available, the Code establishes that lack of scientific information should not be used as a reason not to protect target species, associated or dependent species and non-target species as well as their environment (Articles 6.5 and 7.5.1). So the Code is intended to change the way in which fishery research and development are approached and undertaken. Without considerable inter-country collaboration in the collection and processing of data, fisheries scientists in each coastal country will find it very difficult to monitor and properly analyze the status of many shared stocks, or the related marine environmental conditions in their seas.

To compare the operational oceanographic data infrastructures of valuable for fisheries and aquaculture with the articles of the Code of Conduct for Responsible Fisheries (CCRF), we searched the Code for articles and clauses mentioning data and information. Articles were classified into four large data categories (general principles, fisheries management, aquaculture management and fisheries research) after their corresponding names in the text of the Code. These categories should be read exactly as they are included in the respective legal clauses of the Code. Fourteen legal clauses have defined the data areas covered by the Code (see Figure 1).

The existence of this Directory of Legislation in Force (the objectives of the Code of Conduct for Responsible Fisheries) is, in fact, reflected in the long-term goals for healthy oceans outlined in NOAA's next generation strategic plan. To capture a key understanding of ecosystems with a view to inform re-

Figure 1. Distribution of the 1995 FAO Code articles in the 4 data categories. The clauses correspond to the 4 categories as represented in the Code by the name of the articles, article 6 "general principles", article 7 "fisheries management", article 9 "aquaculture management" and article 12 "fisheries research", respectively.

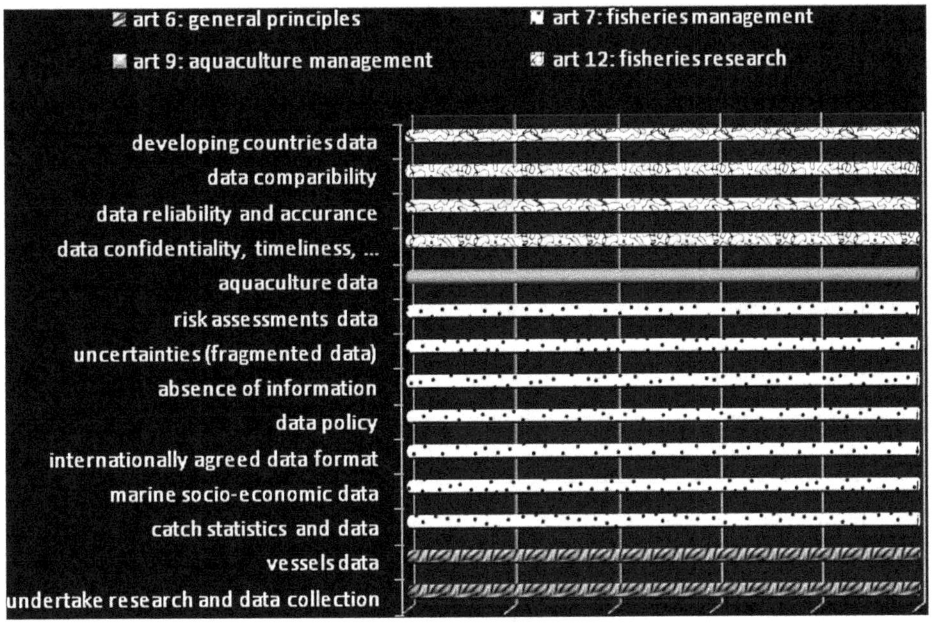

source management decisions, the Code promotes research and relevant environmental factors (Article 2.i) on fisheries as well as on associated ecosystems. And there is an equivalence between recovered and healthy marine and coastal species, and an Article to promote protection of living aquatic resources and their environments and coastal areas (Article 2.g). This rule also takes into account healthy habitats that sustain resilient and thriving marine resources and communities. Sustainable fisheries and safe seafood for healthy populations and vibrant communities are also stipulated in the Code, that promotes the contribution of fisheries to food security and food quality, giving priority to the nutritional needs of local communities (Article 2.f). These images from fisheries policy set out the key principles for data collection: accuracy; reliability and timeliness; avoidance of duplication through improved coordination; safe storage in database systems; improved availability of data; compliance with laws on personal data protection; access enabling to check the availability and quality of data and the methodology used to collect them.

DATA SYSTEMS STUDIED

Different observational data sets have been studied along this analysis (Table 1). Data collected by the meteorological and oceanographic communities in support of operational oceanography should drive pooling of results from multiple marine databases. Standard extraction and transformation tools together with common analytical tools should be part and parcel of each data infrastructure. All these views have been taken into consideration when approaching both the various data sources and the aggregated data generated by the systems studied.

Table 1. Breakdown of the 13 major operational oceanographic data infrastructures studied

Name	Year	Geographic Site	Data Tipology	Web Portal Address
MyOcean	2009	France (Toulouse)	sst, sss, sea level, winds, optics, etc.	www.myocean.eu
ODP	2013	Russia (Obninsk)	geologics, biological, physical, chemistry	www.oceandataportal.org
SeaDataNet	2004	France/Netherlands (Brest/The Hague)	long term ocean phenomena data	www.seadatanet.org
ICES	1957	Denmark (Copenhagen)	catch data, biological, chemical, oceanographic	ecosystemdata.ices.dk
ARGO	2001	France/USA (Brest/Monterey)	robotics buoys data	www.argo.net
AVISO	1992	France (Toulouse)	ocean altimetry data	www.aviso.oceanobs.com
JCOMM	1999	France (Toulouse)	marine climate	www.jcommops.org/old
EMODnet	2009	Belgium (Oostende)	geologics, chemical, biological, physical	www.emodnet.eu
EuroGOOS	1994	Belgium (Brussels)	operational oceanography	www.eurogoos.org
JRC	2000	Italy (Ispra)	fisheries data	datacollection.jrc.ec.europa.eu
NOAA	1961	USA (Silver Spring (Maryland))	physical, chemical, biological, geological	www.nodc.noaa.gov
Roffer	1986	USA (Miami)	fisheries forecasts	www.roffs.com
CatSat	2001	France (Toulouse)	fisheries forecasts	www.catsat.com

As the objective of this chapter is to develop an insight on operational oceanographic data applied to fisheries and aquaculture, and the threshold criteria used for each data is the agreement with the UN Code of Conduct for Responsible Fisheries, we will try to answer two questions: (1) what is the potential of the data products for describing cross-domain operational oceanographic perspectives, and (2) how can such products be applied to visualize the data policies in 1995 FAO Code. In pursuit of these objectives, a key aspect of our work is focusing on agreed procedures adopted by coastal countries in compliance with the requirements laid down in Article 7.4.7 of that Code. In accordance with the procedure referred data policy concerns the following issues: compilation, availability, confidentiality and timeliness. This is needed to ensure that fishers operate under economic conditions that promote responsible fisheries, as Article 7.6.3 of the 1995 FAO Code says. Maintaining sustainable fisheries is the main concern of this Code. And what is to be sustained includes aquaculture data, as an important asset on a cultivated natural capital information source (Leadbetter et al, 2017).

Input uncertainty and organizational coordination in emergency scenarios design the research agenda of marine data infrastructures (Van Lacker et al., 2017). The operational oceanography approach has to cope with uncertainty estimates and is impacted by emergency conditions related to oil spills, discharges of untreated or improperly treated water from treatment plants, intensive shipping and secondary pollution. While the Code acknowledges the statistical significance of uncertainty regarding the illegal, unreported and unregulated (IUU) fishing scores (Article 6.5), the requirement of viable contingency plans to restrict fishing in the event of an unforeseen emergency caused by excess fishing has focused on defining ways to validate data. Article 7.5.5 requires an almost completely satisfactory plan with triggers in place, and defined ways to validate trigger data. In the Code the arrangements for the evaluation of fisheries data sets and services provided for in clauses 7.1.9, 7.4.2, 7.4.4, 7.4.5, 7.5.1, 7.5.3, 7.5.5 and 12.4 are designed to be compatible with competitive advantages for States and relevant international organizations. An evaluation of the costs and benefits of the various oceanographic operational systems must determine the benefits to be derived from data management. The appropriate approach for this evaluation, to fill identified data gaps and make provision for data management, should ensure that, in the absence of scientific data, fisheries research be initiated as soon as possible (clause 12.3).

MyOcean, a project of the European Commission's 7th Framework Program (FP7), provides the main variables needed to depict the daily state of the ocean (real-time), its short-term evolution (forecast for the upcoming days, up to 10-14 days), and its history over the past 20-50 years (hindcast and reanalysis). Users access the information through a centralized service desk (Bahurel et al., 2009). From May 2015, it has let the floor to the Copernicus Marine Environment Monitoring Service (CMEMS) (Pouliquen et al., 2013).

The IODE Ocean Data Portal (ODP), is the data access and discovery service inside the International Oceanographic Data and Information Exchange (IODE) network. It was presented in 2007 and from 2013 it is maintained and developed from the Russian meteorological data centre RIHMI-WDC. Europe, USA and Australia agreed to contribute to this portal. The Ocean Data Interoperability Platform (ODIP), the initiative to establish an EU/USA/Australia/IOC-IODE coordination platform, is closely associated to the activity of ODP (Belov et al., 2013; Glaves, 2017).

SeaDataNet started in 2004, is the pan-European infrastructure for managing, indexing and providing access to marine data, metadata and products from all coastal states (Schaap, 2017). It has developed a NetCDF format able to serve operational oceanography. Temperature and Salinity (TS) historical data collections from 1900 have been created for meet operational oceanography and fisheries communities that need time series of in situ observations to study long term ocean phenomena and influences on the environment (Simoncelli et al., 2013).

From 1957, the International Council for the Exploration of the Sea (ICES) has stimulated exchange for marine data acting as a focal point for the oceanographic community. Scientists from ICES, as identified through the Working Group on Operational Oceanographic products for Fisheries and Environments (WGOOFE), require the following top five data products: temperature, salinity, currents, primary productivity, and algal blooms (Pinto & Holdsworth, 2013). The group addresses the lack of communication among fisheries, environmental, and oceanography researchers and views it as a core cause of the possible failure of fisheries management.

ARGO is the ocean international observing system and was established in 2001 by JCOMM (IODE/WMO Joint Technical Commission for Oceanography and Marine Meteorology). In 2007, Argo achieved its original goal of deploying 3000 robotic buoys all over the world in waters clear of ice located between each other at a distance of 300 km. Argo has two data websites: the IFREMER Argo Server and the U.S. GODAE Argo Server. Argo data are being used to aid the monitoring of environmental conditions that affect fish stocks and biological productivity, as they are ideally suited to monitor intra-seasonal and longer time scale variations in ocean temperature, heat content, mixed layer depth, and circulation (Folorunsho, 2006).

AVISO stands for Archiving, Validation and Interpretation of Satellite Oceanographic data (Rosmorduc et al., 2014). The central component of its interactive visualization device is a consideration of operational oceanography as a way to forecast sea state. Between the worldwide altimetric products distributed by Aviso are the multi-mission altimeter-data developed during GODAE (Global Ocean Data Assimilation Experiment, from 1997), Ssalto/Duacs (Data Unification and Altimeter Combination System).

JCOMM, officially started in 1999. One of the primary initial priorities for JCOMM (WMO-IOC) is the development and implementation of operational oceanography, to develop climate services for fisheries and aquaculture based on the available marine climate data. Its Web portal offers different mechanisms for accessing marine and oceanographic data depending on the application, needs, or source of the data.

European Marine Observation and Data Network (EMODnet) (Green Paper on EU Maritime Policy, 2006), the new integrated maritime policy for the European Union, develops from its five thematic portals a catalogue of data products. EMODnet biology is linked to IODE through the coordination by VLIZ[1], and assembling the participation of several OBIS nodes and NODC's including HCMR[2], IFREMER[3] and IEO[4]. EMODnet Seabed Habitats interactive mapping portal (or webGIS), is linked to the ICES habitat mapping metadata catalogue.

EuroGOOS (European Global Ocean Observing Systems), was created in 1991 in response to the desire of many nations to improve forecasts of climate change, and management of marine resources (Ehlers, 2013). It is a partner in the physical parameters group of EMODnet, that promotes data tools in support of operational management of fisheries, as much as data products and services for the aquaculture industry. Seven international conferences on EuroGOOS have contributed to the initiative. EuroGOOS data can be accessed through five regional operational oceanographic web portals.

Joint Research Centre (JRC) fisheries data collection web site is in place since 2000 (Marine Knowledge 2020, 2013). This EU data form the basis for scientific opinions and recommendations formulated in Scientific, Technical and Economic Committee for Fisheries (STECF) Annual Reports.

By 2012, the US National Oceanic and Atmospheric Administration (NOAA) has adopted an implementation plan of the FAO Code of Conduct for Responsible Fisheries, by which use key innovations have been identified in National Marine Fisheries Service's (NMFS) jurisdiction. NOAA Fisheries Service has provided the first nation-wide compilation of by-catch estimates in U.S. commercial fisheries. These data are monitored for its quality as they establish performances measures for improvements in protecting living marine resources.

Fishing operational oceanography data systems developed commercially have been scrutinized. Catsat provides dimensional oceanographic maps corresponding to fishing areas, and Roffer's Ocean Fishing Forecasting Service Inc. provides fisheries forecasts from SRS since 1987.

DATA POLICIES FOR NATIONAL REGULATIONS

In order to produce national fishing regulations and analyze their implementation, in particular for shared stocks and protected maritime areas, scientists need global data use, the integration of data gathering, experimentation, organization and assessment as an effective way to the attainment of sustainable fishing activities. As States should compile data on catch and fishing effort in a consistent manner (Article 7.4.4), data gathering (the compilation of data) should be a part of any working policy as a good tool that will, among others, enable the interested public to get closer to the data industry associated with the 1995 FAO Code. Table 2, (below) lists links to the data policies of the infrastructures studied regarding compilation criteria.

Table 2. Operational oceanographic data policies exposed by compilation criteria

Name	Data Policy
MyOcean	marine.copernicus.eu/web/27-service-commitments-and-licence.php
ODP	www.iode.org/index.php?option=com_content&view=article&id=423&Itemid=100036
SeaDataNet	www.seadatanet.org/Data-Access/Data-policy
ICES	www.ices.dk/marine-data/guidelines-and-policy/Pages/ICES-data-policy.aspx
ARGO	www.argodatamgt.org/
AVISO	www.aviso.altimetry.fr/fileadmin/documents/data/tools/hdbk_duacs.pdf
JCOMM	www.jcommops.org/dbcp/data_policy
EMODnet	www.emodnet.eu/what-emodnet
EuroGOOS	eurogoos.eu/download/other_documents/eg99_37datapolicy_2.pdf
JRC	ec.europa.eu/jrc/en/publication/eur-scientific-and-technical-research-reports/jrc-data-policy
NOAA	www.nodc.noaa.gov/about/datapolicy.html

Information was gathered regarding MyOcean practices associated with data from drifting buoys, from 28 country and 129 product categories and their source fisheries. Data compilation is the key of the cooperation between MyOcean in-situ TAC and SeaDataNet, contributing to the achievement of conclusive results in the areas of quality control, "best copy", archival, and re-analysis datasets.

Significant efforts are required in compilation of ocean data labelled with Digital Object Identifiers (DOIs). Indeed, IODE is implementing Published Ocean Data (POD), a parallel e-repository to its OceanDocs system targeted at datasets rather than documents. But POD is not fully scalable being the provision of potential storage by IODE beyond expectations (Leadbetter et al., 2013). Encouraging the development of a new marine data infrastructure (including theirs compilation and compilation) IODE has sponsored ODP (the Ocean Data Portal). The data gathered by ODP include operational in-situ/ satellite data stream, delay mode and historical data sets, and merged data sets from different sources.

SeaDataNet insures the long term archiving of a large number of multidisciplinary data (i.e. temperature, salinity current, sea level, chemical, physical and biological properties) collected by many different sensors installed on board of research vessels, satellite and the various platforms of the marine observing system. These large and diverse data sets collected by oceanographic fleets and automatic observation systems, integrate data resources from 40 National Oceanographic Data Centers and Satellite Data Centers.

The ICES portal is operating as a starting point where existing products are grouped by dataset, by ICES areas, by sub-basins, by regions, by species, and by parameter. ICES Data portal gathers all the fishing and aquaculture contents for the user, and it also means oceanographic data. ICES data policy strives to rescue and archive valuable data relevant to the ICES mission that are at risk of being lost, including those residing in reports and documents.

Argo collects salinity/temperature profiles from a sparse array of robotic floats in ice-free oceans (deeper than 2000 m), and also information on the surface and subsurface currents. Data are collected on the vessels through automatic devices, then scientifically validated and documented with metadata. And a central repository of information on quality control supports consistent data compilation and effective data formatting.

Aviso altimetry and operational oceanography data comes from seven past satellite missions, four current missions (Jason-2, Cryosat, Saral, HY-2) and 5 future missions. Main processing steps involving in data compilation inside Aviso are product homogenization, data editing, orbit error correction, and reduction of long wavelength errors (Le Traon et al., 2009).

JCOMM Network assembles data from surface drifting and open ocean moored buoys, the ships of opportunity and aerological programmes, ocean time-series and reference station network, and from the satellite data Argo network of sub-surface profiling floats. EMODNET is compiling marine data and information from a variety of European organizations. EuroGOOS, as a regional alliance of GOOS, collects and integrates marine in-situ data.

The JRC data are collected on the basis of National Programmes in which the MS (Member States) indicate which data is collected, the resources they allocate for the collection and how data is collected. At NOAA, first in 2005 a national bycatch data compilation was reached, and public reviews of Agency stock assessments (e.g., SEDAR) are available from NMFS science centres.

CatSat compiles data on sea level anomalies, plankton concentration, sea surface (and sub-surface) temperature, salinity, thermocline, sea surface (and sub-surface) currents, and weather forecast. Roffer's data from at least nine high resolution satellites (U.S. and European, twice a day, infrared and ocean color) along with other oceanographic and meteorological data are integrated to provide a comprehensive fishing product.

DATA MEETS THEIR NEEDS

It become apparent that a mismatch exists between user (aquaculture and fisheries scientists) requirements and the perceived requirements identified by the producers of oceanographic data products (Berx et al., 2011). By interpreting operational as being service oriented, delivered on demand, and generic, the producers of operational oceanographic products are investing in the development of tools to deliver data in real time and at high resolution, and have accordingly built large systems to handle these kinds of requests. However, researchers in fisheries and aquaculture indicate that they also need high-quality time series (historic data) that are regularly updated and flexible in terms of spatial and temporal limits and resolution.

Many physical oceanographic products are coming online, but in addition to the biophysical products researchers also need integrated coupled analysis from chemical and biological oceanography. Within the fishery science user community, fish-related data products, like ichthyoplankton abundance and distribution, are considered as part of operational oceanography. Users do not seem to understand the sheer size of the data available. So they need to familiarize with scientific data formats, like netCDF, or providers will need to incorporate new Web applications to preselect and download data in a variety of formats. The quick growth of spatial planning tools to support aquaculture and fisheries has led to the frequent production of new products, with older versions being difficult to use. Data citation and publication practices, currently promoted by the International Oceanographic Data and Information Exchange and the Scientific Committee on Oceanic Research (IODE-SCOR) project on data citation could help prevent this phenomenon (Leadbetter et al., 2013; Manzella, 2017).

To monitor and control fishing activity in the oceans from vessel trajectories, clause 6.11 ensures that vessels flying their flags fulfill their obligations concerning the collection and provision of data relating to their fishing activities. Control data from fishing vessels, such as Vessel Monitoring Systems (VMS) data, provide detailed information on fishing effort (e.g. tonnage, power, trajectories) at high spatial and temporal resolution, and on fishing distribution at different spatial and temporal scales. Other reasons why VMS data can be useful are that catch statistics are evaluated by cross validation and verification. In fact, fishing footprint based on Flag states submission is validated by VMS data. But VMS data open access is not guaranteed, and flags of convenience (FOC) data has the lowest priority, because FOC in the long-run means higher real costs and freight rates. Because flags of convenience, lead to the formation of a regime of immunity in the shipping world where the intellectual property rights may not even reside with those who have collected the data or those who are processing it. In fact, such intellectual rights may belong to the employer or to the organization that has paid for the data collection.

Moreover, engaging with the fisheries sector and potential user groups is essential to communicate the rationale for the way in which such datasets are released and to identify a level of detail for the VMS data that does not breach confidentiality concerns but still meet user needs. The EMODnet usefully makes public geological and hydrological data, but the availability of the data held by the VMS is not yet available from EMODnet. Indeed, data with information that can be traced back to a particular individual - for instance fishing vessel tracks - can generally be made anonymous with no loss of usefulness, by removing those parts (e.g. fishing vessel name) that may be connected to an individual or by aggregating data.

Within the EU framework for the collection and management of fisheries data in place, statistical data on catches and landings transmitted by Member States to the Commission are less detailed than the data collected under the Joint Research Council (JRC) Data Collection Regulation. For instance, they are not disaggregated by vessel type, although the data are free of any restrictions on use. JRC began work in fisheries in 1999 with research on the potential use of Synthetic Aperture Radar (SAR) to detect fishing vessels from satellites. As VMS cannot be used to monitor vessels whose system is switched off, in 2003 JRC proposed a Vessel Detection System (VDS), a satellite remote-sensing system that relies on Synthetic Aperture Radar (SAR) instruments. FISHREG is the EU Action which coordinates this process, and it contributes to the fight against IUU fishing (including research based on forensic genetics and chemistry for product traceability). Fishing effort and discards data are available from the JRC Data Collection Web Site. In addition, FISHREG does research in the spatial and temporal variability of pelagic fish habitats, it maintains an advisory role for action EMODNET by DG MARE on common data infrastructure, and its benefits include offshore aquaculture. FISHREG is part of the Maritime Affairs Unit inside the Institute for the Protection and Security of the Citizen (IPSC) of the Commission's Joint Research Centre (JRC).

So in Europe, public debate and stakeholder participation takes place with these reduced data, and Member States (MS) may refuse to provide data on vessels activity based on information from vessel satellite monitoring to end-users for the purposes of public debate. But the rebuilding of depleted fish stocks is explicitly covered by Article 7, clause 7.6.10, and clause 6.5 prescribes a precautionary approach to all fisheries (proportionate to the degree of uncertainty, risk and reversibility of impacts). By way of consequence, uncertainty and provision of scientifically sound and sufficient data and information are potential problems in tracing seafood products. Exacerbated when vessels routinely switches off VMS before entering neighboring nations' ports or transships catches to Flags of Convenience vessels (Pramod et al., 2014). A 2001 European Parliament report estimated that over 10 percent of its fleet (i.e. 3.721 vessels) under EU beneficial ownership was registered under flags of convenience (FOCs) (Hosch, 2009), and that the environmental impact of pirate fishing vessels is virtually impossible to evaluate, as, by definition, data on catches are non-existent or, at best, unreliable.

Figure 2. Needs of data concerning IUU (FOC related data). "VMS" indicates vessel monitoring system data which are collected by EMODnet, ICES, SeaDataNet, AVISO and NOAA, "VDS" vessel detection system is a specific method of work at JRC, fishing effort and the level of "discard" data are provided by JRC, "catch" composition threshold are detailed by SeaDataNet, sea surface height anomalies ("SSHA") is an altimeter product available from AVISO. Whenever the observed IUU is within the commercial area, the data are offered by the satellite remote sensing ("SRS") facility integrated in Roffer.

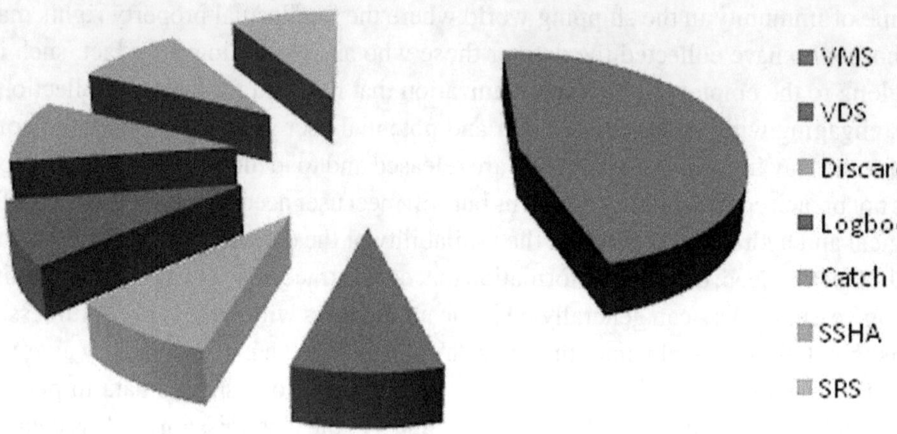

The operational uses of VMS data (identified under Article 6) are relevant for different end-users of marine observation data products and forecasts concerning fisheries, catching and fish farming. ICES aims at the further development of an operational forecasting system based upon VMS data. Emphasis is placed on the needs of data exploring the spatial distribution of fisheries, and fishers behaviour at sea. ICES is mandated to request VMS and logbook information as a basis for its use in spatial analysis to facilitate advice to fisheries management (by Article 18(1)(a) of regulation (EC) No 199/2008 (Council, 2008)). An specific ICES e-infrastructure (SGVMS) has been established for interpolating Vessels routes. On the basis of 2015 data call, ICES will provide a secure (https) folder, and a unique password for each country to upload their data to. Moreover ICES has an agreed VMS/logbook access and use policy for this type of data, which governs the process of who is given access and what they can do with the data.

SeaDataNet provides a coarse-grained inventory for tracking oceanographic data collected by research vessels, the Cruise Summary Report (CSR). Most marine disciplines are represented in the CSR (originally developed by IOC), including VMS data and catch composition thresholds (Schaap & Lowry, 2010). From AVISO have been downloaded delayed time and merged SSHA (sea surface height anomalies) products, and matched with the VMS-derived fishing positions, and sampled in ArcGIS 9.2 (Saitoh et al., 2011).

In 2011, NOAA estimates that 90% of seafood consumed in the United States was imported, and that about half of total imports are from aquaculture (Pramod et al., 2014). Otherwise illegal fisheries is known to occur in U.S. domestic waters with some estimates as high as 10-20%, so the general principles of Article 6.11 makes it necessary to obtain data to track illegal fishing or sources of fish (Pitcher et al., 2008). To address illegal, unreported and unregulated (IUU) fishing, NOAA has put into operation data workshops from its stock assessment agency. As the importance of a complex web of diverse data flowing into scientific advice for resource managers is recognized specifically in Article 12 of the Code, NOAA web site has a GIS Data section for fisheries, and a commercial fisheries statistics division web access to rapidly summarize U.S. commercial fisheries landings. And the implicit goal of Article 12 is to increase the scientific information available to allow for greater certainty and confidence and to reduce risk (National Oceanic and Atmospheric Administration, 2012). The NOAA Vessel Monitoring System (VMS) allows to monitor compliance (the use of flags of convenience to circumvent regulations), track violations, and provide substantial evidence for prosecution. This VMS data is subject to strict confidentiality requirements.

CatSat (the French worldwide satellite service to help fishing) has had no opportunity to be used concerning FOC fishing countries. Apparently, the Miami-based Roffer's Ocean Fishing Forecasting Service, Inc., providing fisheries forecasts derived from SRS (satellite remotely sensed information) since 1987, does not monitor vessels fishing activities under flags of convenience.

As a general comment on the needs of information and data concerning IUU, the point being to give expression to the Article 6 of the Code, it could be said that a global database of large-scale fishing vessels and vessels authorized to fish beyond the area of national jurisdiction of the flag State of the vessel should be established. A study of this kind could not be carried out without direct contact with, and assistance from, a wide spectrum of individuals and institutions concerned. Indeed, Article 7.5.2 stipulates that States should take into account fragmented data in relation to stock conditions (reference points, levels and distribution of fishing mortality and the impact of fishing activities (discards)). In particular, a reference point indicates a particular state of a fisheries indicator corresponding to a situation considered as desirable ("target reference point"), or undesirable and requiring immediate action ("limit reference point" and "threshold reference point") (Caddy and Mahon, 1995; Garcia, 1996). In addition, on closer inspection, Article 7.5.3 looks at how States, inter alia, should determine stock specific target/ limit reference points (also referred to as "reference value"). From the side of a responsible fisheries based information system (for instance, such a system can be the Fisheries Information System (FIS), built by Ifremer and available from SeaDataNet), these are the problems of a common statistical understanding, of a common web of harmonized data and information, a common knowledge framework specifically recognized by Article 12.

Two fundamental pitfalls or caveats of cost-led strategies are to be found in Articles 2 and 7. The costs incurred by governments could be calculated on the basis of factors such as the additional cost of research (as promoted by Article 2.i) resulting from scientific uncertainties (as examined by the Article 7.5.2) that arise from lack of good information on the catch and data of the species caught in IUU fisheries, the increased cost of monitoring, surveillance and enforcement at sea, and the extra cost of port and market based inspection schemes needed to combat IUU fishing. Longer-term costs could include the loss of long-term benefits to the economy due to the lower productivity of overfished stocks as a result of IUU fishing, and the loss of tax revenue or income to the State. Using Lloyd's data, Taiwan, Honduras, Panama, Spain, and Belize are the top five countries where companies that own or operate fishing vessels flagged to one of the top 14 FOC fishing countries are based.

HOW TO DISCOVER DATA?

Operational oceanography data centres have implemented web portals to introduce and browse their products. Data are available, through text search at one of the web sites or by combining metadata-oriented search engines and catalogues of long-term data sets, archives or operational products. Catalogues offering discovery services to coordinate activities are especially included for products facing responsible fisheries criteria. Could we do better? In spite of the fact that Article 7.4.2 states that the research in support of fishery conservation and management should be disseminated to interested parties, the question could simply be posed in these worlds: why won't people use data from ocean models in their fisheries model?

People won't use data from present day data infrastructure in their fishery models. And this is why, end-to-end models are less data-conditioned, key data are hard to integrate in assessment models, often there are no actual data on key processes, and models generally include too many parameters to unambiguously estimate all anyway. The Code also underlines, in Article 7.4.5, that sufficient knowledge of social, economic and institutional factors should be developed through data gathering, analysis and research. However, the availability of consistent, comparable and accessible marine socio-economic data is limited. For example, EMODnet provides data access to marine data across discipline based themes, scientific data (bathymetry, geology, seabed habitats, chemistry, biology, physics) and anthropogenic data (human activities). But no single source for socio-economic data. The importance to find and discover data related to aquaculture activities is recognized by Article 9.2.4, this Article states that appropriate databases and information networks to collect, share and disseminate aquaculture data should be established to facilitate cooperation on planning for aquaculture development at the national, sub-regional, regional and global level. This point is possibly not relevant for EuroGOOS.

Figure 3. Breakdown of the 3 major articles on data dissemination, sufficiency and discoverability in the 1995 FAO Code of Conduct for Responsible Fisheries. The infrastructures corresponding to the principles prescribing that research in support of fishery conservation should be disseminated are determined in the Article 7.4.2; those executing the precept of producing sufficient social knowledge concern Article 7.4.5; and for the principles inspiring better discoverability of data the implementation of the mandate guiding the planning for aquaculture is informed by the Article 9.2.4.

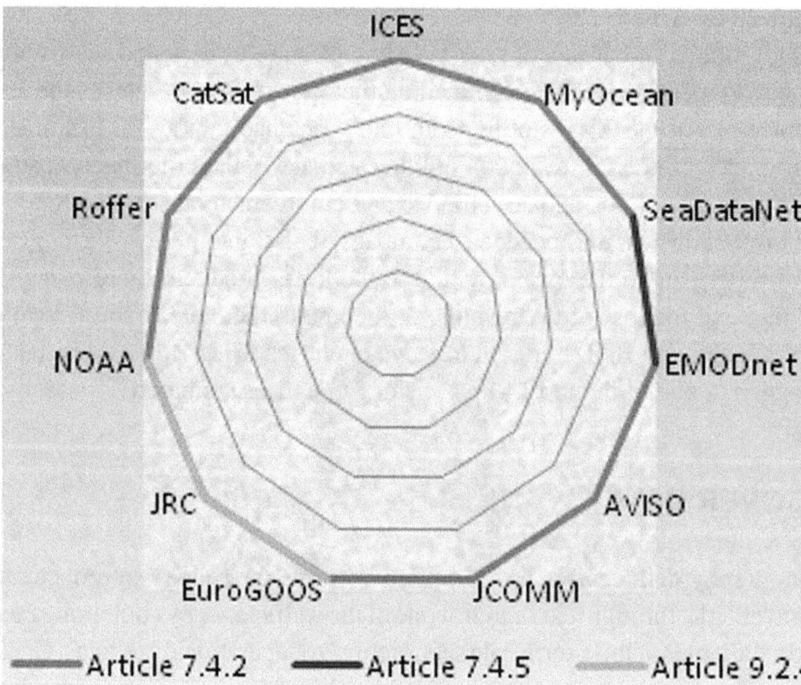

The main source of operational oceanographic data for fisheries is the ICES working group WGOOFE. WGOOFE Portal products are linked in such a way that it should be easy for users to find the kind of oceanographic data that they are looking for. The ICES database of trawl surveys (DATRAS) has been developed to make available the exchange data for all species, and for the catch-per-unit-effort distribution data. As part of an ICES initiative an inventory details the availability of data: oceanographic; contaminants, biological effects and biological community; fish trawl survey; eggs and larvae database; fish predation and historical plankton. Argo floats data and metadata are available through a series of web services, from WMO Number, and also from Argo interoperability services. MyOcean Insitu data are available via ftp and they are also ready via http. In IODE Ocean Data Portal data are available from a data search platform, and through a metadata search tool. SeaDataNet offers maximized availability through a SeaDataNet Common Data Index (CDI), in two versions: 'Quick search', 'Extended search'. Aviso long time-series are available with more than twelve years of gridded sea level anomalies over the whole ocean. JCOMM data collection activity is monitored from a common data section (buoys, ships, floats, tide gauges, GTS, data centres).

EMODnet access to data is available from seven portals. GOOS provides access to data and information products on a sustainable basis. JRC data availability concerns: EU Aquaculture, Fisheries Dependent Information, Fleet Economic Performance, Mediterranean and Black Sea, Fish Processing Industry and the EMFF context indicators.

From NOAA there is ample 2005 data on 81 federal commercial fisheries (480 fish, 54 marine mammals, 19 sea turtles, and 28 seabird by catch estimates). A 2015 NOAA Data Catalog prototype offers: 63.764 datasets; although the availability and completeness of the data are not guaranteed.

Catsat provides maps corresponding to a fishing area, software to be installed onboard for download and visualization of data. In the commercially available data sector, Roffer provides fisheries forecasts from SRS (satellite remote-sensing), by integrating infrared, ocean color chlorophyll, altimeter, radar, current meter, and meteorological data and produces a detailed oceanographic analysis.

UNDERSTANDING THE PRODUCTS

To make a valid analysis many different types of data and information are used in fisheries management. The importance of this complex web of diverse data flowing into scientific advice for resource managers is recognized specifically in Article 12 of the FAO Code. From climatic and oceanographic information to species-specific catch and population trends data, the knowledge of the entire context on the data products involves the provenance of events for operational production (Ma et al, 2017). These data come from a wide variety of sources and are synthesized and managed through multiple processes to make the information available and useful to decision-makers. And the understanding of the product's attributes involves: what is produced, the ocean region covered and its scale resolution, how it has been produced, who produced it, when and where it is made available, for how long, with what accuracy, delivery format and network delivery services, data and network service policy, fitness for purpose, qualification process, historical events with respect to time series, etc.

Product understanding for the general public is incorporated in the Article 12.3 of the Code. Common knowledge is required from States, because data generated by research should be analyzed and be published in a readily understood fashion. On the behalf of all who will be sharing it, this valuable insight must integrate the Code principles as a contribution to fisheries conservation, management and development. An effective way to attain the best scientific evidence, data gathering identify means by which ecosystem considerations can be included in capture fisheries management, pointing out future challenges in relevant strategies. Of particular relevance, nevertheless, is a desire to promote common efforts toward research and to utilize the knowledge required for the responsible management of living marine resources even in the absence of adequate scientific information. Data-less marine management must not be an obstacle to research.

As far as the best scientific knowledge is based on data, the use of the resource should respect confidentiality where appropriate, and be distributed in a timely way. These are common practices for data management for implementing the Code amongst countries. In this respect, States may limit public access to the catch data sets, where such access would adversely affect the confidentiality of commercial and industrial information, including the public interest in maintaining statistical confidentiality. The fact that data is not copyright-protected does not mean that the government may not restrict access to those data on the basis of other mechanisms (e.g. rules of secrecy). Otherwise, at the oceanographic operational data level, when you want to work with the data provided by services it is important that the services meet certain criteria regarding response time. Article 7.4.4 states that the complete and reliable statistics on catch and fishing effort should be updated regularly; and that States should compile and disseminate such data in a manner consistent. In moving the observational data (often via satellites) collected at sea to the appropriate archive or data distribution centres, the data are exchanged in real-time (up to 30 days old for oceanographic data, a few hours old for meteorological data) or in a less timely exchange referred to as delayed-mode. Real-time data exchange normally uses the GTS (Global Telecommunications System) with relatively few and well controlled formats. In delayed-mode, there are many formats and the communications channel is becoming increasingly the Internet.

To understand the status of data confidentiality and timely provision/access of the operational oceanographic products, we interpret both criteria focusing on the systems studied. MyOcean data are free for the purposes of public tasks, and its temporal resolution is approximately of 3 hours, also offering offline products. The Ocean Data Portal (IODE), by ensuring seamless access to ocean data and in order to fulfill its reporting obligations, establishes a timely access to quality data. The data sets provided by SeaDataNet can be retrieved by electronic download, and their use is free of charge (from the license), the arrangements for their provision includes that the timeliness of data dissemination is both in real-time and delayed mode. Argo data are publicly available within hours after collection. The altimetric data from Aviso are available from a Live Access Server and the timeliness of data is both delayed and near-real time. Operational oceanographic data that can be used from ICES portal are of free access except commercial fisheries data and confidentiality rules involving VMS data; ICES data are available in a timely way according to established principles. While JCOMM Network data are in public access, and are updated daily for Argo floats, data buoys and ship data are updated monthly and real-time maps and statistics are disseminated. The arrangements for the sharing of marine real-time data from EMODNET portals, make them freely available for browsing and downloading (older data require credentials). EuroGOOS data are accessible to the public and researchers in common simple methods, as near real-time measurements. Once published in STECF reports, the processed and aggregated fisheries data from the EU JRC becomes publicly available for further studies and analyses; these datasets are updated annually when the relevant STECF reports are published. A 2015 NOAA Data Catalogue is making data publicly released without requiring a login or password, the timeliness of these bycatch data is still under active development. Both CatSat and Roffer data are commercial so they are not publicly available, the timeliness of their transmission is real-time and near real-time, respectively.

DATA FORMATS FOR INTEROPERABILITY

Access to research data requires a broad international agreement on interoperability. The first requirement for interoperability is that the standards employed should be explicitly mentioned, because incompatibility of technical and procedural standards can be the most serious barriers to multiple uses of data sets. As the marginal costs of transmitting data through the Internet are close to zero, it is indeed in digital format that the greatest potential lies for improvements in the efficient distribution of data. Besides, relevant international data documentation standards, and research data formats are a key consideration in technological and semantic interoperability.

Research data are principally aimed at the computer-readable format. And catch data, altimetric and biological data are formatted and validated to integrate previously-developed ocean information systems. As the tasks of the fisheries data infrastructure are the indexing, retrieval and sharing of many data resources, it is becoming widely accepted that data content capturing, integration, and management require the development of detailed, formal interoperability protocols. Data formats are required to share and to be used by any number of different systems and also to achieve better indexing of resources, better retrieval of resources, and increased interaction within the fishery community (see Table 3 below). According to that, the concept of interoperability has various meanings such as uniform business processes, controlled terminology, communication protocols, reference implementations, or data standards allowing data exchange.

Fitting data together is a challenge. Interoperability means the possibility for data sets to be combined, and for services to interact, without repetitive manual intervention, in such a way that the result is coherent and the added value of the data sets and services is enhanced (Directive INSPIRE, 2007). The conceptual heterogeneity in the fishery information systems requires something beyond simple keyword-based searching. Data searches based on tagged content or sophisticated natural-language techniques require some conceptual structuring of the linguistic content of data, it requires conceptual interoperability. But it also requires syntactic interoperability. Syntactic operability refers to the common underlying data structure which enables exchange of data between different sites. Specific data formats

Table 3. Available fisheries and aquaculture data formats

Format	NetCDF/CF	OBIS	DwC	SDN	OGC	ODV	ASCII	XML	GIF	GF3	IC	Excel
EMODNET		X	X	X	X							
SeaDataNet	X			X		X						
JCOMM	X											
MyOcean	X											
ODP	X						X	X				
AVISO	X						X		X			
Argo	X									X		
ICES										X	X	
EuroGOOS	X					X						
JRC												X
NOAA	X											

are among the tools of syntactic interoperability, both for ensuring data storage and data transport. The notions of semantic and syntactic interoperability refer to a common understanding required to interchange information (Diviacco & Leadbetter, 2017).

The problems regarding the interoperability of marine and fishery data are common to a large number of policy themes and are experienced across the various levels of public authority. Typically, a considerable part of the effort involved in scientific stock assessments consists in assembling data from different sources. Although experience for benthic data is available for the immediate coast but not much further. And fisheries data is generally not publicly available (they do not include illegal landings and discards). Since 1995, the implementation – and thus impact - of the FAO Code of Conduct for Responsible Fisheries, has not resulted in an universal widely adopted format standard for data interoperability. But as its principles and provisions have been endorsed and adopted in almost unanimous fashion by the countries covered (Hosch, 2011), a wealth of data lying in fisheries and aquaculture resources involves a considerable need in interoperability. Relevance of Code themes always takes into account that status of data, particularly because while fundamental changes in the fisheries sector has remained few, farm and environmental management has carried a rather important degree of change to the aquaculture sector.

So, as far as 1995 Code Article 7 covers information on the adoption and implementation of fisheries management principles; clause 7.4.6 states that an internationally agreed format is mandatory for States compiling fishery-related and other supporting scientific data. Otherwise, clause 7.4.7 involves an effort in satisfying public data availability by addressing a recommendation to compile data and make them available, in a manner consistent with any confidentiality requirements. In this view, this could be achieved by determining an agreed data format to all members of the fisheries organizations. For efficient data sharing, fisheries handling systems make efforts to define and describe their products, and to adopt a common standard to format data. netCDF is the primary data storage and transport format which has been focused to ensure interoperability between systems. Climate and Forecast (CF) metadata conventions are a second component of this common low layer architecture: CF-compliant data files are self-describing (no external tables are needed to interpret the file). OPeNDAP is the network data access protocol permitting ocean science applications to read files in formats with which they were previously incompatible, and it also offers aggregation capabilities (Blower et al., 2009).

The EMODnet portal is able to provide links to metadata, data and data products for different systems by establishing direct machine-to-machine connections. The portal manages different data protocols for handling marine data by implementing different data formats. For this interoperability several possible data flows can be identified: IPT (DIGIR), custom services, OGC web services and file harvesting (CDI). The majority of the partners use the IPT (Integrated Publishing Toolkit), to publish their marine data. IPT is used to publish and share biodiversity datasets and is specifically designed for interoperability: it enables the publishing of content in databases, Microsoft Excel spreadsheets, or text files using open standards (namely the Darwin Core). The international Darwin Core standard on exchanging data is the data format behind the IPT exchange mechanism and allows the data to be made easily available through the EMODnet data portal. Specifically, the portal objective is to format the data and to perform taxonomic and data standardizations to improve interoperability within the portal and with other data systems. Once this has been done, quality control procedures can run on all the data and the results of these can help to determine the suitability of the data in the creation of data products. To improve interoperability, several data formats are possible: OBIS scheme, DarwinCore Occurrence Scheme (IPT) (extension: DwC MeasurementOrFact), Biological data exchange format (BIODEF - SDN), Shape files (OGC compliant), and web services. The mechanism through which the data become accessible in the

EMODnet portal are: IPT, SeaDataNet (SDN) format, OGC (WFS), its own portal web services or a combination of protocols. In general the data flow is largely ready.

A specific data format is available to enable National Oceanographic Data Centers (NODC's) to make marine data accessible by using the SeaDataNet infrastructure. The format is a general and high-level one focusing on common information elements for marine data. At the same time it is sufficiently flexible/extendable to be applicable for the variety of marine data NODC's are managing and it is possible to derive OBIS or Darwin Core compatible datasets from the format. Otherwise SeaDataNet updates all the information, thus making possible the interoperability between the diverse data systems like metadata, vocabulary standards and the data formats used by the quality control procedure. Indeed, SDN general data analysis and visualization software, Ocean Data View (ODV), offers a seamless connection to SeaDataNet that is upgraded to the latest version for SDN ODV data formats, and to the quality flag scales. SeaDataNet format standards and interoperability tools have expanded to other EU-projects: Black Sea Scene datasets can be queried by using the SDN Common Data Index (CDI), so using the SDN-metadata CDI format (ISO 19115 compliant). Apart from that, the data are provided in SDN data file format: ODV for data profiles (e.g. time-series), NetCDF-CF for gridded data (e.g. ADCP). CASPINFO project, Geo-Seas project, and EUROFLEETS are all governed by the SeaDataNet metadata and data formats and access protocols (Data Transport Formats for data sets delivery, OGC standards for mapping and viewing services, SOAP Web Services). SeaDataNet standards and interoperability protocols have been adopted as core elements in EMODNET.

Interoperability is a central issue for JCOMM because of the many contributors to data collection, processing and distribution. As a strong data management architecture JCOMM is essential to operate Global Data Centres for these networks: Argo profiling floats, OceanSITES long term moorings, and GOSUD surface underway observations from ships. All these systems have elected to base their data management on netCDF and CF data format conventions (Blower et al., 2009).

MyOcean extensive standardization and interoperability results in that users can download products according to their needs in a unique format, netCDF. Data can be created and understood by the basic netCDF utilities, and the products are delivered in NetCDF format via OPeNDAP and ftp. MyOcean includes tasks to implement interoperability with other systems, with the Weather Information System (WIS) or GEO Portal (HMA).

The ODP transport data file formats that are available are NetCDF (E2E convention), ASCII and XML. The ODP interoperability specifications include typical solutions for common dictionaries, formats, protocols and other means for heterogeneous data unification. The IODE Ocean Data Portal identifies and recommends standards to provide interoperability with IODE data centres to allow shared use of metadata, data and products. Interoperability is one of ODP major underlying principles, and SeaDataNet can be searched through ODP. Thus ODP adopts and supports major standards from ISO and the Open Geospatial Consortium (OGC) to provide access from many platforms within a single portal. As one of the major components within ODP, the interoperability package is very critical to reduce or avoid duplication of efforts when required data and information already exist.

Data format standardization was progressing from 1995, when AVISO, the showcase of CNES (National Centre for Space Studies) activities in altimetry, initially was distributed through Internet. Aviso ocean altimetry data on internet uses a choice of file formats (gif, ascii, NetCDF). Since 1992, Aviso manages 10 satellite missions, making radar altimetry more usable and web-interoperable in the coastal ocean. Also, the relationship between AVISO and MyOcean is guaranteed by standards for Ocean Data Interoperability. From 2003, Aviso utilizes OPeNDAP, a standard for Ocean Data Interoperability, thus enabling access to remote data sets through familiar data analysis and visualization packages (Rosmorduc et al., 2006).

Argo profiling floats data User's Manual has designed and documented a set of metadata conventions for netCDF. And within the CF community the decision has been made to extend the Argo format to be CF compatible and benefit from the wider availability of its software tools (Hankin et al., 2009). Inside SeaDataNet, Argo data have served to harmonize the CF and IOC GF3 parameter codes and implement gateways between the two vocabularies.

Data collected on behalf of ICES are archived in ICES data format, which is only intended for ICES internal data handling. Otherwise, as the ICES Data Centre accepts a wide variety of marine data and meta-data types, the RNODC Formats (ICES) follow the formatting practices (GF3) and advice proposed in the guidelines on how oceanographic profile data should be submitted written down by the IOC's Group of Experts on the Technical Aspects of data Exchange (GETADE). Catch data (submitted in aka exchange format spreadsheet) are extracted in the InterCatch (IC) format at ICES level. In terms of marine metadata interoperability, ICES uses SeaDataNet codes for ship names.

Within EuroGOOS, OceanSITES NetCDF, and SeaDataNet ODV, are used for data exchange between ROOSes' portals. These formats are revised periodically by the DataMEQ (Management, Exchange and Quality) working group to benefit from technical progress. DATAMEQ Working Group recommends definition of a common strategy to handle duplicates and to get interoperability between real-time and delayed mode data streams. SeaDataNet maintains the European Directory of Ocean Observing Systems (EDIOS) and improves access to real-time data.

Fisheries and aquaculture data in place by JRC can be downloaded in standard formats for more advanced statistical analyses and modeling, during the Scientific, Technical and Economic Committee for Fisheries (STECF) working groups meetings. The DCF platform for data exchange uses EXCEL templates formats. JRC fisheries data interoperability follows INSPIRE.

NOAA agreed upon netCDF-CF as the base file format for the entire collection of mission data. NOAA recognized the cost savings and efficiencies gained with these standards, by making the altimetry data stream easier to utilize and because it delivers the data in various output formats as preferred by different user communities. The Group for High Resolution Sea Surface Temperature (GHRSST) mandates the use of CF-compliant netCDF format. Interoperability for gridded data (satellite products, and climatological analyses) uses the OPeNDAP data access protocol. Since 2009 NOAA's National Data Buoy Center has served these data via OPeNDAP, providing both DAP and OGC/WCS access. Argo floats observations are available from a NOAA OPeNDAP long term archive (Hankin et al., 2010).

CatSat provides compressed format to reduce the costs of satellite communication. Roffer's Ocean Fish Forecasting Service, Inc. interoperable data sets are available among GCOOS (Gulf of Mexico Coastal Ocean Observing System) partners. ROFFS™ derives a complete analysis in an easy-to-use format.

CONCLUSION

This brief review of data approaches under the 1995 FAO Code of Conduct for Responsible Fisheries has pointed to recent developments in cross-domain operational oceanography that require higher visibility of the data products for fisheries and aquaculture. The data systems studied illustrate the relevance of the Code's clauses recommending specific types of measurements with diagnostic value to support fisheries and aquaculture data management was (Caddy, 2007). In particular a capacity to meet temporal resolution, time frame, reporting frequency, and interoperability formats requirements were the criteria used to identify the fulfillment of the reporting obligations. It was pointed out how useful is to adhere to the Code standards in data affairs, especially in view of the uncertainty derived from the impacts of fishery and aquaculture exploitation. Uncertainty is made explicit in the research agenda of marine data infrastructures concerning fisheries, catching and fish farming. The resulting data indicate that the development of marketing practices and improvement of consumer information can enhance the move to more responsible fisheries (OECD, 2000, p.7). As the data (particularly information on targets and threats) need to be fed back on an ongoing basis for purposes of managing fisheries and aquaculture, all stakeholders provide aggregated data services. But the informational issues that need to be addressed include high-priority problems that need to be solved with raw data. It is important to have a methodology to display and breakdown the major articles in the Code on data dissemination, sufficiency and discoverability. EMODNET and SeaDataNet are more closely based than the other data infrastructures on the articles of the Code, in that social, economic and institutional factors related to sustainability evaluated with data are clear. A broad approach based on disseminating research data in support of fishery conservation is satisfied by the main data sources, such as ICES, MyOcean and NOAA, and it is also facilitated by private companies (CatSat and Roffer). Tools based on better discoverability of data for aquaculture are prioritized by European resources, like the JRC and EuroGOOS.

Regarding their purpose of curtailing illegal catches and showing compliance with the international legal instrument, the data infrastructures studied are consistent with the objectives of the VMS and SRS data. Thus, they contribute to improve the chain of custody of documents and to traceability checks by adding greater relevance to operational fisheries oceanography.

We have compared code compliance with other relevant developments of satellite-based operational oceanography applied to fisheries (including aquaculture). The costs of enforcement can be prohibitive, and people won't use data from present data infrastructures in their fisheries and aquaculture model. Especially to answer the issue of uncertainty for the estimation of the impacts of fishing on marine ecosystems, and of the socioeconomic stakeholders' demands, the vanguard of the data systems analyzed implement the 1995 FAO Code by merging satellite/environment data with catch data and biological data.

REFERENCES

Bahurel, P., Adragna, F., Bell, M. J., Jacq, F., Johannessen, J. A., Le Traon, P.-Y., . . . She, J. (2009). Ocean monitoring and forecasting core services, the European MyOcean example. In Oceanobs 2009.

Belov, S., Mikhailov, N., & Belova, K. (2013). Ocean Data Portal: from data access to integration platform. In *IMDIS 2013: International Conference on Marine data and Information Systems*.

Berx, B., Dickey-Collas, M., Skogen, M. D., De Roeck, Y. H., Klein, H., Barciela, R., & Schrum, C. et al. (2011). Does operational oceanography address the needs of fisheries and applied environmental scientists? *Oceanography (Washington, D.C.)*, *24*(01), 166–171. doi:10.5670/oceanog.2011.14

Bijma, J., Pörtner, H.-O., Yesson, Ch., & Rogers, A. D. (2013). Climate change and the oceans - What does the future hold? *Marine Pollution Bulletin*, *74*(2), 495–505. doi:10.1016/j.marpolbul.2013.07.022 PMID:23932473

Blower, J. D., Blanc, F., Clancy, M., Cornillon, P., Donlon, C., Hacker, P., & Srinivasan, A. et al. (2009). Serving GODAE data and products to the ocean community. *Oceanography (Washington, D.C.), 22*(3), 70–79. doi:10.5670/oceanog.2009.67

Caddy, J. F. (2007). *Using questionnaires based on the Code of Conduct for Responsible Fisheries as a diagnostic tool in support of fisheries management.* Roma: FAO.

Caddy, J. F., & Mahon, R. (1995). Reference points for fisheries management.[Rome: FAO.]. *FAO Fisheries Technical Paper*, 347.

Council. (2008). Council Regulation (EC) No 199/2008 of 25 February 2008 concerning the establishment of a Community framework for the collection, management and use of data in the fisheries sector and support for scientific advice regarding the Common Fisheries Policy. *Official Journal of the European Union, 51*(L 60), 1-12.

Directive INSPIRE. (2007). Directive 2007/2/EC of the European Parliament and of the Council of 14 March 2007 establishing an Infrastructure for Spatial Information in the European Community (INSPIRE). *Official Journal of the European Union, 50*(L 108), 1-14.

Diviacco, P., & Leadbetter, A. (2017). Balancing Formalization and Representation in Cross-Domain Data Management for Sustainable Development. In *Oceanographic and Marine Cross-Domain Data Management for Sustainable Development.* Hershey, PA: IGI Global.

Ehlers, P. (2013). The need for sustained operational oceanographic services. In *Sustainable operational oceanography:Proceedings of the Sixth International Conference on EuroGOOS.*

Folorunsho, R. (2006). The objectives of the Argo workshop. In *Argo Capacity Building for the Atlantic Countries Workshop.* University of Ghana.

Food and Agriculture Organization of the United Nations. (2000). *Code of Conduct for Responsible Fisheries* (Orig. 1995. Repr. 2000). Rome: FAO. Available: ftp://ftp.fao.org/docrep/fao/005/v9878e/v9878e00.pdf

Garcia, S. M. (1996). The Precautionary Approach to Fisheries and its Implications for Fishery Research, Technology and Management: An Updated Review. In *FAO.- 1996. Precautionary approach to fisheries. Part 2: scientific papers. FAO Fisheries Technical Paper, 350(2).* Rome: FAO.

Glaves, H. (2017). Developing a Common Global Framework for Marine Data Management. In *Oceanographic and Marine Cross-Domain Data Management for Sustainable Development.* Hershey, PA: IGI Global.

Green Paper on EU Maritime Policy. (2006). *Background paper No. 4a on the European Marine Observation and Data Network.* COM(2006) 275 final Volume II – Annex, 7.6.2006.

Hankin, S., Blower, J. D., Carval, T., Casey, K. S., Donlon, C., Lauret, O., . . . Harlan, J. (2010). netCDF-CF-OPENDAP: Standards for ocean data interoperability and object lessons for community data standards processes. In Oceanobs 2009.

Hosch, G., Ferraro, G., & Failler, P. (2011). The 1995 FAO Code of Conduct for Responsible Fisheries: Adopting, implementing or scoring results? *Marine Policy*, *35*(2), 189–200. doi:10.1016/j.marpol.2010.09.005

Le Traon, P. Y., Larnicol, G., Guinehut, S., Pouliquen, S., Bentamy, A., Roemmich, D., & Breivik, L. A. et al. (2009). Data assembly and processing for operational oceanography: 10 years of achievements. *Oceanography (Washington, D.C.)*, *22*(3), 56–69. doi:10.5670/oceanog.2009.66

Leadbetter, A., Cheatham, M., Shepherd, A., & Thomas, R. (2017). Linked Ocean Data 2.0. In *Oceanographic and Marine Cross-Domain Data Management for Sustainable Development*. Hershey, PA: IGI Global.

Leadbetter, A., Raymond, L., Chandler, C., Pikula, L., Pissierssens, P., & Urban, E. (2013) Ocean Data Publication Cookbook. Paris: UNESCO.

Ma, X., Fu, L., Fox, P., Di Stefano, M., & West, P. (2017). Documenting Provenance for Reproducible Marine Ecosystem Assessment in Open Science. In *Oceanographic and Marine Cross-Domain Data Management for Sustainable Development*. Hershey, PA: IGI Global.

Manzella, G., Bartolini, R., Bustaffa, F., D'Angelo, P., DeMattei, M., Frontini, F., & Novellino, A. et al. (2017). Semantic Search Engine for Data Management and Sustainable Development. In *Oceanographic and Marine Cross-Domain Data Management for Sustainable Development*. Hershey, PA: IGI Global.

Marine Knowledge 2020. (2013). *From seabed mapping to ocean forecasting*. Green Paper COM. Available at: http://ec.europa.eu/maritimeaffairs/pdf/com_2006_0275_en_part2.pdf

National Oceanic and Atmospheric Administration. (2012). NOAA fisheries implementation plan of the FAO code of conduct for responsible fisheries. U.S. Department of Commerce National Oceanic and Atmospheric Administration, National Marine Fisheries Service.

Organisation for Economic Co-operation and Development. (2000). *Transition to Responsible Fisheries. Government Financial Transfers and Resource Sustainability: Case Studies. AGR/FI(2000)10/FINAL.* Paris: OECD.

Pinto, C., & Holsworth, N. (2013). ICES Data portal, a single entry point to marine knowledge. In *IMDIS 2013: International Conference on Marine data and Information Systems.*

Pitcher, T., Pramod, G., Kalikoski, D., & Short, K. (2009). Not honouring the code. *Nature*, *457*(7230), 658–659. doi:10.1038/457658a PMID:19194431

Pitcher, T. J., Kalikoski, D., Pramod, G., & Short, K. (2008). *Safe conduct? Twelve years fishing the UN Code*. WWF.

Pontecorvo, G. (2003). Insularity of scientific disciplines and uncertainty about supply: The two keys to the failure of fisheries management. *Marine Policy*, *27*(1), 69–73. doi:10.1016/S0308-597X(02)00053-2

Pouliquen, S., Carval, T., Guillotin, D., Loubrieu, T., Coatanoan, C., Grouazel, A., … Manzella, G. (2013). MyOcean In Situ TAC: A new in situ service for operational and research communities. *Mercator Ocean—Coriolis Quarterly Newsletter*, *48*, 3-9.

Pramod, G., Nakamura, K., Pitcher, T. J., & Delagran, L. (2014). Estimates of illegal and unreported fish in seafood imports to the USA. *Marine Policy*, *48*, 102–113. doi:10.1016/j.marpol.2014.03.019

Rosmorduc, V., Bronner, E., Guinle, T., Maheu, C., Morrow, R., Nino, F., & Birol, F. (2014). AVISO+, the new reference web portal for altimetry. *Geophysical Research Abstracts*, 16.

Saitoh, S. I., Mugo, R., Radiarta, I. N., Asaga, S., Takahashi, F., Hirawake, T., & Shima, S. et al. (2011). Some operational uses of satellite remote sensing and marine GIS for sustainable fisheries and aquaculture. *ICES Journal of Marine Science*, *68*(4), 687–695. doi:10.1093/icesjms/fsq190

Schaap, D. (2017). SeaDataNet – Towards a Pan-European Infrastructure for Marine and Ocean Data Management. In *Oceanographic and Marine Cross-Domain Data Management for Sustainable Development*. Hershey, PA: IGI Global.

Schaap, D. M. A., & Lowry, R. K. (2010). SeaDataNet – Pan-European infrastructure for marine and ocean data management: Unified access to distributed data sets. *International Journal of Digital Earth*, *3*(sup1), 50–69. doi:10.1080/17538941003660974

Simoncelli, S., Coatanaon, C., Bäck, Ö., Sagen, H., Scory, S., Tezcan, D., & Tonani, M. et al. (2013). Temperature and salinity historical data collections for the european marginal seas: aggregation and quality assessment procedures. In *IMDIS 2013: International Conference on Marine data and Information Systems*.

Van Lancker, V., Francken, F., Kint, L., Terseleer, N., Van den Eynde, D., De Mol, L., & van Heteren, S. et al. (2017). Oceanographic Data Management; Quills and Free Text to the digital age and 'Big Data. In *Oceanographic and Marine Cross-Domain Data Management for Sustainable Development*. Hershey, PA: IGI Global.

World Bank, Food and Agriculture Organization of the United Nations. (2009). *The Sunken Billions: The Economic Justification for Fisheries Reform*. Washington, DC: The World Bank.

ENDNOTES

1. The Flanders Marine Institute (Dutch: Vlaams Instituut voor de Zee, VLIZ)
2. The Hellenic Centre for Marine Research (HCMR)
3. IFREMER; English: French Research Institute for Exploitation of the Sea
4. Spanish Institute of Oceanography (IEO)

Chapter 14
Integrated Monitoring in the Voordelta, The Netherlands:
Monitoring and Data Management for Evaluation of Nature Compensation Measures

Niels Kinneging
Deltares, The Netherlands

Meinte Blaas
Deltares, The Netherlands

Arjen Boon
Deltares, The Netherlands

Kees Borst
Rijkswaterstaat, The Netherlands

Gerrit Hendriksen
Deltares, The Netherlands

Gerard Van der Kolff
Deltares, The Netherlands

Theo Prins
Deltares, The Netherlands

Willem Stolte
Deltares, The Netherlands

ABSTRACT

Monitoring of the environmental effects of a harbour extension and the compensation measures is a very complex task. The Voordelta area has high natural values, but is also of high economic importance. To implement a monitoring strategy for this area a multidisciplinary consortium has been formed, consisting of a number of institutes and companies. A central data management facility was set up for data storage and management. This chapter illustrates the data management approach using the Voordelta monitoring programme for the years 2004 to 2013. A central data management facility was set up for data storage and management. A repository gives access to raw data files to all team members. From the analysis of the raw data a number of information products have been developed and disseminated to the authorities and the public through Google Earth. It will be shown, that the presence of a strong multidisciplinary team and good collaboration is the key to success in this complex programme. The way the data have been managed supports this process enormously.

DOI: 10.4018/978-1-5225-0700-0.ch014

INTRODUCTION

The last decades there is an increased awareness of the need for the protection of the marine environment. The European Commission (EC) has issued a number of directives that require EU member states to protect and restore marine biodiversity and habitats, to assess the impact of human activities on the marine environment and coastal and transitional waters and to regulate the human activities leading to deterioration of the marine environment. These directives are the Birds Directive (1979, amended in 2009) (European Parliament and Council, 2009), the Habitats Directive (Council of the European Communities, 1992), the Water Framework Directive (2000) and the Marine Strategy Framework Directive (European Parliament and Council, 2008). Furthermore the EC has established a network of protected areas, the Natura 2000 network.

The implementation of marine environmental policies requires a lot of data and information on the state of the environment and on trends in various parameters that describe the ecosystem. This is widely acknowledged in the last decades and initiatives for professional data management are being implemented. The same data dependence holds for marine projects in general, and data management should be an integral part of those projects.

The project taken below to illustrate the importance of data management in large monitoring projects revolved around the effectiveness of various nature compensation measures to compensate for the significant loss of marine habitat and foraging area protected under the Habitats Directive. This loss was caused by the second coastal expansion of the Port of Rotterdam.

To evaluate this compensation measure a Monitoring and Evaluation Plan (MEP) was setup and monitoring was carried out. A part of the MEP was dedicated to the complex management of the streams, quality, formatting and access of data both during the project and afterwards.

This kind of MEP is multidisciplinary by nature and involves experts like ecologists, biologists, environmental chemists and physicists. A lot of data is gathered and must be combined to answer the evaluation questions. Key to successful co-operation in a multidisciplinary consortium is the generic and open availability of the data and project results for the consortium members. Data management is therefore indispensable in complex monitoring projects, like the one discussed in this chapter.

BACKGROUND

European Policy Drivers

The major challenge of the current society is to protect our natural environment and still support the growing population and increasing economic activities. A lot of attention in this respect is given to the marine environment and especially the coastal regions. Currently, marine protected areas are predominantly found in near shore and coastal waters (Reker et al, 2015). The pressure on these areas from economic activities is growing rapidly.

Therefore the European Commission launched the Blue Growth strategy within the Europe 2020 strategy for smart, sustainable and inclusive growth. Blue Growth is the long term strategy to support sustainable growth in the marine and maritime sectors as a whole. The 'blue' economy represents roughly 5.4 million jobs and generates a gross added value of almost €500 billion a year.

The European Union has adopted a number of directives, which try to regulate the sustainable development of the marine environment. A thorough and complete description of all directives is beyond the scope of this chapter. We will limit ourselves to some major drivers of the European policy.

Some directives have the specific goals to protect wild flora and fauna, like the Habitat Directive (Council of the European Communities, 1992) and the Birds Directive (European Parliament and Council, 2009). Under these directives protected areas can be designated as Natura 2000 sites, with specific objectives to protect natural values. Human activities that have a negative effect on the conservation objectives of a Natura 2000 site require compensation measures, if there are reasons of overriding public interest to continue those activities.

Other directives give a framework for member states to develop a marine policy, like the Marine Strategy Framework Directive (MSFD) (European Parliament and Council, 2008) and the Maritime Spatial Planning Directive (MSP) (European Parliament and Council, 2014).

Instead of protecting specific species the latter directives, especially the MSFD, adopted an ecosystem approach. The aim is to improve the protection of the marine environment by integrating environmental protection and sustainable use. Member states must define Good Environmental Status (GES) and targets for the marine environment. These GES and targets are assessed every six years on the basis of environmental monitoring. The MSFD also specifically addresses the role of data and information in the process. Article 19(3) requires that all data and information resulting from the assessments and the monitoring programmes are made available. In order to streamline the data flows for the MSFD a special working group (WG on Data, Information and Knowledge Exchange, DIKE) has been established.

Information Needs

The information need should be the driver for all monitoring activities. The information need can be identified for different spatial scales (local, national, regional) and for different time scales (near real time, years, decades, centuries).

From the above a clear information need has been identified on a European level. The European Commission recognizes that information is needed to establish a marine policy and to monitor and assess the programmes, that are defined to protect and improve the marine environment. The same kind of information exists on a regional and national scale. On a regional scale the regional sea conventions, like OSPAR, HELCOM, Barcelona and Black Sea conventions, also make environmental assessments of the regionals seas and have an information need to do so. The same holds for the individual member states on a national scale. The time scale for this type of monitoring is usually one or more decades.

Not all information needs arise from environmental requirements. For operational oceanography a lot of data have to be gathered in (near) real time. These data are used for tidal predictions, flood warning systems and for the management of various activities, like shipping and dredging. For these monitoring activities international co-operation has been established within the context of the Global Ocean Observing System of the Intergovernmental Oceanographic Commission of UNESCO (IOC GOOS).

Research on climate change has a very different information need. This research requires both very large spatial and long temporal scales. The behaviour of the oceans must be studied on a global scale and preferably data should be available for a number of centuries. The Intergovernmental Panel on Climate Change (IPCC) co-ordinates global climate research.

On the opposite spectrum of climate change research is the information need for marine projects. Projects can have various goals and requirements, but usually a project is limited in time and space.

Furthermore most projects require a combination of data sources to fulfil the information need. The subprojects dedicated to data and information typically have a multidisciplinary character.

A number of categories for project can be distinguished:

- In relation to infrastructural projects as well as the development of new marine activities usually some preparatory studies are carried out, a feasibility study and an environmental impact assessment.
- The aim of a feasibility study is to find out whether a new activity can be carried out economically. Local conditions, like wind, waves and composition of the sea floor, need to be determined.
- An Environmental Impact Assessment (including an Appropriate Assessment) is a legal obligation to evaluate beforehand the impact this new activity has on the environment. Different scenarios have to be investigated and usually mitigation and compensation measures are evaluated at this stage. The compensation measures for the coastal extension of the Port of Rotterdam led to the study that is described as a case in this chapter.
- During the implementation of infrastructural projects also a lot of information is needed. This information is usually related to operational requirements of the building process, but also as an obligation to the legal authorities.
- After a project is completed and compensation measures have been taken, the outcome of those measures have to be evaluated and it has been assessed whether the goals for the measures are met.
- A final category of projects are research projects. Here scientific questions are the central drivers for the information needs. Especially for larger research projects a good data management is essential to its success. A good example is the Sand Engine project (see http://www.dezandmotor.nl/en/home/). In this project a large sand nourishment is deposited on the Dutch coast between Rotterdam and The Hague. The goal is to realize a sustainable enforcement of the Dutch coast. The project involved various studies by scientists from all over the world. The data management of the project has been structured using the OpenEarth building blocks described in this chapter.

With regard to data management, projects are faced with three problems.

First, in many cases a project needs data beyond the boundaries of the project. In order to retain data useful for a BACI approach ((Before-After Control-Impact, e.g., Green, 1979) a proper assessment of the baseline ('before') conditions has to be established by T_0-monitoring. In the daily practice of many projects there is, however, a lot of pressure to start directly after decisions (and budget) have been formalized, which typically leaves very little room for monitoring of the baseline. Therefore the project has to obtain data from other sources, like previous projects and historic, longer term monitoring. The 'control' data imply that the information needs extend beyond the spatial scale of the impact region, such that data from neighbouring monitoring stations must be incorporated.

Secondly due to the fixed time of projects the data and information generated by a project is often not available for future use. Project funding in many cases does not incorporate the long-term consolidation of data. This attitude prevents projects setting up a proper data management system.

Thirdly, there is an increasing demand for open accessibility of data, in particular when data are collected in monitoring programs funded by public authorities and when data are used for assessments under legal obligations.

Types of Data

As remarked, projects require data and information from various sources. Often the character of the data imposes different technical solutions for the data management and therefore a summary of the types of data is given, that can be encountered in projects:

- High resolution spatial data are generally obtained by remote sensing measurement systems. Basically the data points are sampled irregularly over the surface, but often for easy handling by GIS systems and visualization the data are resampled to a regular grid. The remote sensing systems can be either satellite imagery (radar, water colour, temperature, etc), air borne measurements (Lidar, photography) or from vessel mounted systems (sonar and multibeam echo sounders).
- In situ measurements know a coarse spatial sampling. Many semi-continuous measurements are done automatically at measurement station on poles, platforms and buoys. Other in situ measurements are done by a probe or by sampling the water column, followed by laboratory analysis.
- A compromise between these two data types are measurements taken along a track or trajectory. Examples of these data are single beam echo soundings or bird measurements using GPS tracking devices.
- A special type of data results from numerical modelling. Numerical modelling can be used to make predictions of the state of the environmental parameters or to obtain data at a much finer spatio-temporal resolution and in locations and times not achievable through measurements. With the use of data assimilation, numerical models can serve as data-interpolation tool and determine derived parameters such as energy or mass fluxes or difficult to observe quantities in the water column such as near-bed shear stresses. The numerical models thus are a powerful tool, but they require an adequate schematization of the area and good initial and boundary conditions.

Data Initiatives

Data are crucial to all marine activities, but the gathering of data is expensive. Therefore it is profitable to share data. The last decades many initiatives were taken to make monitoring data publicly available. These initiatives focus on either the data acquisition or the data distribution.

Although in most cases the United States of America plays a front role in the field of open data, the focus of this chapter is on the European initiatives. The need for a transnational approach is primarily present for operational fields like meteorology and operational oceanography. In these fields international co-operation is well developed and infrastructures are present. Other fields in oceanography and environmental sciences followed later and infrastructures are under development. The European Commission has set up some major programmes for the development of infrastructures for data handling. For the acquisition of satellite remote sensing the Copernicus programme is leading and for data management and distribution for marine data the EMODnet programme plays a central role.

Copernicus

Since 1998 the European Union used the GMES (Global Monitoring for Environment and Security) programme to develop systems for earth observation. Within GMES the successful ESA Envisat satellite mission has been performed. Although Envisat's main goal was research and development, the data were often used within longer-term monitoring programmes. In 2012, the Copernicus programme was launched as a successor to GMES. Copernicus specifically targets operational and continuous services with the Sentinel satellite missions.

For the purpose of this chapter especially the services of the Copernicus programme are important. Copernicus provides -amongst others- the Copernicus Marine Environment Monitoring Services (CMEMS), in which satellite imagery data are available as open data. This core service may be augmented by bespoke downstream services targeted at a specific region or user sector. The amount and quality of data in the Copernicus services will be growing in the coming years. Even if some of these satellite data products may have limited resolution or accuracy given the projects information needs, the data are useful as background or as boundary condition for more detailed data and models.

EMODnet

In 2008 the EMODnet (European Marine Observations and Data Network) has been launched by the Directorate General of Maritime Affairs (DG MARE) of the EC. EMODnet is the most important initiative of DG MARE in relation to marine data and should become the central infrastructure for the sharing of marine data. The goal of EMODnet is to make marine data publicly available and thus stimutate the economic developments (Blue Growth). Another purpose is to harmonize marine data and perform quality control on the data. More information can be found in Giorgetti et al (2015).

At this moment the major work of EMODnet is done within 7 thematic lots:

- Chemistry
- Biology
- Bathymetry
- Physics
- Geology
- Habitat mapping
- Human activities

Within these 7 lots data are gathered and information products (thematic maps of different European sea regions) are made.

At this moment EMODnet has access to a lot of governmental data and data from research institutes, but data from private companies and projects are difficult to disclose for EMODnet. Therefore a new initiative was taken to facilitate the uploading of these data to some central oceanographic data centres, the EMODnet Ingestion initiative. This project will start in the first half of 2016.

SeaDataNet

The EMODnet initiative has been built largely on the SeaDataNet infrastructure. SeaDataNet is a project under the Seventh Framework Programme of the EU, in which many marine institutes and data centres across Europe co-operate. Schaap et al. (2017, chapter 7 this book) give more information on the SeaDataNet infrastructure.

CASE STUDY: THE *VOORDELTA* PROJECT

The *Voordelta* is the south-western part of the Dutch coastal zone of the North Sea where the rivers Rhine, Meuse and Scheldt discharge into the North Sea (see Figure 1). The total area is about 900 square kilometers. The outer delta of this originally estuarine system is a relatively shallow and highly dynamic area with many subtidal and some intertidal sandbanks, with a high ecological value. Therefore, this area has been assigned as a Natura 2000 site (a Special Area of Conservation under the Habitats Directive and a Special Protection Area under the Birds Directive). The relevant natural values protected within this MEP project were the habitat type H1110 (Sandbanks which are slightly covered by sea water all the time), and three bird species, common scoter (*Melanitta nigra)*, common tern (*Sterna hirundo)* and sandwich tern (*Sterna sandvicensis)*. Other natural values in this area are its function as a nursery area for juvenile fish, and an important foraging and resting habitat for seals and for other bird species. Next to its natural value, it has an important economic value for fishing, shipping and recreation.

Since 2009, a second expansion of the Port of Rotterdam was realized by land reclamation within this Natura2000 site: the Maasvlakte 2 of approximately 2500 ha. Rotterdam is one of the most important harbours in Europe and the extension was considered necessary to consolidate the international position of the Port of Rotterdam. Two major targets were formulated by the government for the Project development Mainport Rotterdam (PMR):

- Reinforce the industrial activities in the Rotterdam harbour
- Improve the environmental conditions for the larger Rotterdam area

As determined in an Appropriate Assessment, Maasvlakte 2 resulted in a significant loss of protected habitat and foraging areas of the three aforementioned bird species in the Voordelta. As required by EU's Habitats Directive this loss of habitat had to be compensated for. Since no quantitative compensation goal could be established (i.e. create additional habitat H1110), a qualitative compensation goal was formulated: to increase the quality of the habitat and to reduce human disturbance.

Two types of compensation measures were taken:

- The implementation of an area where bottom trawling for demersal fish (by vessels with an engine power of more than 260 hp) was banned (seafloor protection area) to increase the biomass of benthic fauna, and
- The assignment of a number of zones within this seafloor protection area where human disturbance was reduced, to improve foraging conditions for common scoters, common tern and sandwich tern (no-entry areas).

Figure 1 also shows these areas in the Voordelta.

At the same time, a Monitoring and Evaluation Plan (MEP) was set up for assessing and evaluating the ecological effects of the compensation measures. This plan was commissioned by Rijkswaterstaat, the governmental agency in The Netherlands for water management. The monitoring of the first period of the MEP (2009 - 2014) is now completed, and all data have been collected.

Figure 1. The Voordelta area with the seafloor protection area (red line) and the protected areas for birds in yellow

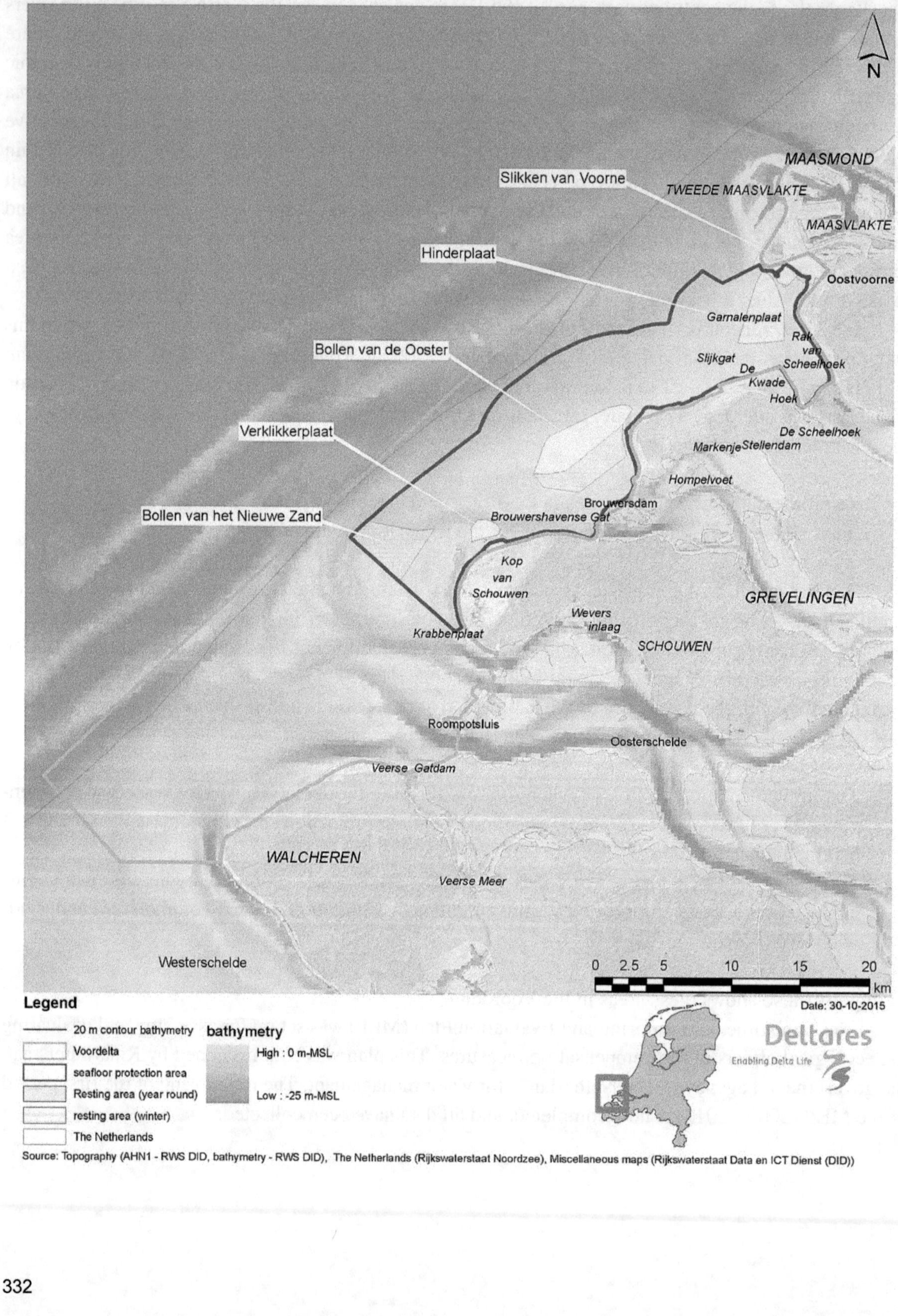

Source: Topography (AHN1 - RWS DID, bathymetry - RWS DID), The Netherlands (Rijkswaterstaat Noordzee), Miscellaneous maps (Rijkswaterstaat Data en ICT Dienst (DID))

MONITORING STRATEGY

The monitoring has been centred around monitoring and evaluation questions that were related to 4 main goals of the compensation measures:

1. Compensation for loss of habitat type H1110
2. Compensation for loss of foraging area for the Common Scoter
3. Compensation for loss of foraging area for the Sandwich Tern
4. Compensation for loss of foraging area for the Common Tern

The monitoring programme was setup according to the monitoring cycle as described by Timmerman et al, 2001 (Figure 2).

Information Needs

The high-level goals of the MEP were further detailed in a set of concrete and testable hypotheses on the expected outcome of the compensation measures. From these hypotheses, the most relevant and measurable parameters were chosen as indicators to be used in testing the hypotheses. These parameters formed the basis for setting up the monitoring programme (information strategy). Central to this approach was that the indicators were chosen according to a cause-and-effect chain set up to follow the effects of the primary drivers (reduction of bottom trawling and human activities) on the benthic assemblages, the fish community and the local populations of the three bird species. The example worked out below relates to

Figure 2. The monitoring cycle as a sequence of related activities
From Timmerman et al., 2001

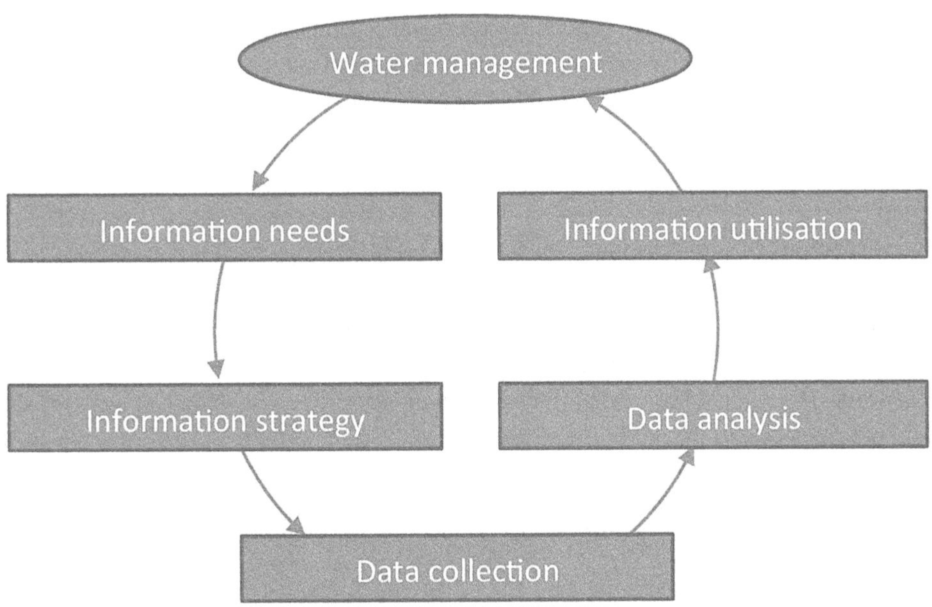

the effects of the compensation measures on the population of the common scoter (main goal 2). This goal needed data from four work packages (birds, benthos, human activities and abiotic conditions) and is therefore a good example of the need for proper data integration in this project (note: data from the work package fish were not relevant for the compensation measure for common scoter).

Information Strategy

It was crucial to isolate the effects of the compensation measures as much as possible from all other (external) effects, like natural variability, autonomous trends, etc. Therefore, the monitoring strategy followed the BACI approach. The strategy involved monitoring in a T_0 period (2004-2007) before the port expansion and the compensation measures were realized, and monitoring in a T_1 period after implementation of the compensation measures. To eliminate autonomous effects the measurements are performed in reference areas as well (control-impact). To investigate long term autonomous trends the duration of the monitoring programme was too short and data were used from the national monitoring programme of Rijkswaterstaat. Additional monitoring of common scoters was carried out in coastal waters in the north of the Netherlands, which is another important wintering area for this species. For the effect of the ban on bottom trawling, monitoring took place in the seafloor protection area and in a reference area within the Natura 2000 site Voordelta.

Data Collection

The information strategy was organized within 5 thematic work packages: Benthos, Fish, Birds, Abiotic Conditions and Human Activities with cross-references between these work packages.

For the data collection a combination of various techniques was chosen:

- An extensive sampling programme for sediment, benthos and (young demersal) fish. Pelagic fish was sampled on an ad-hoc basis together with observed bird (tern) densities. This delivered data on densities, biomass, species diversity and distribution of the organisms.
- In-situ measurements on the distribution of breeding birds (terns), and their breeding success in breeding colonies.
- Airborne observations of resting and foraging birds (terns and common scoters) on the water and sand banks.
- Local observations of bird (common scoter) behaviour in near-coastal waters.
- Remote sensing via GPS loggers to track flights of sterns, providing information on bird distribution and foraging flights, and via vessel tracking systems to track shipping.
- Numerical modelling of abiotic parameters like temperature, salinity, bottom shear stress induced by waves and currents. These modelled data were used for assessing the relationship between physical parameters and the studied species groups, and the effects of the compensation measures on physical parameters.
- Data from long-term national monitoring programmes served as background information on the trends and variability of benthos, fish and birds.

To perform the above mentioned monitoring programme a multidisciplinary team was formed for the MEP, consisting of a number of research institutes and consultants. The monitoring targets were further detailed to determine how the effect of the compensation measures was assumed to work out on the benthos and the different bird species. This consortium and its activities were led by Deltares, which was also responsible for the integration of the results from the 5 thematic working packages. Deltares also developed the central data management facility for this project.

Data Management

Data management is very important for any monitoring project, but this is especially the case for a complex, long-term and multidisciplinary program that involves many different research groups as was the case in the Voordelta monitoring. Data management tools support the exchange of the data between consortium members, keep track of versions and facilitate the final dissemination of the results to the authorities and to the public. Traceability and archiving of project results is crucial for future reuse of the monitoring data, including other purposes than the initial project. This requires a solid data management. Therefore, a central data management facility was set up for data storage and management. OpenEarth has been chosen as the platform for the data management of this project. OpenEarth is a free and open source initiative to setup a basic infrastructure for data and knowledge management in various projects. The initiative was taken as a reaction to the ad-hoc approaches that are often chosen in projects. De Boer et al (2012) give an introduction to the OpenEarth approach which has evolved from a philosophy of working with data to a method that provides well tested methods and building blocks to gather, host and disseminate data in data intensive projects.

Monitoring is costly and monitoring data and information are valuable assets. Unfortunately a lot of project data is gathered for the benefit of the project and is not available for re-use. OpenEarth aims for methods to lower the obstacles for re-use by storage and dissemination of data using international standards. This extra investment on the storage pays off during the re-use of the data, but has also an effect on efficiency because of the use of the same building blocks in other projects. Users and project members will benefit from the efforts of their predecessors. The more re-users, the larger the benefit.

The OpenEarth philosophy assumes that data can be best reused, when some criteria are met:

- They are described using well maintained and accepted vocabularies (in this case the Dutch vocabulary AQUO (mapping with SeaDataNet is available);
- They are accessible via international standards for data exchange (i.e. OGC standards);
- They are applicable on all systems;
- They are accessible at different levels, from raw data to derivative images;
- They are scalable and
- They are easily discoverable through well described catalogues.

Figure 3 shows the different levels of data and information. Researchers usually work with the raw data (left) in order to see all details necessary to answer their questions whilst professionals try to extract information form the data in a more generic/standardized way to be able to meet their clients who like to use a higher level of aggregation of data and information.

The usual access to data from a user perspective is from the top-right corner in Figure 3. A data user starts by examining the availability of data ('discovery') and how they look ('viewing'). Finally the user can decide to use the data ('downloading').

Figure 3. Stages of data and information in OpenEarth data management

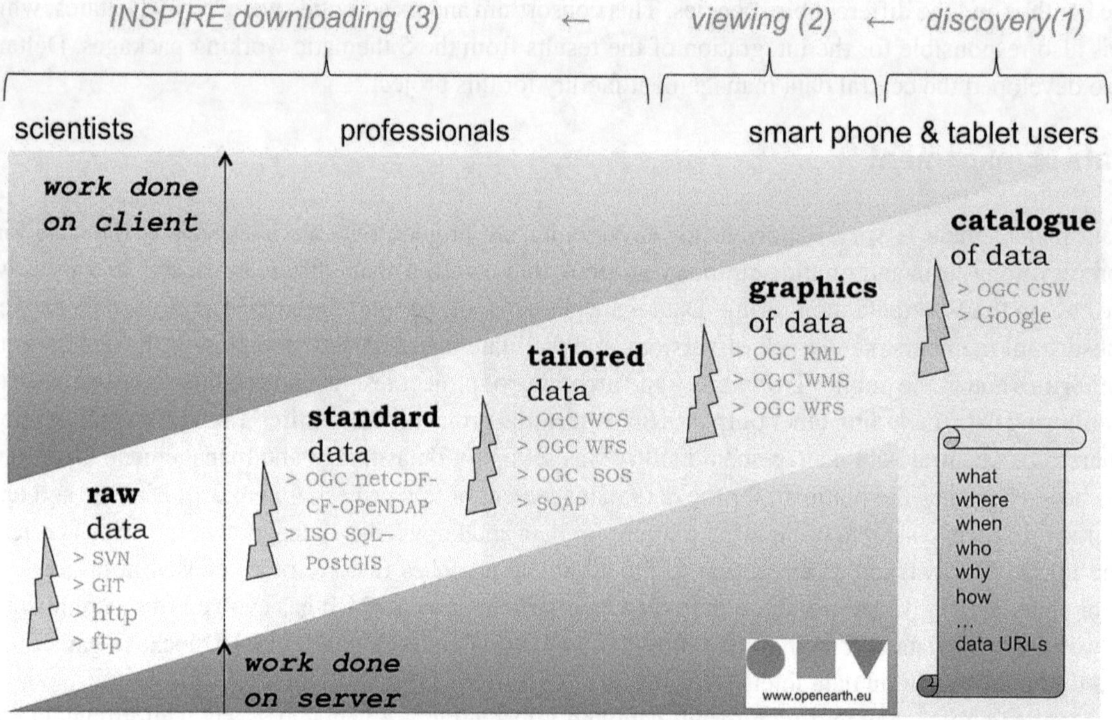

Figure 3 shows the internal processing steps applied to the data from bottom to top and the ease of use of the data from left to right. In the diagonal the syntactic standards are denoted. International standards for data exchange that are used are from the OGC (Open Geospatial Consortium), CF (Climate and Forecasting) and ISO. Apart from the syntactic standard also standards on the data content should be followed, the so called semantic standards.

Mapping of the internal semantic standard to other standards are made at the level of services. As part of the EMODnet Biology project, Voordelta data have been provided as a service to EMODnet biology using the OBIS (Ocean Biogeographic Information System) semantic standard. With the INSPIRE Marine Pilot project (JRC, EC), a pilot is running to provide INSPIRE datasets as a service, based on a data system with similar architecture, using SeaDataNet (BODC) semantic standards for parameters.

From the analysis of the raw data a number of information products have been developed and disseminated to the authorities and the public through web services. These services follow internationally accepted standards, such that information can be disseminated through Google Earth and to portals that support these standards for data and information exchange

Deltares OpenEarth Software Stack

The various building blocks in the processing sequence are depicted in Figure 4. The OpenEarth stack is a suite of open source software products. The suite consists of a repository, where raw data and scripts are stored (version controlled). From this repository aggregated data can be transferred to a structured geo-database and disseminated through a Geoserver and made discoverable by Geonetwork which

Figure 4. OpenEarth Software Stack Deltares (PMR implementation)

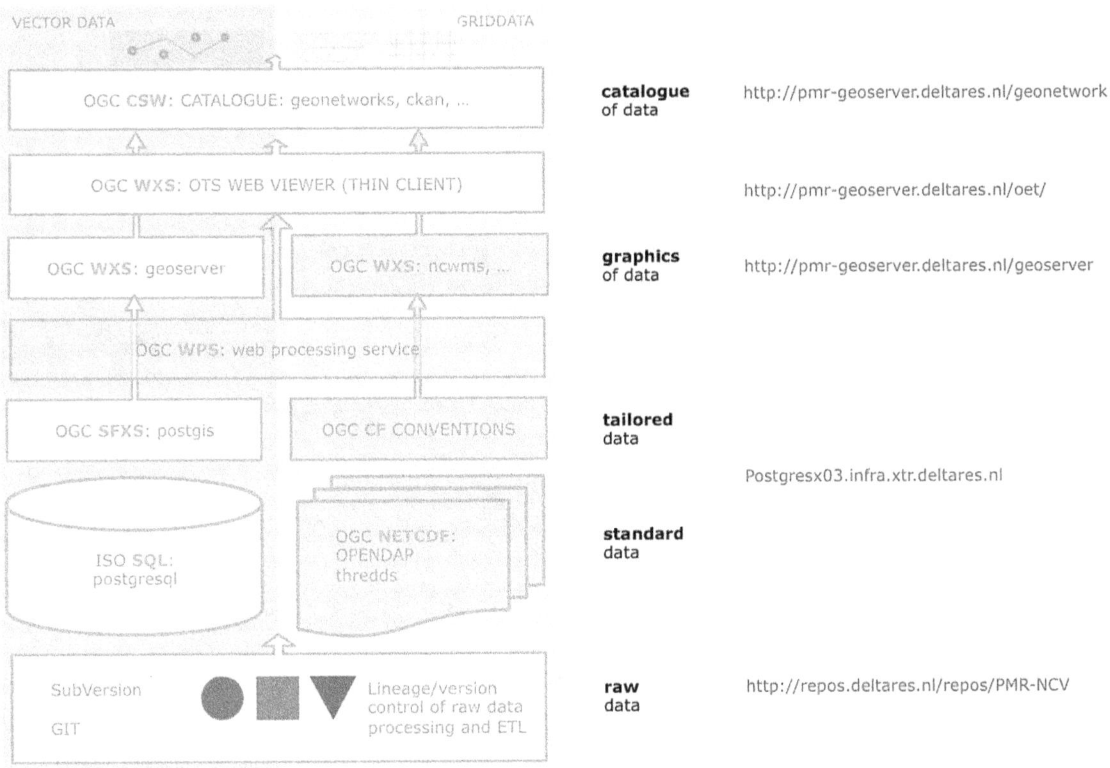

implements the OGC-CSW (Catalogue Service for the Web). A separate implementation can be chosen based on the type of data. In this chapter we focus on the implementation of the left column in Figure 4, which addresses mainly point data. The right column is used for grid data.

At this moment several millions of data entries have been stored and are available for analysis to answer the questions within the Voordelta monitoring project.

Data Analysis, Results and Interpretation

Goals

The objective of the project was to determine the effect of the compensation measures in the Voordelta in relation to the extension of the Port of Rotterdam, Maasvlakte 2. For this chapter we limited ourselves to the intermediate analysis carried out for the second evaluation question on the compensation for loss of foraging area for the common scoter. Furthermore, it is not in the scope of this chapter to discuss the results of the analysis in detail. For this we refer to Prins et al. (2014). Here we discuss which analyses were carried out, which databases were used for the different analyses and how the data management system contributed to the analyses.

Figure 5. Protected areas and Reference areas in the Voordelta, used for the BACI analysis

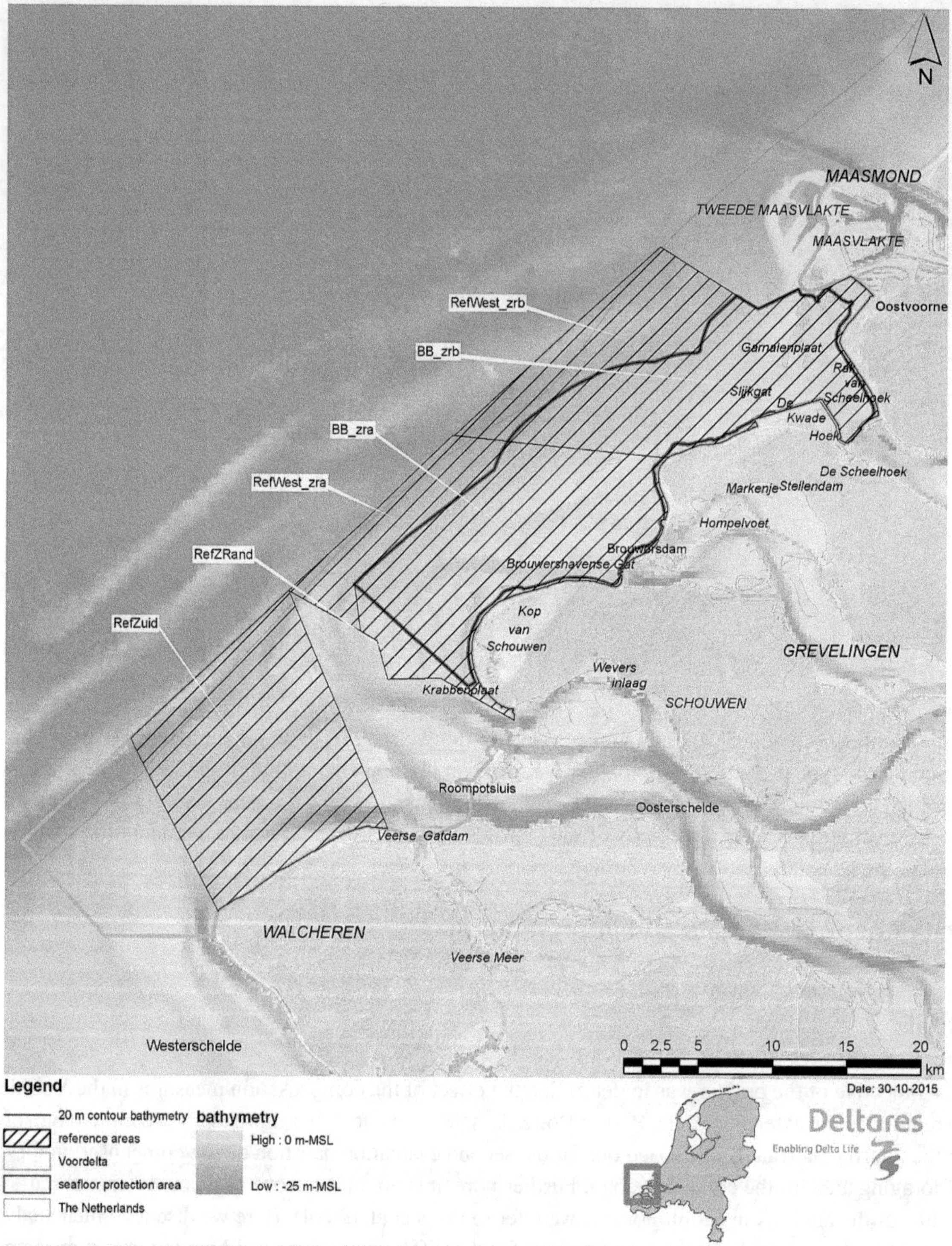

From the above-mentioned goal of compensation for the common scoter, two sub questions were derived:

1. Does the implementation of the protected area result in maintaining food availability in the Voordelta for common scoters?
   a. Are there changes in bird days and distribution pattern?
   b. Are changes in bird days and distribution due to food availability and/or due to other possible causes?
2. Does the implementation of the no-entry areas result in maintaining the Voordelta as a foraging and resting area for the common scoters?
   a. Are there changes in bird days and distribution pattern? (same as 1a)
   b. Are the changes in bird days and distribution due to the no-entry areas and/or other possible causes?[1]

Data

From the above, it followed that data were needed from four different work packages:

- WP Benthos: data on potential food availability for the common scoter, consisting of different species and specific sizes of shellfish. These data were collected in the autumn sampling survey for benthos with a box corer and a benthic sledge.
- WP Birds: data on numbers and distribution of birds: numbers of scoters in different areas of the Voordelta were counted bi-weekly or monthly by airborne surveys. Food preferences (species, size) of the common scoters were assessed by collecting dead scoters that washed up on the shore (over the whole of the Dutch coast) and examining stomach contents
- WP Human activities: data on the human activities, mainly shipping and coastal recreation. Data were collected, as much as possible, simultaneously with the bird surveys.
- WP Abiotic conditions: data on depth, water currents, and wave action. Data on currents and wave action were derived from modelling exercises mainly, based on external forcing (larger southern North Sea model) and based on the most recent depth measurements of the area.

The following data were analysed:

- Benthos surveys

Two types of benthos sampling were carried out: sampling with a Reineck boxcorer, and with a benthic sledge. The first was directed at sampling the relatively abundant and small benthic animals such as polychaetes, the second was used for sampling the less abundant and larger organisms such as shellfish. Depending on species and size, final abundance was based on the most suitable sampling method or on a combination of both methods. The available food mass was compared to the T_0 measurements. For common scoters, shellfish are the staple food and abundance of shellfish is assumed to be one of the two main determining factors regulating the abundance and distribution of the scoters (next to disturbance).

- Bird surveys

The numbers and distribution of common scoters were assessed from airborne surveys of the Voordelta area. Numbers and distribution were also assessed in surveys of reference areas in the outer Delta of the Wadden Sea, because this area is the most important wintering area of common scoters in The Netherlands. This area was especially important as a reference for the autonomous interannual variation in the numbers of common scoters. Also, behavioural observations were made during the day and the twilight periods. These observations were used for assessing in more detail the reactions of scoters to local disturbances by human activities.

For food preference, data were collected from stranded dead scoters along the whole of the Dutch coast (volunteer programme). The collected birds were dissected, studied for condition (weight, diseases) and for stomach contents. The species composition of the shellfish found in the stomach was determined and shell length of the individual shells was assessed.

- Shipping traffic

The intensity of shipping was monitored using AIS (Automated Identification System) data. A careful quality assessment of the AIS data was needed to remove artefacts from the data set. These can consist of discontinuity of AIS pings in time and unrealistic vessel speeds. These effects were removed from the data to obtain more more 'clean' shipping data.

- Human disturbance

For assessing disturbance from other human activities than shipping, airplane surveys concurrently with the bird surveys in the Voordelta were used. Activities assessed were fisheries, recreation, military exercises, sediment extraction, small airplanes, monitoring activities, and maintenance and disaster remediation.

Analyses

In order to answer the above questions, the following main analyses were conducted:

1. Number of bird days in the T_0 (2004-2007) and in the years following the compensation measures (2009-2014)
2. Comparison of numbers and distribution of scoters in protected areas and outside those areas.
3. Analysis of the food preference of the common scoters, the spatial distribution and the abundance of food within the Voordelta.
4. Analysis of shipping traffic in the Voordelta.
5. Statistical analysis of the spatial distribution of the common scoters and the spatial relationship with food (shellfish), abiotic parameters (bottom current speed, depth), human disturbance (shipping traffic).

Results

The results of the analyses are discussed shortly.

Conclusions were based on combinations of observational data (Figure 6). A comprehensive report describes the results of this intermediate analysis (Prins et al. 2014). A final evaluation of the effects of the compensation measures is foreseen for 2020.

The numbers and distribution of common scoters, within and outside of the protected zones were compared between T_0 and T_1 years. Results showed large interannual variability, which overshadowed any temporal trends. A comparison with numbers of birds in the northern coastal waters of the Netherlands showed that numbers in the Voordelta were only a small fraction of the total number of the birds overwintering in Dutch coastal waters.

A comparison with long term monitoring data indicated that numbers in the Voordelta in the T_0 and T_1 years were much lower than in previous decades (1980's and 1990's). Comparison of these three data sets indicated that there was a long term decreasing trend in numbers of overwintering birds, unrelated to management measures in the Voordelta.

The inventory of abundance and biomass of shellfish suitable as food for common scoters showed large variability in the availability and distribution of suitable food items for the birds. The interannual variation in bird numbers was not correlated with the interannual variation in total food biomass.

Observations on shipping traffic, combined with observations of the distribution of birds, indicated that the common scoters were found both inside and outside the protected areas, but always in areas with no or very low shipping traffic.

Figure 6. Diagram showing the contribution of data and information from the various work packages, relevant for the questions with respect to the common scoter

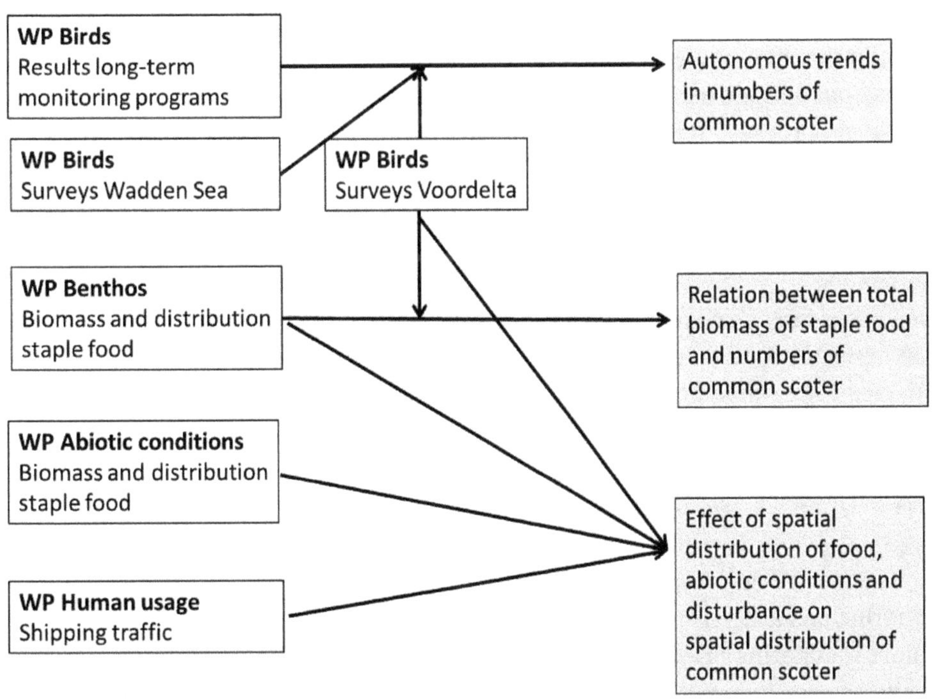

An extensive statistical analysis was carried to relate the spatial distribution of the common scoter with the spatial distribution of environmental conditions, food (biomass of shellfish of suitable sizes) and human disturbance by shipping traffic. This analysis indicated that the availability and accessibility of food (determined by biomass of food and the physical factors depth and current speed that influence the foraging opportunities of the diving ducks) and disturbance by shipping are factors influencing the distribution of the birds within the Voordelta. This result was used to optimize the delineation of the protected areas.

CONCLUSION AND LESSONS LEARNED

The monitoring of the Voordelta area is a very complex task, that requires a strong multidisciplinary team. It is a typical example of a marine project, which is required to evaluate the effects of infrastructural activities and the compensations measures that accompany these activities. The monitoring programme should be able to handle data from various sources, like data collected within the programme and data from national and international programmes. Furthermore these data covered a number of disciplines with their specific standards and formats.

The goals of the data management were to facilitate the exchange of data between researchers, enable version control of data sets, facilitate quality assurance and archiving of the data that underpin the results and conclusions of the monitoring project.

At this moment (2015-2016) the monitoring programme is still active and this chapter describes some findings up to now. Good collaboration and frequent interaction was the key to success. The way the data have been managed supported this process.

First of all, data management assured that all team members had direct access to all data. Quality assurance of the data could be handled professionally. Furthermore, the data management systems optimally supported data dissemination to the authorities and to stakeholders.

In practice, in this first phase of the monitoring programme exchange of data between researchers did not always occur through the central data management facility. It appears that with smaller data sets and for short term purposes, direct exchange of data between researchers is perceived as being easier to deal with and less time consuming. It is expected that, with increasing size of the data sets, longer duration of the monitoring program and personnel changes in the research team, the role of the central data management facility will grow and ultimately will be of vital importance to secure data exchange and data availability.

The sound procedures that are involved in the data management system support transparency, logic and cohesion in the set up and implementation of any monitoring strategy.

Finally the data are available for re-use in other projects. Thus future projects will benefit.

OUTLOOK

The approach chosen in this monitoring project provides not only a good basis for collaboration during the monitoring and analysis stages, but also forms a basis for future reuse and extension of the data sets. Any future monitoring project interested in these data for the Voordelta region or ongoing (inter) national monitoring programmes will benefit from the open and generic structure to retrieve the -by then historic- information on meta level (what has been measured, when and how) as well as on data level. The chosen structure can be extended to cover additional quantities or geographic regions.

The data management philosophy and systems described in this chapter are also used in other complex monitoring projects, like the Sand Engine.

For reuse of data from marine projects the data management systems should be disclosed through a general infrastructure, like the EU initiative EMODnet. EMODnet is promoted by the EU as the central portal for marine data and steps are taken to facilitate the storage and disclosure of marine project data. The data from the Voordelta project are integrated in the EMODnet Biology lot. Also on national level various initiatives are present to make data available to marine specialists and the public. In The Netherlands the Digital Delta (Van Oord, 2011) initiative will now be extended to the marine environment.

ACKNOWLEDGMENT

The authors would like to thank Rijkswaterstaat for permission to publish this paper. The other members of the consortium were IMARES (benthos, fish), CSO (human activities), Arcadis/Alkyon (abiotic modelling), Bureau Waardenburg and INBO (birds) and NIOZ-Yerseke (benthos). Without their effort this project would not have been possible.

REFERENCES

Council of the European Communities. (1992). *Council Directive 92/43/EEC of 21 May 1992 on the conservation of natural habitats and of wild fauna and flora*. Brussels, Belgium: Council of the European Communities.

De Boer, G. J., Baart, F., Bruens, A., Damsma, T., Van Geer, P., Grasmeijer, B., & Santinelli, G. et al. (2012). OpenEarth: using Google Earth as outreach for NCK's data. In *Jubilee Conference Proceedings NCK-days 2012* (pp. 97-103) Enschede, The Netherlands: University of Twente. doi:10.3990/2.177

European Commission. (2006). *Green paper, Towards the future Maritime Policy for the Union: a European vision for the oceans and seas*. Brussels, Belgium: European Commission.

European Commission. (2007). *Interpretation manual of European Union habitats*. Brussels, Belgium: European Commission.

European Parliament and Council of the European Union. (2008). *Directive 2008/56/EC of the European Parliament and of the Council of 17 June 2008 establishing a framework for community action in the field of marine environmental policy*. Strasbourg, France: European Parliament.

European Parliament and Council of the European Union. (2009). *Directive 2009/147/EC of the European Parliament and of the Council of 30 November 2009 on the conservation of wild birds*. Brussels, Belgium: European Parliament.

European Parliament and Council of the European Union. (2014). *Directive 2014/89/EC of the European Parliament and of the Council of 23 July 2014 establishing a framework for maritime spatial planning*. Brussels, Belgium: European Parliament.

Giorgetti, A., Vinci, M., Schaap, D., Fichaut, M., & Barth, A. (2015, June) *The EMODnet and SeaDataNet infrastructure: a challenge towards a European multidisciplinary interoperability*. Paper presented at 10th Baltic Sea Science Congress, Riga, Latvia.

Green, R. H. (1979). *Sampling design and statistical methods for environmental biologists*. Chichester, UK: Wiley-Interscience.

Prins, T. C., Van der Kolff, G. H., Boon, A. R., Reinders, J., Kuijper, C., Hendriksen, G., . . . Adema, J. (2014). *PMR Monitoring natuurcompensatie Voordelta, Eindrapport 1e fase 2009-201*. Retrieved from http://kennisonline.deltares.nl/product/30731

Reker, J., Annunziatellis, A., Mo, G., Tunesi, L., Globevnik, L., Snoj, L., . . . Korpinen, S. (2015). Marine protected areas in Europe's seas. An overview and perspectives for the future (EEA report No 3/2015). European Environment Agency.

Timmerman, J. G., De Boer, J., Hisschemöller, M., & Mulder, W. H. (2001). Specifying information needs: Improving the working methodology. *Regional Environmental Change*, *2*(2), 77–84. doi:10.1007/s101130100030

Van Koningsveld, M., De Boer, G. J., Baart, F., Damsma, T., Den Heijer, C., Van Geer, P., & De Sonneville, B. (2010). OpenEarth - Inter-Company Management of: Data, Models, Tools & Knowledge. In *Proceedings WODCON XIX Conference*. Beijing, China: Central Dredging Association.

Van Oord, K., Nijhoff, A., Prins, M., & Van de Giesen, N. (2011). Water verdient het. Advies Topsector Water (Report of Topteam Water). The Hague, The Netherlands: Topteam Water. (in Dutch)

ENDNOTE

1 Note that for answering these questions properly, more detailed hypotheses were formulated (see Prins et al. 2014)

Chapter 15

Analysis of Ocean in Situ Observations and Web-Based Visualization:
From Individual Measurements to an Integrated View

Alexander Barth
University of Liege, Belgium

Charles Troupin
Balearic Islands Coastal Ocean Observing and Forecasting System (SOCIB), Spain

Sylvain Watelet
University of Liege, Belgium

Aida Alvera-Azcárate
University of Liege, Belgium

Jean-Marie Beckers
University of Liege, Belgium

ABSTRACT

The sparsity of observations poses a challenge common to various ocean disciplines. Even for physical parameters where the spatial and temporal coverage is higher, current observational networks undersample a broad spectrum of scales. This situation is generally more severe for chemical and biological parameters because such sensors are less widely deployed. The present chapter describes the analysis tool DIVA (Data-Interpolating Variational Analysis) which is designed to generate gridded fields from in situ observations. DIVA has been applied to various physical (temperature and salinity), chemical (concentration of nitrate, nitrite and phosphate) and biological parameters (abundance of a species). The chapter also shows the technologies used to visualize the gridded fields. Visualization of analyses from in situ observations provide a unique set of challenges since the accuracy of the analysed field is not spatially uniform as it strongly depends on the location of the observations. In addition, an adequate treatment of the depth and time dimensions is essential.

DOI: 10.4018/978-1-5225-0700-0.ch015

INTRODUCTION

In situ measurements of ocean properties are generally sparsely distributed and thus undersample the ocean variability. A good knowledge of the scales and dynamics of the ocean is essential to make the best use of this limited amount of data. The data distribution is often very inhomogeneous, both in space and time. Accuracy is also inhomogeneous and point measurements (e.g., moorings, coastal stations) are not necessarily representative of typical ocean conditions. Therefore it is necessary to take all these aspects into account when analysing in situ measurements.

Various applications and interpretations of marine data require a gridded field covering the complete domain as a first step. These applications include, for example:

- Computation of budgets (e.g., heat, salt content),
- Identification of long-term trends,
- Identification and characterization of oceanographic features,
- Determination of derived variables from parameters not necessarily measured at the same location (for example density which is a function of temperature and salinity),
- Model initialization and validation.

In principle, some of these applications could be also performed directly on the measured observations. For example, the average ocean temperature could be naively computed as the mean of all observations. However in practice the data distribution is very inhomogeneous in space (with a higher density of observations near the surface and close to the coast) and in time (e.g., more ship-based observations are generally available during summer than during winter and the density of observations has increased substantially during the last decades). Hence computing the mean using directly the observations would result in a biased estimation of the average temperature.

BACKGROUND

Most analysis methods are linear techniques, which means that the analysed field (or its anomaly versus a reference state) at a given location is a weighted linear combination of the observations. Different methods exist to compute those weights. The most obvious way to compute them is to use a monotonic decreasing function of the distance between the location of the analysed field and the observations: observations farther away from a given location have less influence than observations nearby. This approach is implemented in the Cressman method (Cressman, 1959). The weighting function usually decreases over a given length-scale and is exactly zero beyond a given search radius. Various weighting schemes have been proposed (Barnes, 1964). The search radius can be made to vary in space depending on data coverage and/or dynamical scales. The Cressman weighting is a very simple and numerically quite efficient method. However, it suffers from some limitations, in particular no estimate can be obtained at locations where no observation is located within the search radius. In regions with very few observations, the method can therefore return a discontinuous field. In addition, the presence of topographic barriers disconnecting different water bodies cannot be taken into account easily. Another limitation is that all observations are assumed to have a similar error variance since the weighting is based only on distance.

The optimal interpolation method (Gandin, 1965; Bretherton et al., 1976) solves some shortcomings of the Cressman method. A background field represents the first guess of the field. Observations are treated as anomalies relative to this background field. Optimal interpolation is a statistical approach in the sense that the result minimizes the total expected error provided the error covariance of the background field and of the observations are known. In practice, it is however difficult to estimate the error covariance of the observations, and even more difficult to estimate the error covariance of the background field. Most approaches use simple analytical correlation functions based on distance to parameterize the background error covariance (Robinson, 1996; Guinehut et al., 2002; Roberts-Jones et al., 2012), but more sophisticated approaches based on empirical orthogonal functions (Kaplan et al., 1997; Rayner et al., 2003; Beckers et al., 2006) or an ensemble of states derived from e.g. a numerical model (Evensen, 2007) have also been used. The analysis derived from optimal interpolation naturally provides results defined over the entire domain over which the first guess is defined. Far away from an observation, the analysed field gradually reverts back to the first guess. Unlike the Cressman weighting, optimal interpolation allows the computation of the relative error variance of the analysed field. The major difficulty of optimal interpolation is to derive a realistic error covariance and its parameterization is an active field of research nowadays (Hyer and She, 2007; Buongiorno Nardelli et al., 2010; Buongiorno Nardelli, 2012).

A different approach to derive a complete field based on in situ observations can be obtained by defining a cost function. Instead of minimizing the total expected error variance of the field as in optimal interpolation, the variational methods minimize the misfit relative to observations while imposing additional constraints. In particular, the resulting field must be smooth over a given length-scale. It has been shown that this approach can be seen as mathematically equivalent to optimal interpolation, with the peculiarity that the background error covariance is not defined explicitly as in optimal interpolation but implicitly in terms of a regularization constraint in the cost function being minimized. Boundary conditions of the regularization constraint naturally decouple basins which are geographically close but not connected due to the topography. While not impossible, it is quite difficult to include such constraints in optimal interpolation. Among the variational methods one can distinguish between techniques similar to 3D-Var where the regularization constraint is defined by a diffusion operator (Bennett et al., 1996; Weaver et al., 2003; Lorenzo et al., 2007; Moore et al., 2011) or spline interpolation where the cost function minimizes a series of derivatives of the field to ensure a smooth result (Wahba and Wendelberger, 1980; Wahba, 1990; McIntosh, 1990; Brasseur and Haus, 1991; Brasseur et al., 1996; Troupin et al., 2012). While the first class of methods generally uses an iterative solver, in particular the conjugate gradient method, the second class of formulations requires an efficient sparse matrix solver (Chen et al., 2008; Davis and Hager, 2009). Because the mathematical formulation of the variational approach can be seen as a special case of optimal interpolation, it is not possible to give a clear-cut distinction between these approaches.

Other related data analysis techniques consist in expressing the analysed field as a combination of a set of normal basis functions. These basis functions can be for example Fourier modes, the eigenvectors of the Laplacian operator (Lipphardt et al., 2000; Kaplan and Lekien, 2007) or empirical orthogonal functions (Beckers and Rixen, 2003; Alvera-Azcárate et al., 2009). These techniques are in fact related to optimal interpolation and variational analysis, since the solution of the latter methods can also be seen as a linear combination of the eigenvectors of the underlying background error covariance matrix (Beckers et al., 2006).

As ocean observations are sparse, gridding these observations by the aforementioned techniques is often a first step to visualize fields of ocean parameters. The DIVA tool has been integrated in the visualization tool Ocean Data View (Schlitzer, 2002) for this purpose. In the following we limit ourselves to web-based visualization. The Reading e-Science center developed ncWMS which is a primarily Web Map Service (de la Beaujardiere, 2006) to visualize netCDF files (Blower et al., 2013). ncWMS is a Java application which is deployed using a standard servlet container such as Apache Tomcat. On top of ncWMS, the browser-based client GODIVA2 has been developed (Blower et al., 2009). Other web-based clients include Lw4nc2 (Lightweight for netCDF viewer) and Leaflet.TimeDimension (https://github.com/socib/Leaflet.TimeDimension) both developed by SOCIB, Spain.

The present chapter describes the web-based visualization tool OceanBrowser. The server component is written in Python using the matplotlib library (Hunter, 2007).

These web-based visualization systems have different strengths and weaknesses. The focus of the development of OceanBrowser has been the ability to visualize along arbitrary vertical sections. Those systems are also compatible to a certain degree thanks to the Web Map Service protocol published by the Open Geospatial Consortium. However, some extensions to the protocol are necessary to cover the specific case of analysing ocean data.

MAIN FOCUS OF THE CHAPTER

This chapter aims to address the following issues in the context of managing marine observations:

- It introduces the problem of the data analysis of in situ observations.
- The challenges in deriving gridded fields and the visualization of the data fields will be discussed.
- Based on a concrete implementation, it will show how web-based tools and open standards can be used to integrate datasets from different oceanographic disciplines and to facilitate the interpretation of the dataset across different domains.

DATA-INTERPOLATING VARIATIONAL ANALYSIS

An analysis tool called DIVA (Data-Interpolating Variational Analysis) has been developed to create gridded fields from sparse in situ observations. The purpose of the analysis is to give the best estimation of a continuous spatial field of a given variable from a set of discrete measurements, taking into account the errors attached to the measurements.

Basic Formulation

The Variational Inverse Method (VIM) implemented in DIVA aims to find an analysed field that satisfies a set of constraints over a given domain, expressed in the form of a cost function. Before elaborating its mathematical formulation, we consider that we are working with data anomalies: a reference (first guess) field is subtracted from the data before any analysis. In DIVA, this reference field can be either the null guess (i.e., data are already considered as anomalies), the data average, a regression plane derived from the data set or even a previous DIVA analysis (on a longer period for example). Once the analysis is performed, the reference field is added in order to return to the original variable. For a two-dimensional problem, the cost function J to minimize is defined as follows (Troupin et al., 2012):

$$J[\varphi] = \sum_{j=1}^{N_d} \mu_j \left[d_j - \varphi(x_j, y_j)\right]^2 + \|\varphi\|^2 \qquad (1)$$

with

$$\|\varphi\|^2 = \int_D \left(\alpha_2 \nabla\nabla\varphi : \nabla\nabla\varphi + \alpha_1 \nabla\varphi . \nabla\varphi + \alpha_0 \varphi^2\right) dD \qquad (2)$$

where the coefficient α_0 penalizes the field itself, α_1 penalizes gradients (no trends), α_2 penalizes variability (regularization), and μ_j coefficients penalize data-analysis misfits (objective analysis). Thus, this cost function takes into account the deviation between analysis $\varphi(x_j, y_j)$ and data d_j (observation constraint) and the smoothness of the analysis (regularization constraint), while it is also possible to take the advection by currents into account (physical constraint; see next subsection).

Introducing the characteristic length *L*, we can define dimensionless space coordinates: $\hat{\nabla} = L\nabla$ and $D = L^2\hat{D}$, and rewrite the equations (1) and (2) as

$$J[\varphi] = \sum_{j=1}^{N_d} \mu_j L^2 \left[d_j - \varphi(x_j, y_j)\right]^2 + \int_{\hat{D}} \left(\alpha_2 \hat{\nabla}\hat{\nabla}\varphi : \hat{\nabla}\hat{\nabla}\varphi + \alpha_1 L^2 \hat{\nabla}\varphi . \hat{\nabla}\varphi + \alpha_0 L^4 \varphi^2\right) d\hat{D}. \qquad (3)$$

Without loss of generality, we can choose $\alpha_2 = 1$. α_0 fixes the characteristic length *L* for which the first and the last terms are equivalent:

$$\alpha_0 L^4 = 1 \qquad (4)$$

The term $\mu_j L^2$ accounts for the weight on each individual observation, depending on the signal (σ^2, the variance of the background field) to noise (ϵ_i^2) ratio (see also Barth et al., 2014),

$$\mu_j L^2 = 4\pi \frac{\sigma^2}{\epsilon_i^2}. \qquad (5)$$

Finally, the term $\alpha_1 L^2$ fixes the relative importance of gradients. In order to penalize the gradients with a similar weight than the second derivatives, we set

$$\alpha_1 L^2 = 2\zeta \qquad (6)$$

where the adimensional shape factor ζ is set to 1 by default in DIVA.

The overall signal-to-noise ratio λ and the overall error variance of the observations ε^2 are defined as follows:

$$\lambda = \frac{\sigma^2}{\epsilon^2};\ \epsilon^2 = \frac{\sum_{j=1}^{N} \epsilon_j^2}{N} \tag{7}$$

This allows one to specify the following relative weights:

$$w_j = \frac{\mu_j L^2}{4\pi\lambda} \text{ with } \sum_{j=1}^{N} \frac{1}{w_j} = N. \tag{8}$$

Given ζ and w_j, the relative importance of each constraint (μ, α_0, α_1, α_2) is thus linked to two main control parameters: the signal-to-noise ratio λ and the correlation length L. Both can be obtained automatically from the datasets using DIVA tools or can be left to the user's choice.

To ensure that the analysis problem is well-posed, derivatives of sufficiently high order have to be included in the cost function. In fact for an analysis in n dimension, the cost-function should include at least a derivative of the order of $1+n/2$ rounded up for an odd number of dimensions. This guarantees that the analysis problem is well-posed and has a continuous derivative (Barth et al, 2014).

If the covariances are chosen accordingly, the VIM is equivalent to the optimal interpolation method. However, the number of operations in the VIM is almost independent of the number of data, leading to a potential gain of computational time for large problems. And since the minimization problem of the VIM is solved by a finite-element method with a mesh (Figure 1) only defined on the ocean, the spurious propagation of information through topographic barriers is easily avoided, unlike most implementations of optimal interpolation.

Figure 1. A finite-element mesh computed by DIVA when using a correlation length of 1.5° over the Black Sea at the surface. Each color corresponds to a different triangular element mesh. Note the smaller mesh size near the coastlines.
For a more accurate representation of this figure, please see the electronic version.

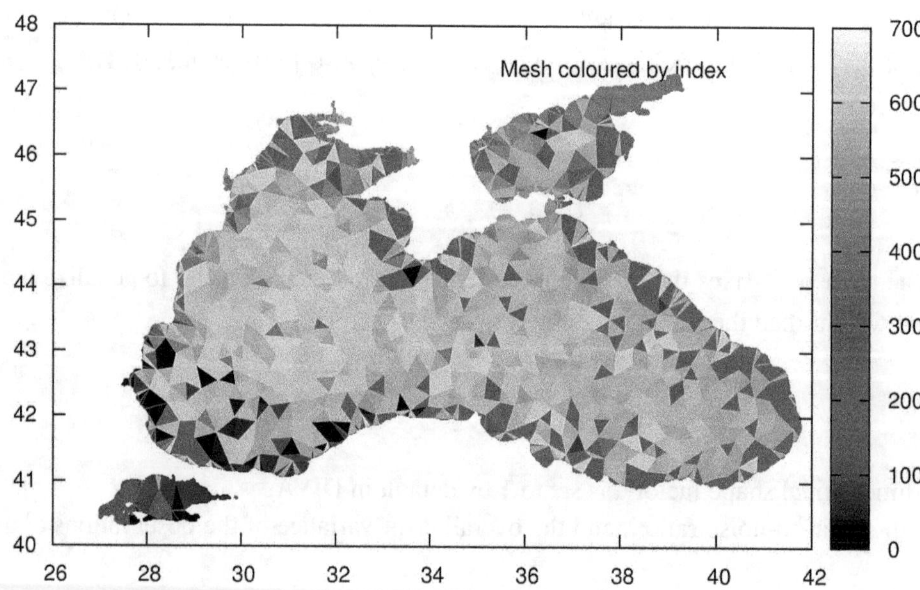

Advection Constraint

The cost function J can be augmented with an optional behaviour constraint, in order to use the advection field as a complementary source of information for the interpolation. Let us consider the general equation for a scalar oceanic quantity φ (Brasseur, 1995):

$$\frac{\delta\varphi}{\delta t} + \mathbf{u}.\nabla\varphi + \nu_3 \frac{\delta\varphi}{\delta x_3} = Q^{\varphi} + \lambda_h \nabla\nabla\varphi + \frac{\delta}{\delta x_3}\left(\lambda_v \frac{\delta\varphi}{\delta x_3}\right) \quad (9)$$

where $\mathbf{u}$ represents the horizontal speed, ν_3 the vertical component of the speed, λ_h and λ_v are horizontal and vertical coefficients of diffusion, and Q^{φ} is the term corresponding to the source/decay of φ.

In equation (9), we can invoke certain simplifying assumptions. In most cases, we can neglect Q^{φ} for instance when analysing temperature or salinity for which the sources are confined at the air-water interface. Further, we ignore the temporal derivative in equation (9) since DIVA aims to deal with slowly evolving situations, such as climatologies. Besides, DIVA performs its 3D analyses via superimposed 2D analyses, so the vertical movement cannot be considered in equation (9). With these assumptions in mind, we get:

$$\mathbf{u}.\nabla\phi \simeq \lambda_h \nabla\nabla\varphi \quad (10)$$

and the resulting advection-diffusion constraint to add to equation (1):

$$C\left[\varphi\right] = \gamma\{\mathbf{u}.\nabla\varphi - \lambda_h \nabla\nabla\varphi\}^2 \quad (11)$$

where γ weights the relative importance of this advection term with respect to the observations and regularization ones, in order to be in the same order of magnitude. In the dimensionless version (equation 3) of the cost function $J\left[\varphi\right]$, the advection constraint is added as follows (Troupin et al., 2012):

$$J_a\left[\varphi\right] = J\left[\varphi\right] + \theta U^2 \int_{\hat{D}} [\mathbf{u}.\nabla\varphi - \frac{A}{L}\nabla\nabla\varphi]^2 \, d\hat{D} \quad (12)$$

where U is the characteristic velocity scale computed from $\mathbf{u}$, A is the horizontal diffusion coefficient, and θ a scaling factor ensuring the same weight as the other constraint when set to 1.

This advection constraint tends to increase the correlation in the direction of $\mathbf{u}$ (Figure 2), which can be derived from isobaths, isopycnals, can be the geostrophic velocity computed from temperature and salinity, or even provided from the output of a hydrodynamic model (Figure 3).

Figure 2. Example of SST analysis over the Black Sea without advection constraint (panel a). The World Ocean Database 2013 was used for data extraction. The analysis was performed between 1980 and 2010 on the month of March. Panel (b) shows the same SST analysis, with the advection constraint enabled. The used velocity field is shown in Figure 3. This advection constraint tends to smooth the short scales gradients where the currents are strong, especially in the central part of the Black Sea (thanks to the enhanced propagation of the information).

For a more accurate representation of this figure, please see the electronic version.

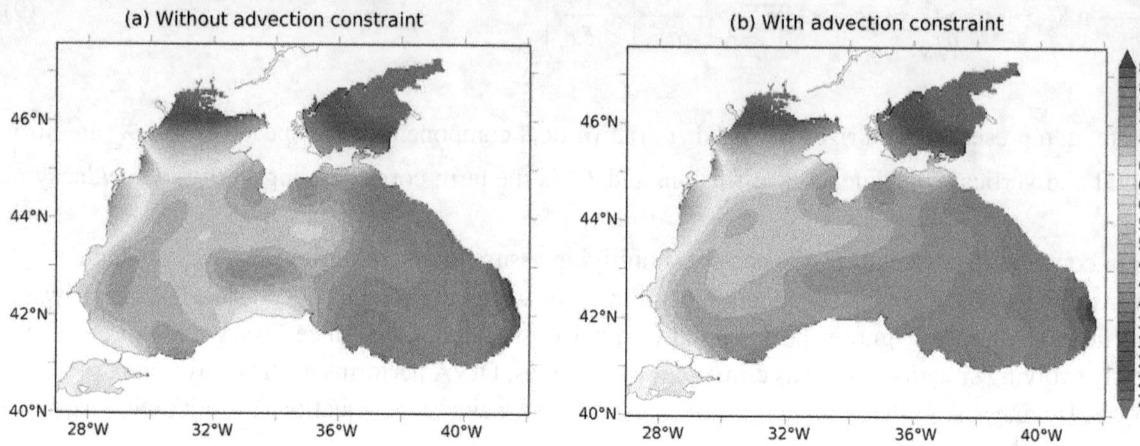

Figure 3. The velocity field we used for this experiment is described in Capet et al. (2012). It consists of a 1980-2010 average of the surface currents output of an eddy-resolving 3D model, for the month of March.

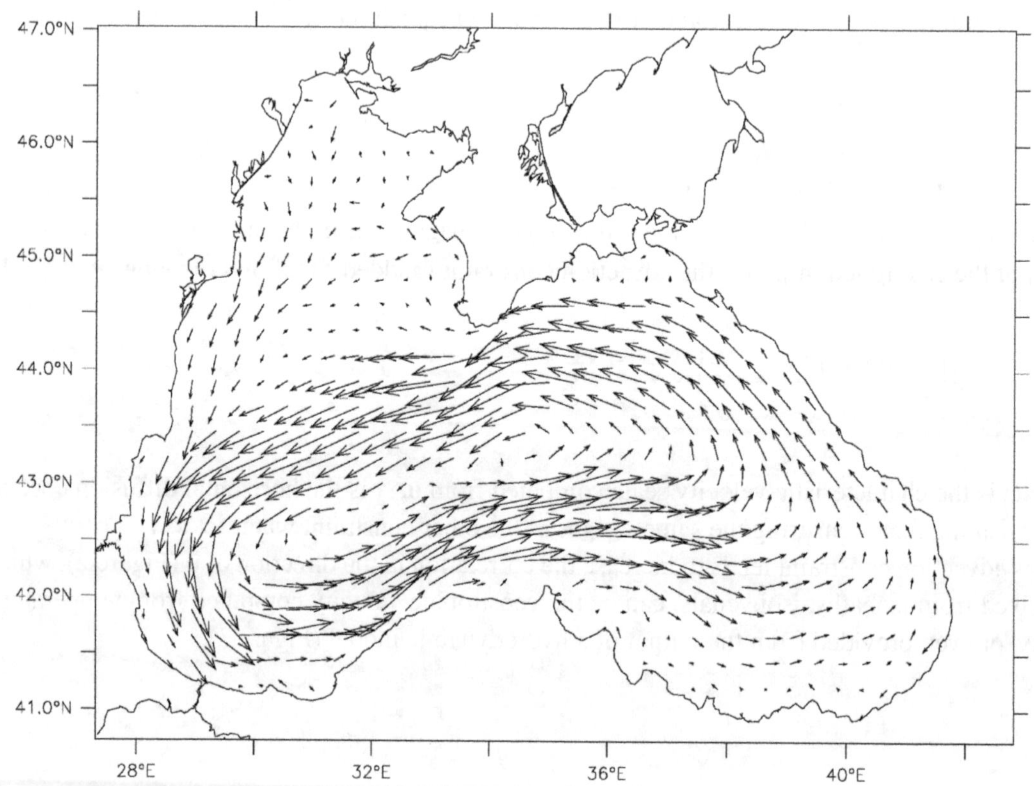

Anamorphosis Transformation

Using a non-gaussian distributed variable can lead to unrealistic values when performing the DIVA analysis. Indeed, DIVA tries to interpolate at best where no data are available, which can imply negative values if the nearest observations are close to zero. Obviously, we have to reject such a result for variables such as salinity, chlorophyll-a concentration, or other concentration values.

An easy way to overcome this issue is to apply a transformation on the data prior to the DIVA analysis. After the analysis, the inverse transformation is applied on the analyzed field. Several transformations have been implemented in DIVA: logarithm, logit and anamorphosis. While the two first transformations are predefined, the last one is directly based on the shape of the original data histogram in view of creating a gaussian-like distribution which favours more realistic analyzed values. An example is shown in Figure 4.

Detrending

One of the most challenging issues in data analysis is the uneven temporal distribution of data, which causes representativity errors. For example, the observation frequency is very often higher in summer than in winter, leading to a positive bias in the temperature analysis.

A detrending method, described in Duchon (1977), has been implemented in DIVA and applied by Capet et al. (2014) in order to recognize and eliminate this bias. Let us define the groups as the various time intervals which can be considered for any trend detection: years, seasons, months... In one of these groups (e.g. the years), each observation corresponds to a unique class c_j e.g. 1952). The data analysis misfit term in equation (1) can be rewritten by including an unknow trend value for each class (dC_1,dC_2,...):

Figure 4. Histogram of nitrate concentration [µmol/l] at the surface of the Black Sea between 1900 and 2015 (panel a) and histogram of nitrate concentration after the anamorphosis transformation (panel b). Its shape is as close as possible to a gaussian distribution.

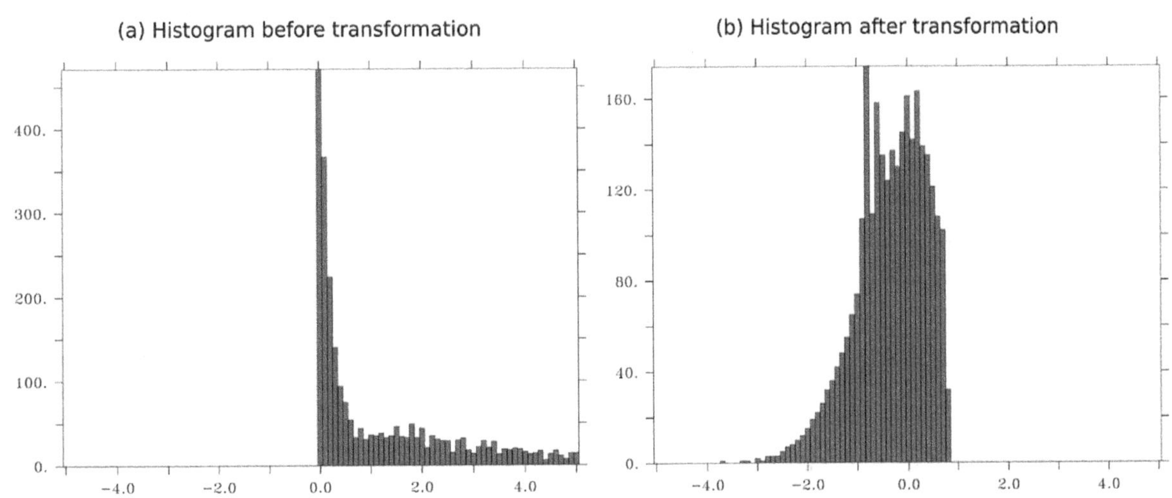

$$\sum_{j\in C_1}\mu_j\left[d_j - dC_1 - \varphi\left(x_j, y_j\right)\right] + \sum_{j\in C_2}\mu_j\left[d_j - dC_2 - \varphi\left(x_j, y_j\right)\right] + \ldots \tag{13}$$

If the function $\varphi\left(x_j, y_j\right)$ were known, we could easily derive the dC_j from a minimization on each class:

$$dC_1 = \frac{\sum_{j\in C_1}\mu_j\left[d_j - \varphi\left(x_j, y_j\right)\right]}{\sum_{j\in C_j}\mu_j} \tag{14}$$

and similarly for the other classes. But since $\varphi\left(x, y\right)$ is unknown too, we have to solve the issue by an iterative process, starting by an analysis without detrending. Using this field $\varphi\left(x, y\right)$, we compute a first guess of the trends in each group and subtract it from the original data. After that, a new analysis is performed, the trends are recalculated, and the iterations continue until a convergence criterion is fulfilled. This method has been successfully used in Capet et al. (2014) for the cold content of the Black Sea cold intermediate layer as well as for its mixed layer depth. In both cases, the method allowed the reduction of the bias caused by the uneven data availability between summer and winter or between cold and warm years.

Analysis in More than Two Dimensions

DIVA is implemented to work on a two-dimensional space. This is generally a horizontal plane at a given depth and for a given time period. But DIVA can also be applied to a vertical section. For multi-dimensional applications, for instance deriving an ocean property as a function of longitude, latitude, depth and time, horizontal analyses are performed independently for various depth and time instances and then assembled in a 4-dimensional analyses.

Maintaining coherency and avoiding abrupt variations in the depth and time dimension is difficult in this case. Sometimes it is necessary to filter the parameters (in particular the correlation length and signal-to-noise ratio) along the depth and time dimensions and to use data from adjacent layers in the vertical. DIVA has been implemented to work in two dimensions only because it uses a triangular mesh and a finite-element method which are increasingly difficult to implement in 3 or 4 dimensions. However, the variational inverse method on structured grids can be easily extended to higher dimensions. The tool DIVAnd uses structured grids and can in fact work with an arbitrary high number of dimensions. The limiting factor are the available resources (in particular memory and cpu time). DIVAnd solves the equations (1) and (2) where the gradient and Laplacian are the n-dimensional operators. The grid must be structured, but it can be any orthogonal curvilinear grid. For instance one could use a cartesian or a spherical grid.

To assess the skill of the analysis method, Barth et al. (2014) used pseudo-observations (extracted from a global NEMO model at about ¼ degree resolution) at the location of the ARGO profiles to reconstruct monthly global surface temperatures. The use of a 3D-analysis versus a 2D-analysis (without advection constraint) resulted in an improvement of 27% while the best results were achieved with a

3D-analysis with advection constraint. This resulted in an RMS error of 0.85°C, a skill improvement of 44% relative to a 2D-analysis without advection constraint.

The more limited the data coverage is, the more it becomes important to efficiently use observations from close-by depth and time instances. One can therefore expect that the benefit of a multidimensional analysis would be even larger for observations from biological or chemical sensors where the data coverage is poorer.

DIVAnd is written in the MATLAB programming language, but it also runs efficiently under GNU Octave which implements a large part of the MATLAB syntax and functions. It can therefore be used running only free software which is an important aspect to facilitate the adoption of this software by new users.

APPLICATION OF DIVA

DIVA supports the ODV4 spreadsheet format (Ocean Data View, version 4) or simple text files as inputs, which can contain either profiles or time series. These time series can be handled by a weighting option, in order to correct the bias caused by the over-abundance of data on a particular zone (temporal and/or spatial). DIVA was originally designed to work in 2D, although it is now possible to solve pseudo 4D problems. In that case, the problem is divided into 2D layers associated with each other by a vertical filtering of two statistical parameters: the signal-to-noise ratio and the correlation length. Among other parameters, the user can choose the periods (years and months), the depths of the 2D layers and the variables. The results are then stored in a combined netCDF file.

For each 2D analysis, DIVA requires a contour usually representing the interface ocean/land at the considered depth. The topography used to create these contours is left to the user's choice, and several possibilities of extraction are proposed. The user has the possibility to set a variable correlation length over the domain, for instance in the case of smaller structures near the coast.

DIVA is mainly released as a command-line tool, although alternative (lighter) distributions include: DIVA for VirtualBox, DIVA for Matlab, Diva-on-Web and DIVA in Ocean Data View. It can be run on various operating systems (Linux, Mac OS X and Windows).

In order to lower the barrier to use the DIVA products, a web-based viewer has been developed which aims to provide an easy access to the interpolated fields as well as to auxiliary variables such as the error field and the number of observations per grid cell.

Interpolation of Glider Sections

The gridding of data along vertical sections (horizontal coordinate vs. depth) is a particular case of DIVA. The contour that delimits the domain of analysis is easily constructed from the bathymetry. The main difference to take into account is the difference of magnitude between the vertical (10-100 m) and horizontal scale (10-100 km). An analysis performed without considering this difference, i.e. assuming that the domain is isotropic, would lead to a gridded field that does not realistically represent the characteristics of the ocean. The solution adopted and implemented in DIVA consists in applying a coordinate change to the data, based on the ratio between the vertical scale and the horizontal scale.

The following example describes an application gridding glider observations in a vertical section located in the Ibiza Channel, between the Ibiza Island and Spain mainland (Figure 5). The vertical

Figure 5. Track of the CANALES mission in the Balearic Sea in September 2013. The blue line is the section that will be interpolated. The data are open and freely available from www.socib.es

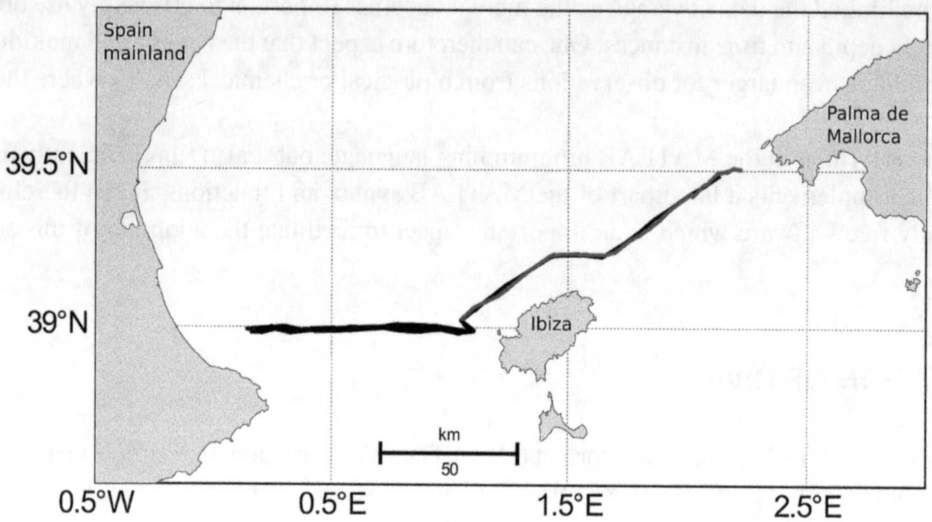

resolution of the data collected by a glider can be very high, thus the use of a gridding tool able to cope with this is often desired.

For this mission, the sampling strategy in the Ibiza Channel was to try to close the section with profiles going from the surface to the bottom (approximately 950 m). The transect (blue line) was repeated 4 times in September-October 2013. Temperature and salinity were acquired for all the transects, while chlorophyll-a concentration was only measured during the first of the four transects, and until a depth of 300 m. The number of data points for the studied transect is as follows: 39575 for salinity and 8452 for chlorophyll-a concentration.

The gridded fields are shown between the surface and 325 m depth for both variables. In this layer, the salinity increases with depth, with the lowest values in the surface layer at the two ends of the Ibiza Channel (Figure 6). Although the interpolation with DIVA has a smoothing effect on the data, the gridded field still displays small scale features characteristic of the studied area.

The chlorophyll-a concentration exhibits a deep chlorophyll-a maximum located between 50 and 100 m depth, with values exceeding 0.7 mg/m^3. Below 150 m, the values are very close to zero (Figure 7).

Post Processing of DIVA Analyses

The final 4D netCDF file includes all useful information about the performed analysis. Eight 4D-variables are available: the analysed field, the absolute error, the relative error, the analyzed field masked by two thresholds (0.3 and 0.5) of relative error (Beckers et al., 2014), the data bins (number of observations in each grid cell over the domain), the outliers bins and the (variable or constant) correlation length field. Three 3D-variables are also available: the analyzed field at the deepest analyzed level, with or without a masking from the relative error field (same two thresholds). The correlation length, the signal-to-noise ratio and the variance of the background field are also available to make the analysis reproducible. Fi-

Figure 6. Vertical section of salinity obtained with DIVA, between the surface and 300 m depth. Black dots indicate data location (only one out of 10 points represented, for the sake of clarity. The bathymetry is extracted from the GEBCO bathymetry with a 30-minute resolution.
For a more accurate representation of this figure, please see the electronic version.

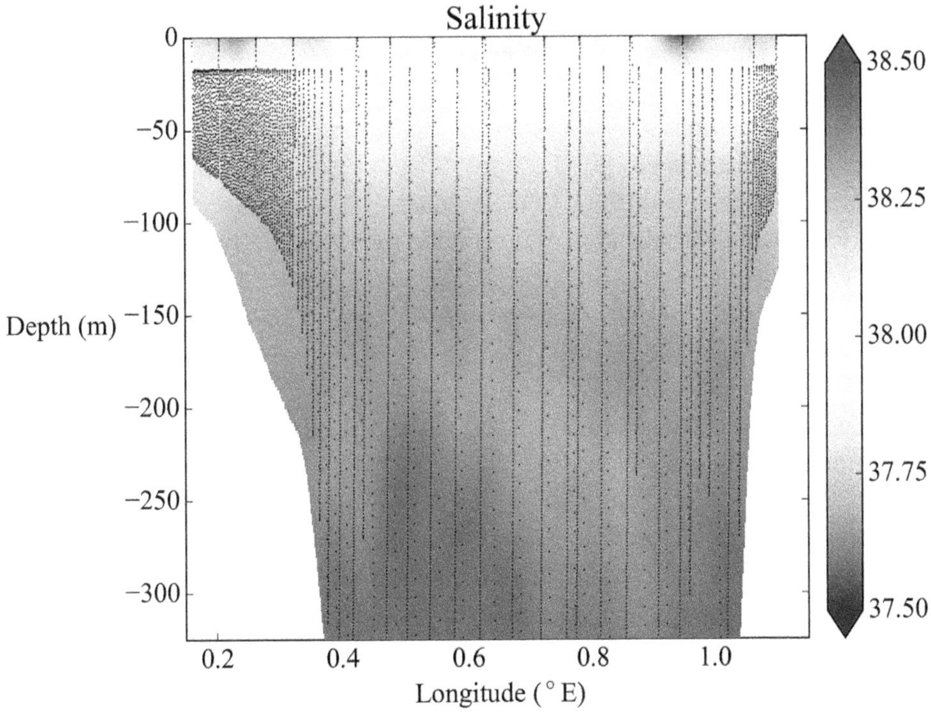

Figure 7. Vertical section of chlorophyll-a concentration obtained with DIVA, between the surface and 300 m depth.
For a more accurate representation of this figure, please see the electronic version.

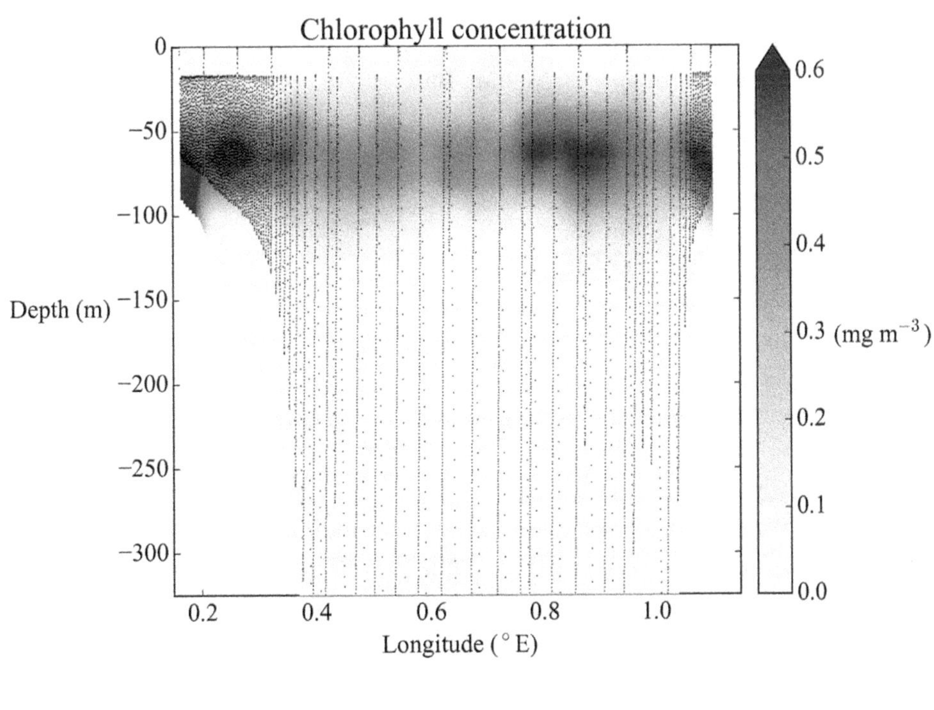

nally, some 1D-variables related to the original observations have been added to this combined netCDF file in order to keep a record of all observations used in an analysis.

On top of that, several global attributes give more metadata information about the creation of the file itself:

- The netCDF conventions followed for its creation,
- The name of the funding project,
- The name of the institution,
- The name of the production group,
- A web link towards the downloading of the products,
- A web link towards the visualization of the products (e.g. OceanBrowser),
- The author email,
- The creation date,
- The title of the netCDF,
- The filename and path,
- The data source(s),
- Some additional comments.

The major part of these global attributes can be specified by the user before running the analysis.

In the event the user wishes to cut the spatial domain of the final netCDF, a script has been implemented in order to avoid reperforming the whole analysis. The user has simply to provide a contour line specifying the new smaller domain. For a quick visualization of the results one can use gnuplot, a command-line utility designed for the visualisation of many different types of data. The command-line implementation makes it perfectly suitable for a combined use with DIVA. A single script generates a set of figures including the contour, the finite-element mesh, the data locations and the analysed field, depending on the files present in the input and output directories. As some users are more familiar with MATLAB or Python, toolkits for plotting the results in these programming environments are also provided (http://modb.oce.ulg.ac.be/mediawiki/index.php/DIVA#Plotting_tools).

To have these tools working, the user simply needs a library for reading of netCDF files and a library for the plotting of data on a map (m_map in Matlab, Basemap in Python). The examples are easily configurable and automatised if one wants to produce a large amount of high-quality figures, suitable for publication in a peer-reviewed journal.

VISUALIZATION OF GRIDDED FIELDS

This section discusses the on-line visualization of DIVA analyses. Contrary to most standard geographic information systems, where there is a strong focus on time-invariant information on the Earth surface (thus no depth and time dimension), the visualization of ocean data sets as generated by DIVA poses several challenges.

The ocean is generally a highly stratified fluid, where the seawater properties vary strongly with depth. While the ocean can be seen as a system made up of individual layers, the vertical movements (upwelling due to winds for instance) and mixing couple the different layers. For instance a vertical section of nutriments is often used to characterize and assess the effect of wind-induced upwelling at the coast. The ocean is also non-stationary as it varies on a broad range of temporal scales. Observational networks generally capture a limited range of these temporal scales. Not all temporal scales are present, but the ocean state is clearly non-stationary. It is necessary to integrate the depth and time dimensions to provide a full 4-dimensional view of the ocean.

A web-based visualization allows the integration with other datasets, which facilitates the interpretation of the dataset in conjunction with results from different domains (for instance physical and biogeochemical variables are strongly related due to the impact of ocean currents). The web interface should allow users to easily find a useful dataset and provide a mechanism to download it (entirely or a subset) and to be interoperable with other visualization platforms.

The web-interface OceanBrowser aims to respond to those challenges and requirements by allowing horizontal and vertical sections. Horizontal sections of the 4-dimensional fields (longitude, latitude, depth and time) can be visualized at a selected depth layer and time slice. The available depth and time slices correspond to the values included in the netCDF file. No additional time and depth interpolation is required for the horizontal sections.

The climatological fields can also be extracted and visualized on arbitrary vertical sections. In this case, the parameters are actually interpolated along a section defined by the user (unlike the horizontal sections). The vertical section can either be drawn by the mouse in the overview map (Figure 8) or generated automatically by using a criteria like a fixed distance from the coast or a fixed ocean depth.

The maps displayed in the browser are created dynamically and therefore several options are made available to the user to customize the graphical rendering of those layers.

Figure 8. To extract the gridded data along a vertical section, the user draws the section in the overview map with the mouse. The section can also be generated automatically by choosing a fixed distance from the coast or isobath level.

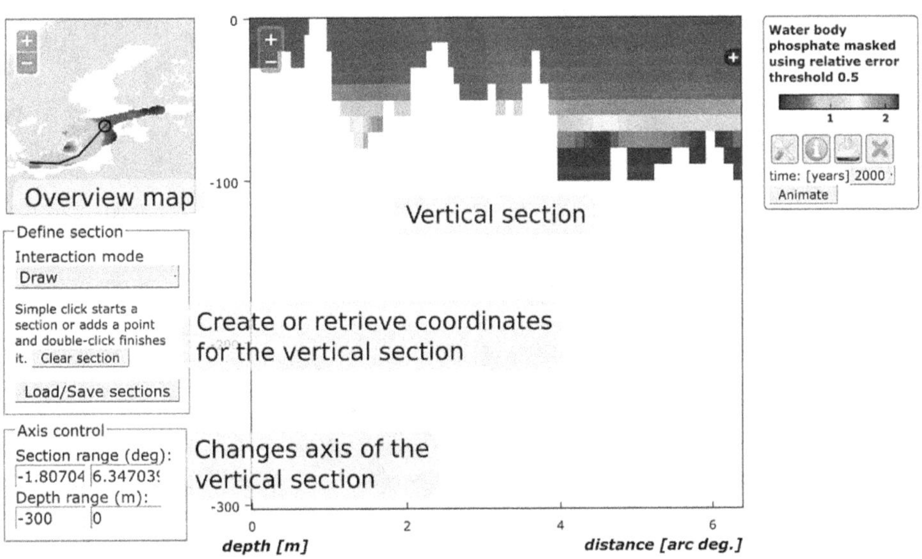

Scalar Plots

Layers representing scalar variables can be displayed either using flat shading, filled contours or simple contours. For flat shading, the color corresponding to a value of a grid cell is uniformly applied to this grid cell (Figure 9). This is the default option. The gridded field can also be displayed using contour lines (with and without filling the color between the contours). While the option flat shading and filled contour result in an opaque layer, the option of simple contours can be used to display different fields from the same domain at the same time and to assess to correspondence between two fields. As the ocean currents and (to a lesser degree) river inputs affect several variables is a similar way, gradients in various fields often align. The similarity of the structures in e.g. salinity and nutrients can for example by easily assessed by combining these variables in a single plot. Several options controlling the color-map are also available.

Vector Plots

Most data files represent scalar fields, but it is also possible to import vector fields into OceanBrowser. The layer can be customized by changing the density of the arrows and by adapting the scaling of the vectors. As an illustration Figure 10 shows the surface currents in the Ligurian Sea derived from the ROMS model (Shchepetkin and McWilliams, 2003, 2005). The spatial distribution of several ocean

Figure 9. The visualization of layer can be customized by using flat shading, filled contours or simple contours

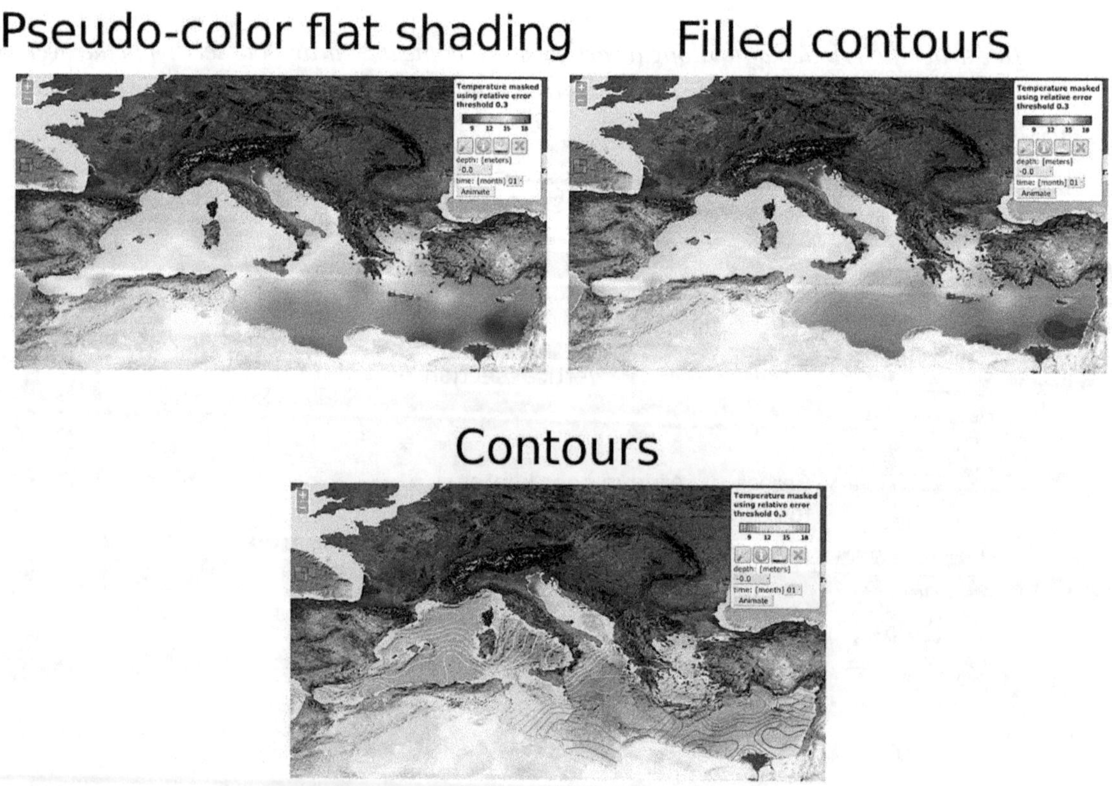

Figure 10. Surface currents generated from ROMS visualized in OceanBrowser

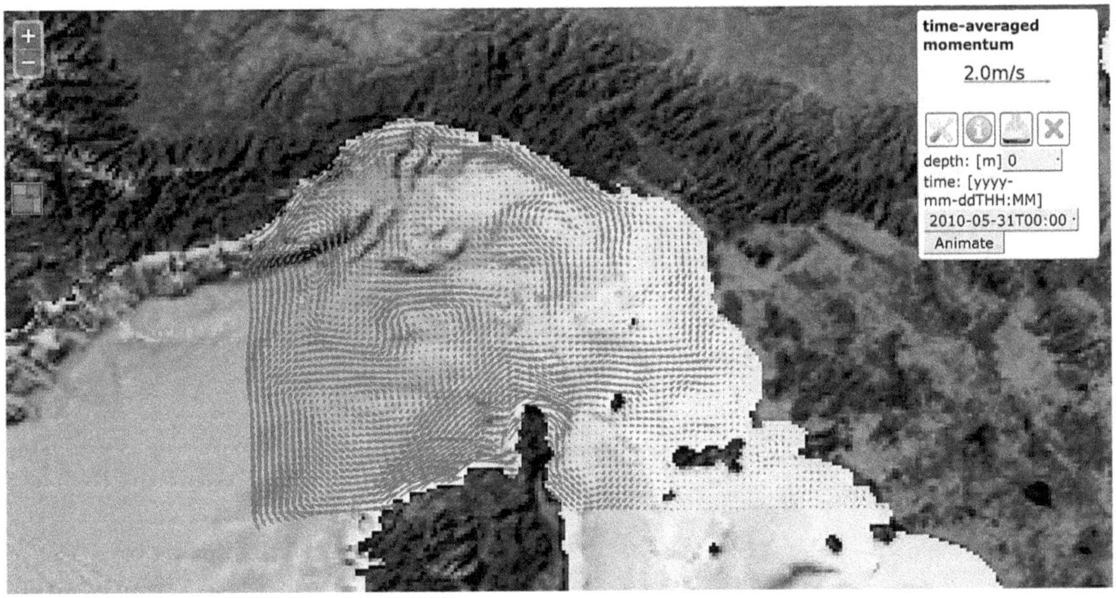

parameters is often related to ocean currents, which is indeed the original motivation to include the advection constraint in the DIVA analysis. A combined view of the ocean currents and the derived analysis thus allows one to visually assess how the information from the observations is spread horizontally.

Location Markers

The primary use of these markers is to show the observation locations used in DIVA. Within the SeaDataNet (Schaap et al., 2017, chapter 7 this book) and EMODNET Chemistry projects, every observation has an identifier which is composed by an institution EDMO code (European Directory of Marine Organisations) and a local identifier. The combination of the EDMO code and local identifier guarantees to uniquely identify a set of observations (usually a vertical profile). Such a layer gives access to the underlying data by linking it to the data download service (Figure 11).

The horizontal and vertical sections can be animated in order to examine the temporal evolution.

The web interface can also import third-party layers by using standard Web Map Service (WMS) requests. The user needs only to specify the URL of the WMS server and its supported version.

Interoperability of OceanBrowser with other web-services and clients depends critically on the use of open standards. Beside the W3C web-standards, OceanBrowser uses OGC WMS and netCDF. The implementation of the WMS protocol for the DIVA data products supports the following requests:

- GetCapabilities

This request is used to provide all layers of the map server. To every parameter and to every region corresponds a different WMS layer. An example of such a request would be: http://ec.oceanbrowser.net/emodnet/Python/web/wms?request=GetCapabilities&service=WMS&version=1.3.0

Figure 11. Location marker of the observations used in DIVA for a given depth and time slice

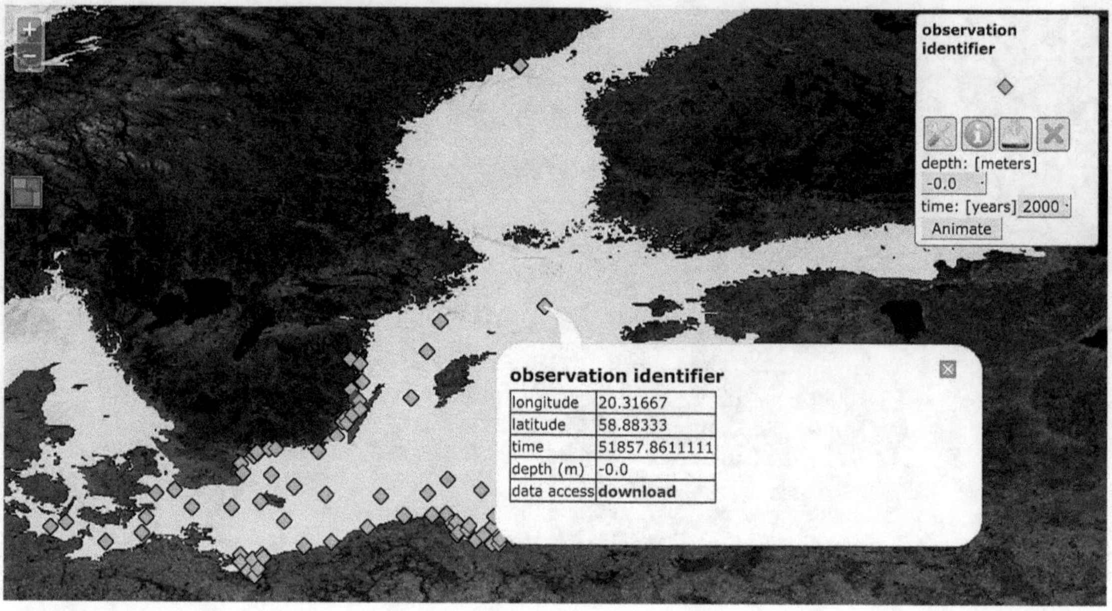

- GetMap

This request allows the extraction of a horizontal section of the 4-dimensional netCDF file at the specified depth and time for a selected latitude/longitude bounding box. By default, the axis are not displayed on a map. This can be activated by setting the non-standard parameter DECORATED to true.

The GetMap can also be used to extract a vertical section. The path of the section is encoded in the SECTION parameter: the longitude and latitude are separated by a comma and the coordinates by the pipe-symbol (|). The x-axis corresponds to the distance in arc degrees along the section (the first point is the origin) and the y-axis in the depth in meters. The parameter RATIO defines the aspect ratio of the vertical section.

Images can be returned in raster (PNG) and vector image formats (SVG, EPS and PDF). They can also be saved as a KML file so that the current layer can be visualized in programs like Google Earth and combined with other information imported in such programs. By providing multiple time instances, the web map server can also return animations in the WebM or MP4 format using this GetMap request. As the animations are generated dynamically, it usually takes a couple of minutes to create them. The frame rate of the animation is controlled through the parameter RATE.

- GetFeatureInfo

This request returns a simple XML file with the underlying value of the analysed field. This information is usually requested when the user clicks on a map.

However, the WMS standards (in version 1.1.1 and 1.3.0) are not completely adequate for ocean analyses. A WMS allows one to represent a data set according to a list of different styles. A legend is

Table 1. Parameters to control the plotting style of WMS layers

Key	Value
method	plotting method (pseudo-color, contours or filled contour)
ncontours	number of contours (ignored for pcolor)
inverted	true if color-map is inverted
cmap	color-map name (http://www.scipy.org/Cookbook/Matplotlib/Show_colormaps)
vmin	minimum of color-map range
vmax	maximum of color-map range

attributed to each style which is a colorbar for scalars. The legend for a given style is represented by a link to an image.

A single legend or colorbar is used for entire data set (for all depth layers and time instances in particular). However, the ocean is strongly stratified and a unique legend does not provide sufficient contrast because the ocean properties at depth are often very different from the properties near the surface. The solution is to make the legend dynamic so that it can be adjusted based on a range of values at a specified depth and time. Instead of using a limited number of styles, the style parameter in OceanBrowser uses a list for key-values pairs. It can contain the information as detailed in Table 1. Key and values are separated by a colon and key-value pairs by a plus sign. The order of the key/values is not significant.

Other WMS servers used for ocean data have developed similar extensions (Blower et al., 2013). For instance the parameters vmin and vmax for the colorbar are called COLORRANGE in ncWMS. Support of this parameter syntax is also implemented in OceanBrowser.

Deployment and Configuration of OceanBrowser

The basic idea was to make the incorporation of data in OceanBrowser as easy as adding a static web page to a web server. Similarly to most web servers, OceanBrowser uses also a "document root" directory under which all netCDF files can be placed, possibly using subdirectories in order to structure the data fields according to e.g. domain, parameters or time range. Internally OceanBrowser holds an inventory of all data fields and the inventory is regenerated when the modification time of the root directory changes (which can be forced using the Linux touch command). For adding a new file to OceanBrowser it is thus not necessary to restart the server application.

Every netCDF file is treated as a virtual folder containing various netCDF variables. If the number of netCDF variables is high, the individual variables can also be grouped in virtual subfolders. This is extensively used for DIVA analysis as several auxiliary netCDF variables are placed in a virtual folder called "Additional fields" which includes for instance the expected error standard deviation of the analysis. The underlying 4-dimensional netCDF file can be either downloaded as a whole from the interface or only as a subset using the OPeNDAP protocol.

While OceanBrowser has been primarily developed to visualize gridded datasets from DIVA it can also be used for satellite datasets which can have a more complex horizontal grid, as shown in Figure 12 which can have a more complex horizontal coordinate as it is sometimes related to the trajectory of the satellite.

Figure 12. L3 Satellite Sea Surface Temperature around the Canary Islands analysed by DINEOF using Medspiration Level 3 data.

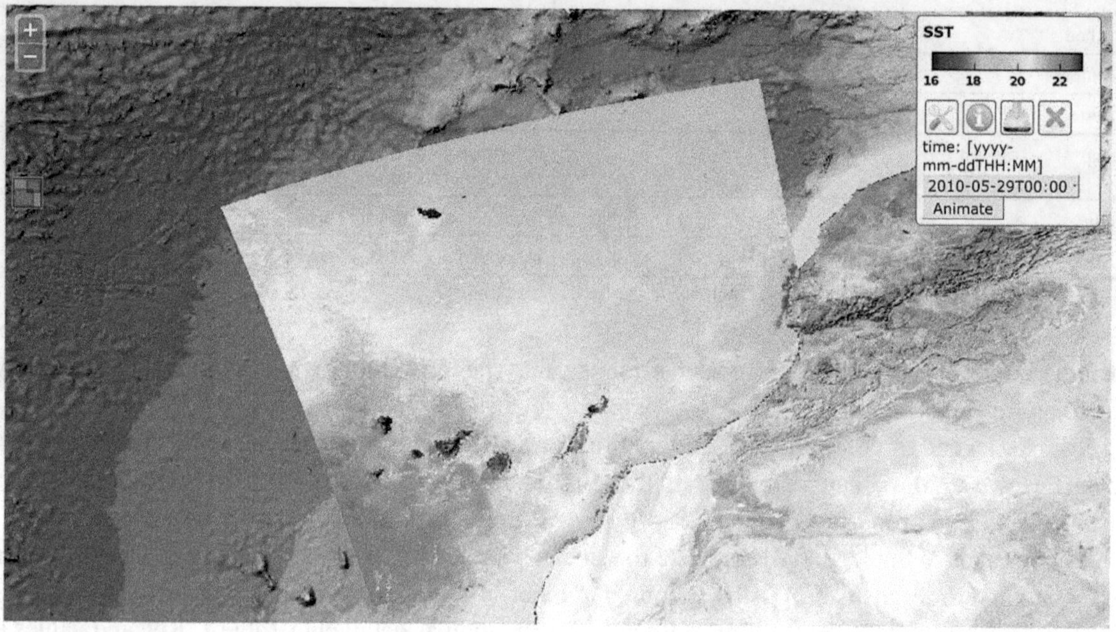

OceanBrowser is composed of a client part running in the web browser written in JavaScript and a server part written in Python. It uses several Python modules and in particular matplotlib and its Basemap plugin for the visualization. The Python server runs on a Linux server with at least 8GB of memory. In total, OceanBrowser requires 13 software packages to be installed on the server. To simplify the deployment, OceanBrowser can be installed using the programme docker. OceanBrowser and all its software dependencies are made available in a software container on the web site called "docker hub" (https://hub.docker.com/r/abarth/oceanbrowser/). Installing and running OceanBrowser can be achieved by running the single Linux command (provided that docker is installed):

```
docker run -p 8080:80 --name my-oceanbrowser-container \
-v /some/netcdf/files/:/var/www/data:ro abarth/oceanbrowser
```

Docker automatically downloads and runs OceanBrowser. OceanBrowser will be available at port 8080 on localhost and servers files in the specified directory. Additional configuration can be done using environment variables (including public URL, name and contact information).

To improve performance, OceanBrowser implements cache control headers for GetCapabilities and GetMap requests. The use of these cache control headers has a significant impact, improving the responsiveness of the web server. HTTP cache control gives a fine grain control about reusing server responses while minimizing the risk to using an out-of-date content.

For example if the client asks for the first time a list of all available layers (GetCapabilities), the server includes the following HTTP headers:

```
Cache-Control: "must-revalidate, max-age=5"
Last-Modified: "Tue, 10 Apr 2015 12:30:18 GMT"
```

This tells the web browser that it must check with the server if a newer version exists (after 5 seconds) and the time-stamp in the "Last-Modified" header reflects the time when the resource was last modified. Any future web requests of the web browser of the same resource are conditional request and include the following header:

```
If-Modified-Since:"Tue, 10 Apr 2015 12:30:18 GMT"
```

The server can either inform the web browser that its cached version is the current version (cache revalidation) or responding with a latest version of the corresponding request (and including the timestamp of the latest version in the Last-Modified header). On OceanBrowser, potential cache revalidation is fast because it requires only to compare the timestamp provided by the web browser and the timestamp of the directory containing all netCDF files.

The caching system has been extended by using in addition server side caching (Figure 13). A caching server (nginx) is placed between the web-browser and OceanBrowser. The web-browser first contacts the caching server which then contacts OceanBrowser. OceanBrowser needs only to generate the response for a given request, if the cached version at the browser and the caching server is outdated (or non-existent). In this case, the new version will be cached at the client (web-browser) and the caching server. Improving the caching system has resulted in a significant improvement of the responsiveness and the download speed to display the list of all layers and map tiles.

VISUALIZATION OF IN SITU OBSERVATIONS

Third-party Web Map Services can be included in OceanBrowser. This is particularly useful to view a climatology in combination with the corresponding in situ observations. Special consideration was given to the integration of the Oceanotron web service developed by Ifremer which provides quality-

Figure 13. To enable server-side caching the nginx server is placed between the client browser and the server software OceanBrowser.

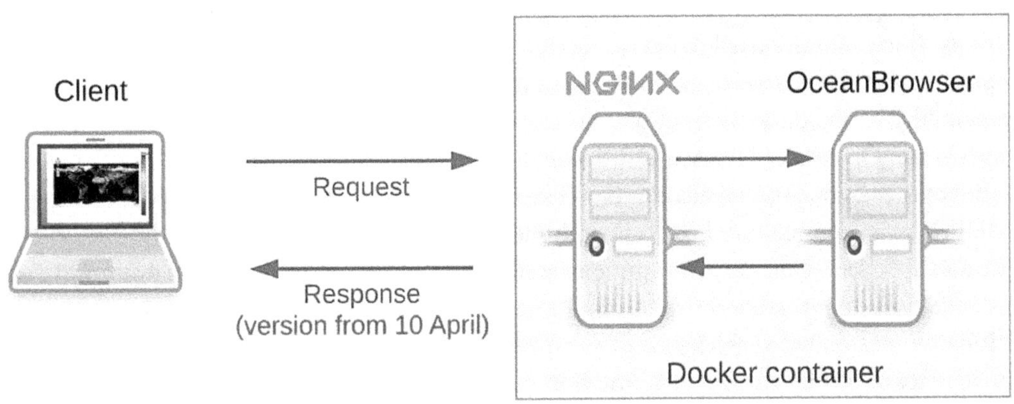

Figure 14. Climatological temperature of the month February and in situ observation of February 2012

controlled temperature and salinity measurements of the Arctic Ocean, North Atlantic Ocean, Baltic Sea, Black Sea, Mediterranean Sea and North Sea. Observations within a specified depth and time range can be queried and displayed on a map. Such observations can be visualized in conjunction with the DIVA analysis (Figure 14). This is useful to validate the climatology in a first step, to identify outliers in the observations, but also to assess anomalous conditions where some observations are different from average conditions.

Oceanotron is based on ncWMS and to realize this integration some of the extensions of the WMS protocol used by ncWMS have been implemented in OceanBrowser in order to allow a customization of the legend style (in particular the colorbar mapping and colorbar range). The way that the depth and the time dimensions are handled for serving in situ observations is also different from the way that those dimensions are represented for gridded climatologies. The depth and time are discrete values for the gridded climatologies, while they must be ranges for in situ observations.

CONCLUSION

DIVA allows the aggregation of in situ observations and the derivation of a coherent spatial view of ocean observations. It is generally used to derive monthly, seasonal or yearly climatological mean estimates.

The accuracy of the analysis can be assessed by taking into consideration the error field which depends essentially on the data distribution, the accuracy of the underlying observations and the correlation length-scale. Originally DIVA was developed to analyse physical parameters like temperature and salinity, but over the time special adaptations (like the anamorphosis transform) have been implemented to make DIVA also suitable for chemical and biological parameters.

Those data fields are generally four-dimensional (longitude, latitude, depth and time) and a web interface was developed to integrate the depth and time dimensions. OceanBrowser allows the visualization of the climatologies either along horizontal sections or arbitrary vertical sections. The time evolution can also be assessed by producing an animation.

DIVA and OceanBrowser are open source and released under the terms of the GNU General Public License and Affero General Public License to promote its use and adoption. Easy deployment of OceanBrowser is possible using Linux containers which allows to bundle all software dependencies and to ease the installation process.

Currently, DIVA does not take into account the possible relationship between variables. In future versions of DIVA we plan to implement a multivariate version. Such analyses can take the advantage of the relationship between e.g. physical parameters and biological parameters as the spatial coverage of biological parameters is much lower than the coverage of physical parameters. The information of the physical parameters would then be used to improve the analysis of the biological parameter. A multivariate extension of DIVA can also be used to make analysis of surface currents as the individual velocity components are strongly coupled through the model dynamics.

ACKNOWLEDGMENT

The development of DIVA and OceanBrowser was possible thanks to funding from the FP7 EU projects SeaDataNet 2 (contract number 283607) and ODIP (contract number 312492), EMODnet Chemistry (contract number SI2.656742) and EMODnet Biology (contract number SI2.657705) from DG MARE. The F.R.S.-FNRS is acknowledged to fund the position of Alexander Barth. We also thank Ifremer for developing Oceanotron (in particular Thomas Loubrieu and Mickael Treguer) and for DIVA products of the Atlantic Ocean (in particular Aurélie Briand). GEBCO project is acknowledged for providing the GEBCO world map, http://www.gebco.net. This is a MARE publication.

Computational resources have been provided by the Consortium des Équipements de Calcul Intensif (CÉCI), funded by the Fonds de la Recherche Scientifique de Belgique (F.R.S.-FNRS) under Grant No. 2.5020.11.

REFERENCES

Alvera-Azcárate, A., Barth, A., Sirjacobs, D., & Beckers, J.-M. (2009). Enhancing temporal correlations in EOF expansions for the reconstruction of missing data using DINEOF. *Ocean Science*, *5*(4), 475–485. doi:10.5194/os-5-475-2009

Barnes, S. (1964). A technique for maximizing details in numerical map analysis. *Journal of Applied Meteorology*, *3*(4), 395–409. doi:10.1175/1520-0450(1964)003<0396:ATFMDI>2.0.CO;2

Barth, A. (2015, September 11). *OceanBrowser*. Retrieved from https://hub.docker.com/r/abarth/ocean-browser/

Barth, A., Beckers, J.-M., Troupin, C., Alvera-Azcárate, A., & Vandenbulcke, L. (2014). divand-1.0: N-dimensional variational data analysis for ocean observations. *Geoscientific Model Development*, *7*(1), 225–241. doi:10.5194/gmd-7-225-2014

Beckers, J.-M., Barth, A., & Alvera-Azcárate, A. (2006). DINEOF reconstruction of clouded images including error maps. Application to the Sea Surface Temperature around Corsican Island. *Ocean Science*, *2*(2), 183–199. doi:10.5194/os-2-183-2006

Beckers, J.-M., Barth, A., Troupin, C., & Alvera-Azcárate, A. (2014). Approximate and efficient methods to assess error fields in spatial gridding with data interpolating variational analysis (DIVA). *Journal of Atmospheric and Oceanic Technology*, *31*(2), 515–530. doi:10.1175/JTECH-D-13-00130.1

Beckers, J.-M., & Rixen, M. (2003). EOF calculation and data filling from incomplete oceanographic datasets. *Journal of Atmospheric and Oceanic Technology*, *20*(12), 1839–1856. doi:10.1175/1520-0426(2003)020<1839:ECADFF>2.0.CO;2

Bennett, A. F., Chua, B. S., & Leslie, L. M. (1996). Generalized inversion of a global numerical weather prediction model. *Meteorology and Atmospheric Physics*, *60*(1-3), 165–178. doi:10.1007/BF01029793

Blower, J., Gemmell, A., Griffiths, G., Haines, K., Santokhee, A., & Yang, X. (2013). A Web Map Service implementation for the visualization of multidimensional gridded environmental data. *Environmental Modelling & Software*, *47*, 218–224. doi:10.1016/j.envsoft.2013.04.002

Blower, J., Haines, K., Santokhee, A., & Liu, C. (2009). GODIVA2: interactive visualization of environmental data on the Web. *Philosophical Transactions of the Royal Society of London A: Mathematical, Physical and Engineering Sciences, 367*(1890), 1035–1039, doi:10.1098/rsta.2008.0180

Brasseur, P., Beckers, J.-M., Brankart, J.-M., & Schoenauen, R. (1996). Seasonal temperature and salinity fields in the Mediterranean Sea: Climatological analyses of an historical data set. *Deep-Sea Research*, *43*(2), 159–192. doi:10.1016/0967-0637(96)00012-X

Brasseur, P., & Haus, J. (1991). Application of a 3–D variational inverse model to the analysis of ecohydrodynamic data in the Northern Bering and Southern Chuckchi seas. *Journal of Marine Systems*, *1*(4), 383–401. doi:10.1016/0924-7963(91)90006-G

Bretherton, F. P., Davis, R. E., & Fandry, C. B. (1976). A technique for objective analysis and design of oceanographic experiment applied to MODE-73. *Deep-Sea Research*, *23*, 559–582. doi:10.1016/0011-7471(76)90001-2

Buongiorno Nardelli, B. (2012). A novel approach for the high-resolution interpolation of in situ sea surface salinity. *Journal of Atmospheric and Oceanic Technology*, *29*(6), 867–879. doi:10.1175/JTECH-D-11-00099.1

Buongiorno Nardelli, B., Colella, S., Santoleri, R., Guarracino, M., & Kholod, A. (2010). A re-analysis of Black Sea surface temperature. *Journal of Marine Systems*, *79*(1–2), 50–64. doi:10.1016/j.jmarsys.2009.07.001

Capet, A., Barth, A., Beckers, J.-M., & Grégoire, M. (2012). Interannual variability of Black Sea's hydrodynamics and connection to atmospheric patterns. *Deep-sea Research. Part II, Topical Studies in Oceanography*, *77*, 128–142. doi:10.1016/j.dsr2.2012.04.010

Capet, A., Troupin, C., Carstensen, J., Grégoire, M., & Beckers, J.-M. (2014). Untangling spatial and temporal trends in the variability of the Black Sea Cold Intermediate Layer and mixed Layer Depth using the DIVA detrending procedure. *Ocean Dynamics*, *64*(3), 315–324. doi:10.1007/s10236-013-0683-4

Chen, Y., Davis, T. A., Hager, W. W., & Rajamanickam, S. (2008). Algorithm 887: Cholmod, Supernodal Sparse Cholesky Factorization and Update/Downdate. *ACM Transactions on Mathematical Software, 35*(3), 22:1–22:14, doi:.10.1145/1391989.1391995

Cressman, G. (1959). An operational objective analysis system. *Monthly Weather Review, 88*, 327–342. doi:10.1175/1520-0493(1960)088<0327:ITEIBF>2.0.CO;2

Davis, T. A., & Hager, W. W. (2009). Dynamic supernodes in sparse Cholesky update/downdate and triangular solves. *ACM Transactions on Mathematical Software, 35*(4), 27:1–27:23. doi:10.1145/1462173.1462176

de la Beaujardiere, J. (Ed.). (2006). *OpenGIS® Web Map Server Implementation Specification Version 1.3.0*. Open Geospatial Consortium Inc.

Duchon, J. (1977). Splines minimizing rotation-invariant semi-norms in Sobolev spaces. In *Constructive theory of functions of several variables* (pp. 85–100). Springer. doi:10.1007/BFb0086566

Evensen, G. (2007). *Data assimilation: the Ensemble Kalman Filter*. Springer.

Gandin, L. S. (1965). *Objective analysis of meteorological fields*. Jerusalem: Israel Program for Scientific Translation.

Guinehut, S., Larnicol, G., & Traon, P. L. (2002). Design of an array of profiling floats in the North Atlantic from model simulations. *Journal of Marine Systems, 35*(1-2), 1–9. doi:10.1016/S0924-7963(02)00042-8

Hunter, J. (2007). Matplotlib: A 2d graphics environment. *Computing in Science & Engineering, 9*(3), 90–95. doi:10.1109/MCSE.2007.55

Hyer, J. L., & She, J. (2007). Optimal interpolation of sea surface temperature for the North Sea and Baltic Sea. *Journal of Marine Systems, 65*(1-4), 176–189. doi:10.1016/j.jmarsys.2005.03.008

Kaplan, A., Kushnir, Y., Cane, M., & Blumenthal, M. (1997). Reduced Space Optimal Analysis for Historical Datasets: 136 Years of Atlantic Sea Surface Temperatures. *Journal of Geophysical Research, 102*(C13C3), 27,835–27,860. doi:10.1029/97JC01734

Kaplan, D. M., & Lekien, F. (2007). Spatial interpolation and filtering of surface current data based on open-boundary modal analysis. *Journal of Geophysical Research: Oceans, 112*(C12), C12007. doi:10.1029/2006JC003984

Lipphardt, B. L. Jr, Kirwan, A. D. Jr, Grosch, C. E., Lewis, J. K., & Paduan, J. D. (2000). Blending HF radar and model velocities in Monterey Bay through normal mode analysis. *Journal of Geophysical Research: Oceans, 105*(C2), 3425–3450. doi:10.1029/1999JC900295

Lorenzo, E. D., Moore, A. M., Arango, H. G., Cornuelle, B. D., Miller, A. J., & Bennett, A. F. et al. (2007). Weak and strong constraint data assimilation in the inverse Regional Ocean Modeling System (ROMS): Development and application for a baroclinic coastal upwelling system. *Ocean Modelling, 16*(3-4), 160–187. doi:10.1016/j.ocemod.2006.08.002

McIntosh, P. C. (1990). Oceanographic data interpolation: Objective analysis and splines. *Journal of Geophysical Research, 95*(C8), 13,529–13,541. doi:10.1029/JC095iC08p13529

Moore, A. M., Arango, H. G., Broquet, G., Powell, B. S., Weaver, A. T., & Zavala-Garay, J. (2011). The Regional Ocean Modeling System (ROMS) 4-dimensional variational data assimilation systems: Part I - System overview and formulation. *Progress in Oceanography*, *91*(1), 34–49. doi:10.1016/j.pocean.2011.05.004

Rayner, N. A., Parker, D. E., Horton, E. B., Folland, C. K., Alexander, L. V., Rowell, D. P., & Kaplan, A. et al. (2003). Global analyses of sea surface temperature, sea ice, and night marine air temperature since the late nineteenth century. *Journal of Geophysical Research*, *108*(D14), 4407. doi:10.1029/2002JD002670

Roberts-Jones, J., Fiedler, E. K., & Martin, M. J. (2012). Daily, Global, High-Resolution SST and Sea Ice Reanalysis for 1985-2007 Using the OSTIA System. *Journal of Climate*, *25*(18), 6215–6232. doi:10.1175/JCLI-D-11-00648.1

Robinson, A. R. (1996). Physical processes, field estimation and an approach to interdisciplinary ocean modeling. *Earth-Science Reviews*, *40*(1-2), 3–54. doi:10.1016/0012-8252(95)00030-5

Schaap, D. (2017). SeaDataNet – Towards a Pan-European Infrastructure for Marine and Ocean Data Management. In *Oceanographic and Marine Cross-Domain Data Management for Sustainable Development*. Hershey, PA: IGI Global.

Schlitzer, R. (2002). Interactive analysis and visualization of geoscience data with Ocean Data View. *Computers & Geosciences*, *28*(10), 1211–1218. doi:10.1016/S0098-3004(02)00040-7

Shchepetkin, A., & McWilliams, J. (2003). A Method for Computing Horizontal Pressure-Gradient Force in an Oceanic Model with a Non-Aligned Vertical Coordinate. *Journal of Geophysical Research*, *108*(C3), 1–34. doi:10.1029/2001JC001047

Shchepetkin, A., & McWilliams, J. (2005). The Regional Oceanic Modeling System: A split-explicit, free-surface, topography-following-coordinate ocean model. *Ocean Modelling*, *9*(4), 347–404. doi:10.1016/j.ocemod.2004.08.002

Troupin, C., Barth, A., Sirjacobs, D., Ouberdous, M., Brankart, J.-M., Brasseur, P., & Beckers, J.-M. et al. (2012). Generation of analysis and consistent error fields using the Data Interpolating Variational Analysis (DIVA). *Ocean Modelling*, *52–53*, 90–101. doi:10.1016/j.ocemod.2012.05.002

Wahba, G. (1990). *Spline models for observational data*. Philadelphia, PA: SIAM Journal of Applied Mathematics.

Wahba, G., & Wendelberger, J. (1980). Some new mathematical methods for variational objective analysis using splines and cross validation. *Monthly Weather Review*, *108*(8), 1122–1143. doi:10.1175/1520-0493(1980)108<1122:SNMMFV>2.0.CO;2

Weaver, A. T., Vialard, J., & Anderson, D. L. T. (2003). Three- and Four-Dimensional Variational Assimilation with a General Circulation Model of the Tropical Pacific Ocean. Part I: Formulation, Internal Diagnostics, and Consistency Checks. *Monthly Weather Review*, *131*(7), 1360–1378. doi:10.1175/1520-0493(2003)131<1360:TAFVAW>2.0.CO;2

KEY TERMS AND DEFINITIONS

Correlation Length: The length-scale over which the error covariance is significantly different from zero.

Error Covariance: Measure how much the error (of the first guess estimate or of the observation) is related to the error at a different location (or time). The error covariance can be decomposed in variance and decorrelation.

netCDF (Network Common Data Form): A binary data format for multidimensional data sets and widely adopted in oceanography and meteorology.

Signal-to-Noise Ratio: A measure of the variance of the background error compared to the variance of the observation errors.

Variational Inverse Method (VIM): Data analysis method based on the formulation of a cost-function to be minimized. While interpolating a gridded field on arbitrary location is generally considered the "direct" or "forward" problem, the inverse problem tries to deduce from observation at arbitrary location a gridded field.

Web Feature Service (WFS): The OGS server-client protocol for discrete features.

Web Map Service (WMS): The OGC server-client protocol to advertise all available layers and to transfer maps, usually as images.

Compilation of References

52 North Initiative for Geospatial Open Source Software GmbH. (2016). *52norrth exploring horizons.* Retrieved February 14, 2016, from http://52north.org/

52 North. (2014). *Sensor Observation Service*. Retrieved from: http://52north.org/communities/sensorweb/sos/

52 North. (n.d.). Retrieved from: http://52north.org/

Alexander, K., & Hausenblas, M. (2009). Describing linked datasets-on the design and usage of void, the vocabulary of interlinked datasets. In *Linked Data on the Web Workshop (LDOW 09), in conjunction with 18th International World Wide Web Conference (WWW 09).*

Alexander, C., Ishikawa, S., & Silverstein, M. (1977). *A pattern language: towns, buildings, construction* (Vol. 2). Oxford, UK: Oxford University Press.

Alvera-Azcárate, A., Barth, A., Sirjacobs, D., & Beckers, J.-M. (2009). Enhancing temporal correlations in EOF expansions for the reconstruction of missing data using DINEOF. *Ocean Science*, *5*(4), 475–485. doi:10.5194/os-5-475-2009

Angles, R., & Gutierrez, C. (2008). Survey of graph database models. *ACM Computing Surveys*, *40*(1), 1–39. doi:10.1145/1322432.1322433

Arko, R., Chandler, C., Stocks, K., Smith, S., Clark, P., Shepherd, A., . . . Beaulieu, S. (2013). Rolling Deck to Repository (R2R): Collaborative Development of Linked Data for Oceanographic Research. *Geophysical Research Abstracts, 15.*

Atkins, J. P., Burdon, D., Elliott, M., & Gregory, A. J. (2011). Management of the marine environment: Integrating ecosystem services and societal benefits with the DPSIR framework in a systems approach. *Marine Pollution Bulletin*, *62*(2), 215–226. doi:10.1016/j.marpolbul.2010.12.012 PMID:21215974

Atlassian. (2016). *Confluence – Team collaboration Software*. Retrieved April 5, 2016, from https://www.atlassian.com/software/confluence/

Bahurel, P., Adragna, F., Bell, M. J., Jacq, F., Johannessen, J. A., Le Traon, P.-Y., . . . She, J. (2009). Ocean monitoring and forecasting core services, the European MyOcean example. In Oceanobs 2009.

Baraniuk, R. G. (2011). More Is Less: Signal Processing and the Data Deluge. *Science*, *331*(6018), 717–719. doi:10.1126/science.1197448 PMID:21311012

Barker, P. M., Dunn, J. R., Domingues, C. M., & Wijffels, S. E. (2011). Pressure Sensor Drifts in Argo and Their Impacts. *Journal of Atmospheric and Oceanic Technology*, *28*(8), 1036–1049. doi:10.1175/2011JTECHO831.1

Barnaghia, P., Ganza, F., & Abangara, H. (2011). *Sense2Web: A Linked Data Platform for Semantic Sensor Networks.* Retrieved from http://www.semantic-web-journal.net/sites/default/files/swj189.pdf

Barnes, S. (1964). A technique for maximizing details in numerical map analysis. *Journal of Applied Meteorology*, *3*(4), 395–409. doi:10.1175/1520-0450(1964)003<0396:ATFMDI>2.0.CO;2

Barry, J., Boyd, S., & Fryer, R. (2010). Modelling the effects of marine aggregate extraction on benthic assemblages. *Journal of the Marine Biological Association of the United Kingdom*, *90*(1), 105–114. doi:10.1017/S0025315409990737

Barth, A. (2015, September 11). *OceanBrowser*. Retrieved from https://hub.docker.com/r/abarth/oceanbrowser/

Barth, A., Beckers, J.-M., Troupin, C., Alvera-Azcárate, A., & Vandenbulcke, L. (2014). divand-1.0: N-dimensional variational data analysis for ocean observations. *Geoscientific Model Development*, *7*(1), 225–241. doi:10.5194/gmd-7-225-2014

Barth, A., Watelet, S., Troupin, C., Alvera-Azcarate, A., & Beckers, J. M. (2017). Analysis of Ocean In Situ Observations and Web-Based Visualization. In *Oceanographic and Marine Cross-Domain Data Management for Sustainable Development*. Hershey, PA: IGI Global.

Barthes, R. (1957). *Mythologies*. London: Ebenezer Baylis & Son Ltd.

Bartolini, R., Lenci, A., Montemagni, S., Pirrelli, V., & Claudia Soria, C. (2004). Semantic mark-up of Italian legal texts through NLP-based techniques. In *Proceedings of LREC 2004*.

Bassett, R., Beard, R., Burnett, W., Crout, R., Griffith, B., Jensen, R., & Signell, R. (2010). Implementing the National Integrated Ocean Observing System (IOOS®)—From the Federal Agency Perspective. *J. Mar. Tech. Soc*, *44*(6), 32–41. doi:10.4031/MTSJ.44.6.8

Beale, T. (2002, November). Archetypes: Constraint-based domain models for future-proof information systems. In OOPSLA 2002 workshop on behavioural semantics (Vol. 105).

Beale, T., Heard, S., Kalra, D., & Lloyd, D. (2005). *Archetype definition language (adl)*. OpenEHR specification, the openEHR foundation.

Beaulieu, S., Fox, P., Maffei, A., Hare, J., Fogarty, M., & Di Stefano, M. (2013). *Traceability, reproducibility, and scalability in Integrated Ecosystem Assessments*. Oral presentation at ESIP Federation Summer Meeting 2013.

Becher, T., & Trowler, P. R. (2001). *Academic tribes and territories intellectual enquiry and culture of disciplines*. Buckingham, UK: The Society for Research into Higher Education and Open University Press.

Beckers, J.-M., Barth, A., & Alvera-Azcárate, A. (2006). DINEOF reconstruction of clouded images including error maps. Application to the Sea Surface Temperature around Corsican Island. *Ocean Science*, *2*(2), 183–199. doi:10.5194/os-2-183-2006

Beckers, J.-M., Barth, A., Troupin, C., & Alvera-Azcárate, A. (2014). Approximate and efficient methods to assess error fields in spatial gridding with data interpolating variational analysis (DIVA). *Journal of Atmospheric and Oceanic Technology*, *31*(2), 515–530. doi:10.1175/JTECH-D-13-00130.1

Beckers, J.-M., & Rixen, M. (2003). EOF calculation and data filling from incomplete oceanographic datasets. *Journal of Atmospheric and Oceanic Technology*, *20*(12), 1839–1856. doi:10.1175/1520-0426(2003)020<1839:ECADFF>2.0.CO;2

Belov, S., & Mikhailov, N. (2012). *Technical Workshop on the IODE Ocean Data Portal*. Retrieved from ftp://ftpdownload:ftpdownload@odp.meteo.ru/5-ODP_V2.pdf

Belov, S., Mikhailov, N., & Belova, K. (2013). Ocean Data Portal: from data access to integration platform. In *IMDIS 2013: International Conference on Marine data and Information Systems*.

Bennett, A. F., Chua, B. S., & Leslie, L. M. (1996). Generalized inversion of a global numerical weather prediction model. *Meteorology and Atmospheric Physics*, *60*(1-3), 165–178. doi:10.1007/BF01029793

Benthall, B., & Gill, S. (2010). SDI Best Practices with GeoNode. In *Proceedings of the 2010 Free and Open Source Software for Geospatial Conference* (FOSS4G 2010).

Bermudez, L., Montgomery, E., Miller, S. P., Neiswender, C., & Isenor, A. (2011). *The Importance of Controlled Vocabularies*. The MMI Guides: Navigating the World of Marine Metadata. Retrieved from http://marinemetadata.org/guides/vocabs/vocimportance

Berners-Lee, T. (1998). *Semantic web road map*. Retrieved from: https://www.w3.org/DesignIssues/Semantic

Berners-Lee, T. (2006). *Linked Data - Design Issues*. World Wide Web Consortium. Retrieved May 29, 2015 from http://www.w3.org/DesignIssues/LinkedData.html

Berners-Lee, T. (2006). *Linked Data*. Accessible at: http://www.w3.org/DesignIssues/LinkedData.html

Berners-Lee, T. (2011). *Linked data-design issues (2006)*. Retrieved from http://www.w3.org/DesignIssues/LinkedData.html

Berners-Lee, T., Hendler, J., & Lassila, O. (2001). The Semantic Web. *Scientific American*, *284*(5), 34–43. doi:10.1038/scientificamerican0501-34 PMID:11396337

Berx, B., Dickey-Collas, M., Skogen, M. D., De Roeck, Y. H., Klein, H., Barciela, R., & Schrum, C. et al. (2011). Does operational oceanography address the needs of fisheries and applied environmental scientists? *Oceanography (Washington, D.C.)*, *24*(01), 166–171. doi:10.5670/oceanog.2011.14

Bijma, J., Pörtner, H.-O., Yesson, Ch., & Rogers, A. D. (2013). Climate change and the oceans - What does the future hold? *Marine Pollution Bulletin*, *74*(2), 495–505. doi:10.1016/j.marpolbul.2013.07.022 PMID:23932473

Bishop, M. A. (1992) Theory-Ladenness of Perception Arguments. In *Proceedings of the Biennial Meeting of the Philosophy of Science Association* (pp. 287-299). Academic Press.

Bizer, C., Heath, T., & Berners-Lee, T. (2009). Linked Data—The Story So Far. *International Journal on Semantic Web and Information Systems*, *5*(3), 1–22. doi:10.4018/jswis.2009081901

Blaas, M., Dong, C., Marchesiello, P., McWilliams, J. C., & Stolzenbach, K. D. (2007). Sediment-transport modeling on Southern Californian shelves: A ROMS case study. *Continental Shelf Research*, *27*(6), 832–853. doi:10.1016/j.csr.2006.12.003

Blažauskas, N., Grigelis, A., Gelumbauskaitė, L. Ž., Gulbinskas, S., Suzdalev, S., & Ferrarin, Ch. (2015). Towards sustainable use of marine resources in the south-eastern Baltic Sea (Lithuania): A review. *Baltica*, *28*(2), 179–188. doi:10.5200/baltica.2015.28.15

Bleier, T., Bozic'Roman, B., da Costa, A., Costes, S., Iosifescu, I., Martin, O., … Simonis, I. (2009). *SANY and open service architecture for sensors networks*. Retrieved October 30, 2015, from http://www.frisia-it.com/assets/images/SANY_Book.pdf

Blower, J., Haines, K., Santokhee, A., & Liu, C. (2009). GODIVA2: interactive visualization of environmental data on the Web. *Philosophical Transactions of the Royal Society of London A: Mathematical, Physical and Engineering Sciences, 367*(1890), 1035–1039, doi:10.1098/rsta.2008.0180

Blower, J. D., Blanc, F., Clancy, M., Cornillon, P., Donlon, C., Hacker, P., & Srinivasan, A. et al. (2009). Serving GODAE data and products to the ocean community. *Oceanography (Washington, D.C.)*, *22*(3), 70–79. doi:10.5670/oceanog.2009.67

Blower, J., Gemmell, A., Griffiths, G., Haines, K., Santokhee, A., & Yang, X. (2013). A Web Map Service implementation for the visualization of multidimensional gridded environmental data. *Environmental Modelling & Software*, *47*, 218–224. doi:10.1016/j.envsoft.2013.04.002

Blythe, J. N., & Dadi, U. (2012). Knowledge integration as a method to develop capacity for evaluating technical information on biodiversity and ocean currents for integrated coastal management. *Environmental Science & Policy*, *19-20*, 49–58. doi:10.1016/j.envsci.2012.01.007

BODC. (n.d.). *NERC controlled vocabulary*. Retrieved from http://www.bodc.ac.uk/products/web_services/vocab/

Boehm, B., Brown, J. R., Kaspar, H., Lipow, M., McLeod, G., & Merritt, M. (1978). Characteristics of Software Quality. North Holland.

Boldrini, E. (2013a). *SeaDataNet metadata profile of ISO 19115*. Retrieved from http://www.seadatanet.org/content/download/20284/140589/file/CDI-profile-V10.0.pdf

Boldrini, E. (2013a). *SeaDataNet metadata profile of ISO 19115. Version 10.0.0*. Retrieved from http://www.seadatanet.org/content/download/20284/140589/file/CDI-profile-V10.0.pdf

Boldrini, E. (2013b). *SeaDataNet metadata profile of ISO 19115 –XML encoding*. Retrieved from http://www.seadatanet.org/content/download/20285/140594/file/CDI-profile-XML-encoding-V10.0.pdf

Boldrini, E. (2013b). *SeaDataNet metadata profile of ISO 19115 –XML encoding. Version 10.0.0*. Retrieved from http://www.seadatanet.org/content/download/20285/140594/file/CDI-profile-XML-encoding-V10.0.pdf

Boldrini, E. (2013c). *SeaDataNet CSR metadata profile*. Retrieved from http://www.seadatanet.org/content/download/20282/140579/file/CSR-profile-V3.0.pdf

Boldrini. (2013d). *SeaDataNet CSR metadata profile of ISO 19115-2 – XML encoding*. Academic Press.

Bond, C. E., Gibbs, A., Shipton, Z. K., & Jones, S. (2007). What do you think this is? 'Conceptual uncertainty' in geoscience interpretation. *GSA Today*, *17*(11), 4–10. doi:10.1130/GSAT01711A.1

Bond, C. E., Lunn, R. J., Shipton, Z. K., & Lunn, A. D. (2012). What makes an expert effective at interpreting seismic images? *Geology*, *40*(1), 75–78. doi:10.1130/G32375.1

Bond, C. E., Shipton, Z. K., Gibbs, A. D., & Jones, S. (2008). Structural models: Optimizing risk analysis by understanding conceptual uncertainty. *First Break*, *26*(1288), 65–71. doi:10.3997/1365-2397.2008006

Booij, N., Ris, R. C., & Holthuijsen, L. H. (1999). A third-generation wave model for coastal regions: 1. Model description and validation. *Journal of Geophysical Research: Oceans (1978–2012)*, *104*(C4), 7649-7666.

Bosma, W. E., Vossen, P., Soroa, A., Rigau, G., Tesconi, M., Marchetti, A., & Aliprandi, C. et al. (2009). Kaf: a generic semantic annotation format. In *Proceedings of the GL2009 Workshop on Semantic Annotation*.

Botts, M., & Robin, A. (2004). *OGC SensorML: Model and Xml Encoding Standard*. Open Geospatial Consortium.

Boyer, T. P., Antonov, J. I., Baranova, O. K., Coleman, C., Garcia, H. E., Grodsky, A., . . . Zweng, M. M. (2013). World Ocean Database 2013. NOAA Atlas NESDIS 72.

Brasseur, P., Beckers, J.-M., Brankart, J.-M., & Schoenauen, R. (1996). Seasonal temperature and salinity fields in the Mediterranean Sea: Climatological analyses of an historical data set. *Deep-Sea Research*, *43*(2), 159–192. doi:10.1016/0967-0637(96)00012-X

Brasseur, P., & Haus, J. (1991). Application of a 3–D variational inverse model to the analysis of ecohydrodynamic data in the Northern Bering and Southern Chuckchi seas. *Journal of Marine Systems*, *1*(4), 383–401. doi:10.1016/0924-7963(91)90006-G

Bray, T. (2013). *XML's 15th Birthday*. Retrieved February 2, 2016 from http://www.tbray.org/ongoing/When/201x/2013/02/10/XML-at-15

Bretherton, F. P., Davis, R. E., & Fandry, C. B. (1976). A technique for objective analysis and design of oceanographic experiment applied to MODE-73. *Deep-Sea Research*, *23*, 559–582. doi:10.1016/0011-7471(76)90001-2

Brickley, D., & Guha, R. V. (2004). RDF Vocabulary Description Language 1.0: RDF Schema. *W3C*. Retrieved from http://www.w3.org/TR/rdf-schema/

Brin, S. (1998). Extracting Patterns and Relations from the World Wide Web. In *Proceedings of the WebDB Workshop at 6th International Conference on Extending Database Technology*.

Broering, A., Stasch, C., & Echterhoff, J. (2012). *OGC Sensor Observation Service Interface Standard, Version 2.0*. OGC.

Brown, J. S., & Duguid, P. (1998). Organising knowledge. *California Management Review*, *40*(3), 90–111. doi:10.2307/41165945

Brundtland Commission Report. (1987). *Lawlor*. Retrieved from https://en.wikisource.org/wiki/Brundtland_Report on 7/10/2015

Buck, J., & Lowry, R. (2016). Oceanographic Data Management: Quills and Free Text to the digital age and 'Big Data. In P. Diviacco, A. Leadbetter, & H. Glaves (Eds.), *Oceanographic and Marine Cross-Domain Data Management for Sustainable Development*. Hershey, PA: IGI Global.

Buck, J., & Lowry, R. (2017). Oceanographic Data Management; Quills and Free Text to the digital age and 'Big Data. In P. Diviacco, A. Leadbetter, & H. Glaves (Eds.), *Oceanographic and Marine Cross-Domain Data Management for Sustainable Development*. Hershey, PA: IGI Global.

Buongiorno Nardelli, B. (2012). A novel approach for the high-resolution interpolation of in situ sea surface salinity. *Journal of Atmospheric and Oceanic Technology*, *29*(6), 867–879. doi:10.1175/JTECH-D-11-00099.1

Buongiorno Nardelli, B., Colella, S., Santoleri, R., Guarracino, M., & Kholod, A. (2010). A re-analysis of Black Sea surface temperature. *Journal of Marine Systems*, *79*(1–2), 50–64. doi:10.1016/j.jmarsys.2009.07.001

Busschers, F. S., Stafleu, J., Maljers, D., Gunnink, J. L., & Sobisch, H.-G. (2011, October). *New geological insights from stochastic 3D voxel models*. Paper presented at the Annual Meeting of the Geological Society of America. Minneapolis, MN.

Caddy, J. F. (2007). *Using questionnaires based on the Code of Conduct for Responsible Fisheries as a diagnostic tool in support of fisheries management*. Roma: FAO.

Caddy, J. F., & Mahon, R. (1995). Reference points for fisheries management.[Rome: FAO.]. *FAO Fisheries Technical Paper*, 347.

Callender, C., & Cohen, J. (2006). There is No Special Problem About Scientific Representation. *Theoria*, *21*(1), 67–85.

Capet, A., Barth, A., Beckers, J.-M., & Grégoire, M. (2012). Interannual variability of Black Sea's hydrodynamics and connection to atmospheric patterns. *Deep-sea Research. Part II, Topical Studies in Oceanography*, *77*, 128–142. doi:10.1016/j.dsr2.2012.04.010

Capet, A., Troupin, C., Carstensen, J., Grégoire, M., & Beckers, J.-M. (2014). Untangling spatial and temporal trends in the variability of the Black Sea Cold Intermediate Layer and mixed Layer Depth using the DIVA detrending procedure. *Ocean Dynamics*, *64*(3), 315–324. doi:10.1007/s10236-013-0683-4

Car, N. J. (2013). The eReefs Information Architecture. In J. Piantadosi, R. S. Anderssen, & J. Boland (Eds.), Proceedings of the 20th International Congress on Modelling and Simulation (pp. 831–837). Adelaide, Australia. Accessible at http://www.mssanz.org.au/modsim2013/C7/car2.pdf

Car, N. J., Stenson, M. P., & Hartcher, M. (2014). A Provenance Methodology And Architecture For Scientific Projects Containing Automated And Manual Processes. In *Proceedings of the 11th Conference on Hydroinformatics*. Accessible at: http://academicworks.cuny.edu/cc_conf_hic/57

Cassard, D., Bertrand, G., Billa, M., Serrano, J.-J., Tourlière, B., Angel, J.-M., & Gaál, G. (2015). ProMine Mineral Databases: New Tools to Assess Primary and Secondary Mineral Resources in Europe. In P. Weihed (Ed.), 3D, 4D and Predictive Modelling of Major Mineral Belts in Europe (pp. 9-58). Mineral Resource Reviews Book Series. Cham, Switzerland: Springer International Publishing.

Cassard, D., Tertre, F., Bertrand, G., Schjøth, F., Tulstrup, J., Heijboer, T., & Vuollo, J. (2014). EURARE IKMS: An Integrated Knowledge Management System for Rare Earth Element Resources in Europe. In *ERES 2014 - 1st European Rare Earth Resources Conference* (pp. 292-300). Retrieved from http://www.eurare.eu/docs/eres2014/fifthSession/GuillaumeBertrand.pdf

Centre for Environmental Data Archival (CEDA). (2016). *Proposals for CF.* Retrieved April 5, 2016, from http://cfeditor.ceda.ac.uk/proposals/1?status=active&namefilter=&proposerfilter=&descfilter=&filter+and+display=filter

CF Community. (2015). *CF Conventions and Metadata.* Retrieved October 30, 2015, from http://cfconventions.org/standard-names.html

CF Community. (2016). *NetCDF Conventions.* Retrieved October 30, 2015, from http://www.unidata.ucar.edu/software/netcdf/conventions.html

Chakravartty, A. (1995). Truth and Representation in Science: Two Inspirations from Art. In R. Frigg & M. Hunter (Eds.), *Beyond Mimesis and Convention: Representation in Art and Science* (pp. 33-50). Berlin: Springer.

Chandler, C., & Arko, R. (2015). Big Data Challenges: Variety and Veracity. Ocean Data Interoperability Platform Workshop Presentation, Paris, France.

Chapelle, O., Metlzer, D., Zhang, Y., & Grinspan, P. (2009). Expected Reciprocal Rank for Graded Relevance. In *Proceedings of the Conference on Information and Knowledge Management.*

Chen, Y., Davis, T. A., Hager, W. W., & Rajamanickam, S. (2008). Algorithm 887: Cholmod, Supernodal Sparse Cholesky Factorization and Update/Downdate. *ACM Transactions on Mathematical Software, 35*(3), 22:1–22:14, doi:.10.1145/1391989.1391995

Chen, N., Liping, D., & Genong Yua, M. M. (2009). A flexible geospatial sensor observation service for diverse sensor data based on Web service. *ISPRS Journal of Photogrammetry and Remote Sensing*, *64*(2), 234–242. doi:10.1016/j.isprsjprs.2008.12.001

Clark, J. (1999). *XSL Transformations (XSLT) Version 1.0.* W3C Recommendation. Retrieved from http://www.w3.org/TR/xslt

Clarke, R. (2005). A proposal for an open content licence for research paper (Pr)ePrints. *First Monday*, *10*(8), 1–11. doi:10.5210/fm.v10i8.1262

Claustre, H., Bishop, J., Boss, E., Bernard, S., Berthon, J. F., Coatanoan, C., . . . Uitz, J. (2009). Bio-optical profiling floats as new observational tools for biogeochemical and ecosystem studies: potential synergies with ocean color remote sensing. *Proceedings of OceanObs'09: Sustained Ocean Observations and Information for Society.* Retrieved February 10, 2016, from http://www.ioccg.org/groups/Claustre_et_al._OO09.pdf

Clough, P., & Sanderson, M. (2013). Evaluating the performance of information retrieval systems using test collections. *Information Research, 18*(2). Retrieved from http://www.informationr.net/ir/18-2/paper582.html

Commons, C. (2001). *"Some Rights Reserved": Building a Layer of Reasonable Copyright.* Retrieved 2005-09-14 from the World Wide Web: http://creativecommons.org

Compton, M., Corsar, D., & Taylor, K. (2014). *Sensor data provenance: SSNO and PROV-O together at last.* In 7th International Semantic Sensor Networks Workshop (October 2014).

Cooper, K. M. (2012). Setting limits for acceptable change in sediment particle size composition following marine aggregate dredging. *Marine Pollution Bulletin, 64*(8), 1667–1677. doi:10.1016/j.marpolbul.2012.05.031 PMID:22721693

Cooper, K. M., Froján, C. R. B., Defew, E., Curtis, M., Fleddum, A., Brooks, L., & Paterson, D. M. (2008). Assessment of ecosystem function following marine aggregate dredging. *Journal of Experimental Marine Biology and Ecology, 366*(1), 82–91. doi:10.1016/j.jembe.2008.07.011

Cornillon, P., Adams, J., Blumenthal, M. B., Chassignet, E., Davis, E., Hankin, S., & Sirot, J. et al. (2009). NVODS and the development of OPeNDAP - an integrative tool for oceanographic data systems. *Oceanography (Washington, D.C.), 22*(2), 116–127. doi:10.5670/oceanog.2009.43

Council of the European Communities. (1992). *Council Directive 92/43/EEC of 21 May 1992 on the conservation of natural habitats and of wild fauna and flora.* Brussels, Belgium: Council of the European Communities.

Council. (2008). Council Regulation (EC) No 199/2008 of 25 February 2008 concerning the establishment of a Community framework for the collection, management and use of data in the fisheries sector and support for scientific advice regarding the Common Fisheries Policy. *Official Journal of the European Union, 51*(L 60), 1-12.

Coward, R., & Ellis, J. (1977). *Language and Materialism: Developments in Semiology and the Theory of the Subject.* London: Routledge & Kegan Paul.

Cox, S. (2012). *Best practice in formalizing a SKOS vocabulary. Solid Earth and Environment GRID.* Commonwealth Scientific and Industrial Organisation (CSIRO). Retrieved from https://www.seegrid.csiro.au/wiki/Siss/VocabularyFormalizationInSKOS

Cox, S. (2013). *Geographic information – Observations and measurements.* OGC and ISO 19156.

Cox, S. (2013, December). *Observations to information.* Abstract IN42A-01 presented at 2013, Fall Meeting, AGU, San Francisco, CA.

Cox, S. (2013, October). An explicit OWL representation of ISO/OGC Observations and Measurements. In SSN@ISWC (pp. 1-18).

Cox, S. (2015). *Ontology alignment – is PROV-O good enough?* Open Geospatial Consortium GeoSemantics Summit, Boulder, CO.

Cox, S., & Taylor, P. (2015). OM-JSON - a JSON implementation of O&M. Presentation to the Open Geospatial Consortium Technical and Planning Committee Sensor Web Enablement Domain Working Group, Nottingham, UK.

Cox, S. (2006). *Observations and measurements. Open Geospatial Consortium Best Practices Document.* Open Geospatial Consortium.

Cox, S. J. D. (2016). (in press). Ontology for observations and sampling features, with alignments to existing models. *Semantic Web.* doi:10.3233/SW-160214

Cox, S., Yu, J., & Rankine, T. (in press). SISSVoc: A Linked Data API for SKOS vocabularies. Submitted to Semantic Web - Interoperability, Usability. *Applicability.*

CRC. (1986). *CRC Handbook of Chemistry and Physics* (67th ed.). CRC.

Cressman, G. (1959). An operational objective analysis system. *Monthly Weather Review, 88*, 327–342. doi:10.1175/1520-0493(1960)088<0327:ITEIBF>2.0.CO;2

Dalyander, P. S., Butman, B., Sherwood, C. R., Signell, R. P., & Wilkin, J. L. (2013). Characterizing wave-and current-induced bottom shear stress: US middle Atlantic continental shelf. *Continental Shelf Research, 52*, 73–86. doi:10.1016/j.csr.2012.10.012

Dambrink, R. M., Maljers, D., Schokker, J., & Stafleu, J. (2015, November). *Tailor-made maps generated from 3D geological voxel models.* Paper presented at the Annual Meeting of the Geological Society of America. Baltimore, MD.

David, P. A., & Bunn, J. A. (1988). The economics of gateway technologies and network evolution. *Information Economics and Policy, 3*(2), 165–202. doi:10.1016/0167-6245(88)90024-8

Davidson, S. B., & Freire, J. (2008). Provenance and scientific workflows: challenges and opportunities. In *Proceedings of the 2008 ACM SIGMOD international conference on Management of data.* doi:10.1145/1376616.1376772

Davis, T. A., & Hager, W. W. (2009). Dynamic supernodes in sparse Cholesky update/downdate and triangular solves. *ACM Transactions on Mathematical Software, 35*(4), 27:1–27:23. doi:10.1145/1462173.1462176

De Backer, A., Van Hoey, G., Coates, D., Vanaverbeke, J., & Hostens, K. (2014). Similar diversity-disturbance responses to different physical impacts: Three cases of small-scale biodiversity increase in the Belgian part of the North Sea. *Marine Pollution Bulletin, 84*(1), 251–262. doi:10.1016/j.marpolbul.2014.05.006 PMID:24889315

De Boer, G. J., Baart, F., Bruens, A., Damsma, T., Van Geer, P., Grasmeijer, B., & Santinelli, G. et al. (2012). OpenEarth: using Google Earth as outreach for NCK's data. In *Jubilee Conference Proceedings NCK-days 2012* (pp. 97-103) Enschede, The Netherlands: University of Twente. doi:10.3990/2.177

de Jong, M. F., Baptist, M. J., van Hal, R., de Boois, I. J., Lindeboom, H. J., & Hoekstra, P. (2014). Impact on demersal fish of a large-scale and deep sand extraction site with ecosystem-based landscaped sandbars. *Estuarine, Coastal and Shelf Science, 146*, 83–94. doi:10.1016/j.ecss.2014.05.029

de La Beaujardière, J. (2009). *IOOS Data Management Activities.* IEEE. Retrieved from http://ieeexplore.ieee.org/xpls/abs_all.jsp?arnumber=5422388

de La Beaujardière, J., Beegle-Krause, C. J., Bermudez, L., Hankin, S., Hazard, L., Howlett, E., ... (2010a). Ocean and Coastal Data Management. In *Proceedings of OceanObs'09: Sustained Ocean Observations and Information for Society.* European Space Agency. doi:10.5270/OceanObs09.cwp.22

de la Beaujardiere, J. (Ed.). (2006). *OpenGIS® Web Map Server Implementation Specification Version 1.3.0.* Open Geospatial Consortium Inc.

de La Beaujardière, J., Mendelssohn, R., Ortiz, C., & Signell, R. (2010b). Building the IOOS® Data Management Subsystem. *J. Mar. Tech. Soc, 44*(6), 73–83. doi:10.4031/MTSJ.44.6.10

De Mauro, A., Greco, M., & Grimaldi, M. (2015, February). What is big data? A consensual definition and a review of key research topics. In AIP Conference Proceedings (Vol. 1644, pp. 97-104).

De Nies, T., Magliacane, S., Verborgh, R., Coppens, S., Groth, P., Mannens, E., & Van de Walle, R. (2013). Git2PROV: exposing version control system content as W3C PROV. In *Posters & Demonstrations Track within the 12th International Semantic Web Conference* (ISWC-2013).

Degraer, S., Verfaillie, E., Willems, W., Adriaens, E., Vincx, M., & Van Lancker, V. (2008). Habitat suitability modelling as a mapping tool for macrobenthic communities: An example from the Belgian part of the North Sea. *Continental Shelf Research*, *28*(3), 369–379. doi:10.1016/j.csr.2007.09.001

Degrendele, K., Roche, M., De Mol, L., Schotte, P., & Vandenreyken, H. (2014). Synthesis of the monitoring of the aggregate extraction on the Belgian Continental Shelf from 2011 till 2014. In: L. De Mol & H. Vandenreyken (Eds.), *Which future for the sand extraction in the Belgian part of the North Sea?Proceedings of the Study Day Sand Extraction* (pp. 3-29). Blankenberge, Belgium: FOD Economie.

Degrendele, K., Roche, M., Schotte, P., Van Lancker, V. R., Bellec, V. K., & Bonne, W. M. (2010). Morphological evolution of the Kwinte Bank central depression before and after the cessation of aggregate extraction. *Journal of Coastal Research SI*, *51*, 77–86.

Delaney, J. (2010, April). *Wiring an interactive ocean* [video file]. Retrieved October 13, 2015 from https://www.ted.com/talks/john_delaney_wiring_an_interactive_ocean?language=en

Di Stefano, M., Fox, P., Beaulieu, S., & Maffei, A. (2012). *The integrated ecosystems assessment initiative—enabling the assessment of impacts on large marine ecosystems: informatics to the forefront of science-based decision support.* Poster Presentation at 2012 ICES Annual Science Conference, Bergen, Norway. Accessible at: http://tw.rpi.edu/media/2012/10/08/cd52/ICES_2012.pdf

Di Stefano, M., Fu, L., West, P., Ma, X., Beaulieu, S., & Fox, P. (2014). *Development of cyberinfrastructure to facilitate collaboration and knowledge sharing for marine Integrated Ecosystem Assessments.* Poster Presentation at ESIP Summer Meeting 2014, Frisco, CO. Accessible at: http://tw.rpi.edu/web/doc/ECOOP-ESIPSummer2014-Poster

Di, L. (2007). (GMU): A General Framework and System Prototypes for the Self-Adaptive Earth Predictive Systems (SEPS). In *Dynamically Coupling Sensor Web with Earth System Models* (AIST-05-0064). ESTO-AIST Sensor Web PI Meeting.

Diesing, M., & Stephens, D. (2015). A multi-model ensemble approach to seabed mapping. *Journal of Sea Research*, *100*, 62–69. doi:10.1016/j.seares.2014.10.013

DiNezio, P. N., & Goni, G. J. (2011). Direct Evidence of a Changing Fall-Rate Bias in XBTs Manufactured during 1986–2008. *Journal of Atmospheric and Oceanic Technology*, *28*(11), 1569–1578. doi:10.1175/JTECH-D-11-00017.1

Ding, L., Michaelis, J., McCusker, J., & McGuinness, D. L. (2011). Linked provenance data: A semantic web-based approach to interoperable workflow traces. *Future Generation Computer Systems*, *27*(6), 797–805. doi:10.1016/j.future.2010.10.011

Directive INSPIRE. (2007). Directive 2007/2/EC of the European Parliament and of the Council of 14 March 2007 establishing an Infrastructure for Spatial Information in the European Community (INSPIRE). *Official Journal of the European Union*, *50*(L 108), 1-14.

Diviacco, P. (2015). Reconciling Knowledge and Collaborative e-Research. In Collaborative Knowledge in Scientific Research Networks. Hershey, PA: IGI Global. doi:10.4018/978-1-4666-6567-5.ch001

Diviacco, P. (2016). *Collaborative toolkit for Scientific Project Development*. Retrieved February 12, 2016, from http://colla.ogs.trieste.it/COLLA.2/

Diviacco, P. Lowry, R., & Schaap, D. (2012). Marine seismic metadata for an integrated European scale data infrastructure: the FP7 Geo-Seas project. *Bollettino di Geofisica Teorica ed Applicata, 53*(2), 243-252. DOI 10.4430/bgta0051

Diviacco, P., & Busato, A. (2013). The Geo-Seas Seismic data viewer: a tool to facilitate control of data access. *Bollettino di Geofisica Teorica ed Applicata, 54*(3), 257-270.

Diviacco, P. (2005). An open source, web based, simple solution for seismic data dissemination and collaborative research. *Computers & Geosciences, 31*(5), 599–605. doi:10.1016/j.cageo.2004.11.008

Diviacco, P. (2012). Addressing Conflicting Cognitive Models in Collaborative E-Research: A Case Study in Exploration Geophysics. In *Collaborative and Distributed E-Research: Innovations in Technologies, Strategies and Applications*. Hershey, PA: IGI Global Press; doi:10.4018/978-1-4666-0125-3.ch012

Diviacco, P. (2013). Addressing Conflicting Cognitive Models in Collaborative e-Research: A Case Study in Exploration Geophysics. In *Collaborative and Distributed e-Research: Innovations in Technologies, Strategies and Applications*. Hershey, PA: IGI Global press.

Diviacco, P., & Busato, A. (2015). Maps, graphs and annotations as Boundary Objects in Knowledge Networks, Distributed Cognition and Collaborative e-Research. In P. Diviacco, P. Fox, C. Pshenichny, & A. Leadbetter (Eds.), *Collaborative Knowledge in Scientific Research Networks*. Hershey, PA: IGI Global. doi:10.4018/978-1-4666-6567-5.ch019

Diviacco, P., DeCawer, K., Leadbetter, A., Sorribas, J., Stojanov, Y., Busato, A., & Cova, A. (2014). *Bridging semantically different paradigms in the field of marine acquisition event logging, Earth Science Informatics*. Springer. doi:10.1007/s12145-014-0192-0

Diviacco, P., & Leadbetter, A. (2016). Balancing Formalization and Representation in cross-domain data management for sustainable development. In P. Diviacco, A. Leadbetter, & H. Glaves (Eds.), *Oceanographic and Marine Cross-Domain Data Management for Sustainable Development*. Hershey, PA: IGI Global.

Diviacco, P., & Leadbetter, A. (2017). Balancing Formalization and Representation in Cross-Domain Data Management for Sustainable Development. In P. Diviacco, A. Leadbetter, & H. Glaves (Eds.), *Oceanographic and Marine Cross-Domain Data Management for Sustainable Development*. Hershey, PA: IGI Global.

Diviacco, P., Lowry, R., & Leadbetter, A. (2011). Collaborative work and tools towards wide scientific community driven metadata model and vocabulary building: The case of the EU GeoSeas project. European Geosciences Union (EGU) General Assembly, Vienna (Austria), April 2010. *Geophysical Research Abstracts*, 12.

Diviacco, P., & Pshenichny, C. A. (2010). Concept-referenced spaces in computer-supported collaborative work. *Geophysical Research Abstracts*, 12.

Diviacco, P., & Pshenichny, C. A. (2011). A case study on the use of event bushes as a formal representation for Computer Supported Collaborative Work in the GeoSciences European Geosciences Union (EGU) General Assembly, Vienna (Austria), April 2010. *Geophysical Research Abstracts*, 12.

Diviacco, P., Pshenichny, C. A., Carniel, R., Khrabrykh, Z., Shterkhun, V., Mouromtsev, D., & Pascolo, P. et al. (2014). Organization of a geophysical information space by using an event-bush-based collaborative tool. *Earth Sci Inform, 10*. doi:10.1007/s12145-014-0182-2

Diviacco, P., Sorribas, J., De Cauwer, K., Sinquin, J. M., Casas, R., Busato, A., & Scory, S. et al. (2017). Repositioning data management near data acquisition. In *Oceanographic and Marine Cross-Domain Data Management for Sustainable Development*. Hershey, PA: IGI Global.

Dublin Core. (2012). *DCMI Metadata Terms*. Retrieved from: http://dublincore.org/documents/dcmi-terms/

Duchon, J. (1977). Splines minimizing rotation-invariant semi-norms in Sobolev spaces. In *Constructive theory of functions of several variables* (pp. 85–100). Springer. doi:10.1007/BFb0086566

Duffy, J. (2006). *Marine ecosystem services*. Retrieved from http://www.eoearth.org/view/article/154472

Dujmović, J. J. (2007). Continuous preference logic for system evaluation. *IEEE Transactions on Fuzzy Systems*, *15*(6), 1082–1099. doi:10.1109/TFUZZ.2007.902041

Dyer, K. R., & Huntley, D. A. (1999). The origin, classification and modelling of sand banks and ridges. *Continental Shelf Research*, *19*(10), 1285–1330. doi:10.1016/S0278-4343(99)00028-X

Eaton, B., Gregory, J., Drach, B., Taylor, K., Hankin, S., Caron, Signell, R., Bentley, P., Rappa, G., Hoeck,. H, Pamment, A., Juckes, M., Raspaud, M. (2011). *NetCDF Climate and Forecast Metadata Conventions, v 1.6*. Retrieved from http://cf-pcmdi.llnl.gov/documents/cf-conventions/latest-cf-conventions-document-1

EC Enterprise and Industry. (2010). *Improving framework conditions for extracting minerals for the EU. Exchanging best practice on land-use planning, permitting and geological knowledge sharing*. Retrieved October 31, 2015 from http://www.google.nl/url?sa=t&rct=j&q=&esrc=s&source=web&cd=1&ved=0CCEQFjAAahUKEwjo4PqEh-3IAhWGXRoKHUh4AZM&url=http%3A%2F%2Fec.europa.eu%2FDocsRoom%2Fdocuments%2F9986%2Fattachments%2F1%2Ftranslations%2Fen%2Frenditions%2Fnative&usg=AFQjCNGOCjuWZYABqUCMw8HqzXXOB43bBw&sig2=sNL54AEbIEG32vmHIfEJ3w

EC. (2000). Directive 2000/60/EC of the European Parliament and of the Council of 23 October 2000 establishing a framework for community action in the field of water policy (Water Framework Directive). *Official Journal of the European Communities,* L327/1.

Eco, U. (1976). A Theory of Semiotics. Indiana University Press. doi:10.1007/978-1-349-15849-2

Ecosystem Assessment Program. (2009). *Ecosystem Assessment Report for the Northeast U.S.* Continental *Shelf Large Marine Ecosystem*. U.S. Department of Commerce, Northeast Fisheries Science Center Reference Document 09-11. Available online at http://www.nefsc.noaa.gov/publications/crd/crd0911/

Ecosystem Assessment Program. (2012). *Ecosystem Status Report for the Northeast Shelf Large Marine Ecosystem - 2011*. U.S. Department of Commerce, Northeast Fisheries Science Center Reference Document 12-07. Available online at http://nefsc.noaa.gov/publications/crd/crd1207/

Eco, U. (1979). *A Theory of Semiotics*. Indiana University Press.

Ehlers, P. (2013). The need for sustained operational oceanographic services. In *Sustainable operational oceanography:Proceedings of the Sixth International Conference on EuroGOOS*.

Elliott, M., Burdon, D., Hemingway, K. L., & Apitz, S. E. (2007). Estuarine, coastal and marine ecosystem restoration: Confusing management and science – a revision of concepts. *Estuarine, Coastal and Shelf Science*, *74*(3), 349–366. doi:10.1016/j.ecss.2007.05.034

Emile-Geay, J., & Eshleman, J. A. (2013). Toward a semantic web of paleoclimatology. *Geochemistry Geophysics Geosystems*, *14*(2), 457–469. doi:10.1002/ggge.20067

Engelhardt, W., & Zimmerman, J. (1982). *Theory of earth science*. Cambridge, UK: Cambridge University Press.

European Commission (2008). *Commission regulation (EC) no 1205/2008 of 3 December 2008 implementing Directive 2007/2/EC of the European Parliament and of the Council as regards metadata*. Author.

European Commission. (2006). *Green paper, Towards the future Maritime Policy for the Union: a European vision for the oceans and seas*. Brussels, Belgium: European Commission.

European Commission. (2007). *Directive 2007/2/EC of the European Parliament and of the Council of 14 March 2007 establishing an Infrastructure for Spatial Information in the European Community (INSPIRE)*. Available at: http://eur-lex.europa.eu/JOHtml.do?uri=OJ:L:2007:108:SOM:EN:HTML

European Commission. (2007). *Directive 2007/2/EC of the European Parliament and of the Council of 14 March 2007 establishing an Infrastructure for Spatial Information in the European Community (INSPIRE)*. Retrieved from http://eur-lex.europa.eu/JOHtml.do?uri=OJ:L:2007:108:SOM:EN:HTML

European Commission. (2007). *Interpretation manual of European Union habitats*. Brussels, Belgium: European Commission.

European Commission. (2008). *Directive 2008/56/EC of the European Parliament and of the Council of 17 June 2008 establishing a framework for community action in the field of marine environmental policy (Marine Strategy Framework Directive)*. Available at: http://eur-lex.europa.eu/legal-content/EN/TXT/PDF/?uri=CELEX:32008L0056&from=EN

European Commission. (2008). *Directive 2008/56/EC of the European Parliament and of the Council of 17 June 2008 establishing a framework for community action in the field of marine environmental policy (Marine Strategy Framework Directive)*. Retrieved from http://eur-lex.europa.eu/legal-content/EN/TXT/PDF/?uri=CELEX:32008L0056&from=EN

European Commission. (2010). *Communication from the European Commission to the European Parliament and the Council: Marine Knowledge 2020: Marine data and observation for smart and sustainable growth COM/2010/0461*. Author.

European Commission. (2010). *Communication from the European Commission to the European Parliament and the Council: Marine Knowledge 2020: marine data and observation for smart and sustainable growth COM/2010/0461*. Retrieved from http://eur-lex.europa.eu/legal-content/EN/TXT/PDF/?uri=CELEX:52010DC0461&from=EN

European Commission. (2010). *European Marine Observation and Data Network: Impact Assessment*. SEC(2010) 999 Final.

European Commission. (2010). *Maritime spatial planning in the EU - Achievements and future developments* [EC COM2010_771]. Publications Office of the European Union, Luxembourg. Retrieved October 31, 2015 from http://ec.europa.eu/maritimeaffairs/documentation/publications/documents/com_2010_771_brochure_en.pdf

European Commission. (2011). *Communication from the EC to the European Parliament, the Council, the European Economic and Social Committee and the Committee of the Regions. Roadmap to a Resource Efficient Europe* [EC COM2011_571]. Publications Office of the European Union, Luxembourg. Retrieved October 31, 2015 from http://ec.europa.eu/environment/resource_efficiency/about/roadmap/index_en.htm

European Commission. (2012). *Green Paper Marine Knowledge 2020*. COM(2012)473Final. Retrieved from http://ec.europa.eu/maritimeaffairs/documentation/publications/documents/marine-knowledge-2020-green-paper_en.pdf

European Commission. (2013). *Galway Statement on Atlantic Ocean Cooperation: Launching a European Union – Canada – United States of America Research Alliance 24 May 2013*. Retrieved from https://ec.europa.eu/research/iscp/pdf/galway_statement_atlantic_ocean_ cooperation.pdf

European Commission. (2014). *Communication from the European Commission: Innovation in the Blue Economy: Realising the potential of our seas and oceans for jobs and growth COM(2014) 254/2 (13/05/2014)*. Author.

European Commission. (2014). *Communication from the European Commission: Innovation in the Blue Economy: realising the potential of our seas and oceans for jobs and growth COM(2014) 254/2 (13/05/2014)*. Retrieved from http://eur-lex.europa.eu/legal-content/EN/TXT/PDF/ ?uri=CELEX:52014DC0254R(01)&from=EN

European Molecular Biology Laboratory European Bioinformatics Institute (EMBL-EBI). (2016). *Chemical Entities of Biological Interest (ChEBI)*. Retrieved April 5, 2016, from https://www.ebi.ac.uk/chebi/

European Parliament and Council of the European Union. (2008). *Directive 2008/56/EC of the European Parliament and of the Council of 17 June 2008 establishing a framework for community action in the field of marine environmental policy*. Strasbourg, France: European Parliament.

European Parliament and Council of the European Union. (2009). *Directive 2009/147/EC of the European Parliament and of the Council of 30 November 2009 on the conservation of wild birds*. Brussels, Belgium: European Parliament.

European Parliament and Council of the European Union. (2014). *Directive 2014/89/EC of the European Parliament and of the Council of 23 July 2014 establishing a framework for maritime spatial planning*. Brussels, Belgium: European Parliament.

European Union. (2008). Directive 2008/56/EC of the European Parliament and of the Council of 17 June 2008: Establishing a framework for community action in the field of maritime environmental policy (Marine Strategy Framework Directive). *Official Journal of the European Union*, L 164/19-40.

Euzenat, J., & Shvaiko, P. (2007). *Ontology matching*. Heidelberg, Germany: Springer.

Evans, B., Wyborn, L., Pugh, T., Allen, C., Antony, J., Gohar, K., . . . Bell, G. (2015). The NCI High Performance Computing and High Performance Data Platform to Support the Analysis of Petascale Environmental Data Collections. In Environmental Software Systems. Infrastructures, Services and Applications (pp. 569-577). Springer International Publishing. doi:10.1007/978-3-319-15994-2_58

Evensen, G. (2007). *Data assimilation: the Ensemble Kalman Filter*. Springer.

Fayyad, U., Piatetsky-Shapiro, G., & Smyth, P. (1996). From data mining to knowledge discovery in databases. *AI Magazine*, *17*(3), 37–54.

Feblowitz, J. (2012). The big deal about big data in upstream oil and gas. *IDC Energy Insights*, 1-11. Retrieved October 31, 2015 from https://www.hds.com/assets/pdf/the-big-deal-about-big-data-in-upstream-oil-and-gas.pdf

Federation of Earth Science Information Partners (ESIP). (2015). *Category: Attribute Conventions Dataset Discovery*. Retrieved February 15, 2016, from http://wiki.esipfed.org/index.php/Category:Attribute_Conventions_Dataset_Discovery

Fellbaum, C. (1998). *WordNet: An Electronic Lexical Database*. MIT Press.

Fernández, J. D., Martínez-Prieto, M. A., Gutiérrez, C., Polleres, A., & Arias, M. (2013). Binary RDF representation for publication and exchange (HDT). *Web Semantics: Science, Services, and Agents on the World Wide Web*, *19*, 22–41. doi:10.1016/j.websem.2013.01.002

Ferrarin, C., Cucco, A., Umgiesser, G., Bellafiore, D., & Amos, C. L. (2010). Modelling fluxes of water and sediment between the Venice Lagoon and the sea. *Continental Shelf Research*, *30*(8), 904–914. doi:10.1016/j.csr.2009.08.014

Ferrarin, C., Umgiesser, G., Cucco, A., Hsu, T.-W., Roland, A., & Amos, C. L. (2008). Development and validation of a finite element morphological model for shallow water basins. *Coastal Engineering*, *55*(9), 716–731. doi:10.1016/j.coastaleng.2008.02.016

G. Fick, & R. H. Sprague, (Eds.). (2013). Decision Support Systems: Issues and Challenges.*Proceedings of an International Task Force Meeting*. Oxford, UK: Pergamon Press.

FOAF. (2014). *FOAF Vocabulary Specification 0.99*. Retrieved from: http://xmlns.com/foaf/spec/

Folorunsho, R. (2006). The objectives of the Argo workshop. In *Argo Capacity Building for the Atlantic Countries Workshop*. University of Ghana.

Food and Agriculture Organization of the United Nations. (2000). *Code of Conduct for Responsible Fisheries* (Orig. 1995. Repr. 2000). Rome: FAO. Available: ftp://ftp.fao.org/docrep/fao/005/v9878e/v9878e00.pdf

Fountain, T., Tilak, S., Hubbard, P., Shin, P., & Freudinger, L. (2009). The Open Source DataTurbine Initiative: Streaming data middleware for environmental observing systems. *International Symposium on Remote Sensing of Environment*.

Fox, P. (2014). *Facilitating Next Generation Science Collaboration: Marine Ecosystems Status Reports and Assessments*. Oral presentation at 2014 IMBER Open Science Conference, Bergen, Norway. Accessible at: http://tw.rpi.edu/web/doc/ECOOP_IMBER_D2_Fox_20140624

Fox, P. (2014). Data Science and Analytics Curriculum Development at Rensselaer (and the Tetherless World Constellation). In *Training Students to Extract Value from Big Data: Summary of a Workshop* (pp. 18–21). Washington, DC: National Academies Press.

Francken, F., Van den Eynde, D., & Van Lancker, V. (2014). On-demand assessment of spatial and temporal variability of sediment transport parameters, Belgian and southern Dutch part of the North Sea. In: L. De Mol & H. Vandenreyken (Eds.), *Which future for the sand extraction in the Belgian part of the North Sea?Proceedings of the Study Day Sand Extraction* (pp. 73-80). Blankenberge, Belgium: FOD Economie.

Fredericks, J. (2015). Persistence of Knowledge Across Layered Architectures. In P. Diviacco, P. Fox, C. Pshenichny, & A. Leadbetter (Eds.), *Collaborative Knowledge in Scientific Research Networks*. Hershey, PA: IGI Global. doi:10.4018/978-1-4666-6567-5.ch013

Fugazza, C., Menegon, S., Oggioni, A., Pepe, M., & Carrara, P. (2014a). The RITMARE Starter Kit: Bottom-up capacity building for geospatial data providers. In *Proceedings of ICSOFT 2014 (9th International Conference on Software Paradigm Trends)*.

Fugazza, C., Pepe, M., Oggioni, A., Tagliolato, P., & Carrara, P. (2015). *Streamlining geospatial metadata in the Semantic Web*. 9th Symposium of the International Society for Digital Earth (ISDE), Halifax, Canada.

Fugazza, C., Pepe, M., Oggioni, A., Pavesi, F., & Carrara, P. (2014b). A holistic, semantics-aware approach to Spatial Data Infrastructures.*3rd International Conference on Data Management Technologies and Applications (DATA)*, Vienna, Austria. doi:10.5220/0004997603490356

Funkquist, L. (2003). A unified model system for the Baltic Sea. In H. Dahlin, N. C. Flemming, K. Nittis, & S. E. Petersson (Eds.), *Elsevier Oceanography Series* (Vol. 69, pp. 516–518). Elsevier. doi:10.1016/S0422-9894(03)80082-X

Gaigalas, A. (1995). *Granulometric classification of clastic deposits and rocks*. Vilnius: Vilnius University Press.

Galiegue, F., Zyp, K., & Court, G. (2013). *JSON Schema: core definitions and terminology*. Retrieved October 12, 2015 from http://tools.ietf.org/html/draft-zyp-json-schema-04

Galkus, A., Jokšas, K., Stakėnienė, R., & Lagunavičienė, L. (2012). Heavy metal contamination of harbor bottom sediments. *Polish Journal of Environmental Studies*, *21*(6), 1583–1594.

Gallant, S. N., Schmidt, R. K., & Car, N. J. (2015). Implementing a glossary and vocabulary service in an interdisciplinary environmental assessment for decision makers. In R. Denzer, M. Argent, G. Schimak, & J. Hrebícek (Eds.), Environmental Software Systems. Infrastructures, Services and Applications. Springer. doi:10.1007/978-3-319-15994-2_38

Gamma, E., Helm, R., Johnson, R., & Vlissides, J. (1994). *Design patterns: elements of reusable object-oriented software*. New York, NY: Pearson Education.

Gandin, L. S. (1965). *Objective analysis of meteorological fields*. Jerusalem: Israel Program for Scientific Translation.

Gangemi, A. (2005). Ontology design patterns for semantic web content. In *The Semantic Web–ISWC 2005* (pp. 262–276). Berlin: Springer Berlin Heidelberg. doi:10.1007/11574620_21

Garau, B., Ruiz, S., Zhang, W. G., Pascual, A., Heslop, E., Kerfoot, J., & Tintoré, J. (2011). Thermal Lag Correction on Slocum CTD Glider Data. *Journal of Atmospheric and Oceanic Technology*, *28*(9), 1065–1071. doi:10.1175/JTECH-D-10-05030.1

Garcia, S. M. (1996). The Precautionary Approach to Fisheries and its Implications for Fishery Research, Technology and Management: An Updated Review. In *FAO.- 1996. Precautionary approach to fisheries. Part 2: scientific papers. FAO Fisheries Technical Paper, 350(2)*. Rome: FAO.

Garde, S., Hovenga, E., Buck, J., & Knaup, P. (2007). Expressing clinical data sets with openEHR archetypes: A solid basis for ubiquitous computing. *International Journal of Medical Informatics*, *76*, S334–S341. doi:10.1016/j.ijmedinf.2007.02.004 PMID:17392019

Gattinoni, P., Pizzarotti, E. M., & Scesi, L. (2014). Geological Risk Management. In *Engineering Geology for Underground Works* (pp. 143–159). Dordrecht, The Netherlands: Springer. doi:10.1007/978-94-007-7850-4_5

GCMD. (2016a). Retrieved from: http://gcmdservices.gsfc.nasa.gov/kms/concept/c5e1c055-768e-4aa3-a0a1-3adfda8ecdca?format=rdf

GCMD. (2016b). Retrieved from: http://gcmdservices.gsfc.nasa.gov/kms/concept/c7112a64-be39-414a-9125-f63ab44ecb5b?format=rdf

GCMD. (2016c). Retrieved from: http://gcmdservices.gsfc.nasa.gov/kms/concept/bd46a0bf-5c06-48af-a6c9-022417b1fffd?format=rdf

GCMD. (2016d). Retrieved from: http://gcmdservices.gsfc.nasa.gov/kms/concept/76943142-e5a9-4ecf-b496-050dd3d97101?format=rdf

GCMD. (2016e). Retrieved from: http://gcmdservices.gsfc.nasa.gov/kms/concept/78f5a84f-1b5b-44a9-97e7-4a1996cd2e36?format=rdf

GCMD. (2016f). Retrieved from: http://gcmdservices.gsfc.nasa.gov/kms/concept/bd24a9a9-7d52-4c29-b2a0-6cefd216ae78?format=rdf

GCMD. (2016g). Retrieved from: http://gcmdservices.gsfc.nasa.gov/kms/concept/ecd03762-df34-49b7-91f2-d8a51acd270e?format=rdf

Gelumbauskaitė, L. Ž. (2010). Interim report of the EMODNET-Geology project. The Nature Research Centre.

Gelumbauskaitė, L. Ž. (2009). Character of sea level changes in the subsiding south-eastern Baltic Sea during Late Quaternary. *Baltica*, *22*(1), 23–36.

Gil, Y. (2014). Intelligent Workflow Systems and Provenance-Aware Software. In *Proceedings of the Seventh International Congress on Environmental Modeling and Software*.

Gil, Y., Michel, F., Ratnakar, V., Hauder, M., Duffy, C., Dugan, H., & Hanson, P. (2015) A task-centered framework for computationally-grounded science collaborations. In *Proceedings of the Eleventh IEEE International Conference on e-Science*. doi:10.1109/eScience.2015.76

Giorgetti, A., Vinci, M., Schaap, D., Fichaut, M., & Barth, A. (2015, June) *The EMODnet and SeaDataNet infrastructure: a challenge towards a European multidisciplinary interoperability*. Paper presented at 10th Baltic Sea Science Congress, Riga, Latvia.

Glaves. (2015). Ocean Data Interoperability Platform (ODIP): Developing a common global framework for marine data management through international collaboration. *Geophysical Research Abstracts*, *17*, EGU2015–EGU12644.

Glaves, H., Schaap, D., Arko, R., & Proctor, R. (2014). Ocean Data Interoperability Platform (ODIP): Supporting the development of a common global framework for marine data management through international collaboration. EGU General Assembly Conference Abstracts, 16, 14366.

Glaves, H. (2017). Developing a Common Global Framework for Marine Data Management. In *Oceanographic and Marine Cross-Domain Data Management for Sustainable Development*. Hershey, PA: IGI Global.

Glaves, H. (2017). Developing a common global framework for marine data management. In *Oceanographic and Marine Cross-Domain Data Management for Sustainable Development*. IGI Global.

Glaves, H., Miller, S., Proctor, R., & Schaap, D. (2013). Mind the Gap: Furthering the development of an international collaboration in marine data management. *Geophysical Research Abstracts*, *15*, EGU2013–EGU12737.

Glaves, H., Schaap, D., Arko, R., & Proctor, R. (2014). Ocean Data Interoperability Platform (ODIP): Supporting the development of a common global framework for marine data management through international collaboration. *Geophysical Research Abstracts*, *16*, EGU2014–EGU14366.

Goecks, J., Nekrutenko, A., Taylor, J., & Team, T. G. (2010). Galaxy: A comprehensive approach for supporting accessible, reproducible, and transparent computational research in the life sciences. *Genome Biology*, *11*(8), R86. doi:10.1186/gb-2010-11-8-r86 PMID:20738864

Gray, J. (2009). Jim Gray on eScience: A Transformed Scientific Method. In The Fourth Paradigm: Data-Intensive Scientific Discovery. Microsoft Research.

Graybeal, J., Isenor, A., & Wang, H. (2009). Moving Between Standards (Crosswalking). In *The MMI Guides: Navigating the World of Marine Metadata*. Retrieved from http://marinemetadata.org/guides/mdatastandards/crosswalks

Graybeal, J., Isenor, A., & Rueda, C. (2012). Semantic mediation of vocabularies for ocean observing systems. *Computers & Geosciences*, *40*, 120–131. doi:10.1016/j.cageo.2011.08.002

Green Paper on EU Maritime Policy. (2006). *Background paper No. 4a on the European Marine Observation and Data Network*. COM(2006) 275 final Volume II – Annex, 7.6.2006.

Green, R. H. (1979). *Sampling design and statistical methods for environmental biologists*. Chichester, UK: Wiley-Interscience.

Grigelis, A. 2012. Report to European Academies of Sciences Advisory Council EASAC. The Lithuanian Academy of Sciences.

Grigelis, A. (2011). Research of the bedrock geology of the Central Baltic Sea. *Baltica*, *24*(1), 1–12.

Gruber, T. R. (1995). Toward principles for the design of ontologies used for knowledge sharing. *International Journal of Human-Computer Studies*, *43*(5-6), 907–928. doi:10.1006/ijhc.1995.1081

Guardian. (2008, July 22). *Will Self* [Author biography]. Retrieved from http://www.theguardian.com/books/2008/jun/11/willself

Guarino, N. (1997). Understanding, building and using ontologies. *International Journal of Human-Computer Studies*, *46*(2-3), 293–310. doi:10.1006/ijhc.1996.0091

Guinehut, S., Larnicol, G., & Traon, P. L. (2002). Design of an array of profiling floats in the North Atlantic from model simulations. *Journal of Marine Systems*, *35*(1-2), 1–9. doi:10.1016/S0924-7963(02)00042-8

Guizzardi, G. (2006, July). The role of foundational ontologies for conceptual modeling and domain ontology representation. In *Databases and Information Systems, 2006 7th International Baltic Conference on* (pp. 17-25). IEEE. doi:10.1109/DBIS.2006.1678468

Habert, G., Bouzidi, Y., Chen, C., & Jullien, A. (2010). Development of a depletion indicator for natural resources used in concrete. *Resources, Conservation and Recycling*, *54*(6), 364–376. doi:10.1016/j.resconrec.2009.09.002

Haines, S., M. Subramanian, E. Mayorga, D. P. Snowden, R. Ragsdale, C. A. Rueda, & M. Howard. (2012). *IOOS Vocabulary and Ontology Strategy for Observed Properties*. IEEE. doi:.10.1109/OCEANS.2012.6405083

Halpern, B. S., Longo, C., Hardy, D., McLeod, K. L., Samhouri, J. F., Katona, S. K., & Zeller, D. et al. (2012). An index to assess the health and benefits of the global ocean. *Nature*, *488*(7413), 615–620. doi:10.1038/nature11397 PMID:22895186

Halpern, B. S., Longo, C., Scarborough, C., Hardy, D., Best, B. D., Doney, S. C., & Samhouri, J. F. et al. (2014). Assessing the Health of the U.S. West Coast with a Regional-Scale Application of the Ocean Health Index. *PLoS ONE*, *9*(6), e98995. doi:10.1371/journal.pone.0098995 PMID:24941007

Hankin, S. C. (2005). *United States National Office for Integrated, Sustained Ocean Observations, United States National Office for Integrated, Sustained Ocean Observations Data Management, and Communications Steering Committee. Data Management and Communications Plan for Research and Operational Integrated Ocean Observing Systems: Interoperable Data Discovery, Access and Archive*. National Office for Integrated and Sustained Ocean Observations. Retrieved from http://www.ioos.gov/library/dmac _plan_2005.pdf

Hankin, S., Blower, J. D., Carval, T., Casey, K. S., Donlon, C., Lauret, O., . . . Harlan, J. (2010). netCDF-CF-OPENDAP: Standards for ocean data interoperability and object lessons for community data standards processes. In Oceanobs 2009.

Hankin, S., Bermudez, L., Blower, J. D., Blumenthal, B., Casey, K. S., & Fornwall, M. et al.. (2010). Data Management for the Ocean Sciences - Perspectives for the Next Decade, 570–79. *European Space Agency, 2010*. doi:10.5270/OceanObs09.pp.21

Hanson, N. R. (1958). *Patterns of Discovery*. Cambridge, UK: Cambridge University Press.

Harnad, S., & Brody, T. (2004). Comparing the impact of open access (OA) vs. non-OA articles in the same journals. *D-Lib Magazine*, *10*(6). doi:10.1045/june2004-harnad

Harris, S., & Seaborne, A. (2013). *SPARQL 1.1 Query Language W3C Recommendation 21 March 2013*. Retrieved from http://www.w3.org/TR/sparql11-query/

Hart, J. K., & Martinez, K. (2006). Environmental Sensor Networks: A revolution in the earth system science? *Earth-Science Reviews*, *78*(3-4), 177–191. doi:10.1016/j.earscirev.2006.05.001

Haslhofer, B., & Klas, W. (2010). A survey of techniques for achieving metadata interoperability. *ACM Computing Surveys*, *42*(2), 1–37. doi:10.1145/1667062.1667064

Havlik, D., Schade, S., Sabeur, Z. A., Mazzetti, P., Watson, K., Berre, A. J., & Mon, J. L. (2011). From Sensor to Observation Web with Environmental Enablers in the Future Internet. *Sensors (Basel, Switzerland)*, *11*(4), 3874–3907. doi:10.3390/s110403874 PMID:22163827

Herman, I. (2008). *W3C semantic web activity*. World Wide Web Consortium.

Hey, T., & Payne, M. C. (2015). Open science decoded. *Nature Physics*, *11*(5), 367–369. doi:10.1038/nphys3313

Hilbert, M. (2013). *Big data for development: From information-to knowledge societies*. Available at SSRN 2205145.

Hitzler, P., Krotzsch, M., & Rudolph, S. (2009). *Foundations of semantic web technologies*. CRC Press.

Hoekstra, R., & Groth, R. (2015). PROV-O-Viz - Understanding the role of activities in provenance. In B. Ludäscher & B. Plale (Eds.), *Provenance and Annotation of Data and Processes* (pp. 215–220). Springer. doi:10.1007/978-3-319-16462-5_18

Holden, E., Linnerud, K., & Banister, D. (2014). Sustainable development: Our common future revisited. *Global Environmental Change*, *26*, 130–139.

Horrocks, I., Patel-Schneider, P. F., Boley, H., Tabet, S., Grosof, B., & Dean, M. (2004). *SWRL: A semantic web rule language combining OWL and RuleML*. W3C Member Submission, 21, 79.

Hosch, G., Ferraro, G., & Failler, P. (2011). The 1995 FAO Code of Conduct for Responsible Fisheries: Adopting, implementing or scoring results? *Marine Policy*, *35*(2), 189–200. doi:10.1016/j.marpol.2010.09.005

Howe, B., Lawson, P., Bellinger, R., Anderson, E., Santos, E., Freire, J., & Silva, C. et al. (2008). End-to-End eScience: Integrating Workflow, Query, Visualization, and Provenance at an Ocean Observatory. In *Proceedings of the IEEE Fourth International Conference on eScience*. doi:10.1109/eScience.2008.67

Howlett, E., R. P. Signell, D. Wilson, D. P. Snowden, & K. R. Knee. (2014). *Data Management Update for the Integrated Ocean Observing System (IOOS®)*. IEEE. doi:10.1109/OCEANS.2014.7003284

Hunter, J. (2007). Matplotlib: A 2d graphics environment. *Computing in Science & Engineering*, *9*(3), 90–95. doi:10.1109/MCSE.2007.55

Hu, Y., Janowicz, K., Carral, D., Scheider, S., Kuhn, W., Berg-Cross, G., & Kolas, D. (2013). A geo-ontology design pattern for semantic trajectories. In *Spatial Information Theory* (pp. 438–456). Springer International Publishing.

Hyer, J. L., & She, J. (2007). Optimal interpolation of sea surface temperature for the North Sea and Baltic Sea. *Journal of Marine Systems*, *65*(1-4), 176–189. doi:10.1016/j.jmarsys.2005.03.008

IBM. (1993). *IBM Dictionary of Computing*. New York: McGraw-Hill.

ICES. (2015). *Interim report of the Working Group on the Effects of Extraction of Marine Sediments on the Marine Ecosystem (WGEXT). ICES CM 2015/SSGEPI:06*. Copenhagen, Denmark: ICES.

ICES. (2016). *Effects of extraction of marine sediments on the marine environment 2005– 2011*. ICES Cooperative Research Report No. 330. ICES.

IDL. (2016). Retrieved from: http://www.exelisvis.com/IDL

Iliffe, J.C., Ziebart, M., Turner, J., Oliveira, J., & Adams, R., (2006). The VORF Project - Joining up Land and Marine Data. *GIS Professional*, (13), 24-26.

Infrastructure for Spatial Information in the European Community. (2013). Retrieved from: http://inspire.ec.europa.eu/index.cfm/pageid/241/documentid/3107

Initial Operating Capability Task Force. (2012). *Technical Guidance for the Implementation of INSPIRE download services*. Author.

Integrated Ocean Observing System (IooS) Implements OGC Standards. (2008). *Another U.S. Government Agency Chooses OGC Standards*. Retrieved from: http://xml.coverpages.org/IOOS-OGC.html

Intergovernmental Oceanographic Commission of UNESCO. (2013). Ocean Data Standards, Vol.3: Recommendation for a Quality Flag Scheme for the Exchange of Oceanographic and Marine Meteorological Data. IOC Manuals and Guides, 54(3).

International Council for the Exploration of the Sea. (2014). Implementing the ICES strategic plan 2014–2018 Linking Science, Advice, Data and Information, and Secretariat. Author.

International Hydrographic Organisation. (2003). Users Handbook on Datum Transformations Involving WGS 84; 3rd Edition, July 2003 (Last correction August 2008) Special Publication No. 60. International Hydrographic Bureau.

International Organization for Standardization (ISO). (2003). *ISO 19115: Geographic Information — Metadata*. Geneva: ISO.

International Organization for Standardization. (2007). *ISO 19139 Geographic information Metadata XML schema implementation. International Organization for Standardization*. Geneva: ISO.

International Organization for Standardization. (2009). *ISO 19115 Geographic Information - Metadata. International Organization for Standardization*. Geneva: ISO.

International Programme on the State of the Ocean. (2016). Retrieved from: http://www.stateoftheocean.org

IOC Task Force for Network Services. (2011a). *Technical Guidance for the Implementation of INSPIRE discovery services*. Author.

IOC Task Force for Network Services. (2011b). *Technical Guidance for the Implementation of INSPIRE view services*. Author.

IPython. (2016). *IPython.Jupyter kernals*. Retrieved from: https://github.com/ipython/ipython/wiki/IPython-kernels-for-other-languages

Irish Marine Institute. (n.d.). *Sensor Observation Service*. Retrieved from: https://github.com/IrishMarineInstitute/sensor-observation-service

Isenor, A., & Lowry, R. K. (2005). *Final report of the ICES/IOC Study Group on the Development of Marine Data Exchange Systems using XML*. Academic Press.

ISO. (2009). *ISO 19115-2: Geographic Information — Metadata — Part 2: Extensions for Imagery and Gridded Data*. ISO.

JCOMM. (2009). *Cookbook for Submitting Data in Real Time and In Delayed Mode*. Joint World Meteorological Organisation-Intergovernmental Oceanographic Commision Technical Commission for Oceanography and Marine Meteorology. Retrieved May 29, 2015 from http://www.jcomm.info/index.php?option=com_content&view=article&id=37

Jena, A. (2015), *A free and open source Java framework for building Semantic Web and Linked Data applications (2015)*. Retrieved October 30, 2015, from https://jena.apache.org/index.html

Jirka, S., Stasch, C., & Broering, A. (2014). *OGC Best Practice for Sensor Web Enablement Lightweight SOS Profile for Stationary In-Situ Sensors - OGC 11-169r1*. Open Geospatial Consortium.

Jokšas, K., Galkus, A., & Stakėnienė, R. (2003). *The Only Lithuanian Seaport and its Environment*. Vilnius: Institute of Geology and Geography.

Jørgensen, F., Møller, R. R., Nebel, L., Jensen, N.-P., Christiansen, A. V., & Sandersen, P. B. E. (2013). A method for cognitive 3D geological voxel modelling of AEM data. *Bulletin of Engineering Geology and the Environment*, *72*(3-4), 421–432. doi:10.1007/s10064-013-0487-2

JSON for Linking Data. (n.d.). Retrieved from: http://json-ld.org/

Judge, M. (2015). *European Marine Observation and Data Network (EMODnet): Making Fragmented Marine Data Relevant and Accessible*. Retrieved October 31, 2015 from http://earthzine.org/2015/10/08/european-marine-observation-and-data-network-emodnet-making-fragmented-marine-data-relevant-and-accessible/

Kaplan, A., Kushnir, Y., Cane, M., & Blumenthal, M. (1997). Reduced Space Optimal Analysis for Historical Datasets: 136 Years of Atlantic Sea Surface Temperatures. *Journal of Geophysical Research*, *102*(C13C3), 27,835–27,860. doi:10.1029/97JC01734

Kaplan, D. M., & Lekien, F. (2007). Spatial interpolation and filtering of surface current data based on open-boundary modal analysis. *Journal of Geophysical Research: Oceans*, *112*(C12), C12007. doi:10.1029/2006JC003984

Kaskela, A., Kotilainin, A., Alanen, U., Stevenson, A., Cooper, R., Green, S., … Verbruggen, K. (2010). Emodnet geology: combining and harmonising sea-bed sediment information. *Marine geological and biological habitat mapping: GeoHab conference: Programme and Abstracts*. Retrieved from http://geohab.org/wellington.html

Kessler, H., Mathers, S., & Sobisch, H. G. (2009). The capture and dissemination of integrated 3D geospatial knowledge at the British Geological Survey using GSI3D software and methodology. *Computers & Geosciences*, *35*(6), 1311–1321. doi:10.1016/j.cageo.2008.04.005

Kimble, C., Grenier, C., & Goglio-Primard, K. (2010). Innovation and knowledge sharing across professional boundaries: Political interplay between boundary objects and brokers. *International Journal of Information Management*, *30*(5), 437–444. doi:10.1016/j.ijinfomgt.2010.02.002

Kogut, P., & Holmes, W. (2001). *AeroDAML: Applying Information Extraction to Generate DAML Annotations from Web Pages*. In First International Conference on Knowledge Capture (K-CAP 2001). Workshop on Knowledge Markup and Semantic Annotation. Retrieved from http://citeseerx.ist.psu.edu/viewdoc/download?doi=10.1.1.21.8180&rep=rep1&type=pdf

Kreps, J. (2014) *Questioning the lambda architecture*. O'Reilly Radar. Retrieved August 28, 2015 from http://radar.oreilly.com/2014/07/questioning-the-lambda-architecture.html

Krzywinski, M., Birol, I., Jones, S. J., & Marra, M. A. (2012). Hive plots—rational approach to visualizing networks. *Briefings in Bioinformatics*, *13*(5), 627–644. doi:10.1093/bib/bbr069 PMID:22155641

Krzywinski, M., Schein, J., Birol, I., Connors, J., Gascoyne, R., Horsman, D., & Marra, M. A. et al. (2009). Circos: An information aesthetic for comparative genomics. *Genome Research*, *19*(9), 1639–1645. doi:10.1101/gr.092759.109 PMID:19541911

Kuhn, T. S. (1962). *The structure of scientific revolutions*. Chicago, IL: University of Chicago Press.

Kumpati, M. (1985). Patent description. US Patent 4774661, filed November 19.

Lark, R. M., Lawley, R. S., Barron, A. J. M., Aldiss, D. T., Ambrose, K., Cooper, A. H., & Waters, C. N. et al. (2015a). Uncertainty in mapped geological boundaries held by a national geological survey: Eliciting the geologists' tacit error model. *Solid Earth*, *6*(2), 727–745. doi:10.5194/se-6-727-2015

Lark, R. M., Marchant, B. P., Dove, D., Green, S. L., Stewart, H., & Diesing, M. (2015b). Combining observations with acoustic swath bathymetry and backscatter to map seabed sediment texture classes: The empirical best linear unbiased predictor. *Sedimentary Geology*, *328*, 17–32. doi:10.1016/j.sedgeo.2015.07.012

Latour, B., & Woolgar, S. (1979). *Laboratory life: The construction of scientific facts*. Princeton, NJ: Princeton University Press.

Lawlor, D. A., Davey Smith, G., & Ebrahim, S. (2004). Commentary: The hormone replace- ment-coronary heart disease conundrum: Is this the death of observational epidemiology? *International Journal of Epidemiology*, *33*(3), 464–467. doi:10.1093/ije/dyh124 PMID:15166201

Lawrence, B., Lowry, R., Miller, P., Snaith, H., & Woolf, A. (2009). Information in environmental data grids. *Philosophical Transactions of the Royal Society of London A: Mathematical, Physical and Engineering Sciences*, *367*(1890), 1003–1014. doi:10.1098/rsta.2008.0237

Le Bot, S., Van Lancker, V., Deleu, S., De Batist, M., Henriet, J. P., & Haegeman, W. (2005). Geological characteristics and geotechnical properties of Eocene and Quaternary deposits on the Belgian continental shelf: Synthesis in the context of offshore wind farming. *Netherlands Journal of Geosciences*, *84*(2), 147–160. doi:10.1017/S0016774600023027

Le Hir, P., Cayocca, F., & Waeles, B. (2011). Dynamics of sand and mud mixtures: A multiprocess-based modelling strategy. *Continental Shelf Research*, *31*(10), S135–S149. doi:10.1016/j.csr.2010.12.009

Le Traon, P. Y., Larnicol, G., Guinehut, S., Pouliquen, S., Bentamy, A., Roemmich, D., & Breivik, L. A. et al. (2009). Data assembly and processing for operational oceanography: 10 years of achievements. *Oceanography (Washington, D.C.)*, *22*(3), 56–69. doi:10.5670/oceanog.2009.66

Lea, D. J., Martin, M. J., & Oke, P. R. (2014). Demonstrating the complementarity of observations in an operational ocean forecasting system. *Quarterly Journal of the Royal Meteorological Society*, *140*(683), 2037–2049. doi:10.1002/qj.2281

Leadbetter, A. (2015). Linked ocean data. In T. Narock & P. Fox (Eds.), The Semantic Web in Earth and Space Science: Current Status and Future Directions. Amsterdam: IOS Press.

Leadbetter, A. M., & Vodden, P. N. (2015). Semantic linking of complex properties, monitoring processes and facilities in web-based representations of the environment. *International Journal of Digital Earth*, 1-25.

Leadbetter, A., & Lowry, R. (2012). *The NERC Vocabulary Server: Version 2.0*. Abstract IN51D-1709 presented at 2012, Fall Meeting, AGU, San Francisco, CA.

Leadbetter, A., Lowry, R., & Clements, D. O. (2012). The NERC Vocabulary Server: Version 2.0. In *Geophysical Research Abstracts 14,Proceedings EGU General Assembly 2012*.

Leadbetter, A., Raymond, L., Chandler, C., Pikula, L., Pissierssens, P., & Urban, E. (2013) Ocean Data Publication Cookbook. Paris: UNESCO.

Leadbetter, A., Raymond, L., Chandler, C., Pikula, L., Pissierssens, P., & Urban, E. (2013). *Ocean Data Publication Cookbook, IOC Manuals and guides No. 64*. Retrieved from http://www.iode.org/index.php?option=com_oe&task=viewDocumentRecord&docID=10574

Leadbetter, A., Raymond, L., Chandler, C., Pikula, L., Pissierssens, P., & Urban, E. (2013). *Ocean Data Publication Cookbook, IOC Manuals and guides No. 64*. Retrieved October 30, 2015, from http://www.iode.org/index.php?option=com_oe&task=viewDocumentRecord&docID=10574

Leadbetter, A. M. (2015). Linked Ocean Data. In T. Narock & P. Fox (Eds.), *The Semantic Web in Earth and Space Science: Current Status and Future Directions* (pp. 11–31). Berlin: IOS Press.

Leadbetter, A., Arko, B., Chandler, C., Shepherd, A., & Lowry, R. (2013). Linked Data: An Oceanographic Perspective. *Journal of Ocean Technology*, *8*(3), 7–12.

Leadbetter, A., Cheatham, M., Shepherd, A., & Thomas, R. (2017). Linked Ocean Data 2.0. In *Oceanographic and Marine Cross-Domain Data Management for Sustainable Development*. Hershey, PA: IGI Global.

Leadbetter, A., Lowry, R., & Clements, O. (2014). Putting meaning into NETMAR – the open service network for marine environmental data. *International Journal of Digital Earth*, *7*(10), 811–828. doi:10.1080/17538947.2013.781243

Lebo, T., Sahoo, S., & McGuinness, D. (2013). *PROV-O: The PROV Ontology*. Accessible at: http://www.w3.org/TR/prov-o/

Lebo, T., Sahoo, S., McGuinness, D., Belhajjame, K., Cheney, J., Corsar, D., ... & Zhao, J. (2013). *Provo: The provontology*. W3C Recommendation, 30.

Lehnert, K. A., Vinayagamoorthy, S., Djapic, B., & Klump, J. (2006). *The digital sample: metadata, unique identification, and links to data and publications*. American Geophysical Union Fall Meeting 2006, San Francisco, CA.

Lenci, A., Montemagni, S., & Pirrelli, V. (2003). CHUNK-IT. An Italian Shallow Parser for Robust Syntactic Annotation. *Linguistica Computazionale*, *16-17*, 353–386.

Levin, P. S., Fogarty, M. J., Murawski, S. A., & Fluharty, D. (2009). Integrated Ecosystem Assessments: Developing the Scientific Basis for Ecosystem-Based Management of the Ocean. *PLoS Biology*, *7*(1), e1000014. doi:10.1371/journal.pbio.1000014 PMID:19166267

Levin, P. S., Kelble, C. R., Shuford, R. L., Ainsworth, C., deReynier, Y., Dunsmore, R., & Werner, F. et al. (2014). Guidance for implementation of integrated ecosystem assessments: A US perspective. *ICES Journal of Marine Science*, *71*(5), 1198–1204. doi:10.1093/icesjms/fst112

Levy, J. M., Snowden, D., Clark, C., Crane, K., Diamond, H. J., Goodrich, D., & Stanitski, D. et al. (2011). The Global Ocean Observing Component of IOOS: Implementation of the Initial Global Ocean Observing System for Climate and the Path Forward. *J. Mar. Tech. Soc.*, *45*(1), 9–18. doi:10.4031/MTSJ.45.1.9

Lezcano, L., Sicilia, M. A., & Rodríguez-Solano, C. (2011). Integrating reasoning and clinical archetypes using OWL ontologies and SWRL rules. *Journal of Biomedical Informatics*, *44*(2), 343–353. doi:10.1016/j.jbi.2010.11.005 PMID:21118725

Liang, B., Li, H., & Lee, D. (2007). Numerical study of three-dimensional suspended sediment transport in waves and currents. *Ocean Engineering*, *34*(11), 1569–1583. doi:10.1016/j.oceaneng.2006.12.002

Lightsom, F. L., Cicchetti, G., & Wahle, C. M. (2015). *Data Categories for Marine Planning*. USGS Open-File Report 2015-1046. USGS. doi:10.3133/ofr20151046

Lipphardt, B. L. Jr, Kirwan, A. D. Jr, Grosch, C. E., Lewis, J. K., & Paduan, J. D. (2000). Blending HF radar and model velocities in Monterey Bay through normal mode analysis. *Journal of Geophysical Research: Oceans*, *105*(C2), 3425–3450. doi:10.1029/1999JC900295

Lorenzo, E. D., Moore, A. M., Arango, H. G., Cornuelle, B. D., Miller, A. J., & Bennett, A. F. et al. (2007). Weak and strong constraint data assimilation in the inverse Regional Ocean Modeling System (ROMS): Development and application for a baroclinic coastal upwelling system. *Ocean Modelling, 16*(3-4), 160–187. doi:10.1016/j.ocemod.2006.08.002

Lott, B. (2012). *Survey of Keyword Extraction Techniques*. UNM Education.

Lowry, J.B.C. (2004). *WGS - AGD - GDA: Selecting the correct datum, coordinate system and projection for north Australian applications*. Department of the Environment and Heritage, Internal Report 473.

Lowry, R. (2003). *EnParDis - Enabling parameter discovery.* Paper presented at 2003 Global Organisation for Earth System Science Portals. Central Laboratory of the Research Councils, Daresbury Laboratory.

Lowry, R., Bird, L., & Haaring, P. (2007): A semantic modelling approach to biological parameter interoperability. In *Proceedings Ocean Biodiversity Informatics:International Conference on Marine Biodiversity Data Management*. VLIZ.

Lowry, R., Fichaut, M., & Schlitzer, R. (2015). *SeaDataNet data file formats - deliverable D8.5*. Retrieved from http://www.seadatanet.org/content/download/16251/106283/file/SDN2_D85_WP8_Datafile_formats.pdf

Lowry, R., Rickards, L., & Brown, J. (2005). *Adding value to oceanographic data at the British Oceanographic Data Centre. In Ensuring Long-Term Preservation and Adding Value to Scientific and Technical Data (PV2005)* (pp. 21–23). Edinburgh, UK: Royal Society of Edinburgh.

Luyten, P. J., Jones, J. E., Proctor, R., Tabor, A., Tett, P., & Wild-Allen, K. (1999). *COHERENS – A coupled hydrodynamical-ecological model for regional and shelf seas: user documentation.* Unpublished Report of the Management Unit of the Mathematical Models of the North Sea. Brussels, Belgium: MUMM.

Maillard, C., Lowry, R. K., Maudire, G., & Schaap, D. (2007). SeaDataNet: Development of a Pan-European Infrastructure for Ocean and Marine Data Management. *Oceans, 2007*. doi:10.1109/OCEANSE.2007.4302435

Maljers, D., Stafleu, J., van der Meulen, M. J., & Dambrink, R. M. (2015). Advances in constructing regional geological voxel models, illustrated by their application in aggregate resources. *Netherlands Journal of Geoscience, 94*(3), 257–270. doi:10.1017/njg.2014.46

Manzella, G. & Manzella, A. (2015). Apeiron: engage students in earth and ocean sciences. *International Journal of Knowledge Society Research, 6*(4), 100-111. DOI: .10.4018/IJKSR.2015100107

Manzella, G., Bartolini, R., Bustaffa, F., D'Angelo, P., DeMattei, M., Frontini, F., & Novellino, A. et al. (2017). Semantic Search Engine for Data Management and Sustainable Development. In *Oceanographic and Marine Cross-Domain Data Management for Sustainable Development*. Hershey, PA: IGI Global.

Marine Community Profile Governance Committee. (2013). *Marine Community Profile Manual Release 1.4*. Retrieved from http://bluenet3.antcrc.utas.edu.au/mcp-doc/index.html

Marine Knowledge 2020. (2013). *From seabed mapping to ocean forecasting*. Green Paper COM. Available at: http://ec.europa.eu/maritimeaffairs/pdf/com_2006_0275_en_part2.pdf

Marinelli, R., & Roventini, A. (2006). The Italian maritime lexicon and the ItalWordNet semantic database. In E. M. Bermudez & L. R. Miyares (Eds.), *Linguistic in the twenty first century* (pp. 173–181). Cambridge Scholar Press.

Marschallinger, R. (1996). A voxel visualization and analysis system based on AutoCAD. *Computers & Geosciences, 22*(4), 379–386. doi:10.1016/0098-3004(95)00093-3

Marz, N., & Warren, J. (2015). *Big Data: Principles and best practices of scalable realtime data systems*. Manning Publications Co.

Ma, X., Fox, P., Tilmes, C., Jacobs, K., & Waple, A. (2014a). Capturing provenance of global change information. *Nature Climate Change*, *4*(6), 409–413. doi:10.1038/nclimate2141

Ma, X., Fu, L., Fox, P., Di Stefano, M., & West, P. (2017). Documenting Provenance for Reproducible Marine Ecosystem Assessment in Open Science. In *Oceanographic and Marine Cross-Domain Data Management for Sustainable Development*. Hershey, PA: IGI Global.

Ma, X., Zheng, J. G., Goldstein, J. C., Zednik, S., Fu, L., Duggan, B., & Fox, P. et al. (2014). Ontology engineering in provenance enablement for the National Climate Assessment. *Environmental Modelling & Software*, *61*, 191–205. doi:10.1016/j.envsoft.2014.08.002

Mayernik, M., DiLauro, T., Duerr, R., Metsger, E., Thessen, A. E., & Choudhury, G. S. (2013). Data conservancy provenance, context, and lineage services: Key components for data preservation and curation. *Data Science Journal*, *12*(0), 158–171. doi:10.2481/dsj.12-039

McCall, J.A., Richards, P.K. & Walters, G.F. (1977). *Factors in software quality.* National Technical Information Services (NTIS), RADC-TR-77-369.

McGuinness, D. (2003). Ontologies come of age. In D. Fensel, J. Hendler, H. Lieberman, & W. Wahlster (Eds.), *Spinning the Semantic Web: bringing the World Wide Web to its full potential* (pp. 171–196). Cambridge, MA: MIT.

McGuinness, D. L. (2003). Ontologies come of age. In D. Fensel, J. Hendler, H. Lieberman, & W. Wahlster (Eds.), *Spinning the Semantic Web: Bringing the World Wide Web to Its Full Potential* (pp. 171–196). Cambridge, MA: MIT Press.

McIntosh, P. C. (1990). Oceanographic data interpolation: Objective analysis and splines. *Journal of Geophysical Research*, *95*(C8), 13,529–13,541. doi:10.1029/JC095iC08p13529

McKee, L. (2015). *OGC Information Technology Standards for Sustainable Development.* OGC® White Paper 14-095. Open Geospatial Consortium.

Mendelssohn, R., & Simons, R. A. (2008, December). ERDDAP-An Easier Way for Diverse Clients to Access Scientific Data From Diverse Sources. In *AGU Fall Meeting Abstracts* (Vol. 1, p. 9). Academic Press.

Merati, N., & Burger, E. (2004). *Implementing Marine XML for NOAA Observing Data.* Paper presented at 2004 User Conference, Esri, Redlands, CA.

Met Office. (2015). *Earth system model development.* Retrieved February 15, 2016, from http://www.metoffice.gov.uk/research/areas/chemistry-ecosystems/earth-system

Mileriené, R., Blažauskas, N., & Gulbinskas, S. (2014). Integration of marine research results into a maritime spatial plan for Lithuania. *Baltica*, *27*(Special Issue), 65–72.

Miles, A., & Bechhofer, S. (2009). *SKOS simple knowledge organization system reference.* World Wide Web Consortium. Retrieved from http://www.w3.org/TR/2009/REC-skos-reference-20090818/

Missier, P., Sahoo, S., Zhao, J., Goble, C., & Sheth, A. (2010). Janus: From Workflows to Semantic Provenance and Linked Open Data. In D. McGuinness, J. Michaelis, & L. Moreau (Eds.), *Provenance and Annotation of Data and Processes* (pp. 129–141). Berlin: Springer. doi:10.1007/978-3-642-17819-1_16

Montgomery, E., & Bermudez, L. (2010). Some Examples of Metadata. In *The MMI Guides: Navigating the World of Marine Metadata.* Retrieved from http://marinemetadata.org/guides/mdataintro/mdataexamples

Mookerjee, M., Vieira, D., Chan, M. A., Gil, Y., Goodwin, C., Shipley, T. F., & Tikoff, B. (2015). We need to talk: Facilitating communication between field-based geoscience and cyberinfrastructure communities. *GSA Today*, *25*, 34–35. doi:10.1130/GSATG248GW.1

Moore, A. M., Arango, H. G., Broquet, G., Powell, B. S., Weaver, A. T., & Zavala-Garay, J. (2011). The Regional Ocean Modeling System (ROMS) 4-dimensional variational data assimilation systems: Part I - System overview and formulation. *Progress in Oceanography*, *91*(1), 34–49. doi:10.1016/j.pocean.2011.05.004

Moreau, L. (2011). Provenance-based reproducibility in the Semantic Web. *Web Semantics: Science, Services, and Agents on the World Wide Web*, *9*(2), 202–221. doi:10.1016/j.websem.2011.03.001

Moreau, L., Clifford, B., Freire, J., Futrelle, J., Gil, Y., Groth, P., & Van den Bussche, J. et al. (2011). The Open Provenance Model core specification (v1.1). *Future Generation Computer Systems*, *27*(6), 743–756.

Morris, C. (1971). *Foundations of the theory of signs*. Chicago, IL: University of Chicago Press.

Mudd, G. M. (2010). The environmental sustainability of mining in Australia: Key mega-trends and looming constraints. *Resources Policy*, *35*(2), 98–115. doi:10.1016/j.resourpol.2009.12.001

Murphy, K. (2012). The social pillar of sustainable development: a literature review and framework for policy analysis. *Sustainability: Science, Practice, & Policy, 8*(1), 15-29. Retrieved from http://www.google.it/archives/vol8iss1/1008-041.murphy.html

Na, A. & Priest, M. (2007). *Sensor Observation Service*. English, (OGC 06-009r6).

Narock, T. W., & Yoon, V. (2015). An Agent-Based Approach for Capturing and Linking Provenance in Geoscience Workflows. *Computers & Geosciences*, *79*, 58–68. doi:10.1016/j.cageo.2015.03.004

NASA. (2015). *Directory Interchange Format (DIF) Writer's Guide, 2015. Global Change Master Directory.* National Aeronautics and Space Administration. Retrieved from http://gcmd.nasa.gov/add/difguide/

NASA. (2016). *Global Change Master Directory*. Retrieved from: http://gcmd.nasa.gov/

National Marine Science Committee. (2015). *National Marine Science Plan: Driving the development of Australia's blue economy.* Retrieved from http://frdc.com.au/environment/NMSC-WHITE/Documents/NMSP_TS_040116%20website%20update.pdf

National Oceanic and Atmospheric Administration. (2012). NOAA fisheries implementation plan of the FAO code of conduct for responsible fisheries. U.S. Department of Commerce National Oceanic and Atmospheric Administration, National Marine Fisheries Service.

National Oceanography Centre. (2015). *SEA-SEARCH*. Retrieved April 5, 2016, from http://www.bodc.ac.uk/projects/european/seasearch/

Nativi, S., & Bigagli, L. (2009). Discovery, mediation, and access services for earth observation data. *IEEE J. Sel. Top. Appl. Earth Observ. Remote Sens.*, *2*(4), 233-240.

Nativi, S., Caron, J., Davis, E., & Domenico, B. (2005). Design and implementation of netCDF markup language (NcML) and its GML-based extension (NcML-GML). *Computers & Geosciences*, *31*(9), 1104–1118. doi:10.1016/j.cageo.2004.12.006

Nativi, S., Craglia, M., & Pearlman, J. (2012). The Brokering Approach for Multidisciplinary Interoperability: A Position Paper: National Research Council of Italy -Institute of Atmospheric Pollution Research(CNR-IIA). *International Journal of Spatial Data Infrastructures Research*, *7*, 1–15.

Nativi, S., Mazzetti, P., Santoro, M., Papeschi, F., Craglia, M., & Ochiai, O. (2015). Big Data challenges in building the Global Earth Observation System of Systems. *Environmental Modelling & Software*, *68*(June), 1–26.

Natural Environment Research Council (NERC). (2015). *Concentration of cadmium {Cd CAS 7440-43-9} per unit dry weight of biota {Mytilus edulis (ITIS: 79454: WoRMS 140480) [Subcomponent: flesh]}*. Retrieved April 5, 2016, from http://vocab.nerc.ac.uk/collection/P01/current/CDDWCF14/

Navigli, R., & Velardi, P. (2008). From Glossaries to ontologies: extracting semantic structure from textual definitions. In P. Buitelaar & P. Cimiano (Eds.), *Ontology learning and population: bridging the gap between text and knowledge* (pp. 71–87). Amsterdam: IOS Press.

NEFSC. (2008). *Examples of Management Jurisdictions of the Northeast Shelf Ecosystem*. Retrieved from: http://www.nefsc.noaa.gov/publications/tm/tm208/images/f7.gif

NERC. (2003). Retrieved from: http://vocab.nerc.ac.uk/collection/P02/current/PPRD/

NERC. (2004). Retrieved from: http://vocab.nerc.ac.uk/collection/P02/current/PSST/

NERC. (2006). Retrieved from: http://vocab.nerc.ac.uk/collection/P02/current/R410/

NERC. (2010). Retrieved from: http://vocab.nerc.ac.uk/collection/P02/current/FCST/

NOAA Environmental Data Management Committee. (2011). *NOAA Data Documentation Procedural Directive Version 1.0*. Available online: https://nosc.noaa.gov/EDMC/PD.DD.php

NOAA. (2015). *NOAA Cruise Level Metadata Template*. Retrieved from https://geo-ide.noaa.gov/wiki/index.php?title=Cruise_Level_Metadata_Template

Nosek, B. A., Alter, G., Banks, G. C., Borsboom, D., Bowman, S. D., Breckler, S. J., & Yarkoni, T. et al. (2015). Promoti. ng an open research culture. *Science*, *348*(6242), 1422–1425. doi:10.1126/science.aab2374 PMID:26113702

O'Brien, K. (2008). *Moving Toward Climate Data Integration: The Observing System Monitoring Center*. Paper presented at the American Geophysical Union, Fall Meeting, San Francisco, CA. http://adsabs.harvard.edu/cgi-bin/nph-bib_query?bibcode=2008AGUFMIN52A.05O

Obama, B. (2013). *Making Open and Machine Readable the New Default for Government Information*. US Government Executive Order, The White House. 2013. Accessible at: http://www.whitehouse.gov/the-press-office/2013/05/09/executive-order-making-open-and-machine-readable-new-default-government

Ocean Observations Panel for Climate. (2015). *Essential variables*. Retrieved October 30, 2015, from http://ioc-goos-oopc.org/obs/ecv.php

Ocean.US. (2002). An Integrated and Sustained Ocean Observing System (IOOS) for the United States: Design and Implementation. Ocean.US.

Oceans Policy Science Advisory Group. (2013). *Marine Nation 2025: Marine Science to Support Australia's Blue Economy*. Retrieved from http://www.aims.gov.au/documents/30301/550211/Marine+Nation+ 2025_web.pdf/bd99cf13-84ae-4dbd-96ca-f1a330062cdf

OGC. (2016). *Ocean Science Interoperability Experiment*. Retrieved from: http://www.opengeospatial.org/projects/initiatives/oceansie

Oggioni, A., Bastianini, M., Carrara, P., Minuzzo, T., & Pavesi, F. (2014). Sensing real-time observatories in marine sites: a proof-of-concept. *Proceedings of the 3rd International Conference on Sensor Networks 2014* (SENSORNETS '14).

Oggioni, A., Tagliolato, P., Fugazza, C., Pepe, M., Menegon, S., Pavesi, F., & Carrara, P. (2017). Interoperability in marine sensor networks through SWE services. In *Oceanographic and Marine Cross-Domain Data Management for Sustainable Development*. Hershey, PA: IGI Global.

Open Geospatial Consortium. (2011). *Observations and Measurements - XML Implementation*. OGC®project document: OGC 10-025r1. Retrieved from http://www.opengis.net/doc/IS/OMXML/2.0

Open Geospatial Consortium. (2013). *OGC® Sensor Web Enablement: Overview and High Level Architecture*. Retrieved October 30, 2015, from http://docs.opengeospatial.org/wp/07-165r1/

Open Science Collaboration. (2012). An open, large-scale, collaborative effort to estimate the reproducibility of psychological science. *Perspectives on Psychological Science*, *7*(6), 657–660. doi:10.1177/1745691612462588 PMID:26168127

Organisation for Economic Co-operation and Development. (2000). *Transition to Responsible Fisheries. Government Financial Transfers and Resource Sustainability: Case Studies. AGR/FI(2000)10/FINAL*. Paris: OECD.

Palmer, M. A. (2012). Socioenvironmental Sustainability and Actionable Science. *Bioscience*, *62*(1), 5–6. doi:10.1525/bio.2012.62.1.2

Papp, Z., & Hakkesteegt, H. (2008). Sensor Web, Sensor Networks: New possibilities and new challenges. In M. Grothe & J. Kooijman (Eds.), *Sensor Web Enablement. Delft: NCG, Nederlandse Commissie voor Geodesie* (pp. 21–39). Delft, The Netherlands: Netherlands Geodetic Commission.

Parra-Arnau, J., Rebollo-Monedero, D., & Forné, J. (2010). *A Privacy-Preserving Architecture for the Semantic Web based on Tag Suppression*. CON-SOLIDER INGENIO 2010 Project. Retrieved from https://pdfs.semanticscholar.org/949a/1cc930d4050a8b2166e58a76a8e244e17cdb.pdf

Peirce, C. S. (1958). *Collected Writings* (C. Hartshorne, P. Weiss, & A. W. Burks, Eds.). Cambridge, MA: Harvard University Press.

Petelin, V. P. (1967). *Granulometric analysis of the marine bottom sediments*. Moscow: Nauka. (In Russian)

Peter-Plag, H., & Pearlman, M. (Eds.). (2009). Global Geodetic Observing System: Meeting the Requirements of a Global Society on a Changing Planet in 2020. Springer Science & Business Media.

Pimentel, J. F. N., Braganholo, V., Murta, L., & Freire, J. (2015). Collecting and analyzing provenance on interactive notebooks: when IPython meets noWorkflow. In *Proceedings of the 7th USENIX Conference on Theory and Practice of Provenance (TaPP'15)*. USENIX Association.

Pinheiro da Silva, P., McGuinness, D. L., & Fikes, R. (2006). A proof markup language for Semantic Web services. *Information Systems*, *31*(4-5), 381–395. doi:10.1016/j.is.2005.02.003

Pinto, C., & Holsworth, N. (2013). ICES Data portal, a single entry point to marine knowledge. In *IMDIS 2013: International Conference on Marine data and Information Systems*.

Pitcher, T. J., Kalikoski, D., Pramod, G., & Short, K. (2008). *Safe conduct? Twelve years fishing the UN Code*. WWF.

Pitcher, T., Pramod, G., Kalikoski, D., & Short, K. (2009). Not honouring the code. *Nature*, *457*(7230), 658–659. doi:10.1038/457658a PMID:19194431

Polanyi, M. (1966). *The Tacit Dimension*. Chicago: University of Chicago Press.

Polanyi, M. (1966). *The tacit dimension*. New York, NY: Anchor Day Books.

Polson, D., & Curtis, A. (2015). Assessing individual influence on group decisions in geological carbon capture and storage problems. In P. Diviacco, P. Fox, C. Pshenichny & A. Leadbetter (Eds.), Collaborative knowledge in scientific research networks (pp. 55-75). IGI Global. doi:10.4018/978-1-4666-6567-5.ch004

Polson, D., & Curtis, A. (2010). Dynamics of uncertainty in geological interpretation. *Journal of the Geological Society*, *167*(1), 5–10. doi:10.1144/0016-76492009-055

Pontecorvo, G. (2003). Insularity of scientific disciplines and uncertainty about supply: The two keys to the failure of fisheries management. *Marine Policy*, *27*(1), 69–73. doi:10.1016/S0308-597X(02)00053-2

Popper, K. (2004). Conjectures and refutations: the growth of scientific knowledge. London: Routledge.

Popper, K. (1963). *Conjectures and refutations*. London, UK: Routledge.

Pouliquen, S., Carval, T., Guillotin, D., Loubrieu, T., Coatanoan, C., Grouazel, A., … Manzella, G. (2013). MyOcean In Situ TAC: A new in situ service for operational and research communities. *Mercator Ocean—Coriolis Quarterly Newsletter*, *48*, 3-9.

Powers, D. M. W. (2011). Evaluation: From Precision, Recall and F-Measure to ROC, Informedness, Markedness & Correlation. *Journal of Machine Learning Technologies*, *2*(1), 37–63.

Pramod, G., Nakamura, K., Pitcher, T. J., & Delagran, L. (2014). Estimates of illegal and unreported fish in seafood imports to the USA. *Marine Policy*, *48*, 102–113. doi:10.1016/j.marpol.2014.03.019

Prins, T. C., Van der Kolff, G. H., Boon, A. R., Reinders, J., Kuijper, C., Hendriksen, G., . . . Adema, J. (2014). *PMR Monitoring natuurcompensatie Voordelta, Eindrapport 1e fase 2009-201*. Retrieved from http://kennisonline.deltares.nl/product/30731

Prior, T., Giurco, D., Mudd, G., Mason, L., & Behrisch, J. (2012). Resource depletion, peak minerals and the implications for sustainable resource management. *Global Environmental Change*, *22*(3), 577–587. doi:10.1016/j.gloenvcha.2011.08.009

Proctor, R., Roberts, K., & Ward, B. J. (2010). A data delivery system for IMOS, the Australian Integrated Marine Observing System. *Advances in Geosciences*, *28*, 11–16. doi:10.5194/adgeo-28-11-2010

Pshenichny, C. A. (2004). Classical logic and the problem of uncertainty. Geological Society, 239, 111-126.

QUDT. (2014). *QUDT – Quanitites, Units, Dimensions and Data Types Ontologies*. Retrieved from: http://www.qudt.org/

Radzevičius, R., Velegrakis, A. F., Bonne, W. M., Kortekaas, S., Garel, E., Blažauskas, N., & Asariotis, R. (2010). Marine aggregate extraction regulation in EU member states. *Journal of Coastal Research SI*, *51*, 15–37.

Rajbhandari, S., & Keizer, J. (2012). The AGROVOC Concept Scheme – A Walkthrough. *Journal of Integrative Agriculture*, *11*(5), 694–699. doi:10.1016/S2095-3119(12)60058-6

Rajman, M., & Besançon, R. (1998). Text mining: natural language techniques and text mining applications. In Data Mining and Reverse Engineering. Springer US.

Ramdeen, S., & Hills, D. J. (2013). *ESIP's Emerging Provenance and Context Content Standard Use Cases: Developing Examples and Models for Data Stewardship*. American Geophysical Union Fall Meeting 2013, San Francisco, CA.

Ratnakar, V., & Gil, Y. (2015). *OntoSoft ontology*. Accessible at: http://ontosoft.org/ontology/software/

Rauber, A., Asmi, A., van Uytvanck, D., & Pröll, S. (2015). *Data Citation of Evolving Data, Recommendations of the Working Group on Data Citation (WGDC), Revision September 24th 2015*. Retrieved October 30, 2015, from https://rd-alliance.org/system/files/documents/RDA-DC-Recommendations_150924.pdf

Rayner, N. A., Parker, D. E., Horton, E. B., Folland, C. K., Alexander, L. V., Rowell, D. P., & Kaplan, A. et al. (2003). Global analyses of sea surface temperature, sea ice, and night marine air temperature since the late nineteenth century. *Journal of Geophysical Research*, *108*(D14), 4407. doi:10.1029/2002JD002670

Reed, C., Botts, M., Davidson, J., & Percivall, G. (2007). OGC® sensor web enablement: overview and high level architecture. In 2007 IEEE Autotestcon (pp. 372–380). IEEE. doi:10.1109/AUTEST.2007.4374243

Reed, C. W., Niedoroda, A. W., & Swift, D. J. (1999). Modeling sediment entrainment and transport processes limited by bed armoring. *Marine Geology*, *154*(1), 143–154. doi:10.1016/S0025-3227(98)00109-1

Reichman, O. J., Jones, M. B., & Schildhauer, M. P. (2011). Science Challenges and Opportunities of Open Data in Ecology. *Science*, *331*(6018), 703–705. doi:10.1126/science.1197962 PMID:21311007

Reid, W. V., Chen, D., Goldfarb, L., Hackmann, H., Lee, Y. T., Mokhele, K., & Whyte, A. et al. (2010). Earth System Science for Global Sustainability: Grand Challenges. *Science*, *330*(6006), 916–917. doi:10.1126/science.1196263 PMID:21071651

Reker, J., Annunziatellis, A., Mo, G., Tunesi, L., Globevnik, L., Snoj, L., . . . Korpinen, S. (2015). Marine protected areas in Europe's seas. An overview and perspectives for the future (EEA report No 3/2015). European Environment Agency.

Rice, J., Arvanitidis, C., Borja, A., Frid, C., Hiddink, J. G., Krause, J., & Norkko, A. et al. (2012). Indicators for sea-floor integrity under the European Marine Strategy Framework Directive. *Ecological Indicators*, *12*(1), 174–184. doi:10.1016/j.ecolind.2011.03.021

Roberts-Jones, J., Fiedler, E. K., & Martin, M. J. (2012). Daily, Global, High-Resolution SST and Sea Ice Reanalysis for 1985-2007 Using the OSTIA System. *Journal of Climate*, *25*(18), 6215–6232. doi:10.1175/JCLI-D-11-00648.1

Robinson, A. R. (1996). Physical processes, field estimation and an approach to interdisciplinary ocean modeling. *Earth-Science Reviews*, *40*(1-2), 3–54. doi:10.1016/0012-8252(95)00030-5

Roemmich, D., Gould, W. J., & Gilson, J. (2012). 135 years of global ocean warming between the Challenger expedition and the Argo Programme. *Nature Climate Change*, *2*(6), 425–428. doi:10.1038/nclimate1461

Rolling Deck. (n.d.). *Dataset Description for Rolling Deck to Repository (R2R) Linked Data*. Retrieved from: http://linked.rvdata.us/.well-known/void

Roos, P. C., & Hulscher, S. J. (2003). Large-scale seabed dynamics in offshore morphology: Modeling human intervention. Reviews of Geophysics, 41(2), 1010. doi:10.1029/2002RG000120

Rosmorduc, V., Bronner, E., Guinle, T., Maheu, C., Morrow, R., Nino, F., & Birol, F. (2014). AVISO+, the new reference web portal for altimetry. *Geophysical Research Abstracts*, 16.

Rossant, C. (2013). *Learning IPython for Interactive Computing and Data Visualization*. Birmingham, UK: Packt Publishing Ltd.

Roventini, A., Alonge, A., Calzolari, N., Magnini, B., & Bertagna, F. (2000). ItalWordNet: a Large Semantic Database for Italian. In *Proceedings of the 2nd International Conference on Language Resources and Evaluation (LREC 2000)*. The European Language Resources Association (ELRA).

Sahoo, S. S., Barga, R., Sheth, A. P., Thirunarayan, K., & Hitzler, P. (2009). *PrOM: A Semantic Web Framework for Provenance Management in Science*. Kno.e.sis Center Technical Report KNOESIS-TR-2009. Accessible at: http://corescholar.libraries.wright.edu/knoesis/445

Saitoh, S. I., Mugo, R., Radiarta, I. N., Asaga, S., Takahashi, F., Hirawake, T., & Shima, S. et al. (2011). Some operational uses of satellite remote sensing and marine GIS for sustainable fisheries and aquaculture. *ICES Journal of Marine Science*, *68*(4), 687–695. doi:10.1093/icesjms/fsq190

Scardamalia, M., & Bereiter, C. (2006). Knowledge building: Theory, pedagogy, and technology. In R. K. Sawyer (Ed.), *The Cambridge handbook of the learning sciences* (pp. 97–115). New York, NY: Cambridge University Press.

Schaap, D. (2014). ODIP - Ocean Data Interoperability Platform - developing interoperability pilot project 1. *Geophysical Research Abstracts*, *16*, EGU2014–EGU2845.

Schaap, D. (2017). SeaDataNet – Towards a Pan-European Infrastructure for Marine and Ocean Data Management. In *Oceanographic and Marine Cross-Domain Data Management for Sustainable Development*. Hershey, PA: IGI Global.

Schaap, D. (2017). *SeaDataNet: Towards a Pan-European Infrastructure for Marine and Ocean Data Management. In Oceanographic and Marine Cross-Domain Data Management for Sustainable Development*. Hershey, PA: IGI Global.

Schaap, D., & Lowry, R. (2010). SeaDataNet - Pan-European infrastructure for marine and ocean data management: Unified access to distributed data sets. *International Journal of Digital Earth*, *3*(sup1Supplement 1), 50–69. doi:10.1080/17538941003660974

Schlitzer, R. (2015). *Ocean Data View User's Guide. Version 4.7.4*. Retrieved from https://odv.awi.de/fileadmin/user_upload/odv/misc/odv4Guide.pdf

Schlitzer, R. (2002). Interactive analysis and visualization of geoscience data with Ocean Data View. *Computers & Geosciences*, *28*(10), 1211–1218. doi:10.1016/S0098-3004(02)00040-7

Schultz-Zehden, A., & Gee, K. (2013). BaltSeaPlan Findings – Experiences and Lessons. Academic Press.

Science and Technology Facilities Council (STFC). (2014). *JASMIN, Petascale storage and cloud computing for big data challenges in environmental science*. Retrieved April 5, 2016, from http://www.jasmin.ac.uk

SeaDas. (2016). Retrieved from: http://seadas.gsfc.nasa.gov

SeaDataNet. (2010). *Data quality control procedures*. Available at: http://www.seadatanet.org/content/download/18414/119624/file/SeaDataNet_QC_procedures_V2_(May_2010).pdf

SeaDataNet. (2015). *SeaDataNet data file formats (D8.5)*. Retrieved from http://www.seadatanet.org/content/download/16251/106283/file/SDN2_D85_WP8_Datafile_formats.pdf

SeaDataNet. (n.d.). Retrieved from: www.seadatanet.org

Searle, J. (1995). *The Construction of Social Reality*. New York: The Free Press.

Semantic Sensor Network Ontology (SSNO). (2009). Retrieved from: https://www.w3.org/2005/Incubator/ssn/ssnx/ssn

Shadbolt, N., & Berners-Lee, T. (2008). Web science emerges. *Scientific American*, *299*(4), 76–81. doi:10.1038/scientificamerican1008-76 PMID:18847088

Shannon, C. E., & Weaver, W. (1949). *A mathematical model of communication*. Chicago, IL: University of Illinois Press.

Shchepetkin, A., & McWilliams, J. (2003). A Method for Computing Horizontal Pressure-Gradient Force in an Oceanic Model with a Non-Aligned Vertical Coordinate. *Journal of Geophysical Research*, *108*(C3), 1–34. doi:10.1029/2001JC001047

Shchepetkin, A., & McWilliams, J. (2005). The Regional Oceanic Modeling System: A split-explicit, free-surface, topography-following-coordinate ocean model. *Ocean Modelling*, *9*(4), 347–404. doi:10.1016/j.ocemod.2004.08.002

Shedroff, N. (1994). *Information interaction design: A unified field theory of design*. San Francisco, CA: Vivid Studios.

Shen, H. (2014). Interactive notebooks: Sharing the code. *Nature*, *515*(7525), 151–152. doi:10.1038/515151a PMID:25373681

Sherman, K. (1991). The Large Marine Ecosystem Concept: Research and Management Strategy for Living Marine Resources. *Ecological Applications*, *1*(4), 350–360. doi:10.2307/1941896

Shneiderman, B. (2008). Science 2.0. *Science*, *319*(5868), 1349–1350. doi:10.1126/science.1153539 PMID:18323442

Signell, R., & Snowden, D. (2010). Advances in a Distributed Approach for Ocean Model Data Interoperability. *J. Mar. Sci. and Eng*, *2*(1), 194–208. doi:10.3390/jmse2010194

Simmhan, Y. L., Plale, B., & Gannon, D. (2006). A framework for collecting provenance in data-centric scientific workflows. In *Proceedings of ICWS'06- IEEE International Conference on Web Services*. doi:10.1109/ICWS.2006.5

Simoncelli, S., Coatanaon, C., Bäck, Ö., Sagen, H., Scory, S., Tezcan, D., & Tonani, M. et al. (2013). Temperature and salinity historical data collections for the european marginal seas: aggregation and quality assessment procedures. In *IMDIS 2013: International Conference on Marine data and Information Systems*.

Skosmos. (2014). *Multispecies Fisheries*. Retrieved from: http://aims.fao.org/skosmos/agrovoc/en/page/c_4987

Skosmos. (2015). *Trophic Levels*. Retrieved from: http://aims.fao.org/skosmos/agrovoc/en/page/c_7969

Snowden, D., Belbeoch, M., Burnett, B., Carval, T., Graybeal, J., Habermann, T., . . . Woodruff, S. (2009). *Metadata Management in Global Distributed Ocean Observation Networks*. Community White Paper for OceanObs09. Retrieved from https://abstracts.congrex.com/scripts/jmevent/abstracts/FCXNL-09A02a-1745513-1-OceanObs09_MetadataManagement_20May2009.doc

Soares, A. (1992). Geostatistical estimation of multi-phase structures. *Mathematical Geology*, *24*(2), 149–160. doi:10.1007/BF00897028

Sobisch, H.-G. (2011, October). *One viewer to publish 3D subsurface models of any modelling origin*. Paper presented at the Annual Meeting of the Geological Society of America, Minneapolis, MN.

Solar, S. V., Demicheli, L., & Wall, P. (2012). Raw Materials Initiative: A contribution to the European Minerals Policy Framework. In R. Sinding-Larsen & F.-W. Wellmer (Eds.), *Non-Renewable Resource Issues: Geoscientific and Societal Challenges* (pp. 21–33). Dordrecht, The Netherlands: Springer. doi:10.1007/978-90-481-8679-2_2

Stafleu, J., Maljers, D., Hummelman, J., & Gunnink, J. L. (2014, April). *Visualization of uncertainty in 3D stochastic voxel models of the Netherlands*. Paper presented at the EGU General Assembly, Vienna, Austria.

Stafleu, J., Sobisch, H.-G., Maljers, D., Hummelman, J., Dambrink, R. M., & Gunnink, J. L. (2013, October). *Visualization and dissemination of 3D geological property models of the Netherlands*. Paper presented at the Annual Meeting of the Geological Society of America, Denver, CO.

Stafleu, J., & Busschers, F. S. (2014). Lithological, grain-size and architectural trends in the Holocene Rhine-Meuse delta - insights from 3D voxel models. In *Proceedings of the 76th European Association of Geoscientists and Engineers Conference and Exhibition: Experience the Energy* (pp. 832-836). Houten, The Netherlands: EAGE. doi:10.3997/2214-4609.20141505

Stafleu, J., Maljers, D., Gunnink, J. L., Menkovic, A., & Busschers, F. S. (2011). 3D modelling of the shallow subsurface of Zeeland, the Netherlands. *Netherlands Journal of Geosciences*, *90*(4), 293–310.

Stakėnienė, R., Galkus, A., & Jokšas, K. (2011). Pollution of Klaipėda port waters. *Polish Journal of Environmental Studies*, *20*(2), 445–459.

Starr, J., Ammann, N., Ashton, J., Barton, A., Elliott, J., Jacquemot-Perba, M., … Ziedorn, F. (2015). *DataCite Metadata Schema for the Publication and Citation of Research Data, Version 3.1 August 2015*. doi:10.5438/0010

Star, S. L., & Griesemer, J. R. (1989). Institutional ecology, translations and boundary objects: Ama- teurs and professionals in Berkeley s museum of vertebrate zoology. *Social Studies of Science*, *19*(4), 387–420. doi:10.1177/030631289019003001

Stelzenmüller, V., Ellis, J. R., & Rogers, S. I. (2010). Towards a spatially explicit risk assessment for marine management: Assessing the vulnerability of fish to aggregate extraction. *Biological Conservation*, *143*(1), 230–238. doi:10.1016/j.biocon.2009.10.007

Stevenson, A., & Glaves, H. (2012). *EMODnet-Geology.5th EMODNET Progress Meeting*, Brussels, Belgium.

Stive, M. J. F., Nicholls, R. J., & de Vriend, H. J. (1991). Sea-level rise and shore nourishment: A discussion. *Coastal Engineering*, *16*(1), 147–163. doi:10.1016/0378-3839(91)90057-N

Stolk, A., & Dijkshoorn, C. (2009). Sand extraction Maasvlakte 2 Project: License, Environmental Impact Assessment and Monitoring. In *Proceedings of the EMSAGG 2009 Conference: A wave of opportunities for the marine aggregates industry*. London, UK: CIRIA.

Sturrock, J. (1986). *Structuralism*. London: Paladin.

Sutherland, J., Townend, I. H., Harpham, Q. K., & Pearce, G. R. (2014). From integration to fusion: the challenges ahead. In A. T. Riddick, H. Kessler, & J. R. A. Giles (Eds.), Geological Society, London, Special Publications: Vol. 408. *Integrated environmental modelling to solve real-world problems*. doi:10.1144/SP408.6

Tang, H. S., Keen, T. R., & Khanbilvardi, R. (2009). A model-coupling framework for nearshore waves, currents, sediment transport, and seabed morphology. *Communications in Nonlinear Science and Numerical Simulation*, *14*(7), 2935–2947. doi:10.1016/j.cnsns.2008.10.012

Taylor, R. B., & Steven, T. A. (1983). Definition of mineral resource potential. *Economic Geology and the Bulletin of the Society of Economic Geologists*, *78*(6), 1268–1270. doi:10.2113/gsecongeo.78.6.1268

Tetherless World Constellation (TWC). (2014). *ECO-OP*. Retrieved from: http://tw.rpi.edu/web/project/ECOOP

Timmerman, J. G., De Boer, J., Hisschemöller, M., & Mulder, W. H. (2001). Specifying information needs: Improving the working methodology. *Regional Environmental Change*, *2*(2), 77–84. doi:10.1007/s101130100030

Tomlinson, M. S., De Carlo, E. H., McManus, M. A., Pawlak, G., Steward, G. F., Sansone, F. J., & Ostrander, C. E. et al. (2011). Characterizing the effects of two storms on the coastal waters of O'ahu, Hawai'i, using data from the Pacific Islands Ocean Observing System. *Oceanography (Washington, D.C.)*, *24*(2), 182–199. doi:10.5670/oceanog.2011.38

Trimonis, E., & Gulbinskas, S. (2000). Bottom sediments of the Klaipėda Strait. *Geologija*, *30*, 20–27.

Trimonis, E., Vaikutienė, G., & Gulbinskas, S. (2010). Seasonal and spatial variations of sedimentary matter and diatom transport in the Klaipėda Strait (Eastern Baltic). *Baltica*, *23*(2), 127–134.

Troupin, C., Barth, A., Sirjacobs, D., Ouberdous, M., Brankart, J.-M., Brasseur, P., & Beckers, J.-M. et al. (2012). Generation of analysis and consistent error fields using the Data Interpolating Variational Analysis (DIVA). *Ocean Modelling*, *52–53*, 90–101. doi:10.1016/j.ocemod.2012.05.002

U.S. Geological Survey. (2016). *Mineral commodity summaries 2016*. U.S. Geological Survey; doi:10.3133/70140094

U.S. Government Accountability Office. (2011). *Key Indicator Systems: Experiences of Other National and Subnational Systems Offer Insights for the United States*. Report to Congressional Addressees GAO-11-396. Available online: http://www.gao.gov/products/GAO-11-396

U.S. IOOS Program Office. (2010). *U.S. Integrated Ocean Observing System: A Blueprint for Full Capability, November 2010*. Retrieved from http://www.ioos.gov/library/us_ioos_blueprint_ver1.pdf

Uhlir, P. F., & Schröder, P. (2007). Open data for global science. *Data Science Journal*, *6*, OD36–OD53. doi:10.2481/dsj.6.OD36

UN resolution A/RES/70/1. (2015). *Transforming our world: the 2030 Agenda for Sustainable Development*. Retrieved from http://www.un.org/en/ga/search/view_doc.asp?symbol=A/RES/70/1 on 12/2/2016

UNEP-WCMC. (2011). Marine and coastal ecosystem services: Valuation methods and their application. UNEP-WCMC Biodiversity Series No. 33.

UNESCO. (1987). GF3: A General formatting system for geo-referenced data. Vol. 2. Technical description of the GF3 format and code tables. Intergovernmental Ocanographic Commission Manuals and Guides.

UNESCO. (1987). GF3: A General Formatting System for geo-referenced data. Volume 2: technical description of the GF3 format and code tables. Paris: Intergovernmental Oceanographic Commission.

UNESCO. (1987). GF3: A General Formatting System for geo-referenced data: Volume 2: technical description of the GF3 format and code tables. Paris: Intergovernmental Oceanographic Commission.

UNESCO. (1993). *Manual of Quality Control Procedures for Validation of Oceanographic Data*. UNESCO, IOC-ICSU, Manuals and Guides No. 26. Retrieved from http://unesdoc.unesco.org/images/0013 /001388/138825eo.pdf

UNESCO. (2013). *Berlin Declaration*. Retrieved from http://www.unesco.org/new/en/social-and-human-sciences/themes/physical-education-and-sport/mineps-2013/declaration/

UNIDATA. (2015). *The NetCDF User's guide*. Boulder, CO: UCAR/Unidata. Retrieved from http://www.unidata.ucar.edu/ software/netcdf/docs/user_guide.html

Unidata. (2015). *THREDDS Data Server (TDS)*. Retrieved February 15, 2016, from http://www.unidata.ucar.edu/software/thredds/current/tds/

Unidata. (2016). *Network Common Data Form (NetCDF)*. Retrieved February 15, 2016, from http://www.unidata.ucar.edu/software/netcdf/

United Nations World Ocean Assessment. (2013). Retrieved from: http://www.worldoceanassessment.org/

Uren, V., Cimiano, P., Iria, J., Handschuh, S., Vargas-Vera, M., Motta, E., & Ciravegna, F. (2006). Semantic annotation for knowledge management: Requirements and a survey of the state of the art. *Web Semantics: Science, Services, and Agents on the World Wide Web*, *4*(1), 14–28. doi:10.1016/j.websem.2005.10.002

US IOOS Program Office. (2010). *Data Integration Framework (DIF) Final Assessment Report, November 22, 2010*. Retrieved from http://www.ioos.gov/library/ioos_dif_assmnt_report_final.pdf

US IOOS Program Office. (2010). *Guidance for Implementation of the Integrated Ocean Observing System (IOOS®) Data Management and Communications (DMAC) Subsystem, March 12, 2010*. Retrieved from http://www.ioos.gov/library/dmac_implementation_2010.pdf

Van den Eynde, D., Giardino, A., Portilla, J., Fettweis, M., Francken, F., & Monbaliu, J. (2010). Modelling the effects of sand extraction, on sediment transport due to tides, on the Kwinte Bank. *Journal of Coastal Research SI*, *51*, 101–116.

van der Meulen, M. J., Doornenbal, J. C., Gunnink, J. L., Stafleu, J., Schokker, J., Vernes, R. W., & van Daalen, T. M. et al. (2013). 3D Geology in a 2D country: Perspectives for geological surveying in the Netherlands. *Netherlands Journal of Geosciences*, *92*(4), 217–241. doi:10.1017/S0016774600000184

van der Meulen, M. J., van Gessel, S. F., & Veldkamp, J. G. (2005). Aggregate resources in the Netherlands. *Netherlands Journal of Geosciences*, *84*(4), 379–387. doi:10.1017/S0016774600021193

van Heteren, S., & Van Lancker, V. (2015). Collaborative seabed-habitat mapping: uncertainty in sediment data as an obstacle in harmonization. In P. Diviacco, P. Fox, C. Pshenichny & A. Leadbetter (Eds.), Collaborative knowledge in scientific research networks (pp. 154-176). IGI Global. doi:10.4018/978-1-4666-6567-5.ch008

van Heteren, S., Vonhögen-Peeters, L., & Maljers, D. (2012). *Multiple querying on geological databases to visualize offshore sand resources in the Netherlands part of the North Sea.* Unpublished TNO report TNO 2012 R10097. Utrecht, The Netherlands: TNO.

Van Koningsveld, M., De Boer, G. J., Baart, F., Damsma, T., Den Heijer, C., Van Geer, P., & De Sonneville, B. (2010). OpenEarth - Inter-Company Management of: Data, Models, Tools & Knowledge. In *Proceedings WODCON XIX Conference*. Beijing, China: Central Dredging Association.

Van Lancker, V., Bonne, W., Uriarte, A., & Collins, M. B. (Eds.). (2010b). *EUMARSAND: European Marine Sand and Gravel Resources.* Special Issue Journal of Coastal Research Royal Palm Beach, FL: CERF.

Van Lancker, V., & van Heteren, S. (2013). Case Study 4: Revisiting the spatial distribution of EUNIS Level 3 North Sea habitats in view of Europe's Marine Strategy Framework Directive. In V. Van Lancker & S. van Heteren (Eds.), *Standardisation and harmonisation in seabed habitat mapping: role and added value of geological data and information* (pp. 86-93). Retrieved October 31, 2015 from http://www.geoseas.eu/documents/geoseas/html_page/d10-5A_gs_wp10_seabed-habitat-mapping_part-Sediment_final.pdf

Van Lancker, V. R. M., Bonne, W., Garel, E., Degrendele, K., Roche, M., & Van den Eynde, D., Bellec, V., Brière, C., Collins, M. BVelegrakis, A. F. (2010a). Recommendations for the sustainable exploitation of tidal sandbanks. *Journal of Coastal Research SI*, *51*, 151–164.

van Ledden, M., Wang, Z. B., Winterwerp, H., & De Vriend, H. (2006). Modelling sand–mud morphodynamics in the Friesche Zeegat. *Ocean Dynamics*, *56*(3-4), 248–265. doi:10.1007/s10236-005-0055-9

Van Oord, K., Nijhoff, A., Prins, M., & Van de Giesen, N. (2011). Water verdient het. Advies Topsector Water (Report of Topteam Water). The Hague, The Netherlands: Topteam Water. (in Dutch)

Van Renssen, A. (2005). *Gellish, A generic extensible ontological language* (Doctoral dissertation). Delft: Delft University of Technology.

Van Saaze, V., & Dekker, A. (2013). Surprising usages of a documentation model: On the notion of boundary objects and beyond. *International Journal of Performance Arts & Digital Media*, *9*(1), 99–113. doi:10.1386/padm.9.1.99_1

Velegrakis, A. F., Ballay, A., Poulos, S. E., Radzevičius, R., Bellec, V. K., & Manso, F. (2010). European marine aggregates resources: Origins, usage, prospecting and dredging techniques. *Journal of Coastal Research SI*, *51*, 1–14.

Vlaams Instituut Voor De Zee (VLIZ). (2016). *WoRMS: World Register of Marine Species.* Retrieved April 5, 2016, from http://www.marinespecies.org/

Vocab. (n.d.). Retrieved from: http://vocab.nerc.ac.uk/.well-known/void

Voigt, T., Tsiftes, N., & He, Z. (2008). Remote Water Monitoring With Sensor Networking Technology. *Ercim News*, 39-40.

Vossen, P., Agirre, E., Calzolari, N., Fellbaum, C., Hsieh, S., Huang, C., . . . van Gent, J. (2008). KYOTO: A System for Mining, Structuring, and Distributing Knowledge Across Languages and Cultures. In *Proceedings of the Fourth International Global Word Net Conference - GWC 2008*. doi:10.4018/978-1-61520-883-8.ch012

Vossen, R. (2011). *DTM exchange format specification*. Retrieved from http://www.emodnet-hydrography.eu/documents/emodnet_hydrography/html_page/euco-0901-002_dtm_exchange_format_specification_v1.5.pdf

W3C. (2004). *SKOS Simple Knowledge Organization System*. Retrieved from: http://www.w3.org/2004/02/skos/

W3C. (2012a). *Web Ontology Language (OWL)*. Retrieved April 5, 2016, from http://www.w3.org/2001/sw/wiki/OWL

W3C. (2012b). *SKOS Simple Knowledge Organization System - Home Page*. Retrieved April 5, 2016, from http://www.w3.org/2004/02/skos/

W3C. (2013). *PROV-O: The PROV Ontology*. Retrieved February 15, 2016, from https://www.w3.org/TR/prov-o/

W3C. (2013). *PROV-O: The Prov Ontology*. Retrieved from: http://www.w3.org/TR/prov-o/#Collection

W3C. (2014a). *RDF Schema 1.1. W3C Recommendation*. Retrieved from: http://www.w3.org/TR/rdf-schema/

W3C. (2014b). *Data Catalog Vocabulary (DCAT)*. Retrieved from: http://www.w3.org/TR/vocab-dcat/

W3C. (2014c). *RDF 1.1 Turtle*. Retrieved from: http://www.w3.org/TR/turtle/

Wahba, G. (1990). *Spline models for observational data*. Philadelphia, PA: SIAM Journal of Applied Mathematics.

Wahba, G., & Wendelberger, J. (1980). Some new mathematical methods for variational objective analysis using splines and cross validation. *Monthly Weather Review*, *108*(8), 1122–1143. doi:10.1175/1520-0493(1980)108<1122:SNMMFV>2.0.CO;2

Wang, Y. (2008). *Software Engineering Foundations, a software science perspective*. Auerbach Publications.

Wannever, B. (1945). As we may think. *The Atlantic Monthly*. Retrieved from http://www.theatlantic.com/magazine/archive/1945/07/as-we-may-think/303881/

Ward, J., & Peppard, J. (2002). *Strategic Planning for Information Systems* (3rd ed.). Chichester, UK: Wiley.

Warner, A. (2002). *A Taxonomy Primer. School of Information*. University of Texas at Austin. Retrieved from http://www.ischool.utexas.edu/i385e/readings/Warner-aTaxonomyPrimer.html

Warner, J. C., Armstrong, B., He, R., & Zambon, J. B. (2010). Development of a coupled ocean–atmosphere–wave–sediment transport (COAWST) modeling system. *Ocean Modelling*, *35*(3), 230–244. doi:10.1016/j.ocemod.2010.07.010

Warner, J. C., Sherwood, C. R., Signell, R. P., Harris, C. K., & Arango, H. G. (2008). Development of a three-dimensional, regional, coupled wave, current, and sediment-transport model. *Computers & Geosciences*, *34*(10), 1284–1306. doi:10.1016/j.cageo.2008.02.012

WCED. (1987). *Our Common Future. World Commission on Environmental and Development*. Oxford, UK: Oxford University Press.

WCED. (1987). *Report of the World Commission on Environment and Development: Our Common Future*. Retrieved from http://www.un-documents.net/our-common-future.pdf

Weaver, A. T., Vialard, J., & Anderson, D. L. T. (2003). Three- and Four-Dimensional Variational Assimilation with a General Circulation Model of the Tropical Pacific Ocean. Part I: Formulation, Internal Diagnostics, and Consistency Checks. *Monthly Weather Review*, *131*(7), 1360–1378. doi:10.1175/1520-0493(2003)131<1360:TAFVAW>2.0.CO;2

Wellmann, J. F., & Regenauer-Lieb, K. (2012). Uncertainties have a meaning: Information entropy as a quality measure for 3-D geological models. *Tectonophysics*, *526-529*, 207–216. doi:10.1016/j.tecto.2011.05.001

Whitley, R. (2000). *The intellectual and social organization of the sciences*. Oxford, UK: Clarendon Press.

Wiberg, P. L., Drake, D. E., & Cacchione, D. A. (1994). Sediment resuspension and bed armoring during high bottom stress events on the northern California inner continental shelf: Measurements and predictions. *Continental Shelf Research*, *14*(10), 1191–1219. doi:10.1016/0278-4343(94)90034-5

Wibisono, A., Bloem, P., de Vries, G. K. D., Groth, P., Belloum, A., & Bubak, M. (2015). Generating Scientific Documentation for Computational Experiments Using Provenance. In B. Ludäscher & B. Plale (Eds.), *Provenance and Annotation of Data and Processes* (pp. 168–179). Springer. doi:10.1007/978-3-319-16462-5_13

Willighagen, E. (2014). Accessing biological data in R with semantic web technologies (No. e185v3). *PeerJ PrePrints*. Retrieved from https://dx.doi.org/10.7287/peerj.preprints.185v3

Woods Hole. (n.d.). Retrieved from: http://www.bco-dmo.org/.well-known/void

Woo, G. (1999). *The mathematics of natural catastrophes*. London, UK: Imperial College Press; doi:10.1142/9781860943867

Woolf, A. (2008). Building the Sensor Web - Standard by Standard. *Ercim News*, 24-25.

World Bank, Food and Agriculture Organization of the United Nations. (2009). *The Sunken Billions: The Economic Justification for Fisheries Reform*. Washington, DC: The World Bank.

World Economic and Social Survey 2013, Sustainable Development Challenges. (n.d.). Retrieved from https://sustainabledevelopment.un.org/content/documents/2843WESS2013.pdf

Yaffe, M. B. (2015). Reproducibility in science. *Science Signaling*, *8*(371), eg5–eg5. doi:10.1126/scisignal.aaa5764 PMID:25852185

Yongguo, J., Zhongwen, G., Keyong, H., Feifei, S., & Feng, H. (2013). Using Sensor Web to Sharing Data of Ocean Observing Systems. In Advances in Wireless Sensor Networks. Berlin: Springer-Verlag.

Yu, J., Car, N. J., Leadbetter, A., Simons, B. A., & Cox, S. J. D. (2015). Towards Linked Data Conventions for Delivery of Environmental Data Using netCDF. Springer.

Yue, P., Gong, J., & Di, L. (2010). Augmenting geospatial data provenance through metadata tracking in geospatial service chaining. *Computers & Geosciences*, *36*(3), 270–281. doi:10.1016/j.cageo.2009.09.002

Zaucha, J. (2014). The key to governing the fragile Baltic Sea. Maritime spatial planning in the Baltic Sea region and way forward. VASAB Secretariat, Riga.

Zemlys, P., Ferrarin, C., Umgiesser, G., Gulbinskas, S., & Bellafiore, D. (2013). Investigation of saline water intrusions into the Curonian Lagoon (Lithuania) and two-layer flow in the Klaipėda Strait using finite element hydrodynamic model. *Ocean Science*, *9*(3), 573–584. doi:10.5194/os-9-573-2013

Zeng, W., Fu, C.-W., Müller Arisona, S., & Qu, H. (2013). Visualizing Interchange Patterns in Massive Movement Data. *Computer Graphics Forum*, *32*(3), 271–280. doi:10.1111/cgf.12114

About the Contributors

Paolo Diviacco is a geophysicist at Istituto Nazionale di Oceanografia e di Geofisica Sperimentale (OGS). He holds a PhD in exploration geophysics, and has 20 years of experience in seismic data processing. He maintained always an interest in computer science and programming, developing software in the field of geophysics and Web-based data management. At the same time, he has always been interested in the philosophical and sociological aspects of scientific production in general and in the geo-sciences in particular. He has been active in developing ideas and Web-based systems that could support scientists while collaborating and foster the collaborative attitude among research institutions. In this context, he worked within several international projects, such as, for example, FP7 ODIP, FP7 Geo-Seas, FP7 Emodnet, FP7 Eurofleets, and the Antarctic Seismic Data Library.

Adam Leadbetter, before joining the Marine Institute as Team Leader for Data Management, worked for ten years at the British Oceanographic Data Centre (BODC). At BODC Adam first had responsibility for managing the ocean data collected by the British Antarctic Survey, and more recently was the Vocabularies Manager. Adam co-chairs the technical committee of the International Coastal Atlas Network project of IOC-IODE and the Vocabulary Services Interest Group of the Research Data Alliance.

Helen Glaves is the Senior Data Scientist at the British Geological Survey with direct responsibility for managing the geoscientific data holdings that support the core science priorities of the organization. She is also co-ordinator of the Ocean Data Interoperability Platform (ODIP) project which is funded by the European Commission, National Science Foundation (USA) and the Australian Government to promote the development of a common global framework for marine data management. Ms. Glaves is also contributing to a number of other national and international initiatives directly related to the sharing, re-use and preservation of earth science data including the Research Data Alliance (RDA) and the Belmont Forum. In 2016 Helen Glaves was awarded the European Geosciences Union's Ian McHarg medal for her contribution to Earth Science and Space Informatics (ESSI).

* * *

Aida Alvera-Azcárate received a PhD in Oceanography in 2004, University of Liège (Belgium). Research Associate at the College of Marine Science, University of South Florida (US), from 2004 to 2007. Working as a researcher in the University of Liège since 2007.

Marcel Bakker is a geologist at TNO - Geological Survey of the Netherlands. He holds a PhD in geography (Queen Mary, University of London, 2004). His main interests include glacial and coastal sedimentary systems, geophysics, neotectonics, groundwater related matters and environmental issues. Currently he works in a broad range of activities and responsibilities including geomodelling, geo(morpho) logical mapping, geophysical interpretation and advisory.

Alexander Barth is a Research Associate of the Belgian National Fund of Scientific Research (Chercheur Qualifié, F.R.S.-FNRS) at the University of Liège. Author of 45 peer-reviewed articles. Guest editor of Ocean Dynamics for the special issue of 47th Liege Colloquium (2015). Co-organizer and member of the scientific committee of 47st Liege Colloquium on "Marine Environmental Monitoring, Modelling and Prediction" (2015). Guest editor of Journal of Marine Systems for the special issue of 41st Liege Colloquium (2009). http://modb.oce.ulg.ac.be/mediawiki/upload/Alex/CV/cv_en.html.

Jean-Marie Beckers is professor of physical oceanography and his research focuses on hydrodynamical modelling, including data assimilation, nesting techniques and statistical assessments. He has authored and co-authored more than 100 peer-reviewed publications. He coordinated EU projects MEDMEX and SANGOMA and acted as Scientific coordinator of SeaDataNet.

Stace E. Beaulieu is a Research Specialist and coordinates the Ocean Informatics initiative at Woods Hole Oceanographic Institution. She holds a Ph.D. in Oceanography from the University of California San Diego and B.S./M.S. in Biological Sciences from Stanford University.

Meinte Blaas is senior adviser/researcher on integrated water quality monitoring and environmental hydrodynamics.

Arjen Boon, Ph.D., is a senior researcher / consultant in the unit Marine and Coastal Systems of Deltares. He is a marine ecologist with a 20 year professional background in fundamental and applied marine ecological research and in consulting for marine policy development. His ecological expert disciplines are benthic ecology, pelagic (process) ecology and ecosystems ecology. He has also well-developed skills in consulting for marine policy, ecosystem impact and risk assessments, and scientific integration and evaluation. He has worked for research institutes in the Netherlands (Royal Netherlands Institute for Sea Research, Netherlands Institute for Fisheries Research (now IMARES)). He has an extensive and in-depth knowledge of process-related ecosystem dynamics and ecological effect relationships regarding the building of offshore wind farms and coastal management Currently, at Deltares he is responsible for the evaluation of scientific monitoring programs, for indicator development at national and European level and for the development and application of the integrated oceanographic and biological knowledge in the shallow coastal environment. He has become responsible for the integration of natural and socioeconomic sciences with CleanSea, an EU FP7 project directed at developing a roadmap for a litter-free marine environment, using an EBM approach (DPSIR, Ecosystem services). His most recent projects involve developing guidance in EU DG ENV projects on the implementation of indicators and MPAs in the MSFD.

Kees Borst is senior consultant at Rijkswaterstaat in the field of monitoring and data management. He is the representative of the Netherlands in the working group DIKE of the MSFD.

Justin J. H. Buck, PhD in polar oceanography and fluid dynamics, is a senior data scientist at the British Oceanographic Data Centre, for 5 years he led the data management for United Kingdom's contribution to the international Argo programme including both real time and climate grade delayed mode activity. A member of the International Argo Data Management Team, he has 8 years of experience in data management protocols for data from robotic ocean observing platforms. He is working with international partners on building global protocols and data infrastructures including; the SenseOCEAN project for data formats and protocols at the sensor level, the EU GROOM project for glider data exchange formats, and the Ocean Data Interoperability Platform (ODIP) for data interoperability at the global level.

Echelle Steel Burns is an undergraduate student pursuing a B.S. in Marine Biology with a minor in English at the University of Hawai'i at Mānoa. Her expected graduation date is in December 2015. She has worked with PacIOOS as a "Marine Data Specialist" throughout her undergraduate career, beginning in August 2012. After graduation, Echelle plans to continue her education by pursuing both an M.S. and PhD in Marine Biology and focus her future work on elasmobranch predators, such as sharks. Eventually, she hopes to hold a faculty position at a university and teach students with similar research interests.

Alessandro Busato is a developer and system administrator at Istituto Nazionale di Oceanografia e di Geofisica Sperimentale (OGS) with 18 years of experience as IT and Web analyst and developer. He has always maintained an interest in e-Science through his previous experiences with NewTech at Trieste Area Science Park, developing Java Web applications, and within several European projects at Trieste Elettra Synchrotron. Currently, he is working within several EU collaborative projects, such as, for example, the FP7 Eurofleet project developing a real time metadata logger, within the GeoSeas project developing high resolution Web viewing and processing facilities for seismic data, and within OGS developing a portal to access legacy geophysical data. Previously, he worked at Insiel Spa developing applications for e-Government. He is skilled in Linux/MacOS system administration, Java, Web technologies and platforms, C, Javascript including kits as JQuery Openlayers, XML, and PostgresQL/Postgis.

Paola Carrara is a researcher at IREA CNR. Her activity focuses on information systems, in particular geographic information on the Web. In RITMARE, the Italian flagship Project on marine research, she coordinates the developing of the interoperable data infrastructure of the Project. In Space4Agri, she is leader of the task regarding the users' requirements collection. She participates in projects related to the integration and sharing of geographic information, also coming from crowdsourcing, with reference to the applications of IREA, such as agriculture, ecology, sea and freshwater.

Raquel Casas, Computer Technical Engineer qualified in 2007 by Universitat de Girona (UDG), and after, in 2013 by Universitat Oberta de Catalunya (UOC) as Computer Scientist Engineer specialized in advanced data management and knowledge, and database management. Since 2006, before to finish her studies in Computer Science Engineering, developed her activity as senior programmer first, and then as analyst in an important Spanish Bank. Her work was design and development custom applications and processes for the proper internal operations of the bank. In the middle of 2013 she had the opportunity to change on a different matter, and join the UTM-CSIC Information Systems Group, and now she is applying her experience in developing and design applications and working with the big data to work in many national and EU projects. One of her tasks is to create a specific profiles for SensorML to define a fixed stations, vessels and its instruments.

Michelle Cheatham is an Assistant Professor at Wright State University in Dayton, OH. She is co-director of the Data Semantics (DaSe) Lab together with Pascal Hitzler. Her research focus in on semantic data integration, including ontology alignment and co-reference resolution, and the privacy implications of Big Data. Additionally, she is exploring the application of machine learning and data mining techniques to problems in the field of binary analysis.

Paolo D'Angelo (MSc in Electronic Engineer) is a software analyst and developer with an experience of more than 10 years; he has also a wide experience in complex databases management. and interoperability instruments and standards In the ETT R&D Department, he is involved in activity related to software design and maintenance, management of the involved human resources in research projects and test plans redaction to verify the correct made of operation of the developed applications.

Karien De Cauwer received her degree of bio-engineer in Environmental Technology in 1995 at the Faculty of Agricultural and Applied Biological Sciences of the University of Ghent. After 2 years research work at the Laboratory of Forest management and Spatial Information Techniques (University of Ghent), she started working at the Belgian Marine Data Centre, OD Nature, a department of the Royal Belgian Institute of Natural Sciences. As database designer and data manager new data management systems were designed and developments coordinated to ensure long term data archival, data quality control and dissemination of wide range of marine data.

Maurizio De Mattei is a graduate in Electronic Engineer and is currently the Research Manager at Delta Progetti. He is involved in national research projects and in the development of innovative information systems in the area of language technologies, marine information systems, big data, analytics and MDA (Model Driven Architecture). Before joining Delta Progetti, he served in a number of Italian companies with several managerial roles and growing responsibilities; he has been COO of a middle-sized software engineering company, managed important (government) projects and programs in Italy and abroad and developed national and European research projects in the field of advanced information and transport systems.

Lies De Mol holds a PhD in Geology (Ghent University, Belgium, 2012). Research focussed on the origin and evolution of cold-water coral mounds in the southern Gulf of Cadiz and the distribution of cold-water coral ecosystems along the French continental margin. After her PhD she switched research to the monitoring and impact of sand extraction in the Belgian part of the North Sea at the Federal Public Service Economy, SMEs, Self-employed and Energy – Continental Shelf Service where she was involved in the TILES project. Currently she is working as geologist at Jan De Nul nv (dredging company).

Robin De Mol received the M.Sc. degree in computer science engineering from the Ghent University, Ghent, Belgium, in 2013. Since then, he has been working on his Ph.D. on information mining and data formatting at the Ghent University, Dept. of Telecommunications and Information Processing under the supervision of Professor Guy De Tré.

Guy de Tré received a M.Sc. in computer science in 1994 and a Ph.D. in engineering in 2000 from Ghent University (Belgium). He is head of the Database, Document and Content Management (DDCM) research group, belonging to the Department of Telecommunications and Information Processing of

Ghent University. He is an expert in soft computing techniques for information management systems. His research comprises fundamental and applied research. The focus of the fundamental research is on the handling of imperfect information. His applied research includes, data quality and big data issues (unstructured data), spatio-temporal modelling, fuzzy querying, and decision support.

Massimo Di Stefano graduated from "Università degli Studi di Napoli "Parthenope"" (Naples, IT) with a Master's Degree in Environmental Science, specializing in Marine Ecosystem. Massimo is now pursuing his Ph.D. in Oceanography at CCOM-JHC (UNH). For the past four years, Massimo worked as Research Assistant III at the Woods Hole Oceanographic Institute (WHOI) for the HABCAM project, and as Software Engineer on staff at the Tetherless World Constellation (RPI) on the ECO-OP project. Massimo's research focus on the development of open-source software infrastructure to facilitate data-science workflows.

Peter Fox is Tetherless World Constellation Chair, Professor of Earth and Environmental Science, Computer Science and Cognitive Science, and Director of the Information Technology and Web Science Program at Rensselaer Polytechnic Institute. Fox has a B.Sc. (hons) and Ph.D. in Applied Mathematics (physics and computer science) from Monash Univsersity. His research and education agenda covers the fields of data science and analytics, ocean and environmental informatics, materials informatics, computational logic, semantic Web, cognitive bias, semantic data frameworks, and solar and solar-terrestrial physics. This research is also applied to large-scale distributed, and data science collaborations. Fox is past-President of the Federation of Earth Science Information Partners (ESIP), past chair of the International Union of Geodesy and Geophysics Union Commission on Data and Information, and serves on the editorial boards of many prominent Earth and space science informatics and Data journals. In 2012, Fox was awarded the European Geoscience Union, Ian McHarg/ Earth and Space Science Informatics Medal, and ESIP's Martha Maiden Lifetime Achievement award for service to the Earth Sciences Information communities. Fox was elected as the first Earth and Space Science Informatics Fellow of the American Geophysical Union in 2015. http://tw.rpi.edu/web/person/PeterFox.

Frederic Francken does research in transport of mud in the coastal zone of the Belgian Continental Shelf in the framework of optimizing the dredging and dumping activities. Research into the impact of sand and gravel extraction on the Flemish Banks. Can these activities occur in a sustainable way? Research into the effects of the offshore wind farms on the marine environment (sediment transport/hydrodynamics). Research into the dispersion of toxic compounds possibly leaking from dumped ammunition from WW I. Scientific Officer for Belgium in the FP7 JERICO Project - http://www.jerico-fp7.eu/.

Francesca Frontini obtained a PhD in corpus linguistics at the University of Pavia; later she joined CNR-ILC in 2010, working on several European projects with a focus on computational lexicography, textual analysis and semantics. In 2014/2015 she was a FMSH "Fernand Braudel" visiting fellow at the Labex Obvil (Paris), where she worked on NLP methods for the analysis of literary texts and literary criticism. She is currently back at CNR-ILC as researcher; her current task is to support the Coordinator of the CLARIN-IT consortium, the national node of the ERIC Common Language Resources and Technology Infrastructure; she is also the representative of CLARIN-IT in the CLARIN Center Committee.

Linyun Fu was a PhD student at the Tetherless World Constellation at Rensselaer Polytechnic Institute. His research focus is Semantic Web Technologies and their applications in provenance modeling and capturing. He joined Twitter, Inc. as a software engineer after graduation.

Cristiano Fugazza is a researcher at the Institute for Electromagnetic Sensing of the Environment of the National Research Council of Italy (CNR-IREA). He is researching in semantic interoperability in the context of project RITMARE. He worked by the Spatial Data Infrastructure Unit (SDIU) of the JRC-IES (Institute for Environment and Sustainability, Joint Research Centre, European Commission) on the creation, harmonization, and exploitation of multilingual thesauri in the SKOS format for the purpose of annotation and discovery of spatial resources. He supported the development of the Global Earth Observation System of Systems (GEOSS) and the European spatial data infrastructure established by the INSPIRE Directive by managing a comprehensive repository of harmonized reference and thematic thesauri. He obtained his Ph.D. in Computer Science by the Dept. of Information Technologies, University of Milan, focusing on Knowledge Management and Semantic Web technologies. He has contributed to, and published in international journals, conferences, and workshops in the domains of spatial data infrastructures, supply chain management, business modeling, digital rights management, personal information protection, and persistent identifiers.

Saulius Gulbinskas is vice rector of Klaipeda University, Geologist.

Vasileios Hademenos is a PhD student at the Renard Centre of Marine Geology (RCMG) at Ghent University. He is a geologist with a MSc in geophysics. Currently, as part of the TILES project, his research focus is on the development of a 3D voxel model of the subsurface of the southern part of the North Sea.

Gerrit Hendriksen is a geo data specialist working in data intensive (marine and coastal) (monitoring) projects. His background is Physical Geography with specialisation Landscape Ecology. The main focus in his work is on data management using OpenEarth as a framework for standardisation, data management and dissemination of data and information in relation to (hydrological) models and monitoring projects. OpenEarth encourages working with open standards for semantics and data/information exchange. Information on OpenEarth can be found at http://openearth.eu.

Niels Kinneging is consultant at Deltares working on information management in relation to monitoring and land subsidence. He has 18 years of experience in the field of monitoring at Rijkswaterstaat.

Lars Kint holds a Master of Science degree (Geology) (Ghent University, Belgium, 2014). He is now a scientific research assistant at the Royal Belgian Institute of Natural Sciences, where he manages and processes data within international projects such as TILES, Transnational and Integrated Long-Term Marine Exploitation Strategies related to the Belgian and southern Netherlands part of the North Sea and EMODnet, the European Marine Observation and Data Network.

Roy Lowry is a part-time Technical Consultant to the British Oceanographic Data Centre in the UK National Oceanography Centre after taking partial retirement in November 2015. Prior to that he was Technical Director, a role he took on in 2000 after working in the organisation for 6 years as a programmer followed by 13 years developing the concept of 'project data management' and running the data management for UK projects such as the NERC North Sea Project, BOFS and LOIS and EU projects such as OMEX and INDIA. During this period, he chaired the international JGOFS Data Management Task Team for nearly nine years. During this time considerable practical experience was gained in the collection and handling of physical, chemical, biological and geological oceanographic data, including participation in more than 10 oceanographic research cruises on the vessels of three nations. In addition to IT management responsibilities within BODC the Technical Director role has a remit to research and develop technologies with the potential to enhance BODC's operational capabilities. This work has focussed on the issues of interoperability between distributed metadata and data repositories through the adoption, and if necessary, development of standards. This has included active participation in two projects, NERC DataGrid and SeaDataNet, building distributed data systems. Within these projects Dr Lowry's work has focussed on facilitating semantic interoperability through development of a well-managed controlled vocabulary infrastructure, including ontologies to support semantic cross-walking.

Xiaogang (Marshall) Ma is an assistant professor of computer science at University of Idaho, USA. Previously he was an associate research scientist of geoinformatics and data science at Rensselaer Polytechnic Institute, USA. He received his Ph.D. degree of Earth Systems Science and GIScience from University of Twente, Netherlands, and D.Eng. and B.Eng. degrees from China University of Geosciences, Wuhan. His research interests include participatory conceptual modeling, data sharing in the semantic web, crowd-sourcing geoinformation, and spatio-temporal analysis of Big and Little Data.

Denise Maljers is a geologist working as a gemodeller and project manager at TNO - Geological Survey of the Netherlands. For 10 years she is working on two- and three-dimensional models of the Dutch subsurface incorporating in these models geometrical properties of geological units as well as lithological and grain-size parameters. These models provide a sound basis for subsurface related questions. She has been working, amongst others, on the application of the models for infrastructural issues and on the calculation of available and exploitable aggregate resources. Further interest lies in the incorporation of uncertainties.

Maurizio Maltese, after classical studies, enrolled at University of Pisa studying for his degree in Digital Humanities and proceeding with an aced Master (second level degree) in Graphics, Interactivity and Virtual Environments; this last experience was very productive since granted the great opportunity to work in research projects in the Edutainment and Serious Gaming fields, as a collaborating student at the Perceptual Robotics Lab of the Sant'Anna Superior Studies School, firstly during a stage and lastly for his final university project for his theses about Digital Storytelling and effective ways to convey narration through new media. After completing his studies, joined Delta Progetti 2000 first for a compensated 6-months stage and then regularly employed to work as a Process Analyst mainly developing solutions in the Natural Language Processing field.

Giuseppe Manzella worked in the National Agency for New Technologies, Environment and Sustainable Development and National Research Council. Actually has a professional collaboration with ETT SpA. His working experiences is in physical oceanography: Modelling, data analysis and information systems. He has participated to national and international oceanographic programs supported by the Italian Ministry of Environment, Italian Ministry of Research, UNESCO IOC, European Commission, European Space Agency. Involved also in the definition of goals and objectives of the Topic Centre on

Marine and Coastal Environment of the European Environment Agency (1995 - 1997). Chairman of the Italian Oceanographic Commission (2008 - 2013) and, in such role, representing Italy in the UNESCO-IOC initiatives and meetings.

John Maurer is a data system engineer at the Pacific Islands Ocean Observing System (PacIOOS) at the University of Hawai'i. He obtained a graduate certificate in remote sensing as well his Masters degree in geography from the University of Colorado at Boulder where he developed software to measure snow from radar on the Greenland ice sheet. He worked with the National Snow and Ice Data Center (NSIDC) for eight years, sparking his interest in Earth science and all things geospatial: an unexpected but comfortable detour from his undergraduate degree in Music, Science, and Technology at Stanford University.

Stefano Menegon is a Research Fellow at the Institute of Marine Science of the National Research Council of Italy. He is an Environmental Engineer and holds a Ph.D. in New Technologies for the City and the Environment. His research has focused on integration of machine learning methods and geographic information, spatial data infrastructures, collaborative geospatial web applications, data sharing strategies for environmental sciences, and smart communities.

Tine Missiaen is a senior marine geophysicist from the Renard Centre of Marine Geology (RCMG) at Ghent University, Belgium. She holds an MSc in physics and PhD in geology and has over 25 years of experience in marine seismics. Her research interest includes ultra high resolution 2D and 3D seismic data acquisition and processing, acoustic techniques for intertidal areas and land-sea boundary studies, acoustics of shallow biogenic gas, and marine geophysical techniques for engineering and environmental studies (a.o. buried chemical warfare). In recent years she has become actively engaged in marine archaeological research and submerged prehistoric landscape studies. She has authored many research papers and (co-)edited several books and special volumes, and was actively involved in numerous EU projects and national research programmes.

Antonio Novellino is the ETT (www.ettsolutions.com) Research Manager and he is coordinating the R&D department activities, and the marine science unit (www.emodnet-physics.eu). He serves on the Board of Directors of Consortium Si4Life (manager of Polo Si4Life - Life Sciences Liguria Region - www.si4life.com), he serves on the board of Consortium Tecnomar (about 130 companies working on maritime and environment technology, www.consorziotecnomar.com) and he is the ETT deputy at regional and national innovation tables. He also serves the techno-scientific board of the Ligurian Cluster of Marine Technology DLTM (www.dltm.it) and the techno-scientific board of Consortium TRAIN (innovation in energy and transport management, www.consorziotrain.org) and member of the EuroGOOS DATAMEQ group for advising (http://eurogoos.eu) on operational oceanography data management procedures. He is co-authors of scientific publications and he participates as reviewer for scientific journals for Elsevier and Frontiers. He is coordinating the activity of MARE/2012/10 - Lotto6 - Knowledge base for growth and innovation in ocean economy: assembly and dissemination of marine data for seabed mapping. He coordinated the activities of MARE/2010/02 - Preparatory Actions for European Marine Observation and Data Network – Physical Parameters.

Alessandro Oggioni, since 2011 researcher at the CNR - Institute for Electromagnetic Sensing of Environment (IREA), is actively involved in the action of data management within the Project Life + EnvEurope and the European Network LTER (Long Term Ecological Research) with aspects relating to the creation of IT tools for the collection, research (through MetaData) and sharing of ecological data; the use of these through the use of geographic services organized for the web (OGC); the realization of workflow processes and web (Web Processing Service - WPS). Since 2012 responsible for the action of realization of the Spatial Data Infrastructure for the project Projects of National Research Program Flag RITMARE, under the sub-project 7.

Fabio Pavesi is a researcher at the Institute for Electromagnetic Sensing of the Environment (IREA) of the National Research Council of Italy (CNR). He graduated in Electronic Engineering at Politecnico di Milano.

Monica Pepe is a researcher at the Institute for Electromagnetic Sensing of the Environment of the National Research Council of Italy. She obtained her Ph.D. in Physical Geography by the Dept. of Earth Sciences, University of Pavia (Italy), focusing on environmental monitoring. She carries out research in optical remote sensing, and focus on automatic mapping of the Earth surface (hydrosphere, cryosphere, biosphere, geosphere). She is involved in Digital Earth research, particularly as regards web services in the framework of Spatial Data Infrastructures and Sensor Web Enablement.

James Potemra is an Associate Professor in the Hawaii Institute of Geophysics and Planetology (HIGP) in the School of Ocean and Earth Science and Technology (SOEST) at the University of Hawaii (UH). Jim has a background in physical oceanography and works on climate-related issues in the western Pacific and eastern Indian Ocean. He is also the data management lead for PacIOOS (http://pacioos.org) and the Asia-Pacific Data Research Center (http://apdrc.soest.hawaii.edu).

Theo Prins is a senior researcher and advisor on marine ecology at Deltares.

Dick M. A. Schaap, MSc Coastal Engineering, Managing Director MARIS, has a background in working as project engineer in Rijkswaterstaat North Sea Directorate (Netherlands Public Works Department), and has extensive experience in coordination of many national and EU projects on marine data management infrastructures. He has been technical coordinator of the SeaDataNet II project. At present he is coordinator of the EU funded EMODnet Bathymetry and EMODnet Data Ingestion projects and the private industry funded SIMORC project, and technical coordinator of EU funded projects EMODnet Chemistry, EMODnet Physics and ODIP II, while partner in EU funded projects EuroFleets 2, CitClops, Micro-B3, EMODnet Biology and CLIPC. He is also invited member of EuroGOOS DATAMEQ group for advising on operational oceanography data management procedures, and member of MODEG, Advisory Group to EMODnet.

Serge Scory, M.Sc. in Engineering (Public works), University of Liège, Belgium, 1982; M.Sc. in Oceanography, University of Liège, Belgium, 1984. Head of the Belgian Marine Data Centre (BMDC) since its creation in 1997. He is and was involved in many national and international projects dealing with marine data and information management. He is currently the Belgian delegate to the International Oceanographic Data and Information Exchange programme (IODE) of the IOC, Head of the Belgian delegation to the WG-DIKE (Data, Information and Knowledge Exchange for the implementation of the Marine Strategy Framework Directive) and Belgian delegate to the Council of the International Council for the Exploration of the Sea (ICES).

Adam Shepherd is a software engineer at the Woods Hole Oceanographic Institution and co-chairs the Vocabulary Services Interest Group for the Research Data Alliance and the Drupal Working Group for the Federation of Earth Science Information Partners. He has been an invited speaker at the American Geophysical Union Fall Meeting, presented at Drupalcon North America, and co-authored a book chapter on oceanographic linked data for IGI Global. Adam holds a BS in Computer Science and a Minor in Business Administration from Northeastern University where he was awarded the 2001 Co-op of the Year for his work at the Woods Hole Oceanographic Institution. With fourteen years of experience building web-enabled software, he concentrates on using science-driven use cases for improving ocean science research capabilities. Currently, his research focus has been on semantically enabled linked data applications for vocabulary alignment, data validation and improved data discovery.

Jean-Marc Sinquin, project manager for marine onboard software manufacturer Oceanographic Research Institute has over 20 years' experience in IT marine solutions. He holds "engineering and IT computer science, degree obtained from H.E.I. (Lille, France) in 1992 and Master's degree in Software Development from ENSTB in 1994. He is currently working on many projects (CARAIBES, GLOBE, …) such as DORIS. For the last eight years, he has been deeply involved under Eurofleets1, Geoseas and EMODnet projects.

Jordi Sorribas, born in Barcelona (Spain) 1965, as a Marine Geologist has been participated in more than 30 oceanographic surveys on-board Spanish and European research vessels. Since 1997 is at the front of the computing and communications team of the Marine Technology Unit of the Spanish National Science Council (Consejo Superior de Investigaciones Científicas, CSIC). During this time has developed scientific software and data acquisition devices for research ships and polar stations using Object Oriented Analysis and development techniques with C++ and Java over Linux, Solaris, MacIntosh and Windows platforms, maintaining a very narrow relationship with scientist that work and uses these systems.

Andrea Spada is CTO at SyO srl.

Jan Stafleu (PhD VU University, Netherlands) is a geologist/geomodeller and project manager at TNO – Geological Survey of the Netherlands (since December 2006). He is involved in the development of three-dimensional geological subsurface models of the shallow subsurface of the Netherlands. In particular, he acts as lead modeller and project manager of the development of the GeoTOP voxel model, which provides a sound basis for subsurface related questions on, amongst others, groundwater extraction and management, land subsidence studies, aggregate resources and infrastructural issues. Before joining the Geological Survey, Jan worked for 11 years at a major ICT company where he gained experience in programming, technical and functional software design, information analysis and project management.

Willem Stolte is a marine plankton ecologist with a background in experimental systems and modelling. Currently he working mostly with (3D) modelling of plankton and cross-project data analysis of marine systems.

Yvan Stoyanov is a senior computer scientist from Royal Belgian Institute of Natural Sciences, with affinity for high-tech innovative environment. He is specialized in computational software for scientific. Her interests broadly are in area of ontology, Web application for scientist data and programming geographical Information Systems.

Sergej Suzdalev has been working at Klaipeda University since 2009. Field of interest: sedimentology, geochemistry, environmental impact assessment.

Paolo Tagliolato is a researcher at the Institute of Marine Science (ISMAR) and the Institute for Electromagnetic Sensing of the Environment (IREA) of the National Research Council of Italy (CNR). He graduated in Mathematics at the State University of Milan. At the same university he completed the Ph.D. degree in Computer Science in the area of multimedia information retrieval. His research focused on information retrieval, spatial analysis of cellular network data for urban applications (at the Politechnic of Milan), semantic spatial data infrastructures and environmental data management with open geospatial standards (at CNR). Since 2015, he is the Italian representative in the European Commission INSPIRE Maintenance and Implementation temporary sub-group on Download Services for Observation Data (MIWP-7a).

Nathan Terseleer is a marine modeller at the Royal Belgian Institute of Natural Sciences, Operational Directorate Natural Environment. He holds a PhD in marine ecosystem modelling (Université libre de Bruxelles, Belgium, 2014) and worked on diatoms and the inclusion of higher functional diversity into biogeochemical models. His research now focuses on seabed dynamics in areas of marine aggregate extraction in the scope of the TILES project. He couples modelling approaches and data analysis to understand the natural variability of the seabed and the potential impact of human activities, aiming at a more informed decision process in the scope of sustainable exploitation. He works on the technical and conceptual aspects of a closer coupling between numerical models and available data, aiming at better validation and interpretation of model results.

Rob Thomas has worked at the British Oceanographic Data Centre (BODC) for over 7 years. At BODC Rob first had responsibility for managing the marine data collected by the Plymouth Marine Laboratory, and more recently the NERC funded UK Ocean Acidification programme. He has been involved with the content governance of the SeaDataNet parameter vocabularies for the last 5 years and currently co-chairs the BODC Vocabulary Management Group. He participates in the Ocean Acidification - International Co-ordination Centre (OA-ICC) international working group on the management of data from biological response to ocean acidification experiments.

Charles Troupin is an Engineer in Physics (University of Liège, 2005) and Doctor in Oceanography (2011). Since March 2014, he is the leader of the Data Centre Facility at the Balearic Islands Coastal Observing and Forecasting System (SOCIB). He has expertise in ocean numerical models and data analysis (time series, spatial and temporal interpolation). He has been involved in several European projects such as SeaDataNet, EMODnet Chemestry, or MyOcean. Recently he worked on the comparison between different sensors and platforms (HF radar, glider, satellite altimeters) in the Ibiza Channel.

Dries Van den Eynde is physical oceanographer, working at the Royal Belgian Institute for Natural Sciences, Operational Directorate Natural Environment since 1988. He obtained his Masters in Civil Engineering at the Catholic University of Leuven, Belgium. He is the team leader of a research group on Suspended Matter and Seabed Monitoring and Modelling since 2009. His main interest is in the development and application of hydrodynamic and wave models, turbulence models, cohesive and non-cohesive sediment transport models and morphodynamic models, including the effect of climate change, as well as executing and analysing measurements of suspended matter and bottom shear stresses. While executing fundamental research is an essential part of his work, the application of the research for the management of the marine environment remains a main concern.

Gerard Van der Kolff is head of the department Software and Data Integration of Deltares Software Center.

Sytze van Heteren (PhD Boston University, USA) is a coastal and marine geologist at the Geological Survey of the Netherlands. For more than 25 years, he has conducted and supervised coastal projects in the Netherlands and in the USA. Key areas of expertise include biogeomorphology of shallow siliciclastic seas, coastal and marine sedimentology, coastal morphodynamics, and applied Quaternary geology. He has represented the Survey in various national and international research consortiums, coordinated the coastal-zone and shallow-marine mapping program, analyzed the impact of storm surges on coastal barriers, and helped to build bridges between marine geology, ecology, and archeology.

Vera Van Lancker (PhD UGent, Belgium) is a marine geologist at the Royal Belgian Institute of Natural Sciences, Operational Directorate Natural Environment. Specialized in geological resources and the impact of human activities, seabed habitats, and sediment dynamics, targeting more sustainable exploitation of marine environments. Coordinator of nationally funded projects and task leader in international projects. Actively involved in the valorization of geological and geophysical data for a wide user community (e.g., EU-FP7 Geo-Seas, DG MARE EMODnet-Geology). Visiting professor at Ghent University. 200+ publications of which 40 peer-reviewed.

Sylvain Watelet is a research assistant at the University of Liège, actively involved in the Gulf Stream research since 2013 and main current developer of the software DIVA.

Patrick West is a principal software engineer with the Tetherless World Constellation at Rensselaer Polytechnic Institute. His current projects are focused on the semantic expression of data science concepts and relationships in various domains, including solar, upper atmosphere, ocean sciences, earth science informatics, as well as computer science areas such as knowledge representation, semantic technologies, distributed semantic data frameworks, robust collaboration and content management environments, and agile software engineering. Previously, Patrick was a senior software engineer at the High Altitude Observatory at the National Center for Atmospheric Research in Boulder, Colorado. Patrick has been a long-term contributor to the Hyrax OPenDAP software project. He has 23 years post-degree experience which includes large and small companies, startups, nonprofits, research organizations, and academia.

Enrique Wulff is a Research Librarian with Spanish National Research Council (CSIC) at the Marine Science Institute of Andalusia - Cádiz. With strong experience as a health sciences, science and social sciences librarian, Mr. Wulff earned a M.Sc. in Biomedicine from University of Cadiz, an M.A. in Hispanic Studies from University of Cadiz, and is a Licentiate in Documentation of the University of Granada.

Index

3D Voxel Modelling 236

A

Action (Event) 199
Action_property (Event) 199
Actor (Event) 199
Aggregates 225
Archetype 34, 42, 44
Argo 3, 13-16, 18-19, 22, 33, 85-86, 135, 307, 309-310, 315-316, 319-320, 354
Autonomous Ocean Observation 13, 17, 22

B

Bathymetry 94, 164, 169-173, 175, 239-240, 289-290, 295, 299, 314, 355, 357
Belgium 233-234
Beneficial Use 294-295, 297
Best practice 22, 31, 55, 57, 62
Big Data 1-3, 85, 92, 94-95, 98, 156, 231, 251
Big Ocean Data 85-87, 98
Blue Economy 48, 67
Boundary Object 36, 42
Brokering 56-57, 59, 218

C

Category (Event) 199
Chemistry 5, 74, 164-165, 167-169, 175, 179, 229, 238, 311, 314, 361, 367
Climatology 164, 365-366
Collaboration Tools 18
Collaborative Work 28, 35, 197
Comment (Event) 199
Common Data Index 53, 55, 57, 63, 135, 138, 140, 162-163, 165, 169, 194, 315, 319
Controlled Vocabularies 2-3, 5-6, 9-10, 12, 18-19, 22, 30-31, 36-37, 53-54, 61-62, 69-71, 89, 115, 188, 208, 211
Controlled Vocabulary 5, 9-10, 14, 17-18, 22, 54, 70, 98, 188, 192, 207, 213
Coordinate systems 51-52, 55
Correlation Length 350, 354-356, 371
Cross-Domain Data Management 23
Cruise 2, 15, 17, 30, 61-62, 68, 80-84, 182, 188-192, 194-199, 312

D

Data Citation 1-2, 14-15, 18, 22, 310
Data Element 199
Data Integration 70, 74-76, 80-81, 200-201, 227, 334
Data Management 1-2, 6-7, 9, 13, 15-16, 23-24, 28-30, 33-34, 37, 47, 49, 55, 64, 69, 71, 73, 82, 116, 127, 150, 155-157, 162, 164, 174, 178-179, 181, 198, 201-203, 219, 254-257, 259, 276, 282-283, 307, 316, 319-320, 325-326, 328-329, 335-337, 342-343
Data Products 18-19, 48, 103, 106, 110, 115-116, 131, 155-156, 164-167, 169, 171, 175-176, 225, 229-232, 234, 239, 242, 252, 271, 303-304, 306-308, 310, 312, 315, 318, 320, 330, 361
Data Quality 2, 18, 155-156, 166-167, 181, 229-230, 232, 235, 241-243, 251
Data Services 56-57, 64, 156, 174, 253-254, 256-257, 259-260, 267, 278, 321
Data-Interpolating Variational Analysis 345, 348
Datum 52, 171, 235
Decision Support System 224-225, 241, 252
Digital Library 140

E

Ecosystem Services 48-49, 67, 227
E-Infrastructure 49, 67, 157-158, 174-176, 312
EMODnet 5, 48, 135, 155-156, 164-165, 167-175, 229, 244, 308, 310-312, 314-316, 319, 321, 329-330, 336, 343, 361, 367
Entropy 237-238, 252

Environmental Impact Models 225, 231, 239-241, 243-244, 252
environmental monitoring 19, 200-202, 327
Error Covariance 347, 371
Eurofleets 61, 63, 110, 162, 181-183, 189, 196-198, 201, 319
European Commission 47-48, 54-55, 67, 128, 136, 155-157, 164, 174, 203, 227, 243, 307, 326-327, 329
event logging 181, 183, 189

G

Geology 74, 179, 203, 225, 229, 244, 267, 284, 286, 289, 314
Geophysics 14, 203, 267
Global Ocean Observing System 15, 22, 85, 162, 254, 280, 327
Global Telecommunications System (GTS) 3, 22
Good Environmental Status (GES) 67, 156, 169, 227, 252, 327
Gravel 225-228, 230, 232-235, 240, 243-244

H

Habitat 101, 225, 231, 234, 292, 308, 326-327, 331

I

In Situ Measurements 255, 257, 271, 278, 346
Information Retrieval 84, 128, 130, 147, 149, 153-154
Information System 115, 129, 131, 134-135, 137-138, 153, 229-230, 234, 244, 313, 319, 336
infrastructure 16-18, 30, 35, 48, 53-55, 57, 61-62, 81, 116-117, 134, 155-157, 164-166, 169, 174, 194, 199, 201, 203-204, 207, 211, 214, 218, 227, 254-255, 260, 278, 285, 287, 294, 305, 307, 309, 311, 314, 317, 319, 330, 335, 343
Integrated Ecosystem Assessment 101, 125
Integrated Ecosystem Assessment (IEA) 125
Integrated Ocean Observing System (IOOS) 63, 201, 254-256, 259, 261, 278, 280
Interoperability 5-6, 9-12, 16, 18, 28, 31-33, 47, 53-57, 59, 61-64, 67, 69, 73, 94, 106, 108, 116-117, 131, 147, 156, 158-159, 164, 200-205, 207, 219, 254, 278, 303, 307, 315, 317-321, 361
IUU fishing 311, 313

K

Knowledge Building 28-29, 36, 129, 131
Knowledge Formalisation 29-30, 33
Knowledge Formalization 33

L

Large Marine Ecosystem (LME) 117, 125
Lemmatisation 153
Linked Data 3, 10, 18-19, 22, 31, 42, 69-71, 74-76, 82, 84-85, 87, 89, 92, 95, 98, 103, 115
Linked Ocean Data 69-73, 85-87, 89, 95, 98

M

marine data 29-31, 33-34, 37, 47-49, 51-57, 60, 64, 69, 128, 131, 149, 155-157, 162, 164-165, 174-176, 218, 229, 244, 285, 306-307, 309-310, 314, 318-321, 329-330, 343, 346
marine environment 47-49, 64, 67-69, 86, 89, 95, 128, 132, 156, 162, 165, 175, 229, 232, 239, 282-285, 298, 307, 326-327, 330, 343
Marine Resources 101, 157, 169, 228, 233, 239, 282-283, 286, 301, 305, 308, 316
marine sciences 29, 33, 86, 117, 184
Marine Strategy Framework Directive 47, 67, 156, 164, 166, 169, 225, 227, 244, 252, 326-327
Message 35-36, 86-87, 89, 95, 199, 276
Metadata 1, 4-5, 9-10, 12-13, 15-18, 22, 29-30, 49, 52-54, 56-63, 69, 73, 103, 106, 110, 112, 116, 128-129, 131, 135, 137, 141-142, 157, 161-166, 174-175, 178, 181-182, 193-194, 197-199, 202-203, 205-208, 211-213, 216, 218-219, 227, 229-230, 234-236, 238, 244, 259, 262-263, 268-269, 289, 307-309, 315, 318-320, 358

N

netCDF (Network Common Data Form) 371
North Sea 7, 165, 225-226, 331, 366

O

Observations and Measurements 18, 32, 63, 83, 94, 116, 179, 201, 203, 205, 230
ocean data 1-2, 12, 32, 34, 37, 47, 53, 55-57, 59, 63-64, 69-73, 85-87, 89, 94-95, 98, 131, 155, 159, 166, 307-309, 315-316, 319, 348, 355, 358, 363
Ontology 10, 12-13, 18, 31-34, 36-37, 42, 45, 54, 69-70, 73-74, 76, 79-80, 83-85, 92, 95, 99-100, 102-103, 106-112, 115-117, 125, 130-131, 135, 137, 149-150, 188-189, 192, 197, 218
Ontology Design Patterns 36, 70, 73-74, 79, 83, 85, 95, 99

Open Access 100, 125, 128, 134-135, 256, 284, 304, 311
Open Geospatial Consortium (OGC) 16, 22, 62, 89, 116, 167, 193, 198, 257, 319
Open Science 100-101, 115, 117, 125
OpenEarth 335-337
Open-Source Project for a Network Data Access Protocol (OPeNDAP) 257-259, 261-262, 264-266, 276, 278, 280, 318-320, 363

P

Pacific Islands Ocean Observing System (PacIOOS) 253-263, 265-268, 270-278, 280
Parameter 3, 5-7, 10, 12, 15, 17, 22, 26, 33-34, 44, 94, 112, 138, 166, 199, 201, 214, 225, 230, 238, 240, 309, 320, 361-363, 367
Parameter Codes 5, 10, 320
Part of Speech 154
Physics 3, 14-15, 135, 179, 229, 314
Program 14, 82, 84, 100, 107, 109-113, 118, 125, 191-192, 197, 199, 252, 254, 259, 265, 278, 307, 335, 342
progressive paradigmatic shift 181, 188
Protocol 53, 57, 103, 184, 199, 257, 271, 280, 318, 320, 348, 361, 363, 366, 371
Provenance 18, 32, 49, 100-112, 115-117, 125, 228, 315

Q

Quality Flag Scheme 68

R

Registry 12, 54, 69, 193, 196, 199
Representation 4, 23, 26, 34-37, 54, 81, 101, 103, 105, 115, 117, 150, 153, 179-181, 198-199, 208, 211, 213, 218, 230, 241-242, 251, 285, 350, 352, 357
Reproducibility 15, 100-101, 117, 125
Research Cruise 2, 15, 17, 68, 82, 182
Resource Description Framework (RDF) 1, 31, 99, 102
Resource Potential 233, 252
Resource Suitability 225

S

Sand 225-228, 230-235, 237, 239-241, 243-244, 270, 287, 291-296, 300, 343
SeaDataNet 5, 10, 30, 49, 53-55, 57, 59, 61-63, 112, 115, 131, 135, 137-138, 145, 155-167, 169, 174-175, 188, 193, 196, 201, 218, 229, 307, 309, 312-313, 315-316, 319-321, 330, 336, 361, 367
Seafloor Integrity 227, 252
Search Engine 8, 127-128, 130-131, 138, 140-143, 146-147, 149-150, 154
Search Engine Indexing 154
Semantic 2, 5-6, 9-13, 17-18, 22, 28, 31-34, 42, 44, 57, 69, 71, 75-76, 79, 81, 83-85, 87, 92, 95, 99-106, 109-110, 112, 115-117, 125, 127, 129-133, 135-136, 140-143, 145-146, 150, 156, 159, 183, 207, 211-212, 214, 216, 218, 317-318, 336
Semantic Interoperability 5-6, 9-11, 31-32, 57, 159, 317
Semantic Modelling 10
Semantic Web 2, 18, 22, 31-32, 34, 42, 69, 71, 84, 87, 95, 99-106, 109-110, 116-117, 125, 211-212
Sensor Observation Service 63, 89, 202
Sensor Web Enablement 1, 3, 16, 22, 31, 62-63, 89, 200, 202, 205
Sensor Web Enablement (SWE) 1, 16, 22, 62-63, 200, 202, 205
SensorML 18, 62-63, 116, 193, 198, 202-203, 205, 207, 212-214, 219
SensorML editor 207
Signal-to-Noise Ratio 349-350, 354-356, 371
SOS client 212-213
Spatial Planning 244, 283, 285-286, 288, 294, 310, 327
Standards 1-2, 4, 9-10, 13, 16-19, 22, 30-31, 52-55, 57, 59, 62-63, 74-75, 81, 89, 95, 100, 102, 104, 116, 135, 146, 155-157, 159, 161, 164-165, 167, 174-175, 182, 195, 201, 203, 206, 216, 229, 257, 261, 280, 295, 317-321, 335-336, 342, 361-362
Streaming Data 99
Subject (Event) 199
Suitability Maps 225, 232, 241-243, 252
survey events 182
Sustainable Development 23-29, 35, 37, 69, 94-95, 127, 129, 131, 178-181, 197, 201-202, 220, 227, 252, 282-283, 286, 294, 327
sustainable fisheries 260, 305-306
Sustainable Resource Management 225, 227, 229, 231-232, 238, 252

T

Tacit Knowledge 28, 37, 42, 129, 188
The Netherlands 225, 234, 236, 238-239, 325, 331, 334, 340-341, 343
Theory Laden-ness 26, 28, 42
Tokenization 154
Tool (Event) 199
Topology 87, 89, 94, 99
Transactional SOS 212-213

U

Uncertainty 14, 74, 225, 229-232, 235-238, 241-244, 252, 304, 306, 311, 321
underway data 17, 63, 178, 181-182, 188, 197
Uniform Resource Identifier (URI) 74, 103, 199
Uniform Resource Locator (URL) 199

V

Variational Inverse Method (VIM) 348, 354, 371
Vocabulary Content Governance 7-9, 18, 22, 159
Vocabulary Technical Governance 9-10, 18, 22
Voxel 225, 232, 235-244, 252

W

Web Feature Service (WFS) 257, 371
Web Map Service (WMS) 137, 257, 265, 348, 361, 371
Web Service 57-58, 62-63, 100, 117, 159, 167, 169, 193, 196-199, 202-203, 206, 315, 318-319, 336, 365
Web-Based Visualization 345, 348, 359
Workflow 86, 101-102, 104-110, 112-115, 117, 125, 137, 178, 207, 212, 225, 232, 235-236, 240

www.ingramcontent.com/pod-product-compliance
Lightning Source LLC
LaVergne TN
LVHW060358300926
849162LV00006B/1298

9781522507000